Advance in Post-harvest Preservation Technology

Advance in Post-harvest Preservation Technology

Editors

Maria Cefola
Bernardo Pace

MDPI • Basel • Beijing • Wuhan • Barcelona • Belgrade • Manchester • Tokyo • Cluj • Tianjin

Editors
Maria Cefola
Institute of Sciences of Food Production (ISPA)
National Research Council (CNR)
Foggia
Italy

Bernardo Pace
Institute of Sciences of Food Production (ISPA)
National Research Council (CNR)
Foggia
Italy

Editorial Office
MDPI
St. Alban-Anlage 66
4052 Basel, Switzerland

This is a reprint of articles from the Special Issue published online in the open access journal *Foods* (ISSN 2304-8158) (available at: www.mdpi.com/journal/foods/special_issues/Advance_in_Post_harvest_Preservation_Technology).

For citation purposes, cite each article independently as indicated on the article page online and as indicated below:

LastName, A.A.; LastName, B.B.; LastName, C.C. Article Title. *Journal Name* **Year**, *Volume Number*, Page Range.

ISBN 978-3-0365-7611-4 (Hbk)
ISBN 978-3-0365-7610-7 (PDF)

Contents

About the Editors

Maria Cefola

Maria Cefola is a researcher of Institute of Sciences of Food Production, National Research Council of Italy (CNR-ISPA). From 2006 to 2010, she attended the Ph.D. School in Innovation Management in Agri-Food Systems in the Mediterranean, researching the post-harvest management of fresh and fresh-cut fruit and vegetable products in the Mediterranean region at the University of Foggia. During her doctorate, she spent six months at the Department of Plant Science of UC Davis. In 2004, she obtained a Master's Degree in Food Science and Technology at the University of Basilicata. She has had significant experience in the post-harvest management of fruits and vegetables within international, national, and regional research projects, and has attended international courses and congress. Her research activities include improving the quality of fresh-cut products through the application of innovative pre-treatment, packaging, and/or storage conditions; innovating the logistic cold chain; and studying innovative non-destructive systems for quality evaluation.

Bernardo Pace

Bernardo Pace is a researcher of the Institute of Sciences of Food Production, National Research Council of Italy (CNR-ISPA). In the last 15 years, he has gained significant experience in the pre-harvest and post-harvest of fresh and fresh-cut fruit and vegetables in international and national projects. He participated in cooperation projects in Tunisia and Albania and in many international congress courses on post-harvest activities. He spent a period at Mann Lab of UC Davis (US), expanding on post-harvest research personal know-how and, with additional support, helped in a new research laboratory on the post-harvest territorial unit of CNR-ISPA in Foggia. He also assumed the role of project leader and led many research projects. He is also involved in the following thematic-based activities: the identification of technologies to improve the quality of whole and fresh-cut products; the application of innovative treatments on vegetables; the validation of new packing materials and storage conditions; the valorisation of by-products of fruit and vegetables; and the application of non-destructive systems that can assess the quality of fruit and vegetables. He is also the author of more than 140 scientific papers, reviews, and book chapters. He serves as a member of the Editorial Board for the MDPI journal *Foods*, and covered the role of Guest Editor for three Special Issues for MDPI.

foods

MDPI

Editorial

Advances Postharvest Preservation Technology

Maria Cefola and Bernardo Pace *

Institute of Sciences of Food Production, National Research Council of Italy (CNR), c/o CS-DAT, Via Michele Protano, 71121 Foggia, Italy; maria.cefola@sipa.cnr.it
* Correspondence: bernardo.pace@ispa.cnr.it; Tel.: +39-0881-630201

Citation: Cefola, M.; Pace, B. Advances Postharvest Preservation Technology. *Foods* **2023**, *12*, 1664. https://doi.org/10.3390/foods12081664

Received: 5 April 2023
Accepted: 11 April 2023
Published: 17 April 2023

Fruits and vegetables are important sources of nutrients such as vitamins, minerals, and bioactive compounds, which provide many health benefits. However, due to suboptimal postharvest management, large quantities of fresh or fresh-cut fruit and vegetables lose their quality and nutritional value before they reach the consumer.

In this Special Issue, *Advances in Post-harvest Preservation Technology*, belonging to the Section "Food Packaging and Preservation", 14 research and 2 review articles are included. Five research topics are addressed: (i) storage and treatment; (ii) packaging and edible coating; (iii) techniques or markers for determining fruit quality; (iv) prediction models; and (v) postharvest technologies and preservation strategies.

Pertaining to the first topic, the article published by Fatchurrahman et al. [1] reports the quality change of fresh goji berry fruit (*Lycium barbarum* L.) stored at different temperatures (0, 5, and 7 °C) for up to 12 d, highlighting significant differences concerning the fruit's phytochemical attributes (vitamin C, soluble solid content, titratable acidity, total polyphenol, 2,2-diphenylpicrylhydrazy (DPPH), and anthocyanin content). This study revealed that goji fruit is optimally stored at a temperature of 5 °C for 9 d. At this temperature, physiological disorders (such as pitting and shriveling) were reduced, while the overall sensorial and nutritional quality attributes, including vitamin C levels, total soluble solid content, and antioxidant activity, were preserved. However, additional technologies such as modified atmosphere packaging may be required to better control decay and extend shelf life. Ma et al. [2] analyzed different forms of walnuts (*Juglans regia* L. cv. Xiangling, with a green husk, shell, and fresh kernels) stored between 0 and −20 °C for 10 months. The results showed that the walnuts with a green husk presented higher-quality kernels compared to the other forms of walnuts analyzed when stored at 0 °C for 3 months. Whereas at a freezing temperature (−20 °C) and during long storage, the walnuts with shells showed advantages with respect to maintaining fatty acid content, total phenols, and total ascorbic acid concentrations compared with other forms of walnuts. To preserve the quality of fresh-cut cauliflower (*Brassica oleracea* L. *botrytis* cv. Xuebai), vegetables were immersed in a solution containing calcium chloride ($CaCl_2$) at a 2% concentration and at different temperatures (0, 20, or 40 °C) for 10 min [3]. While stored at 4 °C for 15 d, changes in sensory quality, firmness, color, ascorbic acid concentration, and the total glucosinolates, polygalacturonase, and lipoxygenase concentrations of cauliflower florets were evaluated. The results showed that a treatment with $CaCl_2$ and high temperature (40 °C) maintained a higher firmness value, preventing the browning and yellowing of fresh-cut cauliflower florets. This treatment was optimal for preventing reductions in ascorbic acid content and total glucosinolates and inhibiting the activity of polygalacturonase and the softening of cauliflower florets compared to the other treatments, particularly with respect to the control sample.

Two papers reported the results of treatments with melatonin (*N*-acetyl-5-methoxytryptamine) applied to preserve and extend the shelf life of mangos and zucchinis. In the first study, a trial measuring the response of on two mango (*Mangifera indica* L.) cultivars to exogenous melatonin (1000 mol L^{-1}) was conducted by Njie et al. [4]. During storage for about one month at 13 °C, physiological parameters, metabolic processes, and relative gene expression levels were measured. For both the cultivars analyzed (Guiqi and Tainong 1), the authors reported that the melatonin treatment did not affect the content of total soluble solids, titratable acidity, or the TSS:TA ratio; instead, it delayed weight loss, increased the period of firmness, lowered respiration rates, and reduced the incidence of decay. Furthermore, the melatonin treatment inhibited the decrease in ascorbic acid, flavonoid, and total phenol content as well as the enzyme activity of polyphenol oxidase. Additionally, for both mango cultivars, a delay in the period during which the malondialdehyde content increased during storage was observed. For both cultivars, in the fruit treated with melatonin, increases in the activity of antioxidant enzymes (superoxide dismutase and ascorbate peroxidase as wells as phenylalanine ammonia-lyase) and their genes' relative expression were noticed compared to the untreated fruit.

Medina-Santamarina et al. [5] reported the efficacy of exogenous melatonin supplementation (1 mM) alone or in combination with 1-methylcyclopropene (1-MCP) 2400 ppbL^{-1} for reducing chilling damage in zucchinis (*Cucurbita pepo* spp. *pepo* cv. Cronos). Zucchinis were stored at 4 °C for 15 d and an additional 2 d at 20 °C. For the zucchinis treated with melatonin + 1 MCP, weight loss and signs of chilling damage were reduced compared to the other treatments and the control, maintaining a higher degree of fruit firmness throughout cold storage. Moreover, for each sampling day, the combined treatment (1-MCP + melatonin) effected the lowest values of malondialdehyde and electrolyte leakage. The combined treatment only reduced the respiration rate for 9 d, while ethylene production was reduced only by the treatment with 1-MCP alone.

While stored at 4 °C, the total soluble solids and total chlorophyll content in the combined treatment were comparable with 1-MCP application. The authors highlighted that the proposed combined treatment could be a promising tool for increasing the storability of zucchinis at suboptimal temperatures.

In another study, to retard peach senescence (*Prunus persica* L. cv. Xiahui 8), treatment with 1-MCP, alone or in combination with nitric oxide, was proposed [6]. At the end of the storage period (8 d at 25 °C), the proposed combined treatment (1-MCP + nitric oxide) yielded the best effects on fruit quality, maintaining good physical characteristics and decelerating the rate at which the fruit firmed, its respiration rate, and ethylene production. During storage, antioxidant enzymes (ROS inhibiting), which affect the ripening and senescence of fruit, were examined. Furthermore, for each compared treatment, twenty phenols were identified and quantified.

The research conducted by Mudalal et al. [7] revealed the preservation effects of different ingredients (fresh onion (*Allium cepa* L.), corn oil, salt, sumac (*Rhus coriaria* L.), and lactic acid) on the quality traits of fresh Za'atar (*Origanum syriacum*) sealed in vacuum bags and stored at 2–4 °C for 42 d. Microbial counts, color, and other sensory attributes were analyzed every 7 d. The study showed that the addition of lactic acid induced a strong preservative effect against aerobic and anaerobic bacteria. Moreover, the addition of sumac improved the preservation of vacuum-packaged oregano stored in refrigerated conditions.

Regarding the second topic, (ii) packaging and edible coating, Vieira et al. [8] reported the results of research on fresh red raspberries (*Rubus idaeus* L. cv. 'Kweli'). To extend the shelf life of this perishable fruit, pads produced with chitosan, green tea, or rosemary ethanolic extracts were incorporated as natural antifungal agents. Sealed fruit trays were packaged under an air atmosphere using polyacid lactic film and stored at 4 °C for 14 d. At the end of the storage period, reductions in the growth and decay of spoiled fruit of 5% and 13% were observed for chitonsan + rosemery and chitonsan + green tea, respectively. Conversely, for the fruit packets without pads, these percentages were equal to 40 and 80% after 7 and 14 d, respectively. The use of polyacid lactic film reduces weight loss, thus preserving the firmness of raspberries, particularly when using pads with green tea or rosemary ethanolic extracts. Total phenols and ascorbic acid content were better preserved in the packages with film pads containing one of each extract. The results highlighted how the application of chitosan pads + rosemary ethanolic extracts could be applied to other soft fruit to increase their marketability.

As an edible coating, chitosan was also used to preserve the quality attributes of fresh date (*Phoenix dactylifera* L. cv Barhi) fruit in combination with extracts of olive cake and orange peel in two different concentrations (1 or 2%) during storage at 4 °C for 28 d [9]. When the chitosan was combined with either of the two extracts, a pronounced effect was observed. In particular, the treatment with olive cake (at 2%) increased the total phenolic content by five-fold and preserved DPPH inhibition to a three-fold greater degree compared to the control at the end of the storage period. For all the tested coatings, an increase in total soluble solids was observed, and the effects on moisture prevention, loss of firmness, and fungal growth did not affect sensory characteristics.

Imeneo et al. [10] studied the effect of an edible pectin-based coating integrated with a lemon extract by-product on the quality attributes of minimally processed carrots (*Daucus carota* cv. Nantes) stored at 4 °C for 14 d. The application of lemon extract preserved the fresh-cut carrots' physiological parameters, thereby delaying their senescence. The carrots showed limited changes in color and white-blush and a good degree of firmness throughout storage due to the presence of calcium chloride in the coating's formulation. The lemon extract by-product improved the microbiological stability of the minimally processed carrots, showing the lowest value of total bacterial activity after 7 d, while a mild increase was observed following the end of the storage period. Total carotenoids, phenolic content, and antioxidant activity values exhibited a similar trend during storage for all treatments, while higher levels were measured in the fresh-cut carrots treated with lemon extract.

Regarding topic (iii) concerning the techniques or markers for determining fruit quality, two papers focused their activity on Candonga strawberries (*Fragaria x ananassa* Duch. cv Sabrosa). The first one, proposed by Cozzolino et al. [11], focused on studying the profiles of volatile and phenolic compounds via HPLC as markers of the strawberries' ripening stage. Strawberries picked at three different harvest times and two ripening stages, namely, when half-red and red, were evaluated. An analysis of polyphenolic compounds revealed that the concentration of anthocyanins increased during the harvest times, whereas the content of flavonoids declined. Overall, fifty-seven volatile compounds were identified, including esters, aldehydes, alcohols, acids, terpenes, furanones, lactones, and others. A multivariate analysis was carried out, for which all the chemical data were considered. The results highlighted that the fruit at the red ripening stage during the first and second harvesting periods were similar, while the red fruit of the third harvesting stage significantly differed. This difference likely indicated overripening or an early stage of fruit senescence. Finally, the authors identified butyl butyrate, ethyl hexanoate, hexyl acetate, nonanal, terpenes, and lactones represented the volatile organic compounds that had a positive impact on consumers' preferences.

The objective of the research conducted by Palumbo et al. [12] was to discriminate two ripening stages (red and half-red) of strawberries, which were harvested at different times during the harvest season, by comparing different techniques, namely, image analysis, the use of an electronic nose, and attenuated total reflection-Fourier transform infrared (ATR-FTIR) spectroscopy. In the principal results concerning a correlation analysis between the data obtained via the e-nose and an analysis of volatile organic compounds, it was revealed that the HS-SPME/MS e-nose experimental data contained a sufficient amount information with which to allow for the discrimination of strawberry samples based on their stage of ripening. Moreover, titratable acidity was correlated with ATR-FTIR and image analysis data. Since titratable acidity usually decreases in strawberries during ripening, its assessment by ATR-FTIR or image analysis might provide a suitable indicator for a fast and non-destructive evaluation of the ripening stages of this fruit.

Hamie et al. [13] studied the possibility of using a non-destructive tool to determine the maturity stage of table grapes (*Vitis vinifera* L. cv. Sugranineteen) during the ripening season. Multiplex® 3 (FORCE–A, Orsay, France) equipped with a portable optical sensor was used for analysis. The study focused on the measurement of skin anthocyanin using two fluorescence indexes, namely, ANTH_RG (chlorophyll fluorescence excited with red and green lights) and FERARI (fluorescence excitation ratio anthocyanin relative index), and other indices concerning the total flavonoid, nitrogen, and chlorophyll content. All measurements were correlated with the principal quality parameters (color, pH, total soluble solids, titratable acidity, total phenols, antioxidant activity, anthocyanin, and flavonoids) obtained by applying common analysis techniques. The results highlighted an important relationship between the total anthocyanins (measured via spectrophotometry) and ANTH_RG and FERARI indices with R^2 values equal to 0.96 and 0.87, respectively. The main result obtained by the researchers was a regression equation developed using the ANTH_RG index (measured by Multiplex® 3) and skin anthocyanin content (obtained via common analysis techniques), which allows one to estimate the previously mentioned parameters directly in-vineyard in rapid mode and without damaging the plant material.

Owoyemi et al. conducted phenotyping analysis of the effects of various preharvest and postharvest quality levels of navel oranges (*Citrus sinensis* L. cv Rustenburg) in order to develop shelf-life prediction models with which to enable the use of a First Expired–First Out (FEFO) logistics strategy [14]. During the analysis, fruit originating from six different orchards was stored for 20 weeks at different temperatures and relative humidities (at 5 °C with 70, 90, or 95% relative humidity or at 2 or 8 °C with 90% relative humidity). Fourteen fruit quality parameters were analyzed every week for each cold storage condition and after one-week of shelf storage at ambient temperature (20 °C). Four different linear and non-linear regression models were tested to determine their ability to predict fruit acceptance scores, namely, multiple linear regression, support vector regression, random forest, and extreme gradient boosting. Based on the obtained data, among the different regression models, extreme gradient boosting combined with a duplication approach provided the most effective approach to predicting fruit quality, yielding an RMSE of 0.217 and an R^2 of 0.891. Thus, in the future, the development of accurate shelf-life prediction models should contribute to optimizing the FEFO logistic management system and thus promote more efficient inventory management and loss reduction. Postharvest storage and preservation strategies for extending the storage life of the king oyster mushroom (*Pleurotus eryngii* D.C.) were described in the review article written by Gou et al. [15]. Many storage preservation techniques, including physical and chemical methods, were described. A suitable storage atmosphere, appropriate packaging, and optimal storage conditions (in terms of temperature and relative humidity) as well as irradiation or drying were some of the preservation methods described in the paper. Other strategies described for preserving the shelf-life of *Pleurotus eryngii* were the use of the lactic acid fermentation process and nanoparticles with which to encapsulate bioactive substances. According to Gou et al. [15], future research on *Pleurotus eryngii* should focus on combining new and traditional technologies to improve postharvest quality. Among such techniques, it was stressed that radiation treatment with 1-MCP alongside a nano-packaging treatment, safe and efficient sterilization, microwaving, and the use of low-pressure electrostatic fields and low-temperature plasma sterilization equipment could be successfully applied during the storage and distribution of king oyster mushrooms. In the study conducted by Palumbo et al. [16], the latest postharvest technologies capable of extending the shelf-life of fruit and vegetables were described. Physical treatments such as microwave heating and the application of high hydrostatic pressure, pulsed electric fields, and cold plasma were reported to reduce microbial load and preserve the freshness and quality characteristics of fruits and vegetables. Moreover, a section about dipping and vacuum impregnation treatment as well as edible active packaging based on natural compounds (including alginate, chitosan, lemon by-product or essential oil, orange peel, and olive cake) was presented. One paragraph described the opportunities offered by selected microbes as control agents since biocontrol is considered one of the more sustainable postharvest approaches for increasing the shelf-life of whole and fresh-cut fruits and vegetables. Examples of innovative, non-destructive techniques for the quality monitoring of fruits and vegetables were described. Among these, it was revealed that image analysis based on traditional imaging in the visible range of the electromagnetic spectrum using a computer vision system is widely used for the in-line grading of many types of fruits and vegetables. Furthermore, it was related that the electronic nose has become one of the most favorable sensing technologies for evaluating the presence and content of specific volatile metabolites corresponding to the presence or loss of the freshness of fresh vegetable products. Finally, near-infrared spectroscopy was described as a technique that can be used to analyze the chemical composition of fresh produce and quality changes during storage. The authors report that this novel technology, which can be applied directly on fields or in industrial lines, is a valid approach for ensuring the traceability and authentication of agricultural produce.

In conclusion, the papers proposed in this Special Issue, which were written within different European and national projects or promoted by private enterprises, report innovative results that extend our knowledge of the postharvest preservation technology applied to fruit and vegetables that are either fresh or minimally processed. The Guest Editors thank all authors that have contributed to enriching the Special Issue and hope that the results presented might constitute a stimulus for further studies.

Author Contributions: M.C. and B.P. equally contributed to organizing, composing, and editing this Special Issue. All authors have read and agreed to the published version of the manuscript.

Acknowledgments: We thank all authors for submitting high-quality manuscripts and the reviewers for their careful evaluation with the common goal of improving the papers.

Conflicts of Interest: The authors declare no conflict of interest.

References

1. Fatchurrahman, D.; Amodio, M.L.; Colelli, G. Quality of Goji Berry Fruit (*Lycium barbarum* L.) Stored at Different Temperatures. *Foods* **2022**, *11*, 3700. [CrossRef] [PubMed]
2. Ma, Y.; Wang, C.; Liu, C.; Tan, J.; Ma, H.; Wang, J. Physiochemical Responses of the Kernel Quality, Total Phenols and Antioxidant Enzymes of Walnut in Different Forms to the Low-Temperature Storage. *Foods* **2021**, *10*, 2027. [CrossRef] [PubMed]
3. Mu, B.; Xue, J.; Zhang, S.; Li, Z. Effects of the Use of Different Temperature and Calcium Chloride Treatments during Storage on the Quality of Fresh-Cut "Xuebai" Cauliflowers. *Foods* **2022**, *11*, 442. [CrossRef] [PubMed]
4. Njie, A.; Zhang, W.; Dong, X.; Lu, C.; Pan, X.; Liu, Q. Effect of Melatonin on Fruit Quality via Decay Inhibition and Enhancement of Antioxidative Enzyme Activities and Genes Expression of Two Mango Cultivars during Cold Storage. *Foods* **2022**, *11*, 3209. [CrossRef]
5. Medina-Santamarina, J.; Serrano, M.; Ruiz-Aracil, M.C.; Ilea, M.I.M.; Martínez-Romero, D.; Guillén, F. A Synergistic Effect Based on the Combination of Melatonin with 1-Methylcyclopropene as a New Strategy to Increase Chilling Tolerance and General Quality in Zucchini Fruit. *Foods* **2022**, *11*, 2784. [CrossRef] [PubMed]
6. Wu, X.; Yuan, J.; Wang, X.; Yu, M.; Ma, R.; Yu, Z. Synergy of Nitric Oxide and 1-Methylcyclopropene Treatment in Prolong Ripening and Senescence of Peach Fruit. *Foods* **2021**, *10*, 2956. [CrossRef] [PubMed]
7. Mudalal, S.; Kanan, D.; Anabtawi, O.; Irshaid, A.; Sabbah, M.; Shtaya, M.; Shraim, F.; Mauriello, G. Application of the Hurdle Technology Concept to the Fresh Za'atar (*Origanum syriacum*) Preservation. *Foods* **2022**, *11*, 3002. [CrossRef] [PubMed]
8. Vieira, T.M.; Alves, V.D.; Moldão Martins, M. Application of an Eco-Friendly Antifungal Active Package to Extend the Shelf Life of Fresh Red Raspberry (*Rubus idaeus* L. cv. 'Kweli'). *Foods* **2022**, *11*, 1805. [CrossRef] [PubMed]
9. Ghafoor, K.; Al-Juhaimi, F.Y.; Babiker, E.E.; Mohamed Ahmed, I.A.; Shahzad, S.A.; Alsawmahi, O.N. Quality Attributes of Refrigerated Barhi Dates Coated with Edible Chitosan Containing Natural Functional Ingredients. *Foods* **2022**, *11*, 1584. [CrossRef] [PubMed]
10. Imeneo, V.; Piscopo, A.; Martín-Belloso, O.; Soliva-Fortuny, R. Efficacy of Pectin-Based Coating Added with a Lemon Byproduct Extract on Quality Preservation of Fresh-Cut Carrots. *Foods* **2022**, *11*, 1314. [CrossRef] [PubMed]
11. Cozzolino, R.; Pace, B.; Palumbo, M.; Laurino, C.; Picariello, G.; Siano, F.; De Giulio, B.; Pelosi, S.; Cefola, M. Profiles of Volatile and Phenolic Compounds as Markers of Ripening Stage in Candonga Strawberries. *Foods* **2021**, *10*, 3102. [CrossRef] [PubMed]
12. Palumbo, M.; Cozzolino, R.; Laurino, C.; Malorni, L.; Picariello, G.; Siano, F.; Stocchero, M.; Cefola, M.; Corvino, A.; Romaniello, R.; et al. Rapid and Non-Destructive Techniques for the Discrimination of Ripening Stages in Candonga Strawberries. *Foods* **2022**, *11*, 1534. [CrossRef] [PubMed]
13. Hamie, N.; Tarricone, L.; Verrastro, V.; Natrella, G.; Faccia, M.; Gambacorta, G. Assessment of "Sugranineteen" Table Grape Maturation Using Destructive and Auto-Fluorescence Methods. *Foods* **2022**, *11*, 663. [CrossRef] [PubMed]
14. Owoyemi, A.; Porat, R.; Lichter, A.; Doron-Faigenboim, A.; Jovani, O.; Koenigstein, N.; Salzer, Y. Large-Scale, High-Throughput Phenotyping of the Postharvest Storage Performance of 'Rustenburg' Navel Oranges and the Development of Shelf-Life Prediction Models. *Foods* **2022**, *11*, 1840. [CrossRef] [PubMed]

15. Guo, Y.; Chen, X.; Gong, P.; Wang, R.; Qi, Z.; Deng, Z.; Han, A.; Long, H.; Wang, J.; Yao, W.; et al. Advances in Postharvest Storage and Preservation Strategies for *Pleurotus eryngii*. *Foods* **2023**, *12*, 1046. [CrossRef] [PubMed]
16. Palumbo, M.; Attolico, G.; Capozzi, V.; Cozzolino, R.; Corvino, A.; de Chiara, M.L.V.; Pace, B.; Pelosi, S.; Ricci, I.; Romaniello, R.; et al. Emerging Postharvest Technologies to Enhance the Shelf-Life of Fruit and Vegetables: An Overview. *Foods* **2022**, *11*, 3925. [CrossRef] [PubMed]

foods

MDPI

Article

Quality of Goji Berry Fruit (*Lycium barbarum* L.) Stored at Different Temperatures

Danial Fatchurrahman *, Maria Luisa Amodio and Giancarlo Colelli

Dipartimento di Scienze Agrarie, degli Alimenti e dell'Ambiente, Università di Foggia, Via Napoli 25, 71122 Foggia, Italy
* Correspondence: danial.fatchurrahman@unifg.it

Abstract: Goji berries are widely known for their outstanding nutritional and medicinal properties; they are usually found in the market as dried fruit or as juice because the fruit has a short shelf-life, and little information is available about its postharvest behavior at low temperatures. This study aimed to determine the storage performance of goji berry fruit by evaluating physicochemical, and sensorial attributes during storage at three different temperatures (0, 5, and 7 °C) for 12 days in a range that has not been extensively studied before. In addition, fruit respiration and ethylene production rates were also measured at the three temperatures. Fruit stored at 0 °C showed the lowest respiration rate and ethylene production (5.8 mg CO_2 $kg^{-1}h^{-1}$ and 0.7 µg C_2H_4 $kg^{-1}h^{-1}$, respectively); however, at this temperature, the incidence and severity of pitting and electrolytic leakage were the highest. In contrast, 5 °C was found to be the best storage temperature for goji berry fruit; the fruit appeared fresh and healthy, had the highest scores during sensory analysis with an acceptable general impression, and had the lowest amount of damage attributable to chilling injury, with 17.1% fruit presenting with shriveling, 12.5% pitting, 6.7% mold, and 35% electrolytic leakage on day 9 of storage. Storage of goji berries at 7 °C resulted in the lowest marketability and the highest incidence of decay. Significant differences were also found in the phytochemical attributes, vitamin C content, soluble solid content (SSC), titratable acidity (TA), SSC/TA ratio, total polyphenol content, 2,2-diphenylpicrylhydrazy (DPPH), and anthocyanin content. This study revealed that a storage temperature of 5 °C for 9 days is recommended to maintain the quality of fresh goji berry. Thus, broadening the existing knowledge of the postharvest behavior of fresh goji berries; our results can help improve the commercial life of goji berries and ensure high-quality attributes throughout distribution.

Keywords: goji berries; shelf-life; postharvest quality; sensorial attributes; freshness; chilling injury

Citation: Fatchurrahman, D.; Amodio, M.L.; Colelli, G. Quality of Goji Berry Fruit (*Lycium barbarum* L.) Stored at Different Temperatures. *Foods* **2022**, *11*, 3700. https://doi.org/10.3390/foods11223700

Academic Editor: Francisca Hernández Garcia

Received: 16 October 2022
Accepted: 16 November 2022
Published: 18 November 2022

1. Introduction

Wolfberries or goji berry fruit (*Lycium barbarum* L.), belonging to the family Solanaceae, are widely recognized for their exceptional health benefits [1]. This shrub is native to Asia, primarily the central north region of Ningxia Hui (China), and was introduced to Europe in the 18th century due to its excellent nutritive and medicinal value [2]. Goji berries are particularly rich in nutraceutical compounds (carotenoids, flavonoids, phenolics, vitamins, and minerals) that exhibit antiaging, antitumor, and antioxidant activities in the human body. Generally, goji berries are cooked and processed as teas, soups, or served with meat and vegetables. They are also utilized for juice, tincture, and wine production [3–5]. The fruit is also consumed in its dried form or processed in powdered form for medicinal purposes [6]. Given their high perishability, high water content, and susceptibility to damage and rot [7], fresh goji berries are generally only available in areas where they are cultivated. This is also due to the lack of information on the goji berry's postharvest behavior and storage recommendations [8]. Dipping in lecithin solution [9] and edible coating applications based on lotus leaf extract [7] have been studied to prolong the shelf-life of goji. A study reported that a mild heat treatment (40 °C for 30 min) combined with a chitosan coating could delay decay occurrence in goji berries for up to 28 days at 2 °C [10]. A test on CO_2-enriched atmospheres concluded that 15–20% of carbon dioxide could help maintain the quality of stored goji berries for up to 14 days [11]. The use of modified atmosphere packaging (MAP), studied by [12], reaching an equilibrium of approximately 10% CO_2 can help preserve berry weight losses and organoleptic quality while avoiding mold occurrence up to day 13 of storage at 7 °C. The postharvest ripening of goji berries has been controlled using salicylic acid treatment, as this acid is used for inhibiting goji berry postharvest decay [13]. Regardless of postharvest technology, temperature and relative humidity are the first storage conditions to be accurately controlled to preserve quality and allow maximum shelf-life [14,15]. Studies on storage temperature of goji berries have only considered low ranges (mostly at around 0 and 2 °C), and high ranges (around 10 or 20 °C). The effect of storage temperatures of −2, 0, 10, and 20 °C on postharvest quality of fresh goji berries was reported by [1] who concluded that 0 °C is the optimum temperature to maintain the berry's phytochemical and sensory qualities. However, the large temperature gap among the tested temperatures is to be considered as there is no information on the temperature range between 0 and 10 °C, especially considering that 5 °C is the temperature frequently used in the cold chain of fresh products for transport and sale. As goji berries belong to the Solanaceae family, they might plausibly be chilling-sensitive, such as tomatoes, bell peppers, and eggplants [16–18]; however, the quality of fruit stored at 0 °C may show a slower degradation compared to when stored at 10 or 20 °C [1]. Therefore, the objective of the present study was to compare the effects of different chilling storage temperatures between 0 and 7 °C to identify the best condition for maintaining goji berry quality and shelf-life, with particular attention to the possible occurrence of chilling injury symptoms.

2. Materials and Methods

2.1. Sample Preparation

Briefly, 3.5 kg of goji berry fruit from 'Sweet Berry' cultivar grown in an open field in Castellaneta, a province of Taranto, Italy was conventionally handpicked with a peduncle at the ripeness level five as defined by a previous report [19] to be the recommended stage of harvest since the fruit are completely red, with close to the highest soluble solids content, and full size, thus benefitting from the highest yield with the least loss caused by the high perishability of the fruit even at harvest. At this stage fruit were harvested with maturity indicators in terms of color (hue°), weight, TA, SSC, and firmness were approximately 34.2 ± 0.85°, 0.83 ± 0.09 g, 0.69 ± 0.03%, 23.9 ± 0.2%, and 0.39 ± 0.04 N, respectively. Damaged fruits were removed, leaving 3.3 kg of healthy fruit with homologous dimensions. Approximately 120 g of fruit was used for the initial evaluation; then, the fruits were split into 27 groups (approximately 120 g each) with three replicates for storage at 0, 5, and 7 °C under 95% relative humidity (RH). Quality attributes of goji berry fruit were determined on the day of harvest and after 5, 9, and 12 days of storage.

2.2. Physical Quality Attributes

The color of the goji berry fruit was measured by extracting images acquired with a spectral scanner (DV s.r.l., Italy). The CIE L*a*b* color space were used for the color parameters. $Hue^\circ = arctg^{\frac{b^*}{a^*}}$ and chromaticity = $\sqrt{a^{*2} + b^{*2}}$ were calculated from the a^* and b^* values [20]. Firmness was assessed on 20 fruits for each replicate by applying a compression test to rupture the fruit between two parallel plates using a texture analyzer (TA-XT2®, Stable Microsystems, Godalming, UK) and a speed of crosshead at 75 mm min^{-1}. The rupture load of the deformation curve was recorded in Newton (N).

2.3. Sensorial Analysis

The sensory evaluation of goji fruit was performed by four trained panel members using a method introduced by [21] for apples, which was then applied to goji berries by [1]. The sensorial properties included firmness, texture, juiciness, sugar–acid ratio, aroma, tastefulness, and general impression. The external properties were shape, size, and color. For each attribute, goji berries were evaluated using a 1–5 anchored scale with 5 = excellent (fresh-like), 4 = very good, 3 = good (limit of marketability), 2 = acceptable (limit of edibility), and 1 = unsatisfying (not edible). The assessment was performed on harvest day using around 200 fruit assessed by each panelist, and 300 fruit each were used from each temperature storage on the evaluation day (after 5, 7, 9, and 12 days of storage). In addition, the incidence of mold and visual damage expressed as a percentage of the total number of fruits for each replicate was recorded.

2.4. Electrolytic Leakage

Relative electrolyte leakage (REL) was evaluated based on a previously published method with some modifications [22]. Instead of water, a mannitol isotonic solution was used to avoid osmotic shock [23]; 0.4 mol L^{-1} mannitol was determined to be the optimal concentration after following the procedure suggested by [24]. Approximately 10 fruit slices were taken from each replicate (approximately 3 g in weight) and placed in a centrifuge tube with 25 mL of 0.4 mol L^{-1} mannitol (Sigma–Aldrich, Steinheim, Germany). The electrical conductivity of the bathing mannitol solution was measured with a conductivity meter (CM35, Crison, Carpi, Italy) after 1 min (C1) and 60 min (C60) of incubation with orbital shaking (DAS12500, Intercontinental Equipment, Rome, Italy) at a speed of 60 cycles min^{-1}. The samples were then frozen at −20 °C for 24 h, and the conductivity (CT) was measured after defrosting for 3 h at 25 °C. The REL was calculated using the following equation:

$$REL\ (\%) = \left[\frac{(C60 - C1)}{CT}\right] \times 100$$

2.5. Physiological and Metabolic Attributes

The respiration rate (mg CO_2/kg/h) of goji berries was measured under static conditions as previously described by [25] with some modifications. Briefly, three replicates of 90 g fruit each were put in 150 mL sealed glass containers with a plastic septum for sampling. The containers were initially left open in a temperature- and humidity-controlled room to acclimate the samples (0, 5, and 7 °C). After closing, around 1 mL of gas samples were collected from each container, after the necessary time had passed to accumulate CO_2 in the headspace reaching a concentration of 0.1–0.2%, and injected into a gas chromatograph (Shimadzu, model 17 A, Kyoto, Japan) equipped with a thermal conductivity detector (200 °C). Separation of CO_2 was performed using a Carboxen 1006 plot (30 m × 0.53 mm, Supelco, Bellefonte, PA, USA) with a flow of the column at 7 mL min^{-1} and an oven temperature of 180 °C. The calculation of the respiration rate was based on the differences in CO_2 concentration, the weight of the sample, the container's free volume, and the elapsed time [26].

Ethylene production (μL C_2H_4/kg/h) was measured using the closed system introduced by [25]. The accumulation of C_2H_4 in the headspace of the sealed containers was measured using gas samples (2.5 mL) that were injected into a gas chromatograph (model 7890 A, Agilent, Santa Clara, CA, USA) installed with a flame ionization detector (FID, Agilent, Santa Clara, CA, USA). The temperature of the detector was set at 300 °C with the hydrogen and air flow at 45 mL min^{-1} and 400 mL min^{-1}, respectively. A metal-packed column 13073-U (Supelco, Bellefonte, PA, USA) was used for the separation of ethylene. Helium as the carrier gas (pressure, 15 psi) was used. The temperature of the oven was set to 120 °C. The concentration of ethylene was then referred to the weight of sample, to the container's free volume, and to the elapsed time.

2.6. Maturity Index

A method introduced by [27] was used for measuring SSC, TA, and pH. Ten goji berries per replicate were homogenized in Ultra Turrax (T18 basic, IKA, Staufen, Germany) and then filtered with two layers of cheesecloth (SWAB 4040, JC NONSTE, Shanghai, China). The obtained juice was used for direct SSC (%) reading using a digital refractometer

(PR32-Palette, Atago N1, Tokyo, Japan). TA and pH were measured in 1 g of juice samples using an automatic titrator (T50 M Terminal, Mettler Toledo, Greifensee, Switzerland). The samples were titrated against a 0.1 mol L^{-1} NaOH solution up to a final pH of 8.1; the results are reported as a percentage of citric acid per 100 g sample.

2.7. Chemical Composition

2.7.1. Determination of Ascorbic Acid, Dehydroascorbic Acid, and Vitamin C Content

Dehydroascorbic acid (DHAA), ascorbic acid (AA), and total vitamin C contents were assessed by homogenizing 5 g of fruit tissue in Ultra Turrax for 1 min with 5 mL of methanol/water (5:95 v/v), citric acid (21 g L^{-1}), ethylenediaminetetraacetic acid solution (EDTA) (0.5 g L^{-1}), and sodium fluoride (NaF) (0.168 g L^{-1}). The homogenate was filtered through cheesecloth, and the pH was adjusted to 2.2–2.4 with the addition of 6 mol L^{-1} hydrochloric acid HCl. Centrifugation at 12,000 rpm for 5 min of homogenate was applied. The supernatant was recovered and filtered through a C18 Sep-Pak cartridge (Waters, Milford, MA, USA) and a 0.2 µm cellulose acetate filter. AA and DHAA were evaluated as described by Zapata and Dufour (1992) with little modifications [28]. HPLC analysis was conducted after the derivatization of DHAA into fluorophore 3-(1,2-dihydroxy ethyl) furol [3,4-b] quinoxaline-1-one (DFQ) with 1,2-phenylenediamine dihydrochloride (OPDA). All the of 20 µL samples were analyzed using an HPLC system (Agilent Technologies 1200 Series; Agilent, Waldbronn, Germany) installed with a binary pump and a DAD detector. DFQ and AA separation were acquired on a Zorbax Eclipse XDB- C18 column (150 mm × 4.6 mm; 5 µm particle size; Agilent Technologies, Santa Clara, CA, USA). Methanol (MeOH)/water (H_2O) (5:95 v/v) was used as mobile phase, containing 50 mmol L^{-1} potassium dihydrogen phosphate and 5 mmol L^{-1} cetrimide at pH 4.5. The rate of the flow was set at 1 mL min^{-1}. The detector was set on wavelengths 348 nm for DHAA and 251 nm for AA. The contents of AA and DHAA are expressed as grams of AA or DHAA/kg f.w.

2.7.2. Determination of Anthocyanin Content

The anthocyanin content was determined using the method introduced by [29]. Two disks (top cut) were taken from fresh goji berries (approximately 1 mm thick). The area of the disks was calculated using the area of the ellipse formula $A = a * b * \pi$. Goji fruit disks were then mixed with 3 mL of acidified methanolic solution (10 mL HCl/L) until submerged and treated for 3 h at 25 °C under dark conditions. The anthocyanin level was measured according to the formula introduced by [30]:

$$Anthocyanin = Absorption_{532\ \mathrm{nm}} - 0.25\ (Absorption_{653\ \mathrm{nm}})$$

Afterward, the molar concentrations of anthocyanins/cm^2 were obtained by dividing the optical density values by the molecular extinction coefficient of cyanidin (2.45 × 10^4) L/mol*cm and then again dividing by the area of the leaf disks [31]. The results are expressed as milligrams of cyanidin per cm^2.

2.7.3. Total Polyphenol Content and Antioxidant Activity

The total polyphenol content was evaluated using 5 g of goji berries homogenized in Ultra Turrax for 1 min in a 30 mL medium containing a 20% water: 80% methanol solution and 2 mmol L^{-1} sodium fluoride. The centrifugation at 9000 rpm for 10 min at 4 °C of homogenate was applied. The method followed a protocol previously used by [32] with slight modifications. Briefly, 100 µL of the extract was mixed with 1.58 mL water, 100 µL of Folin–Ciocalteu reagent, and 300 µL of sodium carbonate solution (200 g L^{-1}). The absorbance was read at 725 nm against a blank using a spectrophotometer (UV-1700, Shimadzu Jiangsu, Suzhou, China) after allowing the solution to stand for 2 h. The total polyphenol content was calculated based on the calibration curve of gallic acid and expressed as milligrams of gallic acid per 100 g of fresh weight (mg GA 100 g^{-1} f.w.). The antioxidant assay was performed according to the procedure described by [33] with minor modifications. Fifty microliters of the same extract, opportunely diluted, were pipetted into 0.950 mL of DPPH solution to initiate the reaction. The absorbance was read at 515 nm after 24 h of incubation. Trolox was used as a standard, and the antioxidant activity is presented as grams of Trolox equivalents per kg of fresh weight (g TE kg^{-1} f.w.).

2.8. Statistical Analysis

The effects of storage duration and storage temperature were analyzed using a two-way ANOVA, and the significance of differences among means was determined using a Tukey's test at $p < 0.05$. All calculations were conducted using the statistical software IBM-SPSS 2019 (1 New Orchard Road, Armonk, NY, USA).

3. Results and Discussion

3.1. Physical and Sensory Aspects

The effects of temperature, storage duration, and their interactions of goji berry quality attributes are shown in Table 1.

Table 1. Effect of storage temperature, storage duration, and interaction on quality attributes of goji berries stored for 12 days at 0, 5, and 7 °C. Mean values of 12 samples are reported (three replicates × four storage durations).

Quality Attributes	Storage Temperature	Storage Duration	Storage Temperature × Storage Duration
Weight loss (%)	****	****	****
Hue (°)	**	**	ns
Chroma	*	***	*
Firmness (N)	***	ns	ns
Mold damage (%)	****	****	****
Shriveling damage (%)	****	****	****
Pitting (%)	****	***	***
Relative Electrolyte Leakage (%)	****	*	*
Soluble Solid Content (%)	****	****	****
Soluble Solid Content Total Acidity	*	*	ns
Total Acidity (%)	*	ns	*
Ascorbic Acid (g/kg)	****	****	ns
Dehydroascorbic Acid (g/kg)	****	ns	ns
Vitamin C (g/kg)	*	****	ns
Anthocyanin (mg cyanidin cm^{-2})	****	****	*
Total polyphenol (g gallic acid/kg)	****	****	****
Antioxidant (g Trolox/kg)	****	****	****
Firmness (score)	***	****	ns
Texture (chewing; score)	****	****	ns
Juiciness (score)	****	****	ns
Sugar–acid ratio (score)	****	****	ns
Aroma (score)	****	****	ns
Tastefulness (score)	****	****	ns
General impression (score)	****	****	ns
Shape (score)	****	****	ns
Size (score)	****	****	ns
Color (score)	****	****	*

Note: (****) $p \leq 0.0001$; (***) $p \leq 0.001$; (**) $p \leq 0.01$; (*) $p \leq 0.05$; ns, not significant.

The goji berry fruit underwent several types of changes during storage, depending on the temperature, including shriveling, pitting with the appearance of black spots, when severe, and the development of decay. Both storage duration and storage temperature significantly affected shriveling, pitting, and mold occurrence (Table 1). It is reported that mold occurrences in fresh goji berries were most likely caused by Alternaria sp. and many other pathogens, including *Penicillium* sp., *Alternaria* sp., *Aspergillus niger*, *Trichoderma*, and *Aspergillus* [34]. As depicted in Figure 1, all types of alterations were already visible after

7 days of storage, and the most severe mold infection was seen at 7 °C, reaching more than 19.6% versus approximately 2 and 4% at lower temperatures of 0 and 5 °C, respectively. As for shriveling, a higher incidence was observed in berries stored at 5 and 7 °C, reaching approximately 5.7% and 10%, respectively. Although in general fruit stored at 0 and 5 °C showed better physical quality in terms of damage, the fruit possessed a higher amount of pitting at 21.2% and 8.2, respectively, compared to temperature storage at 7 °C which was only around 4.6%. The incidence of damaged fruit gradually increased over time, being a maximum of 30% mold on day 12 of storage at 7 °C, against approximately 5% at 0 and 5 °C.

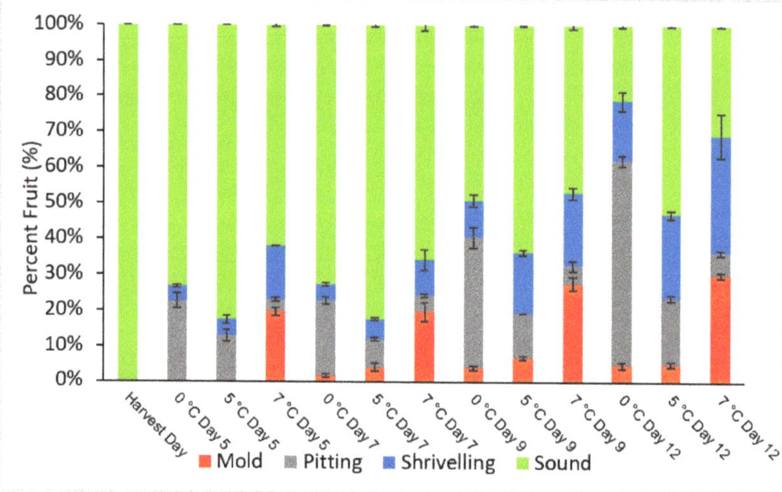

Figure 1. Distribution of fruit alteration types during the storage of goji berries.

The information from Figure 1 and the results on sensory attributes reported in Table 2, revealed that goji berries could be stored maintaining a good quality for about 9 days at 5 °C, although 6.7% of the berries had a very early decay appearance indicated by an early spot of mold on the surface of the fruit's skin. Furthermore, since the highest incidence of pitting was found in goji berries stored at 0 °C where only 49% of the fruit remained sound after 9 days of storage compared to fruit stored at 5 °C with 64% of sound berries, whereas fruit stored at 7 °C, although showing the lowest incidence of pitting, only showed 47.1% of sound fruit. These results indicate that the storage temperature of 5 °C should be preferred over 0 °C. One further consideration is that while berries stored at 0 and 7 °C were characterized by high amounts of alterations which are considered to be unacceptable for marketing (i.e., pitting and mold) in the case of fruit stored at 5 °C the most represented type of alteration was shriveling damage, which could probably be attributed to water loss during storage, and which (unless it is very severe) might not affect the fruit marketability. This is supported by results for sensory evaluation after 9 days of storage (Table 2). As seen in Table 2, the fruit stored at 0 and 5 °C had a reasonable marketable acceptance with a score higher than three (good) for general impression. Among the three storage conditions, after 9 days of storage fruit stored at 5 °C had the highest scores for general impression, tastefulness, aroma, and juiciness reaching 3.3, 3.2, 3.2, and 3.3, respectively. In contrast to fruit stored at 7 °C, which had scores between 2 and 2.5. Although the sensorial attributes of fruits at 0 °C were acceptable, fruits stored at 5 °C showed the highest scores (Table 2). In this experiment, we considered 9 days as the probable shelf life of goji fruit since in the sampling at day 12 the fruit showed scores lower than two for practically all quality attributes, which is very much below the limit of marketability and, in many cases, below the limit of edibility (data not shown). Furthermore, our result is not in accordance with a study by [1] which indicates that the best storage temperature for fresh goji berries is 0 °C when compared to −2, 10, and 20 °C, although in their study storage temperatures between 0 and 7 °C were not explored.

Table 2. Sensory evaluation of goji berries stored at three different temperatures for 9 days. A 1–5 anchored scale was used with 5 = excellent (fresh-like), 4 = very good, 3 = good (limit of marketability), 2 = acceptable (limit of edibility), and 1 = unsatisfying (not edible). In each line, values marked with the same letter are not significantly different according to a Tukey's test ($p < 0.05$).

Sensory Attributes	0 °C	5 °C	7 °C
Firmness	3.2a	3.3a	2.3b
Texture	3.0a	3.2a	2.3b
Juiciness	3.0b	3.3a	2.2c
Sugar–Acid Ratio	3.2a	3.2a	2.3b
Aroma	2.8b	3.2a	2.3c
Tastefulness	2.7b	3.2a	2.5c
General Impression	3.0b	3.3a	2.3c
Shape	3.2a	3.2a	2.3b
Size	3.2a	3.2a	2.3b
Color	3.0a	3.0a	2.0b

Color is an important parameter to assess the quality of produce; a bright red–orange color in fresh goji berries is provided by carotenoids [35]. A study reported that fresh mature goji berries contain as much as 321.1 $\mu g.g^{-1}$ β-carotene [9]. Additionally, among β-carotene types, the content of zeaxanthin esters in goji fruit can exceed 77.5% of βCE/g f.w., and especially zeaxanthin palmitate is found to be abundant, reaching 31–56% of mg βCE/g f.w. [36]. The effect of temperature on accelerating ripening, senescence, color changes, and other attributes is very well known [37]. However, in this experiment, hardly any color change was observed on the berries throughout the storage duration for most of the color primary and secondary attributes (i.e., L*, a*, b*, chroma, and ΔE) (data not shown) although some slight differences in hue angle values were observed at day 12 when fruit held at 7 °C showed a lower hue angle than those stored at 0 °C. Moreover, [1] did not report any difference in color during the storage of goji berries for 12 days where storage temperature varied between −2 and 20 °C.

In a previously published work [19], goji berries were classified into six developmental stages according to their features in terms of size, color, composition, respiration, and ethylene production rates. While full color was reached already at stage four, fruit size at that stage was still less than 60% of its final size, and many other maturity indicators further evolved during development. Fruits in this experiment were harvested at stage five when they had practically completed their growth and almost reached the maximum SSC. Beyond that stage, the fruit had a limited increase in SSC and no further changes in size and color, although, at stage six the berries resulted very soft and slightly overripe with very poor attributes to go through harvest and postharvest handling procedures. The results of this experiment confirm that ripening completion and beginning of senescence did not affect color changes in the berries nor, as it will be shown below, changes in SSC and TA.

Fruit firmness showed a limited decrease during storage, although there were some statistically significant differences among treatments (Figure 2). Most softening happened in the second part of storage as values maintained almost constant during the first week of storage. After that fruit stored at 0 °C showed the highest softening rate always presenting significantly lower values than samples at 5 °C. Fruits stored at 7 °C showed a somewhat intermediate behavior but with values that were always closer to samples at 5 °C. While behavior at 5 °C was confirmed by sensorial analysis, as fruit stored at that temperature obtained the highest score for texture, those stored at 7 °C were perceived as the softest by the evaluators, at least after 9 days of storage. Fruit softening is caused by cell wall and middle lamella hydrolysis, where more than 50 genes related to the cell wall structure show variation in expression, involving complex quantitative trait loci [38]. Expression is often regulated by the ripening process and senescence although in many cases fruit texture, as instrumentally determined, is also strongly influenced by tissue turgidity [39]. In a previously published work [19], goji berries showed an important loss of firmness during their development from stage one (green) to stage three (partially red); after that stage fruit firmness remained more or less constant until the overripe stage (six) although the fruit size increased until stage five. This indicates that the loss of firmness during storage observed in this experiment may not be attributed to ripening and senescence. Further, the weight loss of the fruit stored at 0 °C was the lowest (as later described in Figure 3) followed by that of the fruit stored at 5 °C and 7 °C and this also excludes the idea that the faster softening that occurred in fruit stored at 0 °C might be attributed to a more severe loss of turgor. A change in the structure of middle lamellae due to the reduction of electron density has been observed in mature-green tomato fruit that had been chilled at 5 °C for 15 days [40]. The report explained that the regions of reduced electron density could be attributed to water absorption by the cell wall and mediated by pectinmethylesterase alteration of middle-lamellar pectin. Translocation of water from the cytosol is directly associated with an increase in hydration sites in the cell wall and chill-induced membrane dysfunction [41]. The resulting turgor loss, combined with increased water absorption or swelling of cell walls, may be responsible for chilling-associated softening. These results were also in accordance with mechanisms of woolliness development in chilled nectarines [42].

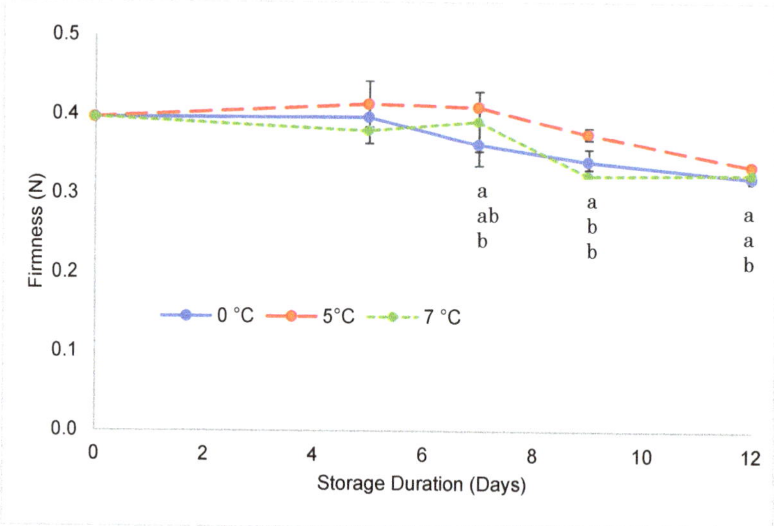

Figure 2. Firmness changes in goji berries during storage at 0, 5, and 7 °C. Values marked with the same letter on the same sampling day are not significantly different, according to a Tukey's test ($p < 0.05$).

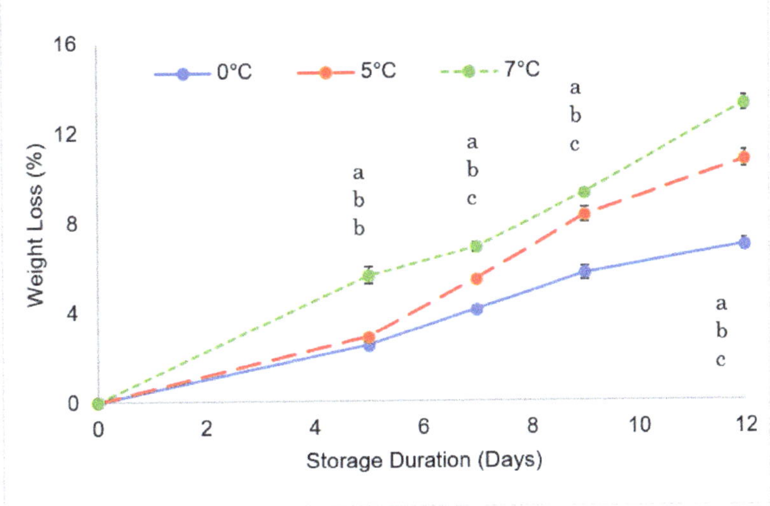

Figure 3. Weight loss changes in goji berries during storage at 0, 5, and 7 °C. Values marked with the same letter on the same sampling day are not significantly different, according to a Tukey's test ($p < 0.05$).

As the storage duration increased the fruit's weight loss, as shown in Figure 3. A significant weight reduction was expected due to the berry's highly perishable nature; however, the rate of weight loss was greatly influenced by storage temperature, probably because thermal conditions strongly influence the water pressure deficit between the fruit tissue and the surrounding air. Fruit stored at 0 °C showed the lowest weight loss (7% on day 12) which was very significantly lower than the weight loss observed for samples stored at 5 °C (11%) and 7 °C (13%). This result confirmed that at a higher temperature storage of 5 and 7 °C we observed higher incidents of shriveling damage due to water loss (Figure 1). Accordingly, [1] reported that goji berries showed a higher weight loss (18%) when stored at 10 °C than when stored at 0 °C (13%) and concluded that a storage temperature of 0 °C was best for retaining fruit freshness for up to 12 days of shelf-life.

Fruit respiration rates increase with storage temperature [25]. Accordingly, in this experiment berries stored at 7 °C showed the highest respiration rate (28.2 mg CO_2/kg/h), while the rates at 5 °C and 0 °C were 13.3 and 5.8 mg CO_2/kg/h, respectively (Table 3). In addition to the effect of temperature on fruit metabolism, the higher respiration values in fruit stored at 7 °C may result from increased cell damage or microbial growth [43]. The rate of ethylene production of goji berry fruit was at a moderate level when stored at 7 °C (3.8 µg C_2H_4/kg*h) and low levels at 0 and 5 °C (0.7 and 0.9 µg C_2H_4/kg*h, respectively). Zhou et al. (2020) reported a respiration rate and ethylene production of 80 mg CO_2/kg/h and 30 µg C_2H_4/kg/h, respectively, for goji berries stored at 4 °C [44]; we observed a lower metabolic activity in comparison to these findings. This difference may be attributed to the different cultivars and growing conditions of goji berries used by Zhou et al. ('Zhongkelvchuan'). Our result is in line with that presented in another study where the respiration rate of goji berries grown in Italy was 27.1 mg CO_2/kg/h at 7 °C [12].

Table 3. Respiration rate and ethylene production in goji berries at different temperatures using five replicates of each temperature measurement.

Temperature	Respiration Rate mg CO_2/kg/h	Standard Error (SE)
7 °C	28.2	3.4
5 °C	13.3	0.8
0 °C	5.8	0.8
Temperature	**Ethylene Production µg C_2H_4/kg*h**	**Standard Error (SE)**
7 °C	3.8	0.1
5 °C	0.9	0.2
0 °C	0.7	0

Figure 4 depicts the electrolytic leakage of the goji berries during storage as being strongly affected by temperature. Electrolytes are contained within membrane-bound compartments in living cells. The proteins and lipids of these membranes are degraded and oxidized under stress (including chilling injury when commodities are sensitive to low temperatures) and during senescence, leading to structural changes that cause loss of integrity and increased membrane permeability [45]. In this case, we may observe a strong effect of storage temperature which influence the changes in REL due to regular senescence. We would have expected that the higher the temperature, the higher the progress in senescence and then the higher the rate of membrane loss of functionality. In this experiment although REL increased for all three treatments after 5 days of storage, the increase was only 2-fold in samples at 7 °C, while samples stored at 5 and 0 °C showed 3- and 4-fold increases, respectively. In addition, while samples stored at 5 and 7 °C showed a further increase in the following 2 days of storage, the value reached in samples stored at 0 °C did not increase with storage time.

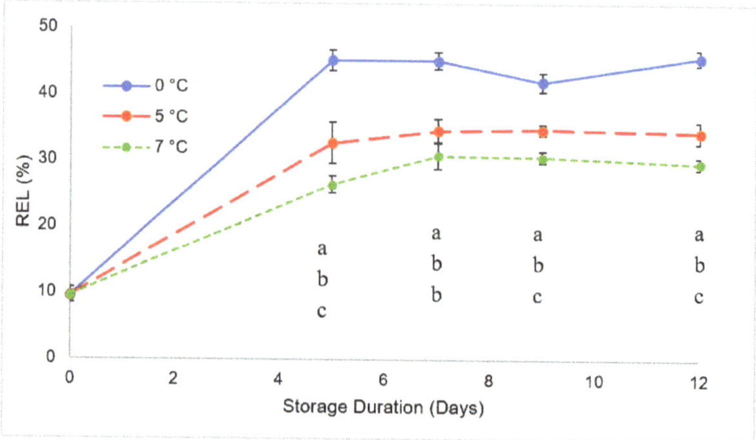

Figure 4. Relative electrolytic leakage of goji berries during storage. Values marked with the same letter on the same sampling day are not significantly different, according to a Tukey's test ($p < 0.05$).

These findings may suggest that while low-temperature damage to membranes was so severe after only 5 days of storage at 0 °C, it may have been caused by biochemical changes in the bilayers which occurred after the primary chilling temperature storage and that may contribute to the development of chilling symptoms reflected by the activated phospholipid catabolism and a consequent accumulation of free fatty acids in the membrane bilayers, contributing to membrane deterioration, and a loss of compartmentation as previously reported in tomatoes [45]. It did not grow any worse as a consequence of tissue senescence, whereas in the other two treatments the damage became more severe after at least 7 days of storage (never reaching the levels showed by berries stored at 0 °C) which indicates it is less likely to be attributed to chilling and more likely to be a result of tissue senescence. Furthermore, it should be taken into consideration that goji berries (as it happens to many other fruits of the Solanaceae family) might be sensitive to temperatures below 10 °C, although the severity of chilling injury symptoms increases with decreasing temperatures. In addition, goji berries are harvested when fully ripe and, at this stage, might better withstand higher ranges of chilling temperatures (e.g., from 5 to 10 °C, such as red tomatoes) as the severity of chilling injury is two times greater in mature green fruits than pink and red fruits at 5 °C [46]. while even a few days of storage at lower ranges might be very critical in terms of quality maintenance.

3.2. Chemical and Nutritional Aspects

Soluble solids slightly decreased during storage (Figure 5A) with the decrease rate at 7 °C significantly higher than that at 0 °C, reaching 22.3% after 12 days, while samples at lower temperatures maintained closer values to the initial values. These results are in line with the findings obtained by [10], who found a decrease in total soluble content from 21 (initial) to 17.32% during storage in the control samples of Chinese wolfberry fruit stored at 2 ± 0.5 °C. The berries used in this experiment had already completed growth, reached full color, and had almost reached maximum values of SSC while on the plant (stage five); a postharvest decrease of SSC in fruit at this stage could be expected due to high respiration and senescence which can be better observed in fruit at 7 °C. On the other hand, harvest at an earlier stage was hardly an option since fruit at stage four have not reached full size yet, thus representing a potentially severe decrease in fruit yield [19]. Nonetheless, the maximum SSC reported for the maturity stage in this study was 23.5%, which remained constant over the 12-day storage period and low storage temperatures delayed senescence, keeping SSC almost unchanged for samples at 5 and 0 °C.

The TA of goji berry fruit changed during storage; namely from 0.7% on harvest day it decreased to about 0.56% at 0 °C and 0.62% at 5 and 7 °C at the end of storage (Figure 5B). Such changes, particularly evident at 0 °C, could be caused by the loss of organic acids and minerals because of chilling stress, as also seen in tomatoes [47]. In addition, the higher percentage of water loss observed in samples at 5 and 7 °C could be the reason for the higher values of TA compared to samples stored at 0 °C, as also observed for SSC in raspberry fruit [48]. These results were reflected in the SSC/TA ratio trend, which at 12 days was highest for fruits stored at 0 °C (41.6) compared with that of fruits stored at 5 and 7 °C (38.2 and 36.2, respectively), as shown in Table 4.

Goji berries are rich in vitamin C, which is beneficial for human health [49]. Figure 6 depicts the vitamin C level as the sum of the contents in AA and DHAA (which are shown in Table 4). Vitamin C decreased during storage for 12 days under all temperature conditions; in particular, it varied from 0.408 to 0.142 g/kg f.w. for fruit stored at 7 °C, which was significantly lower than that for the sample stored at 5 °C (0.175 g/kg f.w.), while samples stored at 0 °C showed an intermediate content (0.163 g/kg f.w.). Goji berries stored at 7 °C showed the lowest vitamin C levels, most probably because of the high levels of oxidation occurring at higher temperatures, as indicated by the higher value of the DHAA content throughout storage. Ascorbic acid is subject to oxidative and enzymatic degradation to dehydroascorbic acid. Ascorbic oxidase is the endogenous enzyme involved in this process [50]. Higher temperature storage might be responsible for the higher oxidation resulting in the higher level of dehydroascorbic acid, which is explained in the results of a study on goji berries which followed the results on kiwi fruits comparing three levels of storage temperature (0, 5, and 10 °C) for 6 days, concluding that the level of DHAA became higher as the temperature storage increased [51].

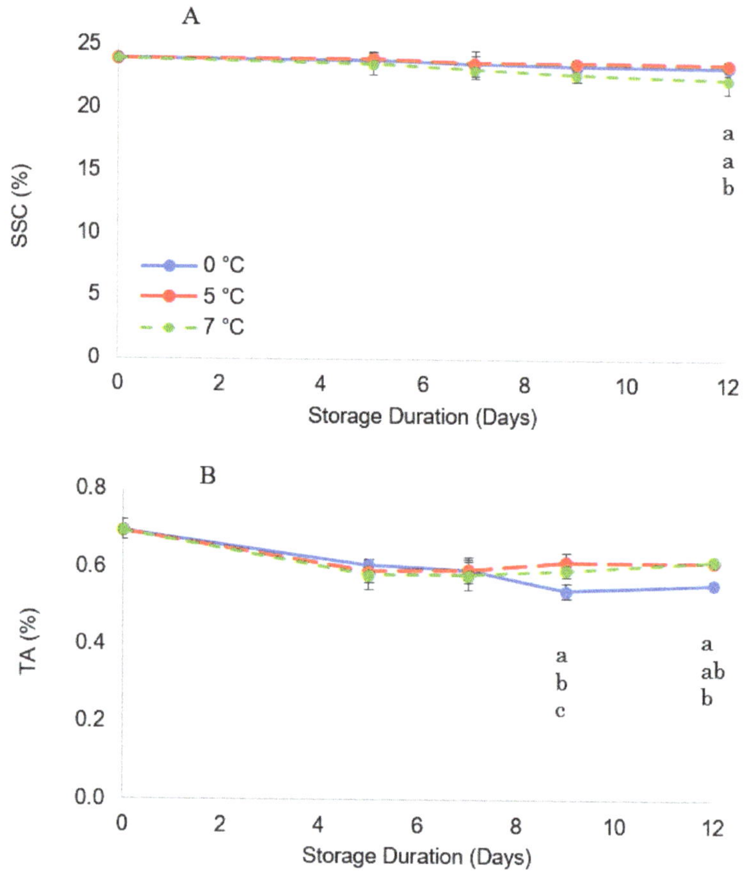

Figure 5. Change in soluble solid content (SSC) (**A**) and titratable acidity (TA) (**B**) during storage. Values marked with the same letter on the same sampling day are not significantly different, according to a Tukey's test ($p < 0.05$).

The higher vitamin C reduction observed at 9 days of storage at 0 °C compared to that at 5 °C may be explained by an intense oxidase activity influenced by the drop of pH (data not shown) as described previously, where the lower pH affects the degradation of vitamin C [52] which may be due to chilling injury, and which is also reported in tomatoes [53].

Our results are consistent with those presented by [54]; they reported that goji berries contain approximately 40 mg/100 g f.w. of vitamin C on the day of harvest, which is comparable with the content of citrus fruits. Therefore, they should be stored in proper conditions since the content is halved after 7 days of storage.

A significant effect of storage temperature was observed in the total polyphenol content of goji berries. Fruit stored at 0, 5, and 7 °C respectively presented values of 2.55, 2.25, and 2.13 g/kg after 12 days (Table 4). The lower levels of total polyphenol at 5 and 7 °C may be attributed to the higher oxidation rates caused by higher temperatures [55]. The range and abundance of phenolic compounds in fruit may vary depending on geographical location, genetic variation, agricultural practices, year of harvest, growth period, or storage conditions [56]. In [54], the authors reported that the content of polyphenolic compounds in *Lycium* spp. was 9.41 g/kg f.w., whereas the authors in [57] reported this value to be 1.42 g/kg.

Table 4. Chemical and nutritional quality of goji berries stored for up to 12 days at 0, 5, and 7 °C. SSC = Soluble solids content; TA = Titratable acidity; AA = Ascorbic acid; DHAA = Dehydroascorbic acid; TP = Total polyphenols; AoxA = Antioxidant activity; AC = Anthocyanin content.

		SSC/TA	AA (g kg^{-1})	DHAA (g kg^{-1})	TP (g gallic Acid kg^{-1})	AoxA (g Trolox kg^{-1})	AC (mg cyanidin-3-glucoside cm^{-2})
Day 0		34.61 ± 1.449	0.254 ± 0.023	0.155 ± 0.015	2.995 ± 0.051	2.157 ± 0.087	1.185 ± 0.03
Day 5	0 °C	39.27 ± 0.45a	0.221 ± 0.01a	0.075 ± 0.006a	2.592 ± 0.014b	1.834 ± 0.039b	1.263 ± 0.389a
	5 °C	40.59 ± 1.11a	0.18 ± 0.01b	0.111 ± 0.043a	2.693 ± 0.071a	1.897 ± 0.025b	0.978 ± 0.412a
	7 °C	40.59 ± 3.67a	0.115 ± 0.01c	0.147 ± 0.029a	2.716 ± 0.009a	2.269 ± 0.08a	1.286 ± 0.441a
Day 7	0 °C	39.73 ± 1.59a	0.193 ± 0.02a	0.067 ± 0.036b	2.352 ± 0.033b	2.106 ± 0.021b	0.853 ± 0.124a
	5 °C	39.76 ± 1.98a	0.14 ± 0.02b	0.105 ± 0.012ab	2.750 ± 0.033a	2.322 ± 0.035a	0.813 ± 0.342a
	7 °C	39.77 ± 2.39a	0.076 ± 0.02c	0.134 ± 0.004a	2.395 ± 0.062b	2.336 ± 0.059a	1.098 ± 0.317a
Day 9	0 °C	43.15 ± 1.68a	0.122 ± 0.032a	0.049 ± 0.008a	2.711 ± 0.063a	2.106 ± 0.07a	0.983 ± 0.203a
	5 °C	38.49 ± 1.76b	0.091 ± 0.023ab	0.112 ± 0.007b	2.374 ± 0.071b	2.322 ± 0.028a	0.857 ± 0.326a
	7 °C	38.33 ± 1.59b	0.052 ± 0.019b	0.133 ± 0.004c	2.258 ± 0.029b	2.336 ± 0.092a	1.159 ± 0.093a
Day 12	0 °C	41.63 ± 0.76a	0.091 ± 0.01a	0.073 ± 0.013b	2.545 ± 0.083a	2.746 ± 0.054b	1.078 ± 0.523a
	5 °C	38.20 ± 0.87b	0.049 ± 0.01b	0.125 ± 0.017a	2.224 ± 0.067b	2.971 ± 0.064ab	1.361 ± 1.028a
	7 °C	36.17 ± 1.46b	0.02 ± 0.01c	0.122 ± 0.007a	2.126 ± 0.054b	3.216 ± 0.082a	0.917 ± 0.168a

In the columns, on the same sampling day, values marked with the same letter are not significantly different according to a Tukey's test ($p < 0.05$).

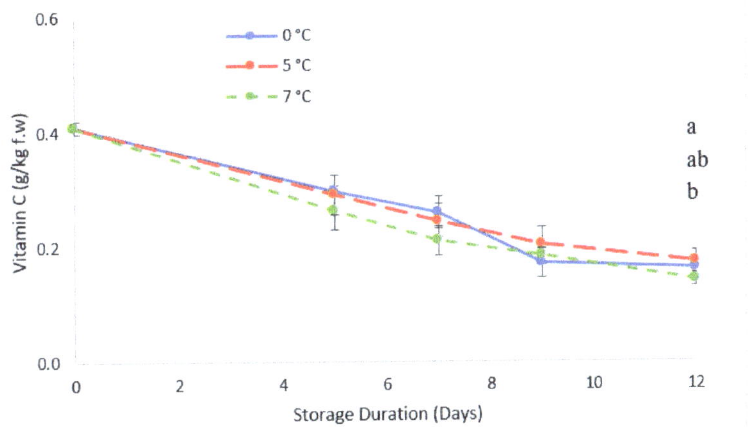

Figure 6. Changes in vitamin C levels during storage. Values marked with the same letter on the same sampling day are not significantly different according to a Tukey's test ($p < 0.05$).

The DPPH antioxidant activity was significantly lower in fruits stored at 0 °C (2.74 g Trolox/kg) than in those stored at 5 °C (2.97 g Trolox/kg) and 7 °C (3.216 g Trolox/kg) on day 12 (Table 4). This may be due to the contribution of the secondary metabolites zeaxanthin and β-carotene—abundant in goji berry fruit which is best conserved at temperatures 7–10 °C [2]. This result is consistent with results presented by [1], where goji berries stored at 10 °C for 12 days showed a higher DPPH antioxidant activity level (2.8 g Trolox/kg) compared with levels in fruits stored at 0 °C (2.3 g Trolox/kg). Furthermore, a significant difference in DPPH activity was observed based on the day of storage, as the level of antioxidant activity increased from the initial day.

Finally, in this study, we did not observe any significant differences in terms of total anthocyanins (expressed as cyanidin) content in goji fruits in relation to storage temperatures which is in accordance with a previous report [1]. Although, it is reported that low temperature induced anthocyanin accumulation on tomato leaves [58] and eggplant during storage at 2 °C [18].

4. Conclusions

Storage temperature is a key factor for the proper storage and handling of fruits, especially for those which are very perishable after harvest such as goji berries. This study complements existing information on the effect of low-storage temperature on goji fruit quality and storability. Our study indicated that 5 °C should be recommended as the appropriate storage temperature over 0 and 7 °C since it induced the lowest level of physiological disorders including pitting, and shriveling while preserving overall sensorial and nutritional quality attributes including vitamin C levels, soluble solids content, and antioxidant activity for 9 days. Although the level of mold was a little higher at 5 °C compared to 0 °C, diffuse chilling symptoms were observed at 0 °C in the form of pitting. These findings indicated that goji fruit may be stored for about 9 days at 5 °C, but additional technologies such as modified atmosphere packaging may be needed to better control decay and to allow safe distribution and consumption.

Author Contributions: Conceptualization, D.F., M.L.A. and G.C.; methodology, D.F. and M.L.A.; software, D.F. and M.L.A.; validation, D.F. and M.L.A.; formal analysis, D.F.; investigation, D.F. and M.L.A.; data curation, D.F. and M.L.A.; writing—original draft preparation, D.F. and M.L.A.; writing—review and editing, D.F., M.L.A. and G.C.; supervision, M.L.A. and G.C.; project administration, M.L.A. and G.C.; funding acquisition, M.L.A. and G.C. All authors have read and agreed to the published version of the manuscript.

Funding: This research was funded by MIUR (Ministero dell'Istruzione, Università e Ricerca) within Program PON ("Dottorati Innovativi con caratterizzazione industriale-2017. Azione: I.1-Dottorati di ricerca innovativi"- CCI: 2014IT16M2OP005 DOT13YISJ8-Borsa: 5).

Data Availability Statement: The data presented in this study are available on request from the corresponding author.

Acknowledgments: We express our gratitude to Masseria Fruttirossi S.r.l, Castellaneta (TA), Italy, for providing samples for this research. We also thank Maria L.V. de Chiara for her contribution related to the formal analysis of vitamin C.

Conflicts of Interest: The authors declare no conflict of interest.

References

1. Jatoi, M.A.; Fruk, M.; Buhin, J.; Vinceković, M.; Vuković, M.; Jemrić, T. Effect of Different Storage Temperatures on Storage Life, Physico-chemical and Sensory Attributes of Goji Berry (*Lycium barbarum* L.) Fruits. *Erwerbs-Obstbau* **2018**, *60*, 119–126. [CrossRef]
2. Kulczyński, B.; Gramza-Michałowska, A. Goji Berry (*Lycium barbarum*): Composition and Health Effects—A Review. *Pol. J. Food Nutr. Sci.* **2016**, *66*, 67–75. [CrossRef]
3. Amagase, H.; Farnsworth, N.R. A review of botanical characteristics, phytochemistry, clinical relevance in efficacy and safety of *Lycium barbarum* fruit (Goji). *Food Res. Int.* **2011**, *44*, 1702–1717. [CrossRef]
4. Benzie, I.; Wachtel-Galor, S. *Herbal Medicine: Biomolecular and Clinical Aspects*, 2nd ed.; CRC Press/Taylor & Francis: Boca Raton, FL, USA, 2011.
5. Potterat, O. Goji (*Lycium barbarum* and *L. chinense*): Phytochemistry, pharmacology and safety in the perspective of traditional uses and recent popularity. *Planta Med.* **2011**, *76*, 7–19. [CrossRef] [PubMed]
6. Zhu, Y.; Zi, G.Q. *Chinese Materia Medica Chemistry, Pharmacology and Applications*; Harwood Academic Publishers: Amsterdam, The Netherlands, 1998.
7. Fan, X.J.; Zhang, B.; Yan, H.; Feng, J.T.; Ma, Z.Q.; Zhang, X. Effect of lotus leaf extract incorporated composite coating on the postharvest quality of fresh goji (*Lycium barbarum* L.) fruit. *Postharvest Biol. Technol.* **2019**, *148*, 132–140. [CrossRef]
8. Ling, L.; Zhao, Y.; Tu, Y.; Yang, C.; Ma, W.; Feng, S.; Lu, L.; Zhang, J. The inhibitory effect of volatile organic compounds produced by Bacillus subtilis CL2 on pathogenic fungi of wolfberry. *J. Basic Microbiol.* **2021**, *61*, 110–121. [CrossRef] [PubMed]
9. Jatoi, M.A.; Jurić, S.; Vidrih, R.; Vinceković, M.; Vuković, M.; Jemrić, T. The effects of postharvest application of lecithin to improve storage potential and quality of fresh goji (*Lycium barbarum* L.) berries. *Food Chem.* **2017**, *230*, 241–249. [CrossRef]
10. Ban, Z.; Wei, W.; Yang, X.; Feng, J.; Guan, J.; Li, L. Combination of heat treatment and chitosan coating to improve postharvest quality of wolfberry (*Lycium barbarum*). *Food Sci. Technol.* **2015**, *50*, 1019–1025. [CrossRef]

11. Kafkaletou, M.; Christopoulos, M.V.; Tsantili, E. Short-term treatments with high CO_2 and low O_2 concentrations on quality of fresh goji berries (*Lycium barbarum* L.) during cold storage. *J. Sci. Food Agric.* **2017**, *97*, 5194–5201. [CrossRef]
12. Palumbo, M.; Capotorto, I.; Cefola, M.; Burbaci, S.; Pace, B. Modified atmosphere packaging to improve the shelf-life of Goji berries during cold storage. *Adv. Hortic. Sci.* **2020**, *34*, 21.
13. Zhang, H.; Ma, Z.; Wang, J.; Wang, P.; Lu, D.; Deng, S.; Lei, H.; Gao, Y.; Tao, Y. Treatment with exogenous salicylic acid maintains quality, increases bioactive compounds, and enhances the antioxidant capacity of fresh goji (*Lycium barbarum* L.) fruit during storage. *LWT* **2021**, *140*, 110837. [CrossRef]
14. Fatchurrahman, D.; Kuramoto, M.; Al Riza, D.F.; Ogawa, Y.; Suzuki, T.; Kondo, N. Fluorescence time series monitoring of different parts of green pepper (*Capsicum annuum* L.) under different storage temperatures. *Comput. Electron. Agric.* **2020**, *179*, 105850. [CrossRef]
15. Mastrandrea, L.; Amodio, M.L.; de Chiara, M.L.V.; Pati, S.; Colelli, G. Effect of temperature abuse and improper atmosphere packaging on volatile profile and quality of rocket leaves. *Food Packag. Shelf Life* **2017**, *14*, 59–65. [CrossRef]
16. Park, M.H.; Sangwanangkul, P.; Choi, J.W. Reduced chilling injury and delayed fruit ripening in tomatoes with modified atmosphere and humidity packaging. *Sci. Hortic.* **2018**, *231*, 66–72. [CrossRef]
17. Fatchurrahman, D.; Kuramoto, M.; Kondo, N.; Ogawa, Y.; Suzuki, T. Identification of UV-fluorescence components associated with and detection of surface damage in Green Pepper (*Capsicum annum* L.). In Proceedings of the 23rd International Conference in Central Europe on Computer, Plzen, Czech Republic, 8–12 June 2015.
18. Tsouvaltzis, P.; Babellahi, F.; Amodio, M.L.; Colelli, G. Early detection of eggplant fruit stored at chilling temperature using different non-destructive optical techniques and supervised classification algorithms. *Postharvest Biol. Technol.* **2020**, *159*, 111001. [CrossRef]
19. Fatchurrahman DAmodio, M.L.; de Chiara, M.L.V.; Mastrandrea, L.; Colelli, G. Characterization and postharvest behavior of goji berry (*Lycium barbarum* L.) during ripening. *Postharvest Biol. Technol.* **2022**, *191*, 111975. [CrossRef]
20. Fatchurrahman, D.; Amodio, M.L.; de Chiara, M.L.V.; Chaudhry, M.M.A.; Colelli, G. Early discrimination of mature-and immature-green tomatoes (*Solanum lycopersicum* L.) using fluorescence imaging method. *Postharvest Biol. Technol.* **2020**, *169*, 111287. [CrossRef]
21. Miller, S.; Hampson, C.; Mc New, R.; Berkett, L.; Brown, S.; Clements, J.; Crassweller, R.; Garcia, E.; Greene, D.; Greene, G. Performance of apple cultivars in the 1995 NE-183 regional project planting: III. fruit sensory characteristics. *J. Am. Pomol. Soc.* **2005**, *59*, 28–43.
22. Navarro-Rico, J.; Martínez-Hernández, G.B.; Artés, F.; Artés-Hernández, F.; Gómez, P.A. Effect of edible coatings and electrolyzed water sanitation on fresh-cut "bimi" broccoli quality. *Acta Hortic.* **2015**, *1071*, 463–469. [CrossRef]
23. Saltveit, M.E. The rate of ion leakage from chilling-sensitive tissue does not immediately increase upon exposure to chilling temperatures. *Postharvest Biol. Technol.* **2002**, *26*, 295–304. [CrossRef]
24. Peng, J.; Tang, J.; Barrett, D.M.; Sablani, S.S.; Powers, J.R. Kinetics of carrot texture degradation under pasteurization conditions. *J. Food Eng.* **2014**, *122*, 84–91. [CrossRef]
25. Kader, A.A. Methods of gas mixing, sampling and analysis. In *Postharvest Technology of Horticultural Crops*, 3rd ed.; University of California: Oakland, CA, USA, 2002; pp. 145–148.
26. Caleb, O.J.; Mahajan, P.V.; Opara, U.L.; Witthuhn, C.R. Modelling the respiration rates of pomegranate fruit and arils. *Postharvest Biol. Technol.* **2012**, *64*, 49–54. [CrossRef]
27. Fatchurrahman, D.; Nosrati, M.; Amodio, M.L.; Chaudhry, M.M.A.; de Chiara, M.L.V.; Mastrandrea, L.; Colelli, G. Comparison performance of visible-nir and near-infrared hyperspectral imaging for prediction of nutritional quality of goji berry (*Lycium barbarum* L.). *Foods* **2021**, *10*, 1676. [CrossRef] [PubMed]
28. Zapata, S.; Dufour, J. Ascorbic, dehydroascorbic and isoascorbic acid simultaneous determinations by reverse phase ion interaction HPLC. *J. Food Sci.* **1992**, *57*, 506–511. [CrossRef]
29. Proctor, J.T.A. Color Stimulation in Attached Apples With Supplementary Light. *Can. J. Plant Sci.* **1974**, *54*, 499–503. [CrossRef]
30. Wells, R. Photosynthetic responses to cutout. In Proceedings of the Beltwide Cotton Conference, San Antonio, TX, USA, 4–7 January 1995; National Cotton Council: Cordova, TN, USA, 1995; pp. 62–64.
31. Siegelman, H.W.; Hendricks, S.B. Photocontrol of anthocyanin formation in turnip and red cabbage seedlings. *Plant Physiol.* **1957**, *32*, 393–398. [CrossRef]
32. Derossi, A.; Mastrandrea, L.; Amodio, M.L.; De Chiara, M.L.V.; Colelli, G. Application of multivariate accelerated test for the shelf life estimation of fresh-cut lettuce. *J. Food Eng.* **2016**, *169*, 122–130. [CrossRef]
33. Capotorto, I.; Amodio, M.L.; Diaz, M.T.B.; de Chiara, M.L.V.; Colelli, G. Effect of anti-browning solutions on quality of fresh-cut fennel during storage. *Postharvest Biol. Technol.* **2018**, *137*, 21–30. [CrossRef]
34. Lan, P.; Gao, F.R.; Chen, C.K.; Wang, W.S.; Han, J.; Ji, H.P.; Yu, J. Separation and identification of pathogenic fungi from the postharvest *Lycium barbarum*. *China Fruit Veg.* **2014**, *34*, 9–12.
35. Zhang, Q.; Chen, W.; Zhao, J.; Xi, W. Functional constituents and antioxidant activities of eight Chinese native goji genotypes. *Food Chem.* **2016**, *200*, 230–236. [CrossRef]
36. Peng, Y.; Ma, C.; Li, Y.; Leung, K.S.-Y.; Jiang, Z.-H.; Zhao, Z. Quantification of Zeaxanthin Dipalmitate and Total Carotenoids in *Lycium* Fruits (Fructus *Lycii*). *Plant Foods Hum. Nutr.* **2005**, *60*, 161–164. [CrossRef] [PubMed]

37. Kasmire, R.F.; Kader, A.A. Handling Tomatoes at Wholesale and Retail: A Guide For Better Quality and Greater Profits. *Outlook* **1978**, *5*, 5–12.
38. Seymour, G.B.; Chapman, N.H.; Chew, B.L.; Rose, J.K.C. Regulation of ripening and opportunities for control in tomato and other fruits. *Plant Biotechnol. J.* **2013**, *11*, 269–278. [CrossRef] [PubMed]
39. Toivonen, P.M.A.; Brummell, D.A. Biochemical bases of appearance and texture changes in fresh-cut fruit and vegetables. *Postharvest Biol. Technol.* **2008**, *48*, 1–14. [CrossRef]
40. Marangoni, A.G.; Jackman, R.L.; Stanley, D.W. Chilling-Associated Softening of Tomato Fruit is Related to Increased Pectinmethylesterase Activity. *J. Food Sci.* **1995**, *60*, 1277–1281. [CrossRef]
41. Jackman, R.L.; Gibson, H.J.; Stanley, D.W. Effects of chilling on tomato fruit texture. *Physiol. Plant.* **1992**, *86*, 600–608. [CrossRef]
42. Dawson, D.M.; Melton, L.D.; Watkins, C.B. Cell Wall Changes in Nectarines (*Prunus persica*). *Plant Physiol.* **1992**, *100*, 1203–1210. [CrossRef]
43. Rodoni, L.M.; Feuring, V.; Zaro, M.J.; Sozzi, G.O.; Vicente, A.R.; Arena, M.E. Ethylene responses and quality of antioxidant-rich stored barberry fruit (*Berberis microphylla*). *Sci. Hortic.* **2014**, *179*, 233–238. [CrossRef]
44. Zhou, Y.; Lai, Y.; Chen, Z.; Qu, H.; Ma, S.; Wang, Y.; Jiang, Y. Evolution of physiological characteristics and nutritional quality in fresh goji berry (*Lycium barbarum*) stored under different temperatures. *J. Food Process. Preserv.* **2020**, *44*, e14835. [CrossRef]
45. Campos, P.S.; Quartin V nia Ramalho J chicho Nunes, M.A. Electrolyte leakage and lipid degradation account for cold sensitivity in leaves of *Coffea* sp. plants. *J. Plant Physiol.* **2003**, *160*, 283–292. [CrossRef]
46. Kalantari, S.; Hatami, M.; Delshad, M. Diverse postharvest responses of tomato fruits at different maturity stages to hot water treatment. *Int. J. Hortic. Sci. Technol.* **2015**, *2*, 67–74.
47. Ibrahim, R.; Rhani, S.A.; Buhri, A. Reduction of chilling injury in tomato (*Solanum lycopersicum*) using different postharvest (pre-storage) treatments. *Acta Hortic.* **2013**, *1012*, 473–478. [CrossRef]
48. Robbins, J.-A.; Sjulin, T.M.; Patterson, M. Postharvest storage characteristics and respiration rates in five cultivars of red raspberry. *HortScience* **1989**, *24*, 980–982. [CrossRef]
49. González-Molina, E.; Domínguez-Perles, R.; Moreno, D.A.; García-Viguera, C. Natural bioactive compounds of Citrus limon for food and health. *J. Pharm. Biomed. Anal.* **2010**, *51*, 327–345. [CrossRef]
50. Saari, N.B.; Fujita, S.; Miyazoe, R.; Okugawa, M. Distribution of ascorbate oxidase activities in the fruits of family cucurbitaceae and some of their properties. *J. Food Biochem.* **1995**, *19*, 321–327. [CrossRef]
51. Lee, S.K.; Kader, A.A. Preharvest and postharvest factors influencing vitamin C content of horticultural crops. *Postharvest Biol. Technol.* **2000**, *20*, 207–220. [CrossRef]
52. Gil-Izquierdo, A.; Gil, M.I.; Conesa, M.A.; Ferreres, F. The effect of storage temperatures on vitamin C and phenolics content of artichoke (*Cynara scolymus* L.) heads. *Innov. Food Sci. Emerg. Technol.* **2001**, *2*, 199–202. [CrossRef]
53. Stevens, R.; Page, D.; Gouble, B.; Garchery, C.; Zamir, D.; Causse, M. Tomato fruit ascorbic acid content is linked with monodehydroascorbate reductase activity and tolerance to chilling stress. *Plant Cell Environ.* **2008**, *31*, 1086–1096. [CrossRef]
54. Donno, D.; Beccaro, G.L.; Mellano, M.G.; Cerutti, A.K.; Bounous, G. Goji berry fruit (*Lycium* spp.): Antioxidant compound fingerprint and bioactivity evaluation. *J. Funct. Foods* **2015**, *18*, 1070–1085. [CrossRef]
55. Réblová, Z. Effect of temperature on the antioxidant activity of phenolic acids. *Czech J. Food Sci.* **2012**, *30*, 171–177. [CrossRef]
56. Dong, J.Z.; Wang, S.H.; Zhu, L.; Wang, Y. Analysis on the main active components of *Lycium barbarum* fruits and related environmental factors. *J. Med. Plants Res.* **2012**, *6*, 2276–2283.
57. Wang, C.C.; Chang, S.C.; Inbaraj, B.S.; Chen, B.H. Isolation of carotenoids, flavonoids and polysaccharides from *Lycium barbarum* L. and evaluation of antioxidant activity. *Food Chem.* **2010**, *120*, 184–192. [CrossRef]
58. Liu, Y.; Tikunov, Y.; Schouten, R.E.; Marcelis, L.F.M.; Visser, R.G.F.; Bovy, A. Anthocyanin biosynthesis and degradation mechanisms in Solanaceous vegetables: A review. *Front. Chem.* **2018**, *6*, 52. [CrossRef] [PubMed]

foods

MDPI

Article

Physiochemical Responses of the Kernel Quality, Total Phenols and Antioxidant Enzymes of Walnut in Different Forms to the Low-Temperature Storage

Yanping Ma [1], Chaoye Wang [1], Chaobin Liu [1], Jiawei Tan [1], Huiling Ma [2] and Jin Wang [3,4,*]

1 College of Forestry, Northwest A&F University, Xianyang 712100, China; myp1273@163.com (Y.M.); 18220183723@163.com (C.W.); liuchaobin@126.com (C.L.); tanjiawei@nwafu.edu.cn (J.T.)
2 College of Life Science, Northwest A&F University, Xianyang 712100, China; hl65@nwafu.edu.cn
3 Key Laboratory of Environmental Medicine and Engineering, Ministry of Education, and Department of Nutrition and Food Hygiene, School of Public Health, Southeast University, Nanjing 210009, China
4 Department of Bioresource Engineering, Faculty of Agricultural and Environmental Sciences, McGill University, Sainte-Anne-de-Bellevue, QC H9X 3V9, Canada
* Correspondence: jinwang_2020@seu.edu.cn or jin.wang6@mail.mcgill.ca

Abstract: Fresh walnut is obtaining high attention due to its pleasant taste and health benefits. This study aimed to evaluate the influence of storage temperatures (0 °C and −20 °C) on the kernel quality, total phenols, and antioxidant enzyme activities of walnuts in three forms (fresh kernels, walnuts with green husk, and walnuts with shell). For a short storage within 3 months at 0 °C, the results revealed that walnuts with green husk provided a better walnut kernel quality resulting from its lower acid value and peroxide value, together with a higher total phenol content and total antioxidant activity, compared with other forms of walnuts. In comparison, frozen storage at −20 °C for a long duration (up to 10 months), found that walnuts with shell showed advantages in improving the kernel quality (fatty acid content, total phenols, and total antioxidant activity) and antioxidant enzyme (peroxidase, catalase, and superoxide dismutase) activities in the kernels, leading to an acceptable range of acid value and peroxide value, compared with other forms of walnuts. Thus, frozen storage at −20 °C showed a potential application in maintaining the walnut kernel quality, especially the walnuts with shell.

Keywords: fresh walnut; total phenols; frozen storage; fatty acid; antioxidant enzyme

Citation: Ma, Y.; Wang, C.; Liu, C.; Tan, J.; Ma, H.; Wang, J. Physiochemical Responses of the Kernel Quality, Total Phenols and Antioxidant Enzymes of Walnut in Different Forms to the Low-Temperature Storage. *Foods* **2021**, *10*, 2027. https://doi.org/10.3390/foods10092027

Academic Editors: Maria Cefola and Bernardo Pace

Received: 31 July 2021
Accepted: 24 August 2021
Published: 28 August 2021

1. Introduction

Walnut (*Juglans regia* L.) has obtained high attention by producers due to its economic values. In 2017, the Food and Agriculture Organization of the United Nations (FAO) reported that walnut production reached 1,819,400 tons and was valued at 7665 million US dollars in China, accounting for 50% of the total global walnut production (FAOSTAT). Furthermore, walnuts are one of the most popular tree nuts attributing to their nutritional values. The United States Department of Agriculture (USDA) Nutrient Database shows that 100 g of dried walnuts contains 16.67 g of protein, 13.33 g of carbohydrates, 6.7 g of dietary fiber, 440 mg of Potassium, and 97 mg of Calcium [1]. Also, walnuts are rich in health-promoting bioactive compounds, including ω-3 fatty acids, plant sterols, polyphenols, and bioactive peptides [2–4]. Many studies have reported that these bioactive compounds can aid against aging, cancers, metabolic syndrome, diabetes, and cardiovascular-related diseases [5–7].

Nowadays, consuming walnuts in fresh form is becoming more common due to their superior flavor, higher Vitamin E, polyphenol and antioxidant content compared to walnuts in dried form [8–10]. However, fresh walnuts have a very short shelf life due to their high moisture content and perishable characteristics. Previous studies reported that modified atmosphere packaging combined with cold storage can extend the shelf life of fresh walnuts for 2 months [9]. ^{60}Coγ-irradiation can improve the shelf life of fresh walnuts up to 3 months [8]. However, the applications of these techniques are limited due to the lack of cold storage room and related equipment in most areas of China. Therefore, more practical methods following less procedures are in need to improve the shelf life of fresh walnuts.

Freezing technology for food preservation has been used by humans for thousands of years. The application of freezing technology has provided many commercial products including broccoli, potatoes, corn, green beans, strawberries, cherries, raspberries, and litchi [11–13]. Studies compared these frozen products with fresh ones and found that no significant differences were observed in the color attributes, taste, and nutrient content [14,15]. Further, the freezing processing is easy to undertake in most families with refrigerators/freezers in the modern world.

However, very limited studies were reported regarding the influences of the freezing technology on the quality of fresh walnuts, especially in different forms of fresh walnuts: fresh kernel, fresh walnuts with shell, fresh walnuts with green husk. Therefore, the objective of the present study is to investigate the physiochemical responses of the kernel quality, antioxidant activity, and protective enzymes, including superoxide dismutase (SOD), peroxidase (POD), and catalase (CAT) to low temperatures under three different forms (fresh kernel, fresh walnuts with shell, fresh walnuts with green husk) during freezing. Hopefully, this study can provide technical support for the preservation of fresh walnuts.

2. Materials and Methods

2.1. Plant Material and Treatments

Fresh walnuts (*Juglans regia* L. cv. *Xiangling*) with green husk were harvested from a local farm (Zhouzhi, GPRS: Lo-108.22207; La-34.16337) in Xi'an, Shaanxi Province, China. After harvest, the walnuts were transferred to the lab and pre-cooled at 0 °C for 24 h. Fresh walnuts were prepared in three forms: walnuts with green husk (Figure 1a), walnuts with shell (Figure 1b), and fresh kernels with seed coat (Figure 1c), and were stored at 0 °C and −20 °C. Each treatment was set in triplicate containing 400 walnuts. During the period of storage, nine sample points were set at the months of 0, 1, 2, 3, 4, 5, 6, 8, and 10. Fifteen walnuts were taken at each point, and their kernels were used for related measurements.

Figure 1. Different forms of fresh walnuts after harvesting: (**a**) walnut with green husk; (**b**) fresh walnut with shell; (**c**) fresh kernels with seed coat; (**d**) fresh kernels without seed coat.

2.2. Relative Electrical Conductivity (REC)

Walnut kernel powder (3 g) was mixed with 30 mL of double distilled water and was incubated at 25 °C for 1 hr. The electrical conductivity (P1) was determined by using a conductometer (DDS-307A, Inesa-Instrument Co., Ltd., Shanghai, China) [16]. Then, the mixed walnut solution was boiled in a water bath for 15 min. After cooling, the value of the electrical conductivity (P2) was recorded. The electrical conductivity of the walnut samples was calculated using the following equation:

$$REC(\%) = \frac{(P_1 - P_0)}{(P_2 - P_0)} \times 100\% \quad (1)$$

2.3. Total Phenol Analysis

According to the method described by Wang et al. (2019), one gram of the walnut sample was mixed with 10 mL of 70% ethanol (v/v), and the mixture was extracted at room temperature for 30 min. After centrifuging at 8000× g for 10 min, the supernatant was stored at 4 °C for the analyses of total phenols and antioxidant activity. The total phenol content of the walnut kernel was determined using the Folin-Ciocalteau assay [17]. Specifically, 0.2 mL of extracts were mixed with 6 mL of double-distilled water, 0.5 mL of Folin-Ciocalteau reagent, and 1.5 mL of a 20% sodium carbonate solution. The mixture was transferred at 75 °C for 10 min, and then the absorbance changes were recorded at 765 nm using a spectrophotometer (UV-3100, Mapada Co. Ltd., Shanghai, China). Gallic acid was used to obtain the standard curve.

2.4. Antioxidant Activity

2.4.1. Ferric Reducing Antioxidant Power (FRAP) Assay

According to the method mentioned by Benzie and Strain (1996), an FRAP working solution was obtained by mixing together an acetic acid buffer (40 mM), a ferric chloride solution (20 mM), and 2,4,6-Tripyridyl-S-triazine (TPTZ) by the rate of 1:1:10 ($v/v/v$) [18]. The walnut kernel extract (0.2 mL) was incubated with 3 mL of FRAP at 37 °C for 30 min. Then, the absorbance of the mixture was recorded at 593 nm. The standard curve was obtained using a 1.0 mol/L $FeSO_4$ solution, and the results were expressed as mmol $FeSO_4$/g FW of walnut kernels.

2.4.2. DPPH Radical Scavenging Activity

DPPH radical scavenging activity was another method used to evaluate the antioxidant activity of the walnut kernels. The walnut kernel extract was incubated with a DPPH solution by the rate of 1:1 at room temperature for 30 min. The color changes were recorded at 517 nm. According to the method described by Wang et al. (2019), the antioxidant activity was calculated by using the following equation:

$$DPPH(\%) = \frac{(Abs_0 - Abs_1)}{Abs_0} \times 100 \quad (2)$$

where Abs_0 is the absorbance value of the blank sample, and Abs_1 is the absorbance of samples [17].

2.5. Activities of Antioxidant Enzymes: CAT, SOD, and POD

One gram of the walnut kernel sample was grounded with 8 mL of a pre-cooled phosphate buffer (0.1 M, pH 6.8) to extract the relevant antioxidant enzymes [10]. After 30 min incubation at 4 °C, the mixture was centrifuged at 10,000× g for 10 min. The supernatant was collected and stored at 4 °C until further analysis.

The CAT activity of the walnut kernels was measured by mixing 400 μL of an enzyme extract, 6 mL of the phosphate buffer (0.1 M, pH 6.8), and 0.2 mL of H_2O_2 (2%). One unit of CAT activity was defined as an increase of 0.01 absorbance unit per min at 240 nm [19]. The activity of SOD was obtained using 3 mL of a working solution containing 100 μL of enzyme extract and nitroblue tetrazolium. One unit of SOD activity is defined as the amount of enzyme that causes 50% inhibition of nitroblue tetrazolium [19]. The absorbance changes of a mixed solution containing 1 mL of enzyme extract, 1 mL of phosphate buffer (0.1 M, pH 6.8), 3 mL of guaiacol (25 mM), and 0.2 mL of H_2O_2 (2%) was recorded at 410 nm for 5 min to analyze the POD activities of the walnut kernels. One unit of POD activity was defined as an increase of 0.01 absorbance units per min [9].

2.6. Changes of O_2^-, H_2O_2, and MDA Contents in Walnut Kernels

The walnut kernels (5 g) were grounded with 15 mL of a pre-cooled phosphate buffer (0.01 M, pH 7.0) and homogenized at 4 °C for 45 min [20]. After centrifuging at 10,000× *g* for 10 min, the supernatant was collected for the analyses of O_2^- and H_2O_2 production. The generation of O_2^- was measured by adding 0.5 mL of the supernatant, 0.5 mL of the pre-cooled phosphate buffer (0.01 M, pH 7.0), and 0.2 mL of hydroxylamine hydrochloride (0.01 M). After incubation at 25 °C for 1 h, 1 mL of p-aminobenzenesulfonic acid (17 mM) and 1 mL of α-naphthyl amine (0.05 M) was added. The color absorbance was measured at 530 nm. Standard curving was obtained using sodium nitrite to calculate the generation of O_2^- and the results were expressed as mmol/g FW [20].

For the determination of the H_2O_2 content, 0.4 mL of the supernatant was mixed with 0.4 mL of the phosphate buffer (0.01 M, pH 7.0), and 0.8 mL of a working solution (potassium dichromate: glacial acetic acid: water = 1:5:15, *w*/*v*/*v*) [9]. After incubating at room temperature for 15 min, the color changes were recorded at 570 nm, and the content of H_2O_2 was expressed as mmol/g. Frozen walnut samples were extracted with 10% of a trichloroacetic acid buffer and incubated at room temperature for 30 min [21]. After centrifuging at 8000× *g* for 20 min, 0.5 mL of the supernatant was mixed with 0.5 mL of thiobarbituric acid (0.5%) and incubated at 100 °C for 15 min, and then the mixture was determined at 450 nm, 532 nm, and 600 nm, respectively. The concentration of MDA was expressed as µmol/g.

2.7. Acid Value (AV) and Peroxide Value (PV)

In the study, 8 g of fresh walnut kernels were dried at 85 °C. Petroleum ether (boiling range, 30–60 °C) was used as a solvent to extract the oil sample at 40 °C for 12 h through a Soxhlet extractor. The oil sample was stored at 4 °C for the analyses of the acid value, peroxide value and fatty acids composition. The standard of GB/T5009.229-2016a and GB/T 5009.229-2016b was used to evaluate the acid value and peroxide value of the walnut kernels, respectively [22,23]. Each analysis was performed in triplicate.

2.8. Fatty Acids Composition

A gas chromatography (6890N, Agilent Technologies, Wokingham, UK) equipped with a DB5 column (0.32 mm × 30 cm) and a flame ionization detector (FID) were used to determine the fatty acid composition in the walnut kernels, according to the method as described by Esteki et al. (2017), with some modifications. The flow rate of the nitrogen gas and H_2 gas was set at 40 mL/min [24]. The injection volume of the oil sample was 0.3 µL, with a split ratio at 30:1. The initial oven temperature was set at 150 °C and held for 3 min. The oven temperature was set to increase to 220 °C at the rate of 5 °C/min and held for 10 min. The injector and detector temperatures were set at 260 °C [24]. Polyunsaturated fatty acid (PUFA) and unsaturated fatty acid (UFA) contents were calculated according to the contents of individual fatty acids. The ChemStation software was used to analyze data obtained from the study.

2.9. Statistical Analysis

All the treatments and analyses were performed in triplicate. Results obtained from the experiment were compared by a one-way analysis of variance (ANOVA) using SPSS 22.0 software (SPSS Inc., Chicago, Illinois, USA). All the data was represented by mean values ± standard deviation (SD), and the significant differences of mean values between samples were evaluated using Duncan's multiple range test ($p \leq 0.05$).

3. Results and Discussion

3.1. Relative Electrical Conductivity (REC)

As shown in Figure 2, an increasing trend in the REC of walnut kernels was observed under different storage conditions. Specifically, no significant difference in the REC was detected in kernels obtained from walnuts with shell and walnuts with green husk on the 2-month storage at 0 °C, while the fresh kernels showed a higher REC. In comparison, a two-fold increase in the REC of kernels present in the walnuts with green husk and walnuts with shell was observed by the end of a 3-month storage of (Figure 2a). The REC was considered as an indicator to evaluate the kernel quality of nuts through testing the concentration of leachates in the kernel soaking solution, reflecting the damage degree and integrity of the cell membrane [25]. The walnuts with shell and green husk showed a lower conductivity compared with fresh kernels, which indicates shell and green husk can prevent kernels from cell membrane damage contributing to a high physiological potential, during the first 2-month storage at 0 °C. Thereafter, the significant increase in the REC might be due to the very limited shelf life when stored at 0 °C [10].

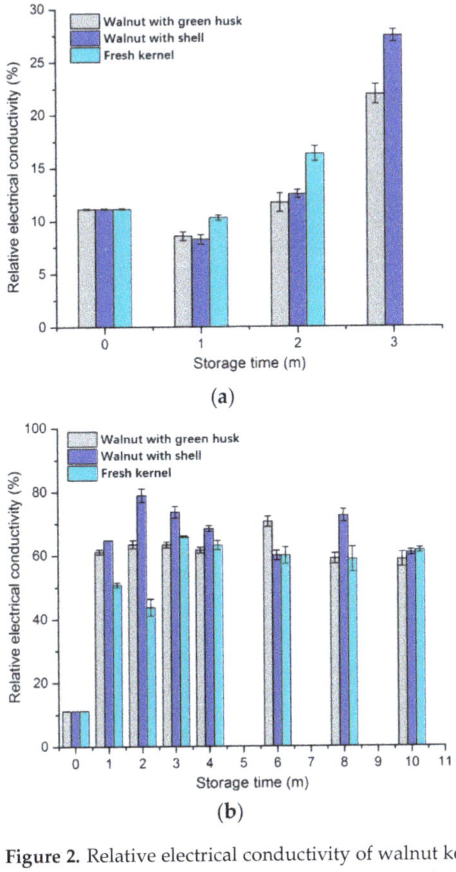

Figure 2. Relative electrical conductivity of walnut kernels under different storage conditions: (**a**) 0 °C and (**b**) −20 °C. Note: m means month in the figure. Data of fresh kernel stored at 0 °C are not available due to the limited shelf life.

In comparison, a six-fold increase in the REC of kernels was observed by the end of the 1-month storage at −20 °C, when compared with the initial level (Figure 2b). Then, for the next 10-month storage, the REC of kernels was maintained at a similar level. It indicates that frozen storage remarkably accelerates the damage to the cell membranes of walnut kernels in all three forms, which is primarily attributed to the growth of ice crystals in the cells, resulting in the breakage of kernel cells [26].

3.2. Total Phenols Content

As shown in Table 1, the total phenol content of the walnut kernels increased during the 3-month storage at 0 °C. The highest phenol content was observed in the walnut kernels with green husk (72.82 mg/100 g FW), followed by the walnuts with shell (65.44 mg/100 g FW) on the third month, while the values of fresh kernels were not available due to the limited shelf life at 0 °C. In comparison, the total phenol content of the walnut kernels stored at −20 °C showed a fluctuating trend. The total phenol content of the fresh kernels increased up to 85.36 mg/100 g from the initial level (44.23 mg/100 g FW) in the second month, and then decreased to 30.55 mg/100 g FW at the end of storage (Table 1). For the walnuts with shell, the total phenols showed a significant enhancement from 44.23 to 77.43 mg/100 g FW after a 1-month storage, and then it was maintained at a stable range of 51.37–75.89 mg/100 g FW during the subsequent duration. While the total phenol content was maintained at a stable level (44.23–53.84 mg/100 g FW) in the walnuts with green husk during the first 6-month storage at −20 °C, it then showed a two-fold increase to 115.98 mg/100 g FW by the end of the 10-month storage.

A higher level of total phenols was observed in the walnuts with shell and green husk compared with the fresh kernels during the storage. After the 10-month storage at −20 °C, the total phenols present in the walnuts with green husk was increased by 2–4 times higher when compared with the initial level. Similarly, in hazelnuts, a study found that a low temperature (−25 °C) can maintain the phenol concentration of kernels during a 12-month storage [27]. This increase in the total phenol content might be strongly associated with the presence of shell and green husk, which can prevent the phenols from oxidation. Studies have reported that a transformation of compounds between the green husk and kernels of walnuts continues after being picked from the trees, which in turn could contribute to the synthesis of phenols in the kernels during storage [9,10]. The dramatic increase in the total phenols of walnuts with green husk in the late storage stage (8–10 months) might be attributed to the closed environmental stresses obtained from the green husk under the frozen conditions.

3.3. Total Antioxidant Activity (TAC)

In the present study, two methods, including FRAP and DPPH assays, were used to quantify the antioxidant activity of the walnut kernels. As shown in Table 1, the results found that the TAC of kernels was improved when stored at 0 °C using FRAP assay. The highest antioxidant activity was observed in the walnuts with green husk (4.90 mmol/g), followed by the walnuts with shell (4.71 mmol/g), while the data of fresh kernels was not available because of their short shelf life. It indicates that walnuts, especially the form with green husk, showed advantages in maintaining or improving the TAC of kernels during a short period of storage at 0 °C. However, no significant difference in the TAC of walnut kernels in three forms was observed when stored at −20 °C. During the 10-month storage, the highest average value of TAC was observed in the walnuts with shell (3.57 mmol/g), followed by fresh kernel (3.39 mmol/g), and walnuts with green husk (2.79 mmol/g). The higher presence of TAC in the walnuts with shell was strongly associated with its higher total phenol content (Table 1). In addition, the results revealed that frozen storage could be beneficial to maintaining the TAC of walnut kernels.

Table 1. Total phenols content and antioxidant activity of walnut kernels during the storage of 0 °C (a) and −20 °C (b). Note: Na means not available. Values with different lowercase letters in the same column 'a–d' and uppercase letter 'A–C' in the same row are significantly different ($p < 0.05$) from each other during storage.

Storage Time	(Month)	Total Phenols (mg/100 g FW)			FRAP (mmol/g FW)			DPPH (%)		
		Fresh Kernel	Walnut with Shell	Walnut with Husk	Walnut Kernel	Walnut with Shell	Walnut with Husk	Fresh Kernel	Walnut with Shell	Walnut with Husk
0 °C	0	44.23 ± 0.62Ab	44.23 ± 0.62Ac	44.23 ± 0.62Ac	2.92 ± 0.26Ab	2.92 ± 0.26Ac	2.92 ± 0.26Ab	16.09 ± 1.25Ab	16.09 ±1.25 Ab	16.09 ± 1.25Ab
	1	48.55 ± 0.55Ba	35.77 ± 1.02Cd	59.62 ± 0.73Ab	5.67 ± 0.24Aa	4.31 ± 0.16Aa	4.16 ± 0.22Aa	16.68 ± 0.85Bb	17.48 ± 1.13Bb	22.42 ± 0.45Aa
	2	46.24 ± 1.32ab	51.35 ± 1.51b	57.52 ± 0.94Abc	4.06 ± 0.33Aab	3.66 ± 0.10Aab	2.92 ± 0.13Ab	21.31 ± 1.14Aa	21.67 ± 1.24Aa	14.30 ± 1.83Bb
	3	Na	65.44 ± 2.11Ba	72.82 ± 0.88Aa	Na	4.71 ± 0.11Aa	4.90 ± 0.15Aa	Na	20.94 ± 0.70Aa	23.48 ± 0.77Aa
−20 °C	1	55.00 ± 1.51Bb	77.43 ± 1.77Aa	53.00 ± 0.88Bc	3.94±0.41Aa	3.97 ± 0.29Aa	2.57±0.35Ab	21.29 ± 0.79Aab	16.27 ± 0.75Ac	18.46 ± 1.72Ab
	2	85.36 ± 1.85Aa	55.98 ± 1.56Bc	53.84 ± 1.26Bc	3.73 ± 0.23Aa	2.94 ± 0.28Ab	3.21 ± 0.10Aa	11.44 ± 1.19Bc	21.04 ± 0.19Ab	10.83 ± 0.69Bc
	3	56.99 ± 1.57Bb	72.85 ± 2.02Aa	49.89 ± 2.05Bc	3.62 ± 0.38Aa	4.15 ± 0.17Aa	3.16 ± 0.15Aa	23.69 ± 0.44Aa	26.39 ± 0.50Aa	21.66 ± 0.24Ab
	6	47.35 ± 1.47Cc	66.29 ± 1.73Ab	52.44 ± 1.54Bc	3.68 ± 0.46Aa	3.84 ± 0.29Aab	2.27 ± 0.29Ab	24.62 ± 0.46Aa	21.26 ± 0.06Ab	18.09 ± 0.36Ab
	8	47.48 ± 2.25Cc	75.89 ± 2.65Ba	96.10 ± 2.10Aa	2.99 ± 0.46Aab	3.19 ± 0.46Ab	3.12 ± 0.13Aa	16.68 ± 0.29Ab	20.72 ± 0.35Ab	19.96 ± 0.18Ab
	10	30.55 ± 1.31Cd	51.37 ± 1.31Bc	115.98 ± 0.77Aa	2.41 ± 0.25Ab	3.34 ± 0.48Ab	2.44 ± 0.17Ab	24.13 ± 1.49Aa	23.38 ± 0.41Aab	26.78 ± 1.30Aa

In regards to the DPPH radical scavenging activity, similar results were observed during the storage at 0 °C (Table 1). The DPPH radical scavenging activity in the fresh kernels and walnuts with shell significantly increased to 21% from the initial level of 16%, while a decrease in the walnuts with husk (14%) was detected by the end of the 2-month storage at 0 °C. During the 10-month frozen storage at −20 °C, the DPPH radical scavenging activity of walnuts with shell gradually increased from 16% to 26%. However, a dramatic decrease (10–11%) in the DPPH radical scavenging activity was investigated in the fresh kernel and walnuts with green husk at the end of the 2-month storage. This fluctuating in the DPPH radical scavenging activity might be attributed to the total phenol content and TAC. Similarly, many studies have reported that a positive correlation between DPPH and total phenols was observed in mangoes, apples, and tomato fruit [28–30].

3.4. SOD, CAT, and POD Activity

SOD, CAT, and POD are three key enzymes which play important roles in catalyzing superoxide radicals, hydrogen peroxides and hydroperoxides into harmless molecules (H_2O_2/alcohol and O_2) [31]. In the present study, these three enzymes showed a fluctuating trend when stored at different conditions (Figure 3). During the first month storage at 0 °C, the SOD activity of the kernels was enhanced and decreased to the initial level in the next month storage, and then a jump enhancement was observed at the end of the 3-month storage (Figure 3a). When frozen storage at −20 °C was applied, walnuts with green husk showed a stable level of SOD with a small peak at the end of the 4-month storage, whereas the peak value of SOD in the walnuts with shell and green husk increased by 2–3 folds compared with the initial level (Figure 3b). After a 6-month storage, the SOD activity of the kernels was maintained at a stable level. A higher level of SOD was observed in the frozen stored walnut samples, which may be due to more SOD being generated in the kernel against low-temperature stress [32,33].

In comparison, an increasing trend in the CAT activity was detected in the kernels when stored at 0 °C during the first 2-month storage, and then a dramatic decrease was observed in the walnuts with shell (Figure 3a). During the storage at −20 °C, the CAT activity of the kernels obtained from the walnuts with shell and green husk significantly increased by 35% from the initial level at the end of the 2-month storage, and then maintained at a stable range (140–160 U min/g FW). In comparison, the CAT activity of the walnuts with green husk started to decrease to a stable range (80–100 U min/g FW) after a 2-month frozen storage at −20 °C. It suggests that green husk played a role in inhibiting the activity of CAT when stored at −20 °C for 10-months, which might be related to the closed environmental condition generated by the green husk. In regards to the POD activity, the results presented a significant decrease (85%) compared with the initial level during the storage at 0 °C and −20 °C (Figure 3a,b). It indicates that low-temperature storage contributed to inhibiting the activity of the POD present in the kernel during the storage.

3.5. MDA, H_2O_2, and O_2^- Production

MDA was considered as an important marker to evaluate the membrane lipid peroxidation during the storage of fruits and nuts [20,34]. As shown in Figure 4, the production of MDA in walnut kernels stored at 0 °C decreased with the rise of storage time, while an increasing trend was observed when stored at −20 °C, especially after the 2-month storage. The highest average value of MDA production was detected in the fresh walnut kernels (6.47 μmol/g FW), followed by the walnuts with shell (6.16 μmol/g FW) and walnuts with green husk (5.78 μmol/g FW) (Figure 4b). The results indicate that green husk could contribute to a lower MDA content in the walnut kernels when stored at −20 °C compared with the other forms of walnuts.

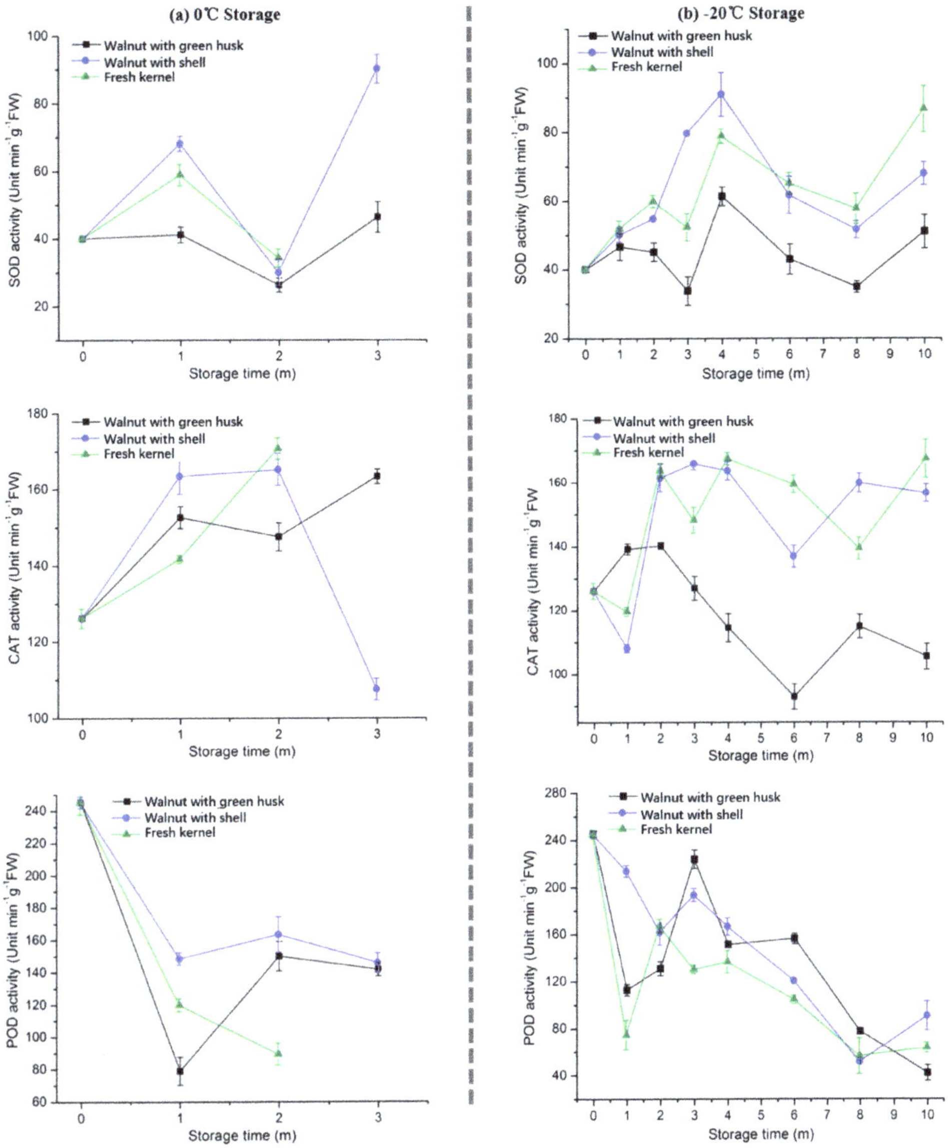

Figure 3. SOD, CAT, and POD activity of walnut kernel under different storage conditions: (**a**) 0 °C and (**b**) −20 °C. Note: m means month in the figure.

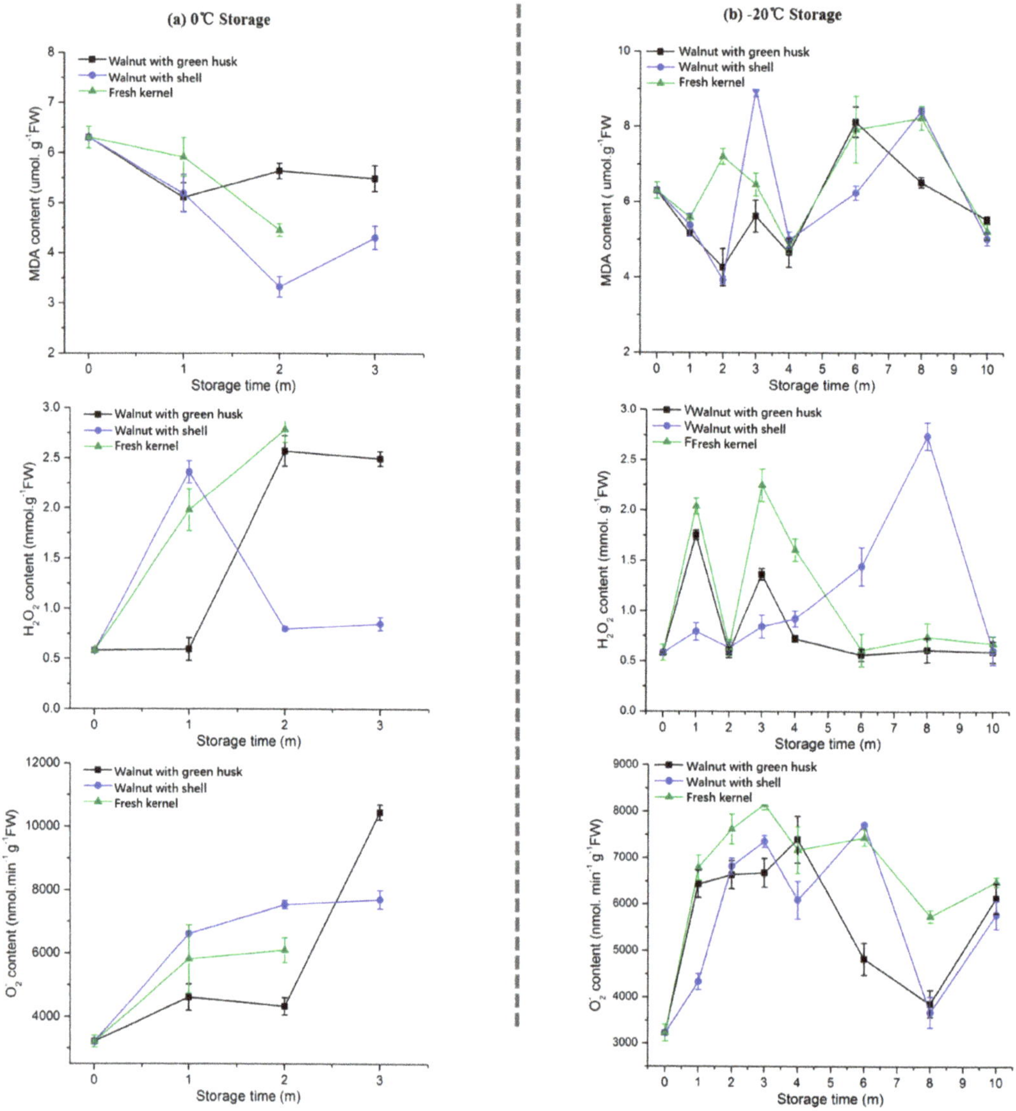

Figure 4. MDA, H_2O_2, and O_2^- production of walnut kernel under different storage conditions: (**a**) 0 °C and (**b**) −20 °C. Note: m means month in the figure.

In comparison, the production of H_2O_2 in the kernels under various conditions showed an increasing trend with the increase of storage time (Figure 3a,b). During the first month of storage at 0 °C, the production of H_2O_2 in the fresh kernels and the kernels obtained from walnuts with shell increased by four-fold compared with the initial level, while the kernels obtained from the walnuts with the green husk maintained a stable level of H_2O_2. During the frozen storage at −20 °C, the production of H_2O_2 in the fresh kernels and walnuts with shell increased sharply during the first 4-month storage and then maintained at the initial level. However, the walnuts with shell delayed the presence of the peak of H_2O_2 maintaining within a stable range (0.6–0.8 mmol/g FW), and then the H_2O_2 was synthesized abundantly from the 4-month to the 8-month storage at −20 °C. Thus, the inhibiting effect on the production of H_2O_2 obtained from the low temperatures was different in various storage conditions. Similarly, an increasing trend was observed in the production of O_2^- in the kernels during the first 3-month storage at 0 °C and −20 °C. After the 4-month frozen storage, the production of O_2^- was reduced to a stable level. Thus, frozen storage can inhibit the production of O_2^- in the kernels of late storage duration.

3.6. Fatty Acid Composition

Fatty acids present in the nuts are considered as the key indicator for evaluating the kernel quality [8]. As shown in Table 2, the changes of fatty acid in the kernels stored at −20 °C were observed, while the relevant data was not available because of the limited shelf life of fresh kernels when stored at 0 °C. Five compositions of fatty acids, including palmitic acid, stearic acid, oleic acid, linoleic acid, and α-linolenic acid, were detected using gas chromatography. Among them, the palmitic acid (16:0) content in three forms of walnut kernels decreased from the initial level (9.21%) to 7.29–7.76% after an 8-month storage at −20 °C. Whereas, the oleic acid (18:0) and unsaturated fatty acid level significantly increased in the kernels obtained from the three forms of walnuts (fresh kernel, walnuts with shell, and walnuts with green husk) compared with their initial concentration. In addition, no significant difference in the stearic acid (18:1), linoleic acid (18:2), linolenic acid (18:3), and polyunsaturated fat content of the kernels was observed during the 8-month storage at −20 °C. As shown in Figure 5, the highest peak area of five compositions of fatty acids was observed in the walnuts with shell, followed by fresh kernels and walnuts with green husk at the end of the 8-month storage at −20 °C. The results indicate that frozen storage maintained the majority of fatty acid content, especially in the kernels of walnuts with shell. In *Canarium* nuts, a significant reduction in the fatty acid of frozen kernels compared with the fresh kernels [35]. In butter, a study showed no significant changes in the fatty acid content during a 24-month frozen storage at −20 °C [36]. Thus, the property changes in fatty acids are different in various food samples.

Table 2. Fatty acid changes of walnut kernels during the storage of −20 °C. Note: Values with different lowercase letters in the same column 'a–c' and uppercase letter 'A–B' in the same row are significantly different ($p < 0.05$) from each other during storage.

Fatty Acid	Storage Time Month	Fatty Acid Composition (%)		
		Fresh Kernel	Walnut with Shell	Walnut with Green Husk
Palmitic acid (16:0)	0	9.21 ± 0.01Aa	9.21 ± 0.01Aa	9.21 ± 0.01Aa
	3	7.86 ± 0.01Ab	8.62 ± 0.06Aab	7.46 ± 0.62Ab
	8	7.29 ± 0.78Ab	7.76 ± 0.85Ac	7.30 ± 0.78Ab
Stearic acid (18:0)	0	2.42 ± 0.16Aa	2.42 ± 0.16Aa	2.42 ± 0.16Aa
	3	2.52 ± 0.01Aa	2.54 ± 0.14Aa	2.41 ± 0.04Aa
	8	2.08 ± 0.01Aa	2.03 ± 0.11Aa	2.18 ± 0.36Aa
Oleic acid (18:1)	0	13.04 ± 0.38Aa	13.04 ± 0.38Aa	13.04 ± 0.38Aa
	3	12.69 ± 0.05Aa	11.18 ± 0.26Aa	13.10 ± 0.48Aa
	8	13.44 ± 0.62Aa	14.54 ± 0.28Aa	13.58 ± 0.62Aa
Linoleic acid (18:2)	0	66.90 ± 0.39Aa	66.90 ± 0.39Aa	66.90 ± 0.39Aa
	3	68.20 ± 0.11Aa	67.93 ± 0.08Aa	68.32 ± 0.11Aa
	8	68.74 ± 0.03Aa	67.35 ± 0.93Aa	67.86 ± 0.45Aa
Linolenic acid (18:3)	0	8.44 ± 0.15Aa	8.44 ± 0.15Aa	8.44 ± 0.15Aa
	3	8.74 ± 0.17Aa	9.73 ± 0.14Aa	8.73 ± 0.31Aa
	8	8.45 ± 0.01Aa	8.32 ± 0.26Aa	9.09 ± 0.01Aa
Polyunsaturated fat (PUFA)	0	75.33 ± 0.92Aa	75.33 ± 0.92Aa	75.33 ± 0.92Aa
	3	76.93 ± 0.48Aa	77.66 ± 0.69Aa	77.05 ± 0.18Aa
	8	77.19 ± 0.38Aa	75.66 ± 0.41Aa	76.95 ± 0.35Aa
Unsaturated fatty acid (UFA)	0	88.37 ± 0.92Aa	88.37 ± 0.92Aa	88.37 ± 0.92Aa
	3	89.62 ± 0.33Aa	88.83 ± 0.43Aa	90.14 ± 1.43Aa
	8	90.63 ± 0.69Aa	90.20 ± 1.66Aa	90.52 ± 1.75Aa

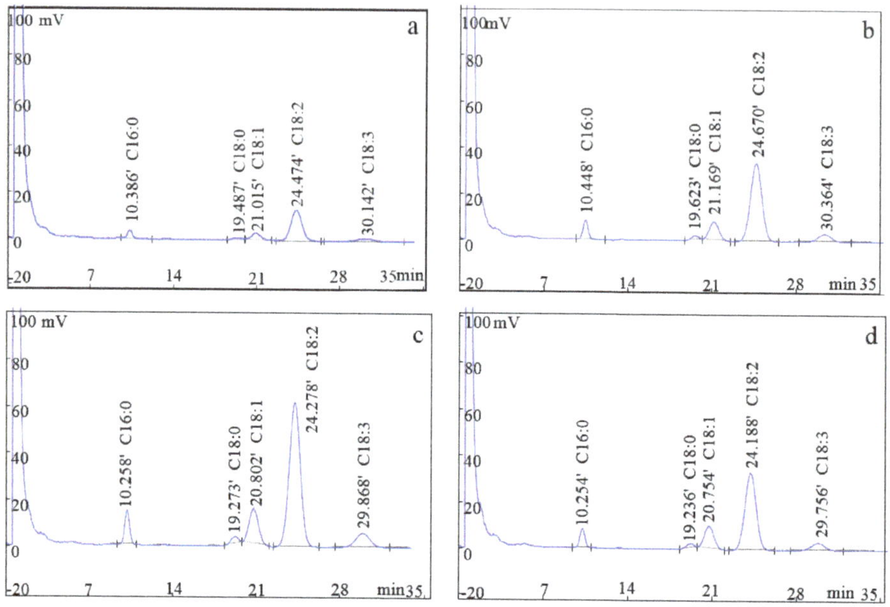

Figure 5. Fatty acid profiles of walnut kernels: (**a**) prior to frozen storage; (**b**) fresh kernel after 8-m frozen storage; (**c**) walnut with shell after 8-m frozen storage; (**d**) walnut with green husk after 8-m frozen storage.

3.7. Total Fat Content, Acid Value, and Peroxide Value of Walnut Kernels

As shown in Table 3, a 20% decrease of fat content in the kernels was observed from walnuts with shell and walnuts with green husk stored at 0 °C, compared with the initial value, whereas only a slight reduction in the fat content was detected during the frozen storage at −20 °C. Similar results have been reported by Ma et al. (2013), where they found walnuts stored at 0 °C still perform energy metabolism, resulting in a degradation of fat in the kernels, while frozen storage decreased this effect [8]. Among the three forms of walnuts, no significant difference in the fat content was observed during the 8-month frozen storage at −20 °C. It indicates that frozen storage could contribute to delaying the degradation of fat present in the kernels. In addition, the high maintenance of fat content may be attributed to the high concentration of fatty acid during storage (Table 2).

The acid value of the kernels was used to evaluate the degree of the degradation in the fatty acids during storage, and the higher acid value contributes to a higher free fatty acid content resulting in a lower fat quality of walnut kernels [37]. As shown in Table 3, the results found a slight increase in the acid value of the walnut kernels after a 3-month storage at 0 °C. The walnuts with green husk (0.67 mg/g) showed a lower acid value compared to the walnuts with shell (0.76 mg/g), which might be strongly associated with the higher total phenol content and total antioxidant activity. During the 8-month storage at −20 °C, the acid value of the walnut kernels increased gradually, and the highest acid value was observed in the walnuts with shell (0.84 mg/g), followed by walnuts with green husk (0.68 mg/g) and fresh kernel (0.61 mg/g). No significant difference was observed between the three forms of walnuts at the end of the 8-month frozen storage at −20 °C. It indicates that low-temperature storage showed a potential advantage in preventing fatty acids from oxidation and degradation.

Table 3. Total fat content, acid value and peroxide value of walnut kernels during the storage at 0 °C and −20 °C. Note: Na means not available. Values with different lowercase letters in the same column 'a–b' and uppercase letter 'A–B' in the same row are significantly different ($p < 0.05$) from each other during storage.

Parameters	Storage Time Month	0 °C			−20 °C		
		Fresh Kernel	Walnut with Shell	Walnut with Green Husk	Fresh Kernel	Walnut with Shell	Walnut with Green Husk
Fat content (%)	0	54.04 ± 2.05A	54.04 ± 2.05Aa	54.04 ± 2.05Aa	54.04 ± 2.05Aab	54.04 ± 2.05Aa	54.04 ± 2.05Aa
	3	Na	44.16± 2.42Ab	44.22± 1.93Ab	56.29 ± 0.86Aa	48.29 ± 2.98Bb	50.84 ± 1.05Ba
	8				50.48 ± 3.66Ab	49.50 ± 0.93Ab	51.33 ± 4.10Aa
Acid value (mg/g)	0	0.57 ± 0.11A	0.57 ± 0.11Ab	0.57 ±0.11Aa	0.57 ± 0.11Aa	0.57 ± 0.11Ab	0.57 ± 0.11Aa
	3	Na	0.76 ± 0.16Aa	0.67 ± 0.03Aa	0.59 ± 0.01Aa	0.67 ± 0.20Ab	0.60 ± 0.05Aa
	8				0.61 ± 0.09Ba	0.84 ± 0.08Aab	0.68 ± 0.02ABa
Peroxide value (mmol/kg)	0	0.59 ± 0.06A	0.59 ± 0.06Aa	0.59 ± 0.06Aa	0.59 ± 0.06Aab	0.59 ± 0.06Ab	0.59 ± 0.06Aab
	3	Na	0.55 ± 0.00Aa	0.49 ± 0.01Aa	0.79 ± 0.14Aa	0.68 ± 0.09Aab	0.84 ± 0.14Aa
	8				0.37 ± 0.06Bb	0.75 ± 0.10Aa	0.44 ± 0.07Bb

In regards to the peroxide value of the walnut kernels, no significant difference was detected at the end of a 3-month storage at 0 °C when compared with the initial level (Table 3). However, the peroxide value of the kernels increased up to 0.84 mmol/kg from the initial level of 0.59 mmol/kg when stored at −20 °C, and then it decreased to a similar value with the initial level. The results indicate that long-term frozen storage (up to 8 months) at −20°C can contribute to maintaining the peroxide value of walnut kernels. Many studies have reported that phenolic compounds can contribute to the lower acid value, peroxide, which in turn could improve the kernel quality of nuts during storage. In fresh walnuts, Wang et al. (2016) and Ma et al. (2020) observed that phenols played a primary role in reducing the lipid oxidation in the kernels, resulting in a negative correlation between the phenol content and the concentration of acid value and peroxide value [10,38]. In hazelnuts, a study found that frozen storage at −25 °C maintained the kernel quality as a result of the high phenolic composition and antioxidant capacity of kernels during the storage [27]. Similarly, in this study, our results observed a higher total phenol content and DPPH in the walnuts during the storage at −20 °C compared with the initial level, resulting in acceptable acid and peroxide values of the kernels (Table 3).

4. Conclusions

In this study, three forms of walnuts, including walnuts with green husk, walnuts with shell, and fresh kernels, were stored at 0 °C and −20 °C. The results found that walnuts with green husk showed a better kernel quality resulted from a lower acid value and peroxide value, and a higher antioxidant activity compared with other forms of walnuts, when stored at 0 °C for a short duration (3 months). In contrast, during the frozen storage at −20 °C for a long duration (up to 10 months), the findings revealed that walnuts with shell showed advantages in improving the fatty acid content, total phenols, and total antioxidant activity compared with other forms of walnuts. Further, the production of H_2O_2 and O_2^- in the kernels was inhibited or delayed because of the higher SOD, CAT, and POD activities, which in turn led to maintaining the acid value (AV) and peroxide value (PV) in an acceptable range. In addition, the walnuts with shell saved space for the storage compared with the walnuts with green husk. Therefore, the walnuts with shell showed a potential to be used for the future frozen storage at −20 °C for long-term storage (up to 10 months) in the food industry. However, the relevant physiochemical mechanisms under the cold storage of fresh walnuts are still not clear, and relevant studies are needed in future research.

Author Contributions: Conceptualization, Y.M. and J.W.; methodology, C.W.; software, C.W.; validation, C.W., C.L. and J.T.; formal analysis, C.W.; investigation, C.L.; resources, J.T.; data curation, C.W.; writing—original draft preparation, C.W. and J.W.; writing—review and editing, Y.M. and J.W.; visualization, H.M.; supervision, Y.M. and J.W.; project administration, Y.M.; funding acquisition, Y.M. and J.W. All authors have read and agreed to the published version of the manuscript.

Funding: The authors would like to acknowledge the financially supported by the funds from Natural Science Foundation of Jiangsu Province [BK20210226], Shaanxi Science Department [2017NY-140], Northwest A&F University, China [A289021414], and the National Innovation Alliance of Walnut Industry (NAWI).

Institutional Review Board Statement: Not applicable.

Informed Consent Statement: Not applicable.

Data Availability Statement: Not applicable.

Conflicts of Interest: The authors declare no conflict of interest.

References

1. US Department of Agriculture. Available online: www.fdc.nal.usda.gov (accessed on 20th May 2018).
2. Jahanbani, R.; Ghaffari, S.M.; Salami, M.; Vahdati, K.; Sepehri, H.; Sarvestani, N.N.; Sheibani, N.; Moosavi-Movahedi, A.A. Antioxidant and Anticancer Activities of Walnut (*Juglans regia* L.) Protein Hydrolysates Using Different Proteases. *Plant Foods Hum. Nutr.* **2016**, *71*, 402–409. [CrossRef]
3. Regueiro, J.; Sánchez-González, C.; Vallverdú-Queralt, A.; Simal-Gándara, J.; Lamuela-Raventós, R.; Izquierdo-Pulido, M. Comprehensive identification of walnut polyphenols by liquid chromatography coupled to linear ion trap—Orbitrap mass spectrometry. *Food Chem.* **2014**, *152*, 340–348. [CrossRef]
4. Sánchez-González, C.; Ciudad, C.; Noé, V.; Pulido, M.L.I. Health benefits of walnut polyphenols: An exploration beyond their lipid profile. *Crit. Rev. Food Sci. Nutr.* **2017**, *57*, 3373–3383. [CrossRef] [PubMed]
5. Estruch, R.; Ros, E.; Salas-Salvadó, J.; Covas, M.-I.; Corella, D.; Arós, F.; Gómez-Gracia, E.; Ruiz-Gutiérrez, V.; Fiol, M.; Lapetra, J.; et al. Primary Prevention of Cardiovascular Disease with a Mediterranean Diet. *N. Engl. J. Med.* **2013**, *368*, 1279–1290. [CrossRef]
6. Lee, J.; Kim, Y.-S.; Lee, J.; Heo, S.; Lee, K.; Choi, S.-W.; Kim, Y. Walnut phenolic extract and its bioactive compounds suppress colon cancer cell growth by regulating colon cancer stemness. *Nutrients* **2016**, *8*, 439. [CrossRef]
7. Shukitt Hale, B.; Thangthaeng, N.; Fisher, D.R.; Bielinski, D.E.; Poulose, S.M. Walnuts improve neuronal and behavioral function in aging. Presented at the Federation of European Nutrition Societies European Nutrition Conference, Boston, MA, USA, 2015.
8. Ma, Y.; Lu, X.; Liu, X.; Ma, H. Effect of 60Coγ-irradiation doses on nutrients and sensory quality of fresh walnuts during storage. *Postharvest Biol. Technol.* **2013**, *84*, 36–42. [CrossRef]
9. Wang, J.; Li, P.; Gong, B.; Ma, H. Phenol metabolism and preservation of fresh in-hull walnut stored in modified atmosphere packaging. *J. Sci. Food Agric.* **2017**, *97*, 5335–5342. [CrossRef]
10. Wang, J.; Liang, S.; Ma, H.; Zhang, P.; Shi, W. Effects of Ethephon on Fresh In-Husk Walnut Preservation and its Possible Relationship with Phenol Metabolism. *J. Food Sci.* **2016**, *81*, C1921–C1927. [CrossRef]
11. Galetto, C.D.; Verdini, R.A.; Zorrilla, S.; Rubiolo, A.C. Freezing of strawberries by immersion in CaCl2 solutions. *Food Chem.* **2010**, *123*, 243–248. [CrossRef]
12. Liang, D.; Lin, F.; Yang, G.; Yue, X.; Zhang, Q.; Zhang, Z.; Chen, H. Advantages of immersion freezing for quality preservation of litchi fruit during frozen storage. *LWT Food Sci. Technol.* **2015**, *60*, 948–956. [CrossRef]
13. Xin, Y.; Zhang, M.; Adhikari, B. Ultrasound assisted immersion freezing of broccoli (*Brassica oleracea* L. var. *botrytis* L.). *Ultrason. Sonochemistry* **2014**, *21*, 1728–1735. [CrossRef] [PubMed]
14. Celli, G.B.; Ghanem, A.; Brooks, M.S.-L. Influence of freezing process and frozen storage on the quality of fruits and fruit products. *Food Rev. Int.* **2015**, *32*, 280–304. [CrossRef]
15. Cheng, L.; Sun, D.W.; Zhu, Z.; Zhang, Z. Emerging techniques for assisting and accelerating food freezing processes: A review of recent research progresses. *Crit. Rev. Food Sci. Nutr.* **2015**, *57*, 769–781. [CrossRef] [PubMed]
16. Aydogdu, A.; Yildiz, E.; Aydogdu, Y.; Sumnu, G.; Sahin, S.; Ayhan, Z. Enhancing oxidative stability of walnuts by using gallic acid loaded lentil flour based electrospun nanofibers as active packaging material. *Food Hydrocoll.* **2019**, *95*, 245–255. [CrossRef]
17. Wang, J.; Wang, J.; Ye, J.; Vanga, S.K.; Raghavan, V. Influence of high-intensity ultrasound on bioactive compounds of strawberry juice: Profiles of ascorbic acid, phenolics, antioxidant activity and microstructure. *Food Control* **2018**, *96*, 128–136. [CrossRef]
18. Benzie, I.; Strain, J. The Ferric Reducing Ability of Plasma (FRAP) as a Measure of "Antioxidant Power": The FRAP Assay. *Anal. Biochem.* **1996**, *239*, 70–76. [CrossRef]
19. Wang, L.; Jin, P.; Wang, J.; Gong, H.; Zhang, S.; Zheng, Y. Hot air treatment induces resistance against blue mold decay caused by Penicillium expansum in sweet cherry (Prunus cerasus L.) fruit. *Sci. Hortic.* **2015**, *189*, 74–80. [CrossRef]
20. Lin, Y.; Lin, H.; Zhang, S.; Chen, Y.; Chen, M.; Lin, Y. The role of active oxygen metabolism in hydrogen per-oxide-induced pericarp browning of harvested longan fruit. *Postharvest Biol. Technol.* **2014**, *96*, 42–48. [CrossRef]

21. Chen, H.; Gao, H.; Fang, X.; Ye, L.; Zhou, Y.; Yang, H. Effects of allyl isothiocyanate treatment on postharvest quality and the activities of antioxidant enzymes of mulberry fruit. *Postharvest Biol. Technol.* **2015**, *108*, 61–67. [CrossRef]
22. National Health and Family Planning Commission of the People's Republic of China. *Food Safety National Standard-Determination of Acid in Food*; GB/T 5009.229-2016a; Chinese Standard Publication House: Beijing, China, 2016. (In Chinese)
23. National Health and Family Planning Commission of the People's Republic of China. *Food Safety National Standard-Determination of Peroxide Value in Food*; GB/T 5009.229-2016b; Chinese Standard Publication House: Beijing, China, 2016. (In Chinese)
24. Esteki, M.; Farajmand, B.; Amanifar, S.; Barkhordari, R.; Ahadiyan, Z.; Dashtaki, E.; Vander Heyden, Y. Classification and authentication of Iranian walnuts according to their geographical origin based on gas chroma-tographic fatty acid fingerprint analysis using pattern recognition methods. *Chemom. Intell. Lab. Syst.* **2017**, *171*, 251–258. [CrossRef]
25. Goneli, A.; Corrêa, P.; Resende, O.; Neto, S.R. Electrical Conductivity for Quality Evaluation of Popcorn Kernels subjected to Mechanical Damage. *Biosyst. Eng.* **2007**, *96*, 361–367. [CrossRef]
26. Kaewtathip, T.; Charoenrein, S. Changes in volatile aroma compounds of pineapple (Ananas comosus) during freezing and thawing. *Int. J. Food Sci. Technol.* **2012**, *47*, 985–990. [CrossRef]
27. Ghirardello, D.; Bertolino, M.; Belviso, S.; Dal Bello, B.; Giordano, M.; Rolle, L.; Gerbi, V.; Antonucci, M.; Spigolon, N.; Zeppa, G. Phenolic composition, antioxidant capacity and hexanal content of hazelnuts (Corylus avellana L.) as affected by different storage conditions. *Postharvest Biol. Technol.* **2016**, *112*, 95–104. [CrossRef]
28. Hoang, N.T.; Golding, J.; Wilkes, M.A. The effect of postharvest 1-MCP treatment and storage atmosphere on 'Cripps Pink' apple phenolics and antioxidant activity. *Food Chem.* **2011**, *127*, 1249–1256. [CrossRef] [PubMed]
29. Liu, C.-H.; Cai, L.-Y.; Lu, X.-Y.; Han, X.-X.; Ying, T.-J. Effect of Postharvest UV-C Irradiation on Phenolic Compound Content and Antioxidant Activity of Tomato Fruit During Storage. *J. Integr. Agric.* **2012**, *11*, 159–165. [CrossRef]
30. Palafox-Carlos, H.; Yahia, E.; Islas-Osuna, M.; Gutierrez-Martinez, P.; Robles-Sánchez, M.; González-Aguilar, G. Effect of ripeness stage of mango fruit (Mangifera indica L., cv. Ataulfo) on physiological parameters and antioxidant activity. *Sci. Hortic.* **2012**, *135*, 7–13. [CrossRef]
31. Ighodaro, O.; Akinloye, O. First line defence antioxidants-superoxide dismutase (SOD), catalase (CAT) and glutathione peroxidase (GPX): Their fundamental role in the entire antioxidant defence grid. *Alex. J. Med.* **2018**, *54*, 287–293. [CrossRef]
32. Jin, P.; Shang, H.; Chen, J.; Zhu, H.; Zhao, Y.; Zheng, Y. Effect of 1-Methylcyclopropene on Chilling Injury and Quality of Peach Fruit during Cold Storage. *J. Food Sci.* **2011**, *76*, S485–S491. [CrossRef] [PubMed]
33. Song, H.; Yuan, W.; Jin, P.; Wang, W.; Wang, X.; Yang, L.; Zhang, Y. Effects of chitosan/nano-silica on post-harvest quality and antioxidant capacity of loquat fruit during cold storage. *Postharvest Biol. Technol.* **2016**, *119*, 41–48. [CrossRef]
34. Scussel, V.M.; Giordano, B.N.; Simão, V.; Manfio, D.; Galvao, S.; Rodrigues, M.N.F. Effect of Oxygen-Reducing Atmospheres on the Safety of Packaged Shelled Brazil Nuts during Storage. *Int. J. Anal. Chem.* **2011**, *2011*, 813591. [CrossRef] [PubMed]
35. Bai, S.H.; Nevenimo, T.; Hannet, G.; Hannet, D.; Jones, K.; Trueman, S.; Grant, E.; Walton, D.; Randall, B.; Wallace, H. Freezing, roasting and salt dipping impacts on peroxide value, free fatty acid and fatty acid concentrations of nut kernels. *Acta Hortic.* **2019**, *1256*, 71–76. [CrossRef]
36. Krause, A.J.; Miracle, R.E.; Sanders, T.H.; Dean, L.L.; Drake, M.A. The Effect of Refrigerated and Frozen Storage on Butter Flavor and Texture. *J. Dairy Sci.* **2008**, *91*, 455–465. [CrossRef] [PubMed]
37. Jiang, L.; Feng, W.; Li, F.; Xu, J.; Ma, Y.; Ma, H. Effect of One-methylcyclopropene (1-MCP) and chlorine dioxide (ClO_2) on preservation of green walnut fruit and kernel traits. *J. Food Sci. Technol.* **2013**, *52*, 267–275. [CrossRef] [PubMed]
38. Ma, Y.; Li, P.; Watkins, C.B.; Ye, N.; Jing, N.; Ma, H.; Zhang, T. Chlorine dioxide and sodium diacetate treatments in controlled atmospheres retard mold incidence and maintain quality of fresh walnuts during cold storage. *Postharvest Biol. Technol.* **2020**, *161*, 111063. [CrossRef]

foods MDPI

Article

Effects of the Use of Different Temperature and Calcium Chloride Treatments during Storage on the Quality of Fresh-Cut "Xuebai" Cauliflowers

Bingyu Mu [1], Jianxin Xue [1,*], Shujuan Zhang [1] and Zezhen Li [2]

1 College of Agricultural Engineering, Shanxi Agricultural University, Jinzhong 030801, China; 13633544838@163.com (B.M.); Z15035658426@163.com (S.Z.)
2 College of Food Science and Engineering, Shanxi Agricultural University, Jinzhong 030801, China; 119834545023@163.com
* Correspondence: sxndxjx@sxau.edu.cn; Tel.: +86-133-1344-0069

Abstract: This study revealed the effect of the use of different temperature and calcium chloride ($CaCl_2$) treatments on the storage quality of fresh-cut "Xuebai" cauliflowers. Fresh-cut "Xuebai" cauliflowers were soaked with 2% $CaCl_2$ solution at different temperatures. The change in the firmness, color, and ascorbic acid (ASA), total glucosinolates (TGLS), polygalacturonase (PG), and lipoxygenase (LOX) content of fresh-cut "Xuebai" cauliflowers during the cold storage period was assessed. In addition, the sensory quality was also evaluated. The results show that the combined treatments with $CaCl_2$ at different temperatures could effectively maintain the storage quality of fresh-cut "Xuebai" cauliflowers. Then, a method based on factor analysis with comprehensive quality evaluation was proposed. A factor analysis with a principal component analysis (PCA) was conducted on nine indicators of cauliflowers. Two principal components were extracted with a cumulative contribution rate of 97.513%. The results demonstrated that the treatment with the best fresh-keeping effect of cauliflowers in storage was the combination treatment at 40 °C with 2% $CaCl_2$ solution, while the optimal storage period was 15 days.

Keywords: fresh-cut cauliflower; storage period; calcium chloride; temperature; factor analysis; quality

Citation: Mu, B.; Xue, J.; Zhang, S.; Li, Z. Effects of the Use of Different Temperature and Calcium Chloride Treatments during Storage on the Quality of Fresh-Cut "Xuebai" Cauliflowers. *Foods* **2022**, *11*, 442. https://doi.org/10.3390/foods11030442

Academic Editors: Maria Cefola and Bernardo Pace

Received: 5 January 2022
Accepted: 31 January 2022
Published: 2 February 2022

1. Introduction

Brassica vegetables are important components of a healthy diet and can help to prevent diseases such as diabetes, cardiovascular disease, and certain cancers; thus, they have attracted the attention of researchers [1–3]. Among cruciferous vegetables, cauliflowers (*Brassica oleracea* L. botrytis) are important components of human diets and have large economic benefits and yields around the world. The demand for cauliflowers has risen sharply due to their high nutritional value and high levels of many bioactive compounds, including glucosinolates, phenolic compounds, and ascorbic acid [4,5].

Fresh-cut fruits and vegetables are packed and sold after fresh raw materials have been selected, cleaned, peeled, and sliced [6]. Nowadays, with changes in lifestyles and consumption habits, the popularity of fresh-cut cauliflowers has risen among consumers due to their ideal nutritional value, excellent sensory qualities, and health-related benefits for the body [7]. As a basic processing method, cutting will cause browning, odor emission, tissue softening, and a loss of nutritional quality in fresh-cut cauliflower, thus shortening its storage time [8]. As a divalent cation nutrient element, calcium (Ca^{2+}) plays a significant role in delaying aging and improving antioxidant capacity in the cell wall and membrane structure. During the cross-linking process between Ca^{2+} and carboxyl groups in pectin, a structure called "egg box" is formed. This helps to increase the strength of the cell wall and the firmness of the tissue structure [9].

Calcium chloride ($CaCl_2$) as a preservative and curing agent has been widely used in the preservation of fresh-cut fruits and vegetables, as it can effectively inhibit the occurrence of diseases, inhibit ethylene production, and delay the aging of fruits and vegetables [10,11]. Aghdam et al. (2013) found that postharvest $CaCl_2$ treatment had a positive effect on the antioxidant capacity and DPPH free radical scavenging capacity of cornelian cherry (*Cornus mas*) fruit. At the same time, the content of active substances such as anthocyanins and ascorbic acid was increased [12]. Zhang et al. (2019) reported that treatment with 2% $CaCl_2$ could reduce the browning of "Nanguo" pear peel during cold storage, maintain a high hardness and polyphenol content, and inhibit the gene expression of phospholipase D and polyphenol oxidase while reducing their activity [13].

In recent years, heat treatment has attracted researchers' attention because of its advantages in terms of safety, lack of contribution to pollution, and easy means of operation. Through heat treatment, the incidence of chilling injury symptoms in fruits and vegetables can be reduced; the occurrence of respiratory peaks can be delayed; losses of water can be reduced; the ripening, decay and yellowing of fruits and vegetables within the storage period can be controlled; the postharvest quality of fruits and vegetables can be maintained, and their shelf life can be prolonged [14]. As a common method of physical pretreatment for fruits and vegetables, cold treatment can reduce the tissue temperature of fruits and vegetables within a short timeframe, cause them to experience low-temperature stress, induce stress resistance, maintain the balance of active oxygen metabolism in fruits, protect the integrity of the cell membrane structure, reduce enzyme activity, delay fruit aging, and maintain the nutritional quality of fruits and vegetables to the greatest extent possible [15]. Duarte-Sierra et al. (2017) reported that heat treatment can lower the respiratory efficiency of broccoli, maintain a good chemical composition, delay the rate of yellowing of the curd, and lead to a significantly higher content of chlorophyll and glucosinolate than is present in untreated broccoli [16]. Yang et al. (2016) found that $CaCl_2$ treatment can increase glutamine biosynthesis in broccoli, promote myrosinase (MYR) gene expression and activity, and improve the total antioxidant capacity and DPPH scavenging capacity [17]. Grzegorzewska et al. (2009) found that, during the storage period, broccoli treated with ice water had a lower rooting rate, lower loss of tightness, and minor color change [18].

Some research reports have shown that $CaCl_2$ treatment combined with other methods is a good method of fruit and vegetable preservation. Wang et al. (2014) reported that the combined treatment with $CaCl_2$ and cold water at 0 °C can inhibit the respiration rate of sweet cherries, help them maintain a brighter luster, enhance their antioxidant system, and reduce the peroxidation of membrane lipids. In the end, the authors achieved their goal of reducing the rate of decay of cherry fruit and delaying its senescence speed [19]. Supapvanich et al. (2012) suggested that the use of hot $CaCl_2$ dips at 40 °C could help to maintain the postharvest quality of fresh-cut sweet leaf bush, delay the decrease in chlorophyll content, delay yellowing, and help to maintain a high nutritional quality, especially with regard to the activities of total phenols, flavonoids, and antioxidant enzymes [20].

Unfortunately, there has been no related exploration of the change in the preservation quality of fresh-cut "Xuebai" cauliflowers through combined treatment with temperature and $CaCl_2$. Therefore, this study aimed to select the optimal temperature of $CaCl_2$ solution for the preservation of fresh-cut "Xuebai" cauliflower florets at 4 °C. The effects of $CaCl_2$ solutions of different temperatures on the color value, firmness, ascorbic acid content (ASA), total glucosinolates content (TGLS), and polygalacturonase (PG) and lipoxygenase (LOX) activities during storage of fresh-cut "Xuebai" cauliflower florets were studied. By measuring multiple indicators, conducting sensory evaluation (SE), combining principal component analysis and factor analysis, and sorting out the scores, the most suitable compound treatment method for fresh-cut "Xuebai" cauliflower preservation was screened. This study provided theoretical support for the research and development of storage and preservation technology for fresh-cut "Xuebai" cauliflower florets and the extension of shelf life.

2. Materials and Methods

2.1. Plant Material and Treatments

"Xuebai" cauliflower (*Brassica oleracea* L. botrytis) was picked in Juxin Agricultural Park (112.49 N and 7.39 E) in Taigu, Shanxi Province of China. After picking, the cauliflower was precooled by vacuum in the field, put into a foam box with small ice cubes, and transported to the laboratory. The selected cauliflower samples were similar in size and without any diseases, insect pests, or mechanical damage. We cut cauliflower into small flower balls (about 50 g) of the same size with a knife sterilized in 0.1% sodium hypochlorite solution, soaked the flower balls in 0.1% sodium hypochlorite solution for 5 min, rinsed them with sterile water, and took them out to dry.

After drying, 21 samples were separated from the freshly cut florets and analyzed. A total of 420 fresh-cut cauliflower samples were selected and divided into four groups (each with 105 samples); then, the following treatments were performed: (1) T20 °C-$CaCl_2$ 0%: immersed in water at 20 °C for 10 min; (2) T0 °C-$CaCl_2$ 2%: immersed in 2% $CaCl_2$ solution at 0 °C for 10 min; (3) T20 °C-$CaCl_2$ 2%: immersed in 2% $CaCl_2$ solution at 20 °C for 10 min; (4) T40 °C-$CaCl_2$ 2%: immersed in 2% $CaCl_2$ solution at 40 °C for 10 min. Then, the fresh-cut "Xuebai" cauliflower florets were controlled in a drain basket, dried, and put into a 0.04 mm PE fresh-keeping bag at a constant temperature of 4 °C and a 90% constant humidity for 16 days. For each treatment, we randomly selected 21 samples (a total of 84 samples) for repeated experiments at 0, 3, 6, 9, 12, and 15 days. Then, we determined the quality index and performed a sensory evaluation.

According to the research of Grzegorzewska et al. (2009) and Xue et al. (2021), the temperature setting, number of storage days, and $CaCl_2$ solution concentration of this experiment were determined [18,21].

2.2. Sensory Quality Scores

After each sampling, a group of volunteers (n = 9) consisting of male and female students were asked to evaluate the following sensory attributes: appearance, aroma, flavor, texture, and overall acceptability. These volunteers received professional training before the assessment and agreed on the assessment criteria. The sensory quality scores for cauliflower ranged from one to nine and were calculated using a weighted method. A score of nine meant the sample was excellent (the color was very good, the flower ball organization was tight, the fragrance was good, and it was not decayed); a score of seven meant it was good (the color and luster were good, tissue, the fragrance was light, and less than one twentieth of the flower buds had spots); a score of five was medium (the color and luster were acceptable, the flower balls were slightly soft, the central organization was loose, there was some fragrance, and one-twentieth to one-fifth of the flower buds had spots); a score of three was poor (the color and luster were not good, more than half of the flower balls were wilted, there was a slight peculiar smell, and one-fifth to one-half of the flower buds had spots); a score of one was extremely poor (the color was very bad, most of the bulbs were wilting, there was a rancid smell, and more than half of the flower buds had spots) [22,23].

2.3. Firmness Measurement

The firmness of the cauliflower floret was measured by the TMS-PRO food physical property analyzer (FTC, Sterling, VA, USA). The cylinder probe used was P / 2 n, the initial speed was 5 mm/s, the test speed was 1 mm/s, the retraction speed was 5 mm/s, the interval between two extrusions was 2 s, the minimum perception force was 0.4 N, and the extrusion depth was 4 mm. The firmness of each sample was measured three times, and the average value was calculated. The maximum firmness value obtained at the first compression was taken as the firmness value of the sample.

2.4. Color Measurement

Referring to the research of Yan et al. (2020), at each sampling time point, the L^*, a^*, and b^* values on the surface of the cauliflower florets were measured using a CR-400 color meter (Konica Minolta, Tokyo, Japan) [23]. Three data points were randomly measured at the top bulge of each sample, and the average value was taken as the final result.

2.5. Ascorbic Acid Content Measurement

Cauliflower samples with weights of 1 g were ground and fixed to 50 mL; the content of ascorbic acid (ASA) was determined by titration with 2-6-dichloro-indophenol. When the ASA was completely oxidized, a drop of the dye caused the oxalic acid solution to immediately appear light pink. This change in color was the end point of titration. According to the titration standard of 2-6-dichloro-indophenol solution, the content of ascorbic acid in cauliflower florets can be calculated. The specific operation method used was based on the method of Chen et al. (2018) [24].

2.6. Total Glucosinolates Measurement

Based on the principle that total glucosinolates (TGLS) are hydrolyzed into glucose by myrosinase (MYR) in cauliflower, the content of TGLS was determined by spectrophotometry. A standard solution of 3-5 dinitrosalicylic acid and glucose was prepared, and the glucose standard curve was drawn. Two 0.5 g grinding samples were accurately weighed and placed in 25 mL calibration tubes, and to each, 0.1 g of sodium fluoride was added. Then, we added 20 mL of boiling water to a test tube, heated it immediately until it boiled, and maintained it there for 10 min. In the other test tube, we added 20 mL of distilled water at 36 °C and kept it in a water bath at 37 °C for one hour. The TGLS was hydrolyzed under the action of MYR. Then, we heated the sample to boiling and kept it there for 10 min. We added six drops of neutral lead acetate to each of the two test tubes, distilled water was added until the solution reached a total of 25 mL, and 0.5 mL of filtrate was used to determine its absorbance.

2.7. Polygalacturonase Measurement

The determination of polygalacturonidase (PG) was performed using visible spectrophotometry. The galacturonidase produced by PG hydrolysis reacted with DNS reagent to produce a brownish red substance with a characteristic absorption peak at 540 nm. The pectinase activity was calculated by measuring the change in the absorbance value at 540 nm.

We weighed 0.1 g of cauliflower, added 1 mL of extract, ground it in an ice bath, centrifuged it for 10 min at 4 °C with sixteen thousand revolutions, diluted the supernatant five times, and followed the steps of the determination instructions (Soleibao Technology Co., Ltd., Beijing, China).

2.8. Lipoxygenase Measurement

The presence of lipoxygenase (LOX) was determined by ultraviolet spectrophotometry (Soleibao Technology Co., Ltd., Beijing, China); LOX can catalyze the oxidation of linoleic acid, and the oxidation product had a characteristic absorption peak at 234 nm. The increasing rate of 234 nm absorbance was measured to calculate the LOX activity.

2.9. Statistical Analysis

For this experiment, we adopted a completely random design. A one-way analysis of variance (ANOVA) was performed on the quality indicators. All indicators were repeatedly measured three times (each repetition included 21 samples), and the results were expressed as mean ± standard deviation (SD). We used the SPSS 17.0 software and Origin 19 software for data analysis and the creation of graphs.

3. Results and Discussion

3.1. *Effect of Different Treatments on the Quality of Fresh-Cut "Xuebai" Cauliflower Florets*

3.1.1. Volunteer Group Sensory Evaluation

The most intuitive indicator that describes the quality of cauliflower and reflects its commercial value is sensory evaluation (SE) [25]. It can be seen from Figure 1 that the sensory score of fresh-cut "Xuebai" cauliflower florets declined during the storage period. The sensory score of the control check (CK) was always lower than that of the other treatment groups. It can be seen from Figure 1 that from 0 d to 6 d, the sensory scores of the four groups of fresh-cut "Xuebai" cauliflower florets did not differ greatly. However, as the storage period grew longer, the decline in the scores of the CK group accelerated. On the 15th day, the CK group had the lowest score. Samples from this group scored only one point, indicating that they had lost their commercial value. These samples received significantly different scores from those of the groups treated with $CaCl_2$ ($p > 0.05$). In general, the effect of the $CaCl_2$ + 40 °C treatment was slightly better than that of the $CaCl_2$ + 0 °C and $CaCl_2$ + 20 °C treatment groups. Our results showed that treatments with $CaCl_2$ can effectively maintain the marketability of cauliflower.

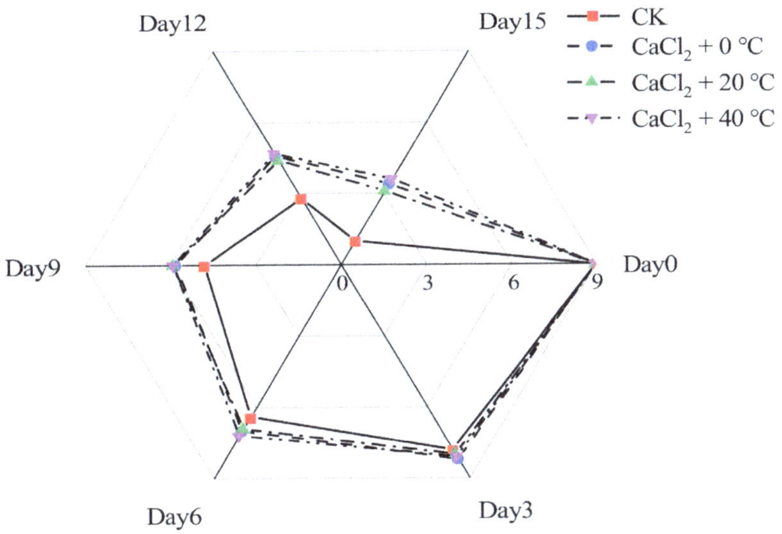

Figure 1. Sensory evaluation of fresh-cut "Xuebai" cauliflower florets given different treatments during storage.

3.1.2. Color

The color of cauliflower is very important for evaluating its value and quality. However, fresh-cut cauliflower will brown and rot in the postharvest storage process, which is visually manifested as a color change [26]. On the zeroth day of storage, the color values L^*, a^*, and b^* of fresh-cut "Xuebai" cauliflower florets were 70.265 ± 0.96, −1.558 ± 0.67, and 9.61 ± 0.35 in sequence. In Table 1, the effects of storage time and treatment methods on the lightness L^* value (light or dark), a^* value (−green to + red), and b^* value (−blue to + yellow) are compared. The results show that the lightness of all samples (CK and treated samples) decreased as the storage time grew longer, while the a^* and b^* values increased over the course of the storage period [27].

Table 1. Effects of different treatments on the color of fresh-cut "Xuebai" cauliflower florets.

Color	Groups	Storage Time/d					
		0	3	6	9	12	15
L^*	CK	70.265 ± 0.96 a	69.207 ± 0.961 a	67.883 ± 0.744 a	66.127 ± 0.736 b	63.467 ± 0.652 c	59.923 ± 0.549 c
	$CaCl_2$ + 0 °C		69.647 ± 0.806 a	67.810 ± 0.902 a	67.510 ± 0.711 ab	66.073 ± 0.347 ab	64.263 ± 0.346 b
	$CaCl_2$ + 20 °C		69.423 ± 0.612 a	68.463 ± 0.728 a	67.290 ± 0.651 ab	65.840 ± 0.517 b	63.733 ± 0.481 b
	$CaCl_2$ + 40 °C		69.873 ± 0.849 a	69.240 ± 0.731 a	68.360 ± 0.580 a	67.370 ± 0.723 a	65.700 ± 0.637 a
a^*	CK	−1.558 ± 0.67 a	−1.517 ± 0.061 a	−1.457 ± 0.098 a	−1.353 ± 0.057 a	−1.187 ± 0.051 a	−0.987 ± 0.080 a
	$CaCl_2$ + 0 °C		−1.540 ± 0.062 a	−1.503 ± 0.062 a	−1.450 ± 0.053 a	−1.357 ± 0.032 b	−1.283 ± 0.049 b
	$CaCl_2$ + 20 °C		−1.533 ± 0.081 a	−1.490 ± 0.070 a	−1.427 ± 0.031 a	−1.333 ± 0.091 b	−1.220 ± 0.123 b
	$CaCl_2$ + 40 °C		−1.547 ± 0.076 a	−1.517 ± 0.100 a	−1.467 ± 0.055 b	−1.397 ± 0.060 b	−1.317 ± 0.075 b
b^*	CK	9.61 ± 0.35 a	10.927 ± 0.787 a	12.877 ± 0.962 a	14.840 ± 0.792 a	17.413 ± 0.751 a	20.317 ± 0.587 a
	$CaCl_2$ + 0 °C		10.433 ± 0.656 a	11.690 ± 0.797 a	12.977 ± 0.290 b	15.393 ± 0.562 b	16.677 ± 0.621 bc
	$CaCl_2$ + 20 °C		10.480 ± 0.579 a	11.877 ± 0.670 a	13.347 ± 0.696 b	15.277 ± 0.535 b	17.430 ± 0.622 b
	$CaCl_2$ + 40°C		10.260 ± 0.321 a	11.470 ± 0.810 a	12.797 ± 0.597 b	14.547 ± 0.621 b	16.100 ± 0.085 c

Note: Different letters in the same column indicate significant differences at the $p < 0.05$ level. L^*—lightness, a^*—redness, b^*—yellowness.

After a prolonged period of storage, the L^* values of $CaCl_2$ treated fresh-cut "Xuebai" cauliflower florets were significantly different from those of the CK group on the 12th and 15th day, indicating that $CaCl_2$ can effectively inhibit the browning of fresh-cut "Xuebai" cauliflower florets and decrease the appearance of spots on the flower bud. Throughout the whole storage period, the L^* value of fresh-cut "Xuebai" cauliflower florets treated with $CaCl_2$ + 40 °C was higher than that of the other three groups. On the 15th day, there were significant differences between cauliflower florets from the $CaCl_2$ + 40 °C treatment group and those from the other three groups ($p < 0.05$). In addition, the a^* and b^* values for cauliflower florets given $CaCl_2$ + 40 °C treatment were the lowest, indicating that $CaCl_2$ + 40 °C treatment can effectively prevent fresh-cut "Xuebai" cauliflower florets from turning yellow and browning. It can be seen that the $CaCl_2$ + 40 °C treatment is better able to maintain the quality of fresh-cut "Xuebai" cauliflower florets.

Consistent with our research results, Zhang et al. (2014) [28] found that after heat treatment, the color value of the heat treatment group was better than that of the other treatment groups [28]. Suzuki et al. (2005) reported that heat treatment can reduce the rate of yellowing of cauliflower [29]. Supapvanich et al. (2012) reported that soaking freshly cut beets with hot $CaCl_2$ solution can effectively lower the activity of chlorophyll-decomposing enzymes, reduce the loss of chlorophyll and protein in vegetables, delay the loss of greenness, and reduce the rate of yellowing [20].

3.1.3. Firmness

In the process of postharvest storage and consumption, firmness is always an important indicator for evaluating the quality of cauliflowers [30]. In this study, the combined treatment with temperature and $CaCl_2$ had a significant effect on the firmness of fresh-cut "Xuebai" cauliflower florets during their period of storage at 4 °C ($p < 0.05$). From Figure 2, there was no significant difference between the treatment groups and the CK group throughout the early and middle stages of storage ($p < 0.05$). However, after storage for a longer period of time, the rate of decrease in the firmness of fresh-cut "Xuebai" cauliflower florets given treatment was significantly lower than that of the CK group (12th and 15th days). By the 15th day, there were significant differences between the CK group and the three treatment groups ($p < 0.05$), indicating that the combined treatment was able to maintain the firmness more effectively and reduce the softening of the "Xuebai" cauliflower florets. The difference between the $CaCl_2$ + 0 °C and $CaCl_2$ + 20 °C treatment groups was not significant ($p > 0.05$), but the $CaCl_2$ + 40 °C treatment group showed a significant difference from the other two groups ($p < 0.05$). A comprehensive comparison showed that the $CaCl_2$ + 40 °C treatment was better able to maintain the firmness of fresh-cut "Xuebai" cauliflower florets.

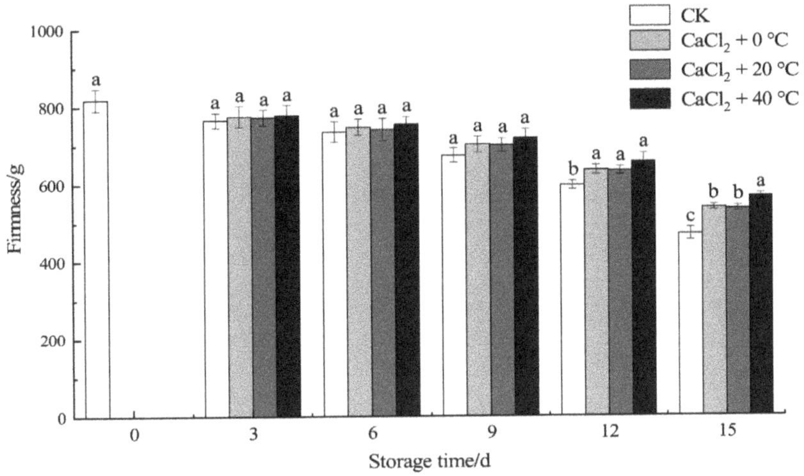

Figure 2. Effect of the different treatments on the firmness of fresh-cut "Xuebai" cauliflower florets. Error bars indicate standard deviation (SD); the statistical significance between the four groups that passed the Tukey significance test is expressed in different lowercase letters, and the significance level is $p < 0.05$.

Duarte-Sierra et al. (2017) studied the effect of heat treatment on the quality of broccoli in storage. The results showed that the use of different temperature treatments can lead to the formation of a high initial CO_2 content, delay yellowing, maintain a higher level of firmness, and lead to a lower weight loss rate of broccoli [16]. Some enzyme activities relating to hydrolyzing cellulose and pectin increased, dissolving the middle gum layer, loosening the structure of the cell wall, and thus causing the softening of cauliflowers over the storage period [31]. The reason for the rapid decrease in the hardness value was the increase in the rate of metabolism, which is related to aging. Wang et al. (2014) showed that calcium pectate was formed after treatment with $CaCl_2$, which increased the rigidity of the cell wall and reduced the content of PG, pectin methyl esterase (PME), B-galactosidase (b-Gal), etc. These substances are located in the middle layer of the cell wall and are closely related to the activity of enzymes related to fruit softening. In addition, calcium can also fix water and maintain the stability of the pressure between cells, delaying the appearance of softening [19].

3.1.4. Ascorbic Acid Content

As shown in Figure 3, with the extension of the storage period of fresh-cut "Xuebai" cauliflower florets, the ASA content in the CK group and treatment groups showed a downward trend. However, the temperature and $CaCl_2$ treatment effectively delayed the decrease in the ASA content. On the 15th day, there was a significant difference between the ASA value of fresh-cut "Xuebai" cauliflower florets given the combined treatments and that of florets in the CK group ($p < 0.05$). Additionally, there was a significant difference between the cauliflowers given the $CaCl_2$ + 40 °C treatment and those in the other two treatment groups ($p < 0.05$). It can be seen that the $CaCl_2$ + 40 °C treatment more effectively inhibited the decomposition of ASA during the storage period of fresh-cut "Xuebai" cauliflower florets.

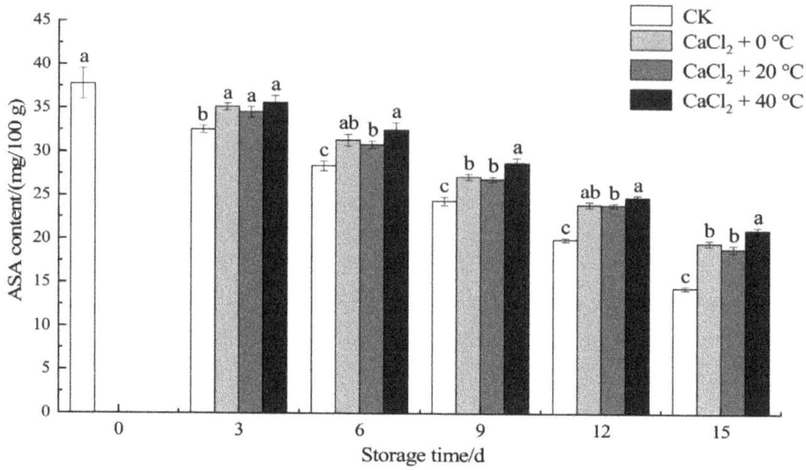

Figure 3. Effect of the different treatments on the ASA content of fresh-cut "Xuebai" cauliflower florets. Error bars indicate standard deviation (SD); the statistical significance between the four groups that passed the Tukey significance test is expressed in different lowercase letters, and the significance level is $p < 0.05$.

Many fruits and vegetables are rich in ASA. As is widely known, ascorbic acid is an important antioxidant for the healthy growth of the human body. Ribeiro et al. (2020), in a study on red pomegranate, and Morteza Soleimani Aghdam et al. (2013), in a study on cornelian cherry fruit, reported the positive effect of calcium treatment on delaying the decrease in ASA content [12,32]. Naser et al. (2018), in a study concerning the quality changes in persimmons, showed that the use of a combined calcium and heat treatment can effectively decrease the reduction in ASA. Calcium treatment can act as a signal to activate the antioxidant system inside the cell; calcium treatment can reduce the rate of free radical degradation at the genetic level and reduce the consumption of ASA [33]. Aguayo et al. (2015) reported that heat treatment was used as an auxiliary method to help increase the absorption of ASA in apple tissues. Combined with calcium treatment, it can delay the decrease in ASA content and improve the quality of apples in storage [34].

3.1.5. Total Glucosinolates Content

Figure 4 shows that during the storage process, the total glucosinolates (TGLS) content of fresh-cut "Xuebai" cauliflower florets followed a trend of increasing first and then decreasing; these results are the same as those of Wei et al. (2016), Duarte-Sierra et al. (2017), and Xue et al. (2020) [21,35,36]. The TGLS content of the control group was 12.7 μmol/100 g on the 9th day, which dropped sharply to 4.2 μmol/100 g on the 15th day. However, the declining trend of the treatment group was not obvious. In the late storage period (12 d and 15 d), the TGLS content of fresh-cut "Xuebai" cauliflowers in the three treatment groups was higher than that in the CK group, showing a significant difference ($p < 0.05$). Treatment with temperature and $CaCl_2$ can effectively delay the degradation of TGLS in fresh-cut "Xuebai" cauliflower tissues and maintain the good nutritional quality of cauliflowers. The content of TGLS in the group treated with $CaCl_2$ + 40 °C was the highest. This shows that the effect of the 40 °C thermal compound treatment was better than that of the 0 °C and 20 °C treatments.

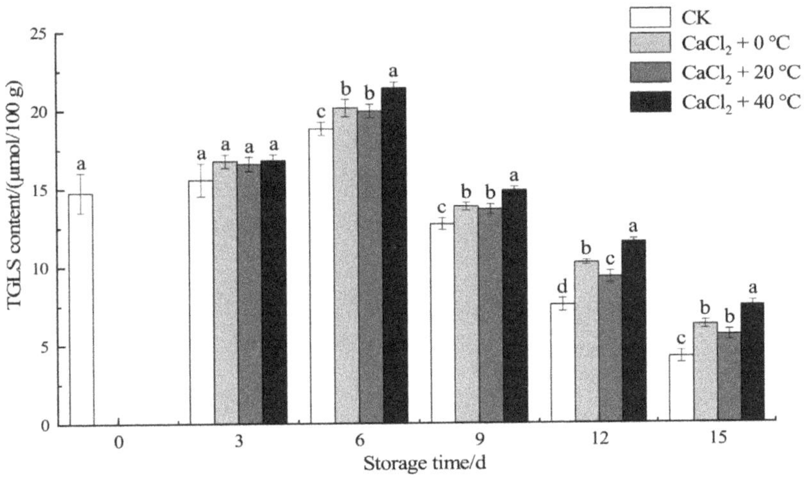

Figure 4. Effect of the different treatments on the TGLS content of fresh-cut "Xuebai" cauliflower florets. Error bars indicate standard deviation (SD); the statistical significance between the four groups that passed the Tukey significance test is expressed in different lowercase letters, and the significance level is $p < 0.05$.

Glucosinolates (GLS) are natural biologically active ingredients in broccoli. The endogenous plant enzyme myrosinase in the plant cell, physically segregated from GLS, is released when chopped or chewed to hydrolyze GLS into various products that help prevent cancer, including isothiocyanates (ITCs), thiocyanates, nitriles, and sulforaphane [35]. GLS are distributed in the vacuoles of plants in the form of salts. In the early stages of storage, the increase in GLS due to the presence of hydroxycinnamic acid helps to maintain the rigidity of the cell walls, which can protect plant tissues from damage. However, in later stages of storage, a decrease in TGLS content occurs due to the degradation of GLS [19]. A.P. Vale et al. (2015) reported that the GLS content of broccoli during cold storage increased first and then decreased. The best consumption time was seven days after harvest. During this period of time, the aliphatic GL of broccoli was at a relatively high level. Sulforaphane was the major decomposition product, and it had the effect of reducing the risk of cancer [36]. Heat treatment can significantly improve the retention of glucoraphanin, maintain the quality of broccoli, and extend the shelf life of products [37]. A study on the effect of $CaCl_2$ treatment on the GLS metabolism of broccoli sprouts showed that it promoted the expression of BrST5b (sulfotransferase 5b), inhibited AOP2 (2-oxoglutarate-dependent dioxygenase 2) expression, and induced the expression of genes that synthesize GLS to promote GLS biosynthesis [17].

3.1.6. Polygalacturonase Activity

Polygalacturonase (PG) is an enzyme that is closely related to softening senescence, and plays an important role in the degradation of cell wall materials [38]. It can be seen from Figure 5 that, as the storage period grew longer, the activity of PG increased, and the activity of PG in the treatment groups was significantly lower than that in the CK group. There was no significant difference between the treatment groups and the control group in the early stage of the storage period ($p > 0.05$). In the late storage period, the PG activity of the fresh-cut "Xuebai" cauliflower florets in the treatment groups was lower than that in the control group, and there was a significant difference ($p < 0.05$). On the 15th day, the PG activity of the group given the $CaCl_2$ + 40 °C treatment was 11.99, which was 23.1% lower than that of the control group. There was a significant difference in the PG activity between the group given the $CaCl_2$ + 40 °C treatment and the other two treatment groups.

The results showed that the treatment with temperature and $CaCl_2$ compound effectively inhibited the activity of PG enzyme during storage and delayed the softening of fresh-cut "Xuebai" cauliflower florets.

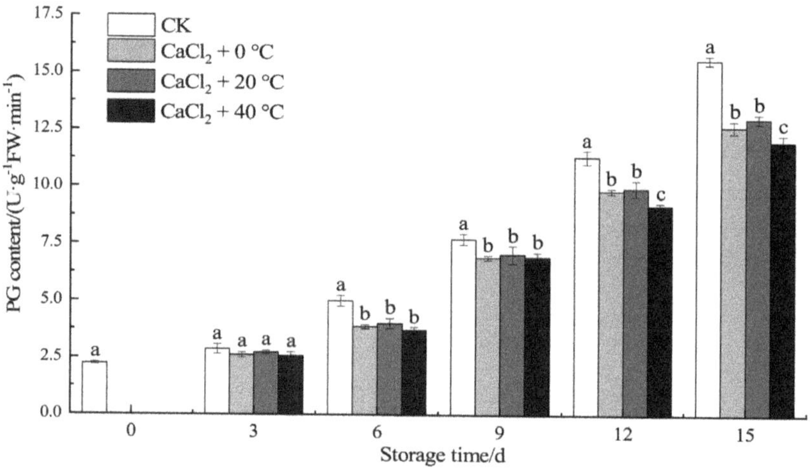

Figure 5. Effect of the different treatments on the PG activity of fresh-cut "Xuebai" cauliflower florets. Error bars indicate standard deviation (SD); the statistical significance between the four groups that passed the Tukey significance test is expressed in different lowercase letters, and the significance level is $p < 0.05$.

During the maturation and senescence of fruits and vegetables, the activity of PG, which is a cell-wall-degrading enzyme, increases. As a result, cell wall components such as pectin are degraded, resulting in the softening of the pulp, increased juice yield, and changes in the textural characteristics. During storage, the de-esterification and long-chain depolymerization of pectin occur, leading to the degradations of insoluble pectin into soluble pectin and pectic acid. The adhesion between the cells is reduced, the cell partitions disappear, and the turgor pressure is lost. This causes the cell viscosity and fruit hardness to decrease and finally leads to the softening of the fruit [39]. Ca+ can interact with pectin to form a calciumpectin gel, causing the cell walls to harden and giving them the ability to resist degradation. The use of calcium reduces the accessibility of cell-wall-degrading enzymes [32], thereby causing the activity of PG enzyme to be lower in the treatment group than in the control group. In general, the inhibition achieved by the $CaCl_2$ + 40 °C treatment was better than that achieved by the other two treatments. We also found that low temperatures do not effectively inhibit the activity of PG enzyme, which is consistent with the results of Tian et al. (2008) [38].

3.1.7. Lipoxygenase Activity

Lipoxygenase (LOX) has a peroxidation effect on lipids. It can cause a large number of free radicals to be produced and participate in the synthesis of ethylene, leading to an increase rate of aging in fruits and vegetables [40]. Figure 6 shows that, with the increase in the storage period, the activity of LOX first increased and then decreased, while the LOX value of the treatment groups was lower than that of the CK group. On the third day, there was a significant difference between the fresh-cut "Xuebai" cauliflower florets in the three treatment groups and those in the CK group ($p < 0.05$). On the ninth day, the LOX enzyme activity of the control group and the treatment groups reached a peak, after which the cauliflower began to show rapid softening and worsening effects of aging. By the 15th day, there was a significant difference between the $CaCl_2$ + 40 °C treatment group and the other two treatment groups ($p < 0.05$). It was observed that the $CaCl_2$ + 40 °C treatment inhibited the LOX activity most significantly.

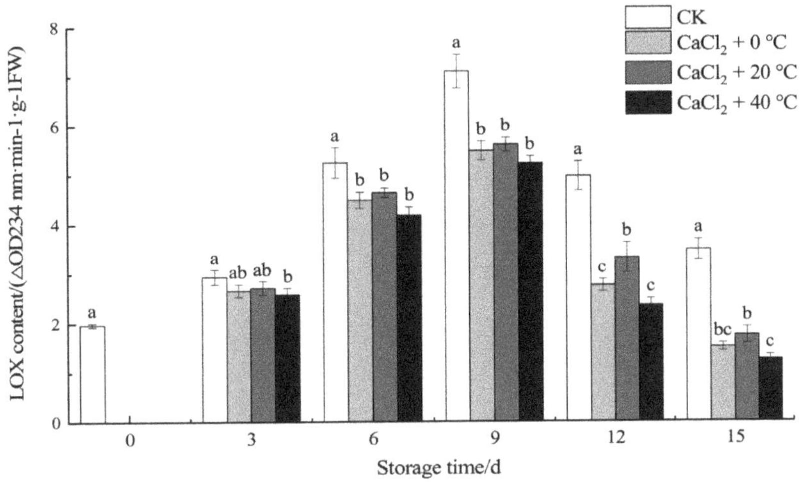

Figure 6. Effect of the different treatments on the LOX activity of fresh-cut "Xuebai" cauliflower florets. Error bars indicate standard deviation (SD); the statistical significance between the four groups that passed the Tukey significance test is expressed in different lowercase letters, and the significance level is $p < 0.05$.

LOX can catalyze the oxidative metabolism of unsaturated fatty acids to generate lipid peroxidation free radicals and other substances, activate membrane lipid peroxidation, and aggravate the damage to the membrane structure. The combination of Ca+ and membranes can protect membrane lipids, reduce plant cell membrane damage, and strengthen the cell wall structure. Ca+ plays an important role in stabilizing the membrane lipid structure, preventing membrane damage and maintaining the membrane integrity. However, exposure to low temperatures will induce Ca+ exudation, increase the Ca+ content in the cytoplasmic matrix, increase the activity of lipoxygenase enzymes in the cell, and aggravate the degradation and peroxidation of membrane lipids [41]. These findings are consistent with our conclusion that the $CaCl_2$ + 40 °C treatment was better than the $CaCl_2$ + 0 °C treatment in this study.

3.2. *Factor Analysis among Cauliflower Quality Indicators*

We carried out an exploratory factor analysis with dimensionality reduction to extract and synthesize the original variables with overlapping information into a small number of common factors, thereby replacing most of the original variable information. This method was used to model and analyze the quality of fresh-cut "Xuebai" cauliflower florets [42–44].

3.2.1. Correlation Analysis of Quality Indicators

During storage, the cell structures of fresh-cut "Xuebai" cauliflower florets were damaged, the permeability of the cell membrane was increased, and the rigidity of the cell wall was decreased. With the degradation of insoluble pectin into soluble pectin and pectic acid, the firmness of fresh-cut "Xuebai" cauliflower florets decreased rapidly and the activity value of the PG enzyme showed an upward trend. However, the ascorbic acid content and L^* value both showed downward trends. From the changing trends of the quality indicators of fresh-cut "Xuebai" cauliflower florets, it could be seen that there was an intrinsic connection between them, so it was necessary to carry out relationship research.

Table 2 shows the results of the relationship analysis carried out on the quality indicators of fresh-cut "Xuebai" cauliflower florets. Very significant positive relationships were observed between the firmness (X1) and L^* value (X2), ASA (X5), TGLS (X6), LOX activity (X8), and SE (X9) ($p < 0.01$). Additionally, negative relationships could be clearly observed between the firmness (X1) and a^* value (X3), b^* value(X4), and PG activity (X7) ($p < 0.01$). We also observed extremely significant positive relationships between the L^* value (X2) and ASA (X5) and TGLS (X6) ($p < 0.01$), as well as an extremely significant negative relationship between the a^* value (X3) ($p < 0.01$) and PG (X7) ($p < 0.05$). The a^* value (X3) had very significant positive relationships with the b^* value (X4) and PG activity (X7) ($p < 0.01$). There was also a clear negative relationship between the a^* value (X3) and ASA (X5), TGLS (X6), LOX activity (X8), and SE (X9) ($p < 0.01$). The b^* value (X4) was extremely negatively associated with the ASA (X5) and TGLS content (X6) ($p < 0.01$), and significantly negatively associated with the LOX activity (X8) ($p < 0.05$). The content of ASA (X5) had a very significant positive relationship with the LOX activity (X8) and SE (X9) ($p < 0.01$). The content of TGLS (X6) had very significant positive relationships with the LOX activity (X8) and SE (X9) ($p < 0.01$). There were also very significant negative relationships between the PG activity (X7) and LOX activity (X8) and SE (X9) ($p < 0.01$).

Table 2. Relationship analysis of the quality indicators of fresh-cut "Xuebai" cauliflower florets during storage.

	Firmness	L^* (X2)	a^* (X3)	b^* (X4)	ASA (X5)	TGLS (X6)	PG (X7)	LOX (X8)	SE (X9)
Firmness (X1)	1.000								
L^* (X2)	0.957 **	1.000							
a^* (X3)	−0.949 **	−0.995 **	1.000						
b^* (X4)	−0.976 **	−0.981	0.978 **	1.000					
ASA (X5)	0.973 **	0.961 **	−0.950 **	−0.991 **	1.000				
TGLS (X6)	0.930 **	0.880 **	−0.877 **	−0.900 **	0.899	1.000			
PG (X7)	−0.988 **	−0.943 *	0.936 **	0.981	−0.985	−0.935	1.000		
LOX (X8)	0.316 **	0.120	−0.131 **	−0.148 *	0.130 **	0.381 **	−0.251 **	1.000	
SE (X9)	0.972 **	0.968	−0.965 **	−0.995	0.993 **	0.899 **	−0.986 **	0.125	1.000

* and ** represent significance at the levels of 0.05 and 0.01. L^*—lightness, a^*—redness, b^*—yellowness.

From Table 2, it can be seen that the correlations between the quality indicators of fresh-cut "Xuebai" cauliflower florets were relatively high. However, if a quality assessment is carried out directly, there will be deviations in the assessment results due to the presence of overlapping information [45]. Thus, a factor analysis was conducted to combine several relevant indicators with uncorrelated indicators [44]. This new set of quality indicators was used for the quality synthesis of fresh-cut "Xuebai" cauliflower florets treated with different temperatures and $CaCl_2$ in order to improve the reliability of the evaluation model.

3.2.2. Principal Component Analysis (PCA)

The raw data of various indexes of fresh-cut "Xuebai" cauliflower florets treated with different temperatures and $CaCl_2$ were statistically analyzed using the Kaiser-Meyer-Olkin (KMO) and Bartlett test methods. It can be seen from Table 3 that the KMO value was 0.794, indicating that the sampling adequacy effect was very good. The significant difference of Bartlett's spheroid test was less than 0.01; this showed that the quality index data of fresh-cut "Xuebai" cauliflower florets were different. We continued to study these data with a factor analysis.

Table 3. KMO and Bartlett's correlation test.

Test Method	KMO Measure of Sampling Adequacy	Bartlett's Test of Sphericity		
		Approx. x2	df	Sig.
Result	0.794	477.288	36	0.000

A principal component analysis (PCA) was carried out in the SPSS 17.0 software to extract feature values from the quality index data of fresh-cut "Xuebai" cauliflower florets and perform a factor analysis. Table 4 shows that the variances of the common factors of all indexes of fresh-cut "Xuebai" cauliflower florets were above 0.917. This indicated that the effect of the factors analysis was good, and that the quality index data of fresh-cut "Xuebai" cauliflower florets were suitable for factor analysis.

Table 4. Factor variance statistics of quality indicators.

Quality Indexes	Initial	Extraction
Firmness(X1)	1.000	0.989
L^*(X2)	1.000	0.969
a^*(X3)	1.000	0.959
b^*(X4)	1.000	0.994
ASA(X5)	1.000	0.983
TGLS(X6)	1.000	0.917
PG(X7)	1.000	0.981
LOX(X8)	1.000	0.993
SE(X9)	1.000	0.991

L^*—lightness, a^*—redness, b^*—yellowness.

The PCA and the quarter rotation method were used to obtain the eigenvalues and rates of contribution of each factor of fresh-cut "Xuebai" cauliflower florets; the results are shown in Table 5.

Table 5. Factor analysis of principal component characteristic values and contributions in fresh-cut "Xuebai" cauliflower florets.

Component	Initial Eigenvalues			Extraction Sums of Squared Loadings			Rotation Sums of Squared Loadings		
	Total	% of Variance	Cumulative	Total	% of Variance	Cumulative	Total	% of Variance	Cumulative
1	7.736	85.960	84.960	7.736	85.960	84.960	7.717	85.744	85.744
2	1.040	11.553	97.513	1.040	11.553	97.513	1.059	11.768	97.513
3	0.113	1.261	98.773						
4	0.085	0.949	99.722						
5	0.013	0.141	99.863						
6	0.008	0.086	99.949						
7	0.002	0.025	99.974						
8	0.002	0.017	99.991						
9	0.001	0.009	100.000						

Table 5 shows the variance contribution rate obtained in the factor analysis of fresh-cut "Xuebai" cauliflower florets. The contribution rates of factor one and factor two to the total variance of fresh-cut "Xuebai" cauliflower floret reached 85.744% and 11.768%, respectively, while the cumulative variance contribution rate reached 97.513%. Nearly 100% of the information in the original dataset was covered by these two principal components, so it was determined that the number of factors in the subsequent factor analysis should be two.

Figure 7 shows a schematic diagram of the factor rotation of the correlation coefficients of the nine original indicators and factor one and factor two. It can be seen from Figure 7 that firmness, TGLS, ASA, L^* value, and SE had higher loads in the positive direction of factor one, while PG, a^* value, and b^* value had higher loads in the negative direction of factor one, which is in line with the actual meaning of each index. It was found that the higher the PG activity, a^* value, and b^* value were, the more significant the deterioration in the quality of fresh-cut "Xuebai" cauliflower florets was. Additionally, the higher the firmness, TGLS, ASA, L^* value, and SE were, the higher the quality of fresh-cut "Xuebai" cauliflower florets was. It can be inferred that factor one comprehensively reflected the main quality indicators, while factor two had a significant positive correlation with the LOX content. This mainly demonstrates the influence of LOX on the storage quality of fresh-cut "Xuebai" cauliflower florets, and its cumulative variance contribution rate was only 11.768%. The impact of this factor on the overall quality was far less important than that of factor one.

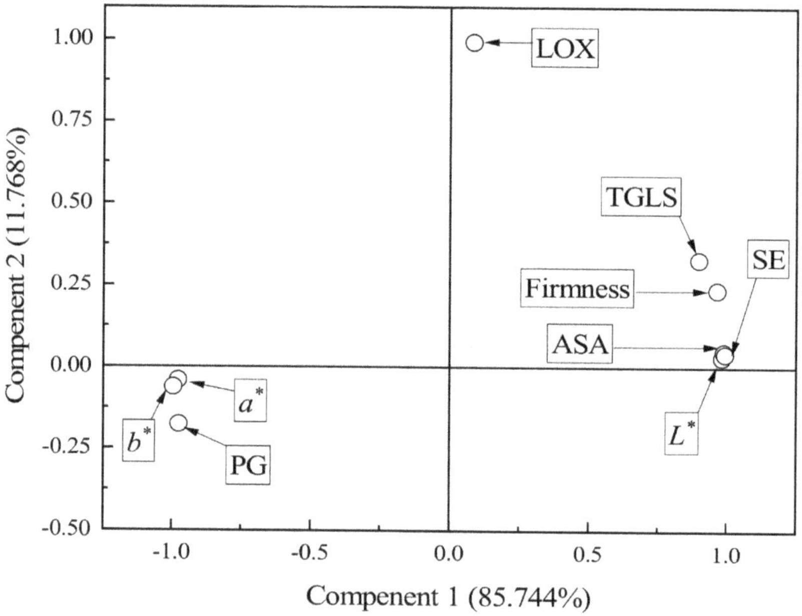

Figure 7. Principal component graph in rotating space. L^*—lightness, a^*—redness, b^*—yellowness.

Table 6 shows the effectiveness of the use of the regression coefficient method to explore the mutual effect of fresh-cut "Xuebai" cauliflower floret quality indicators under $CaCl_2$ treatments at different temperatures. After the data were rotated using the maximum variance method, the expressions of factor one and factor two were obtained.

$$Y1 = 0.123 \times X1 + 0.132 \times X2 - 0.131 \times X3 - 0.133 \times X4 + 0.133 \times X5 + 0.111 \times X6 - 0.126 \times X7 - 0.020 \times X8 + 0.133 \times X9 \quad (1)$$

$$Y2 = 0.091 \times X1 - 0.106 \times X2 + 0.096 \times X3 + 0.079 \times X4 - 0.090 \times X5 + 0.189 \times X6 - 0.032 \times X7 + 0.933 \times X8 - 0.098 \times X9 \quad (2)$$

Table 6. Quality component coefficient matrix of fresh-cut "Xuebai" cauliflower florets.

Quality indexes	Component	
	1	2
Firmness (X1)	0.123	0.091
L^* (X2)	0.132	−0.106
a^* (X3)	−0.131	0.096
b^* (X4)	−0.133	0.079
ASA (X5)	0.133	−0.090
TGLS (X6)	0.111	0.189
PG (X7)	−0.126	−0.032
LOX (X8)	−0.020	0.933
SE (X9)	0.133	−0.098

L^*—lightness, a^*—redness, b^*—yellowness.

These two common factors reflected the effects of different treatments on the storage quality of fresh-cut "Xuebai" cauliflower florets in terms of different aspects. Then, the variance contribution rate corresponding to each common factor was weighted to obtain the following comprehensive score calculation Equation Y:

$$Y = (85.744 \times Y1 + 11.768 \times Y2) \div 97.513 = 0.119 \times X1 + 0.103 \times X2 - 0.104 \times X3 - 0.107 \times X4 + 0.106 \times X5 + 0.120 \times X6 - 0.115 \times X7 + 0.095 \times X8 + 0.105 \times X9 \quad (3)$$

We normalized the original data and substituted them into Equation Y. Table 7 lists the order of the comprehensive quality index values of fresh-cut "Xuebai" cauliflower florets treated with different temperatures and $CaCl_2$.

Table 7. Ranking of the comprehensive quality index values of fresh-cut "Xuebai" cauliflower florets treated with different temperatures and $CaCl_2$.

Method	Storage Time/d	Value	Rank
CK	3	103.285	4
	6	99.066	8
	9	89.918	12
	12	78.230	16
	15	60.603	20
$CaCl_2$ + 0 °C	3	104.983	2
	6	101.324	6
	9	94.059	10
	12	84.302	14
	15	70.889	18
$CaCl_2$ + 20 °C	3	104.545	3
	6	100.455	7
	9	93.628	11
	12	83.878	15
	15	70.091	19
$CaCl_2$ + 40 °C	3	105.410	1
	6	102.522	5
	9	96.307	9
	12	87.251	13
	15	74.803	17

It can be seen from the comprehensive score comparison of $CaCl_2$ treatments at different temperatures in Table 7 that the storage period of fresh-cut "Xuebai" cauliflower florets could be significantly prolonged after treatment with temperature and $CaCl_2$. The fresh-cut "Xuebai" cauliflower florets had the highest score after being treated with $CaCl_2$ + 40 °C, but as the storage time grew longer, their quality continued to decline. After 15 days of storage, the comprehensive quality score of the cauliflowers fell below the pass line and the curd softened and browned, causing then to lose their storage value.

4. Conclusions

This experiment studied the effect of the use of different temperatures and $CaCl_2$ treatments on the preservation of fresh-cut "Xuebai" cauliflower florets. Through a comprehensive evaluation of our factor analysis, the optimal processing method for cauliflowers subjected to a storage period was determined. The test results showed that as the storage period grew longer, the fresh-cut "Xuebai" cauliflower florets developed a peculiar smell; darkened and yellowed in color; exhibited decreased hardness; decreased SE, ASA, and TGLS content, and showed increased PG and LOX activities. The quality of cauliflowers in the treatment groups was significantly better than that of the cauliflowers in the CK group, and they maintained a higher firmness value, higher ASA and TGLS content, and a lower activity of related enzymes. The factor analysis of indicators of fresh-cut "Xuebai" cauliflower florets during the storage period found that the rate of contribution of the first two principal component factors was 97.513%. These two factors were able to represent the change in the cauliflowers' quality. We also established a comprehensive evaluation model for fresh-cut "Xuebai" cauliflowers. The $CaCl_2$ + 40 °C treatment was determined to be the best preservation method for fresh-cut "Xuebai" cauliflower florets. This study provides a theoretical reference for extending the storage time of fresh-cut "Xuebai" cauliflower florets and improving their quality during storage.

Author Contributions: Conceptualization, B.M.; methodology, B.M.; data curation, B.M.; writing—original draft preparation, B.M.; writing—review and editing, B.M.; supervision, S.Z. and Z.L.; project administration, J.X. All authors have read and agreed to the published version of the manuscript.

Funding: This research was funded by the National Natural Science Foundation of China, Grant/Award Number: 31801632.

Institutional Review Board Statement: Not applicable.

Informed Consent Statement: Not applicable.

Data Availability Statement: Not applicable.

Conflicts of Interest: The authors declare no conflict of interest.

References

1. Dos Reis, L.C.; de Oliveira, V.R.; Hagen, M.E.K.; Jablonski, A.; Flôres, S.H.; de Oliveira Rios, A. Effect of cooking on the concentration of bioactive compounds in broccoli (*Brassica oleracea var. Avenger*) and cauliflower (*Brassica oleracea var. Alphina F1*) grown in an organic system. *Food Chem.* **2015**, *172*, 770–777. [CrossRef] [PubMed]
2. Kapusta-Duch, J.; Szeląg-Sikora, A.; Sikora, J.; Niemiec, M.; Gródek-Szostak, Z.; Kuboń, M.; Leszczynska, T.; Borczak, B. Health-Promoting Properties of Fresh and Processed Purple Cauliflower. *Sustainability* **2019**, *11*, 4008. [CrossRef]
3. Samec, D.; Urlic, B.; Salopek-Sondi, B. Kale (*Brassica oleracea var. acephala*) as a superfood: Review of the scientific evidence behind the statement. *Crit. Rev. Food Sci. Nutr.* **2019**, *59*, 2411–2422. [CrossRef] [PubMed]
4. Avato, P.; Argentieri, M.P. Brassicaceae: A rich source of health improving phytochemicals. *Phytochem. Rev.* **2015**, *14*, 1019–1033. [CrossRef]
5. Garcia-Ibanez, P.; Nicolas-Espinosa, J.; Carvajal, M. Plasma membrane vesicles from cauliflower meristematic tissue and their role in water passage. *BMC Plant Biol.* **2021**, *21*, 30. [CrossRef] [PubMed]
6. Zhang, H.; Yamamoto, E.; Murphy, J.; Locas, A. Microbiological safety of ready-to-eat fresh-cut fruits and vegetables sold on the Canadian retail market. *Int. J. Food Microbiol.* **2020**, *335*, 108855. [CrossRef] [PubMed]
7. Yousuf, B.; Qadri, O.S.; Srivastava, A.K. Recent developments in shelf-life extension of fresh-cut fruits and vegetables by application of different edible coatings: A review. *LWT Food Sci. Technol.* **2018**, *89*, 198–209. [CrossRef]
8. Ma, L.; Zhang, M.; Bhandari, B.; Gao, Z.X. Recent developments in novel shelf life extension technologies of fresh-cut fruits and vegetables. *Trends Food Sci. Technol.* **2017**, *64*, 23–38. [CrossRef]
9. Nguyen, V.T.B.; Nguyen, D.H.H.; Nguyen, H.H. Combination effects of calcium chloride and nano-chitosan on the postharvest quality of strawberry (*Fragaria x ananassa Duch.*). *Postharvest Biol. Technol.* **2020**, *162*, 111103. [CrossRef]
10. Hou, Y.; Li, Z.; Zheng, Y.; Jin, P. Effects of $CaCl_2$ Treatment Alleviates Chilling Injury of Loquat Fruit (*Eribotrya japonica*) by Modulating ROS Homeostasis. *Foods* **2021**, *10*, 1662. [CrossRef]

11. Martín-Diana, A.B.; Rico, D.; Frías, J.M.; Barat, J.M.; Henehan, G.T.M.; Barry-Ryan, C. Calcium for extending the shelf life of fresh whole and minimally processed fruits and vegetables: A review. *Trends Food Sci. Technol.* **2007**, *18*, 210–218. [CrossRef]
12. Aghdam, M.S.; Dokhanieh, A.Y.; Hassanpour, H.; Rezapour Fard, J. Enhancement of antioxidant capacity of cornelian cherry (*Cornus mas*) fruit by postharvest calcium treatment. *Sci. Hortic.* **2013**, *161*, 160–164. [CrossRef]
13. Zhang, L.; Wang, J.-W.; Zhou, B.; Li, G.-D.; Liu, Y.-F.; Xia, X.-L.; Xiao, Z.G.; Fei, L.; Ji, S.-J. Calcium inhibited peel browning by regulating enzymes in membrane metabolism of 'Nanguo' pears during post-ripeness after refrigerated storage. *Sci. Hortic.* **2019**, *244*, 15–21. [CrossRef]
14. Zhang, M.; Liu, W.; Li, C.H.; Shao, T.T.; Jiang, X.; Zhao, H.Z.; Ai, W.T. Postharvest hot water dipping and hot water forced convection treatments alleviate chilling injury for zucchini fruit during cold storage. *Sci. Hortic.* **2019**, *249*, 219–227. [CrossRef]
15. Cao, X.; Zhang, F.; Zhao, D.; Wang, Z.; Zhu, D.; Li, J. Effects of Cold Shock Treatment on Freezing Characteristics of Grape. *J. Chin. Inst. Food Sci. Technol.* **2019**, *19*, 201–207. [CrossRef]
16. Duarte-Sierra, A.; Forney, C.F.; Michaud, D.; Angers, P.; Arul, J. Influence of hormetic heat treatment on quality and phytochemical compounds of broccoli florets during storage. *Postharvest Biol. Technol.* **2017**, *128*, 44–53. [CrossRef]
17. Yang, R.Q.; Hui, Q.R.; Gu, Z.X.; Zhou, Y.L.; Guo, L.P.; Shen, C.; Zhang, W.H. Effects of $CaCl_2$ on the metabolism of glucosinolates and the formation of isothiocyanates as well as the antioxidant capacity of broccoli sprouts. *J. Funct. Foods* **2016**, *24*, 156–163. [CrossRef]
18. Grzegorzewska, M.; Kosson, R. The Influence of Postharvest Treatment and Type of Packaging on Quality and Storage Ability of Cauliflower (*Brassica oleracea* L. *var. botrytis*). *J. Fruit Ornam. Plant Res.* **2009**, *71*, 133–142. [CrossRef]
19. Wang, Y.; Xie, X.; Long, L.E. The effect of postharvest calcium application in hydro-cooling water on tissue calcium content, biochemical changes, and quality attributes of sweet cherry fruit. *Food Chem.* **2014**, *160*, 22–30. [CrossRef]
20. Supapvanich, S.; Arkajak, R.; Yalai, K. Maintenance of postharvest quality and bioactive compounds of fresh-cut sweet leaf bush (*Sauropus androgynus* L. *Merr.*) through hot CaCl2 dips. *Int. J. Food Sci. Technol.* **2012**, *47*, 2662–2670. [CrossRef]
21. Xue, J.X.; Wang, K.; Li, Z.Z.; Zhang, S.J.; Mu, B.Y.; Li, Z.H.; Huang, L.; Zhao, H.M.; Sun, H.X. Influences of post-harvest melatonin treatment on preservation quality and shelf life of fresh-cut cauliflower. *Trans. Chin. Soc. Agric. Eng.* **2021**, *37*, 273–283. [CrossRef]
22. Zhang, N.; Yan, R.X.; Guan, W.Q.; Wang, C. Effects of red light-emitting diode (LED) on the postharvest yellowing change of Broccoli. *Spectrosc. Spectr. Anal.* **2016**, *36*, 955–959. [CrossRef]
23. Yan, Z.C.; Shi, J.Y.; Gao, L.P.; Wang, Q.; Zuo, J.H. The combined treatment of broccoli florets with kojic acid and calcium chloride maintains post-harvest quality and inhibits off-odor production. *Sci. Hortic.* **2020**, *262*, 109019. [CrossRef]
24. Chen, X.L.; Zhang, L.H.; Yan, S.L.; Mei, X.; Shi, J.B.; Cai, S.; Sui, Y.; He, J.J. Effect of packaging materials on storage quality of fresh-cut broccoli. *Food Sci.* **2018**, *39*, 246–250. [CrossRef]
25. Xu, D.Y.; Liu, J.; Zuo, J.H.; Gao, L.; Wang, Q. Effect of Methyl Jasmonate Treatment on Quality of Postharvest Pepper Subjected Vibration during Transportation. *Mod. Food Sci. Technol.* **2018**, *34*, 70–76. [CrossRef]
26. Liu, Z.S.; Shi, J.Y.; Zuo, J.H.; Gao, L.P.; Wang, Q.; Meng, D.M. Effect of combined UV-C and red light emitting diode irradiation on storage quality of broccoli. *Food Sci.* **2020**, *41*, 238–245. [CrossRef]
27. Hazbavi, I.; Khoshtaghaza, M.H.; Mostaan, A.; Banakar, A. Effect of postharvest hot-water and heat treatment on quality of date palm (cv. Stamaran). *J. Saudi Soc. Agric. Sci.* **2015**, *14*, 153–159. [CrossRef]
28. Zhang, N.; Yang, Z.; Chen, A.G.; Zhao, S.S. Effects of intermittent heat treatment on sensory quality and antioxidant enzymes of cucumber. *Sci. Hortic.* **2014**, *170*, 39–44. [CrossRef]
29. Suzuki, Y.; Asoda, T.; Matsumoto, Y.; Terai, H.; Kato, M. Suppression of the expression of genes encoding ethylene biosynthetic enzymes in harvested broccoli with high temperature treatment. *Postharvest Biol. Technol.* **2005**, *36*, 265–271. [CrossRef]
30. Wang, L.; Wang, Q.; Liu, H.Z.; Liu, L.; Du, Y.; Zhang, J.S. Research Process on Peanut Processing Characteristics and Quality Evaluation. *J. Chin. Cereals Oils Assoc.* **2011**, *26*, 122–128.
31. Hu, X.Y.; Zhou, G.Y. Preservation Effect of Four Natural Preservatives on Cherry Tomatoes. *Food Sci.* **2012**, *33*, 287–292.
32. Ribeiro, L.R.; Leonel, S.; Souza, J.M.A.; Garcia, E.L.; Leonel, M.; Monteiro, L.N.H.; Silva, M.D.; Ferreira, R.B. Improving the nutritional value and extending shelf life of red guava by adding calcium chloride. *LWT Food Sci. Technol.* **2020**, *130*, 109655. [CrossRef]
33. Naser, F.; Rabiei, V.; Razavi, F.; Khademi, O. Effect of calcium lactate in combination with hot water treatment on the nutritional quality of persimmon fruit during cold storage. *Sci. Hortic.* **2018**, *233*, 114–123. [CrossRef]
34. Aguayo, E.; Requejo-Jackman, C.; Stanley, R.; Woolf, A. Hot water treatment in combination with calcium ascorbate dips increases bioactive compounds and helps to maintain fresh-cut apple quality. *Postharvest Biol. Technol.* **2015**, *110*, 158–165. [CrossRef]
35. Wei, L.; Liu, C.; Zheng, H.; Zheng, L. Melatonin treatment affects the glucoraphanin-sulforaphane system in postharvest fresh-cut broccoli (*Brassica oleracea* L.). *Food Chem.* **2020**, *307*, 125562. [CrossRef]
36. Vale, A.P.; Santos, J.; Brito, N.V.; Marinho, C.; Amorim, V.; Rosa, E.; Oliveira, M.B.P.P. Effect of refrigerated storage on the bioactive compounds and microbial quality of *Brassica oleraceae* sprouts. *Postharvest Biol. Technol.* **2015**, *109*, 120–129. [CrossRef]
37. Reddy, Y.V.R.; Marcy, J.E.; Bratsch, A.D.; Williams, R.C.; Waterman, K.M. Effects of Packaging and Postharvest Treatments on the Shelf-Life Quality of Crown-Cut Broccoli. *J. Food Qual.* **2010**, *33*, 599–611. [CrossRef]
38. Tian, S.; Zhao, H.; Xue, X.; Zhao, Z.; Zhou, J. Effects of LOX and PG Enzymes in Chinese Jujube Fruit (*Zizyphus Jujuba* mill. cv. Dong) on Its Softening and Senescence. *Food Sci.* **2008**, *29*, 446–448.

39. Zhang, P.; Chen, F.; Yang, H.; Li, L.; Gong, B.; Wang, L. Research advances on cell wall disassembly in fruit ripening and softening. *Food Sci. Technol.* **2010**, *35*, 62–66. [CrossRef]
40. Sheng, J.; Luo, Y.; Shen, L. Effects of PG and LOX on the softening and cell ultrastructure changes of postharvest tomato fruit. *Acta Hortic.* **2000**, *27*, 276–281.
41. Wu, J.; Wu, B.; Huang, S.; Chen, C.; Yan, L.; Xu, H.; Lin, J. Phospholipase D and Lipoxygenase of Young Loquat Fruits in Response to Low Temperature Stress. *Plant Sci. J.* **2015**, *33*, 203–209. [CrossRef]
42. Zhu, L.; Ling, J.; Shang, H.; Chen, S.; Cui, Y.; Kang, M. Factor analysis of the effects of controlled freezing-point storage in combination with 1-MCP treatment on quality of 'Yulu' juicy peaches during cold storage. *J. Fruit Sci.* **2016**, *33*, 1164–1172. [CrossRef]
43. Wu, X.; Gao, J.; Li, Y.; Wu, C. Development of A Safety Climate Scale for Geological Prospecting Projects in China. *Int. J. Environ. Res. Public Health* **2019**, *16*, 1082. [CrossRef] [PubMed]
44. Xue, J.; Huang, L.; Zhang, S.; Sun, H.; Gao, T. Study on the evaluation of carboxymethyl-chitosan concentration and temperature treatment on the quality of "Niuxin" persimmon during cold storage. *J. Food Process. Preserv.* **2020**, *44*, 14560. [CrossRef]
45. Olasege, B.S.; Zhang, S.; Zhao, Q.; Liu, D.; Sun, H.; Wang, Q.; Ma, P.; Pan, Y. Genetic parameter estimates for body conformation traits using composite index, principal component, and factor analysis. *J. Dairy Sci.* **2019**, *102*, 5219–5229. [CrossRef] [PubMed]

foods

Article

Effect of Melatonin on Fruit Quality via Decay Inhibition and Enhancement of Antioxidative Enzyme Activities and Genes Expression of Two Mango Cultivars during Cold Storage

Alagie Njie [1,2], Wen'e Zhang [1], Xiaoqing Dong [1], Chengyu Lu [1], Xuejun Pan [1,*] and Qingguo Liu [3,*]

1 College of Agriculture, Guizhou University, Guiyang 550025, China
2 School of Agriculture and Environmental Sciences, University of The Gambia, Kanifing P.O. Box 3530, The Gambia
3 Institute of Subtropical Crops, Guizhou Academy of Agricultural Sciences, Fenglindong Road, Guiyang 562400, China
* Correspondence: xjpan@gzu.edu.cn (X.P.); lqingguo8@gmail.com (Q.L.); Tel.: +86-138-8509-4631 (X.P.); +86-135-9598-4098 (Q.L.)

Abstract: The postharvest deterioration of mango fruits is a critical issue limiting mango storage and preservation due to its climacteric nature. This study evaluated the storage behavior of two mango cultivars and their response to exogenous melatonin (MT, 1000 μmol L^{-1}) treatment in attenuating fruit decay and enhancing fruits' physiological and metabolic processes and gene relative expression subjected to cold storage. MT treatment in both mango cultivars significantly delayed weight loss, firmness, respiration rate, and decay incidence. However, MT did not influence the TSS, TA, and TSS:TA ratio regardless of the cultivar. Moreover, MT inhibited the decrease in total phenol and flavonoid content and AsA content while delaying the increase in the MDA content of mango during storage in both cultivars. In addition, MT dramatically inhibited the enzyme activity of PPO. In contrast, an increase in the activities of antioxidant enzymes (SOD and APX) and PAL and their genes' relative expression was noticed in MT-treated fruits versus control in both cultivars. However, MT treatment was cultivar dependent in most parameters under study. These results demonstrated that MT treatment could be an essential postharvest treatment in minimizing decay, maintaining fruit quality, and extending mango fruits' postharvest shelf life by enhancing the physiological and metabolic processes during cold storage.

Keywords: mango fruit; melatonin treatment; cold storage; decay; physiological and metabolic processes

Citation: Njie, A.; Zhang, W.; Dong, X.; Lu, C.; Pan, X.; Liu, Q. Effect of Melatonin on Fruit Quality via Decay Inhibition and Enhancement of Antioxidative Enzyme Activities and Genes Expression of Two Mango Cultivars during Cold Storage. *Foods* **2022**, *11*, 3209. https://doi.org/10.3390/foods11203209

Academic Editor: Lorenzo Zacarías

Received: 9 August 2022
Accepted: 2 October 2022
Published: 14 October 2022

Publisher's Note: MDPI stays neutral with regard to jurisdictional claims in published maps and institutional affiliations.

1. Introduction

Mango (*Mangifera indica*) is one of the most vital tropical fruits globally due to its unique features: its pleasant flavor, attractive color, distinct taste and aroma, and rich nutritional value (carbohydrates, minerals, fibers, vitamins, polyphenols) [1,2]. In addition, mango fruits are categorized as a climacteric type with relatively high respiration. Mangoes are usually harvested in July and August at very high temperatures, and the harvested mango fruits ripen within a few days under ambient conditions. These factors lead to a relatively short postharvest shelf life due to rapid fruit decay and deterioration, a limiting factor in the long-distance transport and marketing of the fruit [1,3].

Most fruits, including mango, that contain very high water content are highly vulnerable to the attack of microbes, which ultimately cause decay in the harvested produce and reduce the shelf life and quality. Therefore, it is crucial to protect them from postharvest decay caused by microbes, bruises, and mechanical damage while improving shelf life and maintaining postharvest quality. Melatonin (N-acetyl-5-methoxytryptamine, MT), a derivative of tryptophan, is a vital indoleamine hormone widely found in various plant species, including apple, strawberry, grape, cherry, banana, kiwi, etc. [4]. MT plays a role in several physiological processes, such as regulating plant growth and development, delaying leaf senescence, and promoting fruit ripening [5,6]. MT, a potent antioxidant, improves the resistance of plants to both abiotic and biotic stresses by eliminating reactive oxygen species (ROS), elevating the antioxidant system, and increasing the efficiency of other antioxidants in plants. MT has shown its effectiveness in controlling decay by suppressing the activities of various fungi, bacteria, and other microbes, and also boosts the physiological performance of fruits against unfavorable conditions [7]. Similarly, MT was also reported to scavenge free radicals, ROS, and reactive nitrogen species (RNS) directly, act as a signaling molecule at the cellular level, and increase the number of antioxidant enzymes, which promotes its activity as an antioxidant [8]. Among the leading causes of fruit decay during storage is the development of rots caused by a range of fungi. Total phenolic content and antioxidant activity were more significant in MT-treated Santa Rosa plums than in the control, substantially reducing the decay rate during cold storage [9].

Recently, exogenous MT treatment has been tested as an effective postharvest treatment in different fruits. Some researchers found that MT was effective in promoting postharvest banana ripening [10], reducing senescence, and improving the quality of peach fruit and strawberry, respectively [11,12], minimizing chilling injury in peach fruit [13]. Rastegar et al. [14] found that MT at 1000 μmol L^{-1} maintained fruit firmness, ascorbic acid content (AsA), the content of phenolic compounds, and the antioxidant capacity of mango during storage. Similar research by Liu et al. [15] stated that the application of MT (0.5 mM, immersion for 1 h) to 'Guifei' mangoes effectively delayed the changes in ripening parameters, including firmness, total soluble solids content (TSS), titratable acidity (TA), and respiration rate. Moreover, MT significantly controlled polyphenol oxidase (PPO) activity and increased the activity of the catalase (CAT) and peroxidase (POD) enzymes during storage [14]. However, a more recent paper by Bhardwaj et al. [16] discovered that MT treatment has a different effect on different cultivars of mango fruits. They found that 'Langra' mangos responded best to MT treatment by increasing their chilling tolerance during storage, while 'Gulab Jamun' mangos did not show any significant effect [16], which demonstrates that the MT effect is cultivar dependent. MT upregulates critical genes' (*PpSODs*, *PpCATS*, *PpAPXs*, and *PpGRs*) expression in peaches' antioxidant defense [17]. In addition, the transcript expression of phenolic biosynthesis genes was upregulated by MT treatment, indicating that MT could activate the phenylpropanoid pathway [18]. MT reduces ethylene production by regulating the expression of the *MaACO* and *MaACS* genes and delays sharp changes in quality indices in bananas [19]. Furthermore, exogenous MT significantly regulated the transcript levels of the expression of critical genes involved in antioxidative defense (*SOD, CAT, APX, POD*) and *PAL* resulting in increased tolerance to drought stress in Chinese hickory plants [20].

'Guiqi' and 'Tainong 1′ used in this research are two of Guizhou province's most cultivated mango cultivars. 'Guiqi', also known as 'Gui re 82′, is a mango cultivar with a thin peel and is rich in aroma content, TSS, and TA [21]. 'Tainong 1′, an early maturing cultivar with a solid pine flavor, is rich in polyphenols and flavonoids and has high antioxidant scavenging inhibition [22]. Zhang et al. [23] discovered that hot water treatment at 55 °C for 10 min effectively enhanced the chilling tolerance in 'Tainong 1′ mango fruit during ambient storage after exposure to low temperature [23]. The above two cultivars have different physiological characteristics. However, it is speculated that the storage characteristics of these two cultivars, especially 'Guiqi', are not evident. Therefore, the current study aims to compare the storage characteristics of two mango cultivars after MT treatment subjected to cold storage. The study also aims to evaluate the effect of exogenous MT treatment on extending the shelf life of these two cultivars by attenuating fruit decay and enhancing fruits' physiological and metabolic processes and genes relative expressions during storage. The results of this study will give an insight into the storage behavior of these two cultivars of mango and their response to MT treatment. It is also assumed that this research will give insight into how MT treatment will help minimize fruit deterioration and improve antioxidative processes and relative gene expression, thereby maintaining quality and extending the shelf life of mango fruits.

2. Materials and Methods

2.1. Plant Materials

Mango fruits of the early maturing 'Tainong 1′ and late-maturing 'Guiqi' were harvested from Guizhou Mountain Mango Base farm in Wangmo county, Guizhou province (Latitude 25°6′ North, and Longitude 106°6′ East), on 21 July and 8 August, respectively. In total, 240 fruits of 'Tainong 1′ and 'Guiqi' each were harvested at the physiological green mature stage and packed in cardboard boxes. The fruit of each layer was separated with soft fabric to avoid compression injury. After sorting to choose fruits of uniform weight with no visible blemishes, they were transported to the laboratory within 3 h using cardboard boxes via a van with air conditioning at 25 ± 1 °C and 50–60 ± 0.5% relative humidity (RH). Upon arrival at the laboratory, the harvested fruits were treated using sodium hypochlorite (5% v/v) by immersing them for 10 min. Then, they were rinsed using distilled water and air-dried.

2.2. Melatonin Treatment

A complete randomized design (CRD) with three replications was used in this research. Each treatment was assigned 56 fruits and replicated three times. According to our pre-experiment, 1000 μmol L^{-1} MT was used as the final concentration in which fruits were immersed for 30 min. Treatment using distilled water for 30 min served as the control. The fruits were then air-dried at room temperature and stored in a refrigerator at 13 ± 1 °C and 80 ± 1% RH. The fruits were evaluated at three-day intervals during a one-month storage period.

2.3. Quality Parameters

2.3.1. Weight Loss

The fruit weight loss was measured using an electronic scale and measured in grams. The weight loss was measured every three days using 20 fruits from each treatment, compared with the initial weight, and expressed in percentage.

$$weight\ loss\ (\%) = \frac{initial\ weight - final\ weight}{initial\ weight} \times 100$$

2.3.2. Firmness

A handheld penetrometer (GY-4, Yueqing Edberg Instrument Co., Ltd., Yueqing, China) was used to determine the firmness of the mango fruits (n = 3). A peel from the mango fruit was removed from both ends, and the firmness was measured in the equatorial region using a 3.5 mm probe. Firmness was expressed as a force in Newton (N).

2.3.3. Decay Incidence Rate

Decay incidence of mango fruits was the number of fruits showing decay symptoms (rot, lesions, or visible fungal growth) relative to the total number of fruits and expressed in a percentage (percentage). Twenty fruits from each treatment were used to measure decay incidence, and a visual inspection was carried out every three days.

$$Decay\ incidence\ (\%) = \frac{number\ of\ fruits\ showing\ decay\ symptoms}{total\ number\ of\ fruits} \times 100$$

2.3.4. Respiration Rate

The respiration rate expressed as mg $CO_2 \cdot kg^{-1} \cdot h^{-1}$ FW was determined as modifications described by [24]. Six (6) fruits of each treatment were enclosed in a 9 L glass jar at room temperature for 30 min, and three biological replicates were performed. The respiration rate was calculated by the concentration of CO_2, which was measured by a CO_2 gas analyzer (Telaire-7001, Goleta, CA, USA).

2.4. Nutritional Parameters

2.4.1. Total Soluble Solid Content (TSS), Titratable Acidity (TA), and TSS:TA

TSS was measured using a refractometer (Hybrid, PAL-BX I ACID 1, ATAGO, TOKYO) at room temperature and expressed as Brix percentage. For the measurement of TA, 20 µL of mango juice was pipetted with 980 µL of distilled water, and the concentration of TA was determined using a refractometer. The results were expressed in percentage (%). The ripening index was calculated as the ratio of TSS:TA according to [25].

2.4.2. Ascorbic Acid (AsA) Content

In total, 0.5 g of pulp sample was weighed and homogenized with 2 mL of 50 g/L trichloroacetic acid (TCA) in a mortar and grounded in ice condition. The homogenate was centrifuged at 12,000 rpm for 20 min, and the supernatant was collected and used for the assay of the AsA content. Briefly, 1 mL of the sample extract was mixed with 1 mL of 50 g/L TCA solution, and 1 mL of anhydrous alcohol was added and mixed well by vortex. Afterward, 0.5 mL of 4% phosphoric acid was added and vortexed again. After 5 min, 1 mL of 5 g/L BP was added and mixed well, and 0.5 mL of 0.3 g/L $FeCl_3$ was then added and mixed well with the vortex. The mixed solution was placed in the dark at 30 °C for 60 min, and the absorbance was recorded at 534 nm against the blank. Ascorbic acid was used as the standard, and the AsA content was expressed in a 100 g sample (fresh weight), i.e., mg·100 g^{-1} FW.

2.4.3. Total Phenol and Flavonoid Content

One gram (1 g) of frozen mango samples was ground using a mortar and pestle and extracted in 5 mL of 95% (*v*/*v*) methanol. The homogenate was sonicated for 30 min, followed by centrifuging at 11,000× *g* for 20 min at 4 °C. Then, the supernatant was transferred to a new tube and stored at −20 °C, used to determine the total phenol and flavonoid content.

The total phenol content was assayed using the method of Liu et al. [12], with modifications. Briefly, 0.5 mL volume of the methanol extract was mixed with 1 mL Folin–Ciocalteu reagent, following which the reaction mixture was allowed to stand for 5 min at ambient temperature. Next, 3 mL of 7.5% sodium carbonate was added and left to react for 30 min. The absorbance was measured at 765 nm. The results were expressed as mg GAE·100 g^{-1} FW against gallic acid as a standard.

The total flavonoid content was determined using modifications by the colorimetric method [12]. Methanol extract (1 mL) was mixed with 5% $NaNO_2$ solution (0.3 mL). After being allowed to stand for 5 min, the mixture was combined with a 10% $AlCl_3$ solution (0.3 mL), while 1 M sodium hydroxide (NaOH) (1.5 mL) and distilled water (1.9 mL) were added after 5 min. The absorbance of the reaction mixture was recorded at 510 nm. Quercetin (QE) was used as a standard, and the results were expressed as mg QE·100 g^{-1} FW.

2.5. Malondialdehyde (MDA) Content

The MDA content was assayed according to the thiobarbituric acid method [26] with slight modification. Briefly, 1 g samples were ground as homogenate using 5 mL cold 10% (*w*/*v*) TCA solution and centrifuged at 10,000× *g* for 20 min at 4 °C. Then, 2 mL supernatant and 3 mL 0.67% (*w*/*v*) thiobarbituric acid (TBA) were mixed and heated for 20 min in a boiling water bath at 50 °C. After cooling, the absorbance of the supernatant was measured at 450, 532, and 600 nm. Finally, the MDA content was calculated using the formula and expressed as nmol·g^{-1} FW basis:

$$MDA\ content\left(nmol\cdot g^{-1}\right) = 6.45 * (OD_{532} - OD_{600}) - 0.56 * OD_{450}$$

2.6. Enzymes Activities

First, 0.5 g mango samples were rapidly homogenized on the ice with 8 mL ice-cold sodium phosphate buffer (PBS, 50 mmol L^{-1}, pH 7.8) containing 1% (w/v) polyvinyl polypyridine (PVP). Then, the homogenate was centrifuged at 12,000× g for 30 min at 4 °C, following which the supernatant was divided into 1.5 mL centrifuge tubes and stored at −70 °C as a crude extract for PPO, SOD, and APX determination.

2.6.1. Polyphenol Oxidase (PPO) Activity

A 4.0 mL of 50 μmol L^{-1} sodium acetate buffer, pH 5.5, was mixed with 0.5 mL of 50 mmol L^{-1} catechol solution, and finally, 100 μL of enzyme extraction solution was added and started timing immediately. The absorbance of the reaction system at 420 nm was recorded at 15 s as an initial value and then recorded at 1 min intervals for 6 min. The PPO enzyme activity was expressed in $U{\cdot}min^{-1}{\cdot}g^{-1}$ FW.

2.6.2. Phenylalanine Ammonia-Lyase (PAL) Activity

PAL activity was assayed using a plant PAL kit (A137-1-1, Nanjing Jiancheng Bioengineering Institute, Nanjing, China) following the manufacturer's instructions and absorbance measurement at 290 nm and expressed in $U{\cdot}min^{-1}{\cdot}g^{-1}$ FW.

2.6.3. Superoxide Dismutase (SOD) Activity

The reaction mixture includes 1.7 mL of 50 μmol L^{-1} sodium phosphate buffer, pH 7.8, 0.3 mL of 130 μmol L^{-1} methionine solution, 0.3 mL of 750 μmol L^{-1} NBT solution, 0.3 mL of 100 μmol L^{-1} EDTA-Na_2 solution, 0.3 mL of 20 μmol L^{-1} riboflavin solution, and 0.1 mL of enzyme extract. The reaction mixture was exposed under a 4000 lx fluorescent lamp for 10 min before measuring the SOD activity at 560 nm. The reaction system's inhibition of the photochemical reduction of NBT was expressed as one SOD activity unit (U) per min per gram of fresh weight at 50%.

2.6.4. Ascorbate Peroxidase (APX) Activity

A 2.6 mL reaction buffer was added to 0.1 mL of enzyme extract in a test tube. Then, 3 mL of 2 μmol L^{-1} H_2O_2 solution was added to initiate the enzymatic reaction and immediately mixed. The absorbance at 290 nm of the initial reaction was recorded from 15 s after then at 30 s interval for 3 min. The APX enzyme activity was expressed in $U{\cdot}min^{-1}{\cdot}g^{-1}$ FW.

2.7. RNA Isolation and Gene Expression Analysis

Total RNA was extracted from frozen mango pulp for gene expression analysis using RNAprep Pure kit (TIANGEN BIOTECH (BEIJING) CO., LTD DP140916, Beijing, China) and further assessed for quantity and quality with a NanoDrop One. Then, 2 μg of extracted RNA were further treated with FastKing-RT SuperMix and further converted into cDNA with the FastKing gDNA Dispelling RT SuperMix according to the manufacturer's instruction (TIANGEN BIOTECH (BEIJING) CO., LTD; KR170801). Specific primers of *PPO*, *PAL*, *MnSOD*, and *APX* were obtained from National Center for Biotechnology Information (NCBI) and designed by their nucleotide sequence using PRIMER 6 software, and then synthesized by a biological company (Tsingke Biotech (Guiyang) Co., Ltd., Guiyang, China). The qRT-PCR analysis of *PPO*, *PAL MnSOD*, and *APX* genes in mango was performed using gene-specific primers (Table 1) in a total volume of 20 μL with Bioer Real-Time PCR Detection Systems (BIOER Technology Ltd., Gene-9660, Hangzhou, China) using TB Green Premix EX Taq II (TAKARA BIO INC); Cat. RR820A, Beijing, China). The Real-Time qPCR reaction was performed with the following thermal profile: 95 °C 30 s, 40 cycles of 95 °C 5 s and 60 °C 34 s, and finishing with a final cycle at 95 °C 15 s, 60 °C 60 s, and 95 °C per 15 s, to define the reaction melting curve. The qPCR data were calibrated relative to *Actin* as a reference gene at zero time for each treatment, following the $2^{-\Delta\Delta Ct}$ method for relative quantification. The values represent the mean of three biological replicates.

Table 1. The sequences of specific primers used for qRT-PCR quantification.

Gene Name	Gene ID	Forward Primer (5′-3′)	Reverse Primer (5′-3′)
MnSOD	123228203	TAACCGAATCCGTCCGCCTTGG	CCTCCGCCGTTGAACTTGATGG
APX	123203234	CTTCTTCTCTGTCGTTCTCT	TCTCGCACTCTTCAACTG
PAL	123193566	CTGGCTGGTATCAGTAGTG	CCTGGATGGTGCTTCAAT
PPO	123193265	TAGCACACGCAGCGGAGTTGAA	CCCAGTTGCCACCTCATCCTCA
ACTIN	123216838	AGACCACCTACAACTCCAT	ATCCTCCAATCCAGACACT

2.8. Statistical Analysis

Microsoft Excel (V. 2010, Washington, DC, USA), OriginPro (V. 2022b, Northampton, MA, USA), and SPSS (V.26.0, New York, NY, USA) were used as the statistical software to analyze the results obtained from the experiment. ANOVA analysis was used to compare the mean difference. $p < 0.05$ was used as the statistical difference between means using Tukey's test. The storage period, cultivar, and treatment were used as the source of variations in this research.

3. Results

3.1. Effects of MT on Quality Parameters

3.1.1. Weight Loss and Firmness

During the storage period, weight loss measured in percentage increased steadily in both cultivars regardless of the treatment (Figure 1A,B). However, at the end of the storage period, MT treatment illustrated a lower mean percentage loss in weight than control fruits in both cultivars (Figure 1). The lowest weight loss for MT treatment was 9.48% for 'Guiqi' and 9.72% for 'Tainong 1′ as compared to the control with 10.62% for 'Guiqi' and 11.24% for 'Tainong 1′ at 24 d and 27 d storage periods, respectively. There was no cultivar difference in terms of weight loss. The difference between MT and control in both cultivars was significant ($p < 0.05$).

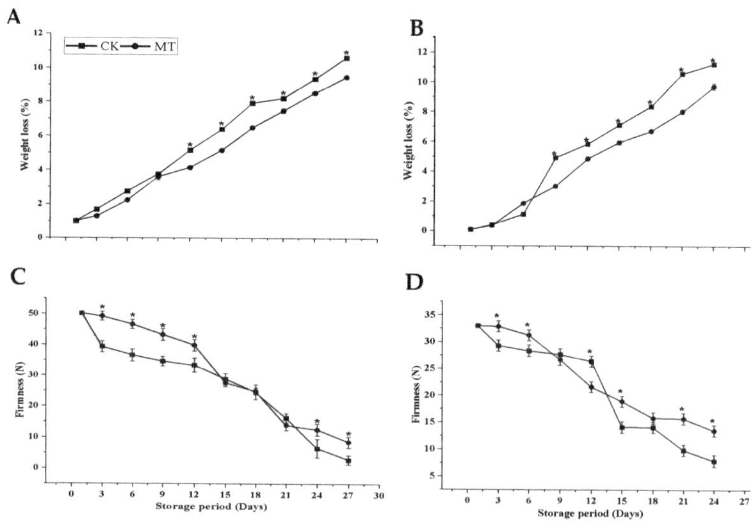

Figure 1. Weight loss (**A**,**B**) and firmness (**C**,**D**) of 'Guiqi' (**A**,**C**) and 'Tainong 1′ (**B**,**D**) fruits treated with 1000 μmol L^{-1} MT stored at 13 ± 1 °C and 80 ± 1% RH. Each value is the mean for three replicates, and vertical bars indicate the standard error (SE). Error bars with an asterisk (*) on the same storage period show a significant difference between the treatments ($p < 0.05$).

The firmness of both cultivars showed a constant decreasing trend, which signifies that the fruits became softer with an increased storage period (Figure 1C,D). However, at the end of the storage period, the MT-treated fruits maintained higher firmness than the control fruits with a significant difference ($p < 0.05$). MT treatment delayed fruit firmness in 'Guiqi' almost thrice that of the control at 27 d (8.47 ± 1.83 and 2.72 ± 1.47), respectively, while in 'Tainong 1′, the delay difference between MT and control at 24 d was almost twice (13.52 ± 1.03 and 7.91 ± 1.09), respectively.

3.1.2. Decay Incidence and Respiration Rate

Fruits started to show signs of decay incidence after 12 d in the 'Guiqi' cultivar and 3 d in 'Tainong 1′, and the trend continued to increase until the end of the storage period (Figure 2A,B). MT treatment showed a lower incidence of decay compared to the control treatment in both cultivars with a statistically significant difference at ($p < 0.05$). The highest decay percentages were 65% and 85% for MT treatment in 'Guiqi' and 'Tainong 1′, and 75% and 95% for the control, respectively, for 24 d and 27 d storage periods, respectively.

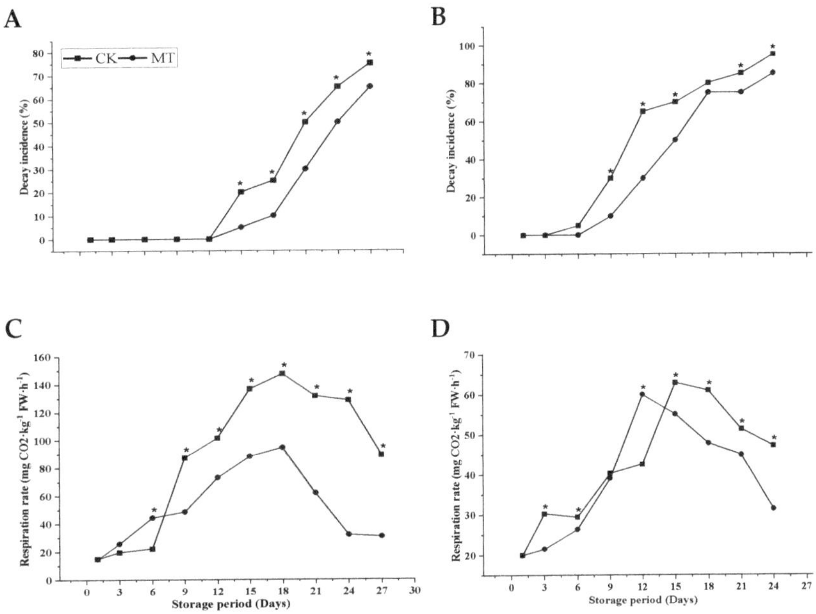

Figure 2. Decay incidence (**A**,**B**) and respiration rate (**C**,**D**) of 'Guiqi' (**A**,**C**) and 'Tainong 1′ (**B**,**D**) fruits treated with 1000 μmol L^{-1} MT. Each value is the mean for three replicates, and vertical bars indicate the standard error (SE). Error bars with an asterisk (*) on the same storage period show a significant difference between the treatments ($p < 0.05$).

Both treatments showed a similar respiration pattern during storage in both cultivars (Figure 2C,D). However, from 9 d to 27 d, MT-treated fruits showed a much lower respiration rate than the control in 'Guiqi', with a significant difference ($p < 0.05$). While in 'Tainong 1′, the difference was significant from 12 d to 24 d, in which MT-treated fruits showed lower respiration at most stages of the storage period. The respiration peak in MT-treated fruits was observed on 18 d (94.56 ± 0.05 mg $CO_2 \cdot kg^{-1}$ $FW \cdot h^{-1}$) in 'Guiqi', while in 'Tainong 1′, it was observed on 12 d (59.96 ± 0.06 mg $CO_2 \cdot kg^{-1}$ $FW \cdot h^{-1}$) as compared to the control with a peak at (147.49 ± 0.14 mg $CO_2 \cdot kg^{-1}$ $FW \cdot h^{-1}$, and 62.91 ± 0.05 mg $CO_2 \cdot kg^{-1}$ $FW \cdot h^{-1}$) in 'Guiqi' and 'Tainong 1′, respectively.

3.2. Effect of MT on Nutritional Parameters of Mango

3.2.1. TSS, TA, and TSS:TA

MT treatment did not affect TSS content in this study due to its higher concentration than the control. At the end of the storage period, the TSS content in MT was 18.23% and 19.01% for 'Guiqi' and 'Tainong 1', respectively. While in control, 17.47% for 'Guiqi' and 18.37% for 'Tainong 1' were observed. However, the difference was not significant ($p < 0.05$) between the treatments in each cultivar at the end of the storage period (Figure 3A,B). TA showed a decreasing pattern in both cultivars regardless of the treatment throughout the storage period (Figure 3C,D). From 6 d in 'Guiqi' and 9 d in 'Tainong 1', MT-treated fruits did not significantly affect the TA content in both cultivars. At the end of the storage period, MT treatment showed a lower TA content (0.92% and 0.74% for 'Guiqi' and 'Tainong 1', respectively) than the control (1.05% and 0.76% for 'Guiqi' and 'Tainong 1', respectively), but the difference was not statistically significant ($p < 0.5$). A continuous increase in the TSS:TA ratio in both cultivars was observed, indicating increasing maturity and ripening during storage. At the end of the storage period, MT treatment did not influence the TSS:TA ratio reduction in both cultivars (Figure 3E,F). However, there was a lower TSS:TA ratio by MT in 'Guiqi' than in the control, while in 'Tainong 1', the TSS:TA ratio was higher for MT than in the control.

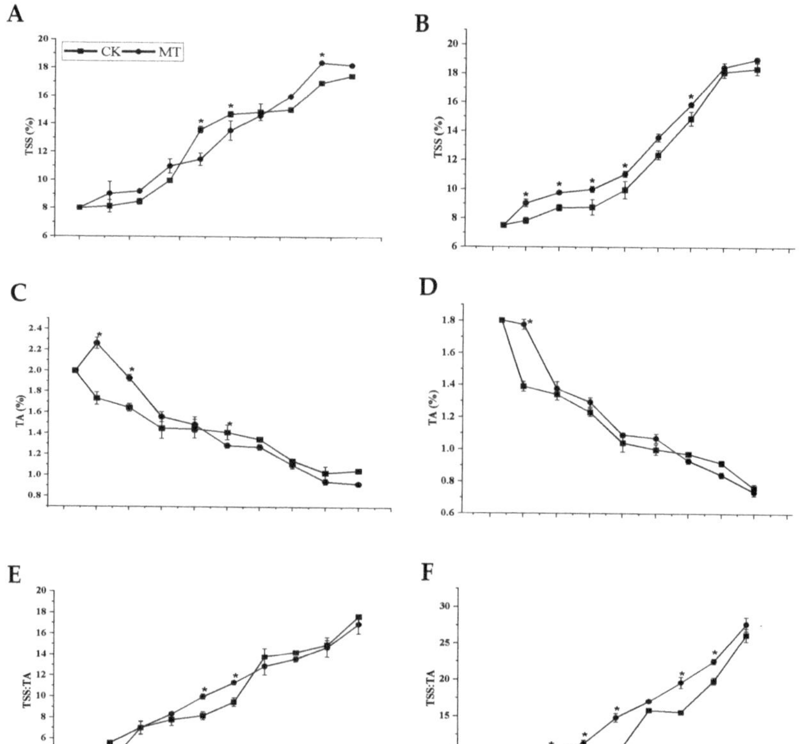

Figure 3. TSS (**A**,**B**), TA (**C**,**D**), and TSS:TA ratio (**E**,**F**) of 'Guiqi' (**A**,**C**,**E**) and 'Tainong 1' (**B**,**D**,**F**) fruits treated with 1000 μmol L^{-1} MT. Each value is the mean for three replicates, and vertical bars indicate the standard error (SE). Error bars with an asterisk (*) on the same storage period show a significant difference between the treatments ($p < 0.05$).

3.2.2. AsA Content

A continuous decrease in the AsA content was observed in both cultivars throughout the storage period regardless of the treatment (Figure 4A,B). However, MT treatment resulted in a higher AsA content than the control fruits at the end of the storage period, and the results were statistically significant ($p < 0.05$). At the 27 d, the AsA in 'Guiqi' was 0.62 ± 0.03 mg/100 g FW for MT treatment as compared to 0.51 ± 0.02 mg/100 g FW for the control, while at 24 d in 'Tainong 1', 0.27 ± 0.01 mg/100 g FW for MT was observed compared to 0.19 ± 0.14 mg/100 g FW for the control. Moreover, the interaction between the treatment and storage period was more significantly affected by MT treatment in 'Guiqi' than in 'Tainong 1' ($p < 0.05$) during most stages of the storage period.

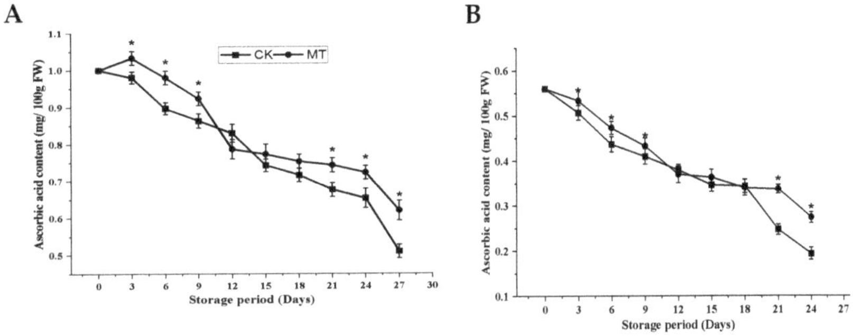

Figure 4. Ascorbic acid content (**A**,**B**) of 'Guiqi' (**A**) and 'Tainong 1' (**B**) fruits treated with 1000 μmol L^{-1} MT. Each value is the mean for three replicates, and vertical bars indicate the standard error (SE). Error bars with an asterisk (*) on the same storage period show a significant difference between the treatments ($p < 0.05$).

3.2.3. Total Flavonoid and Phenol Content

Figure 5A–D illustrated a decline in the mean values of the total flavonoid and phenol content in both the control fruits and the MT-treated fruits throughout the storage period, regardless of the cultivar. However, MT treatment showed higher significant values in the total flavonoid content at the end of the storage period than the control treatment ($p < 0.05$) in both cultivars. The lowest content of total flavonoid in the MT treatment was (22.07 ± 1.08 and 18.82 ± 0.26 mg QE/100 g FW) compared to the control (17.13 ± 1.03 and 14.52 ± 0.75 mg QE/100 g FW) in 'Guiqi' and 'Tainong 1', respectively. A similar trend was noticed in the total phenol content, in which the MT treatment showed higher mean values (32.38 ± 1.04 and 21.80 ± 1.06 mg GAE/100 g FW) than in the control (23.78 ± 2.76 and 18.13 ± 1.06 mg GAE/100 g FW) in 'Guiqi' and 'Tainong 1', respectively. The interaction between the treatment and the storage period was similar in both cultivars but slightly more pronounced in 'Tainong 1' than in 'Guiqi'.

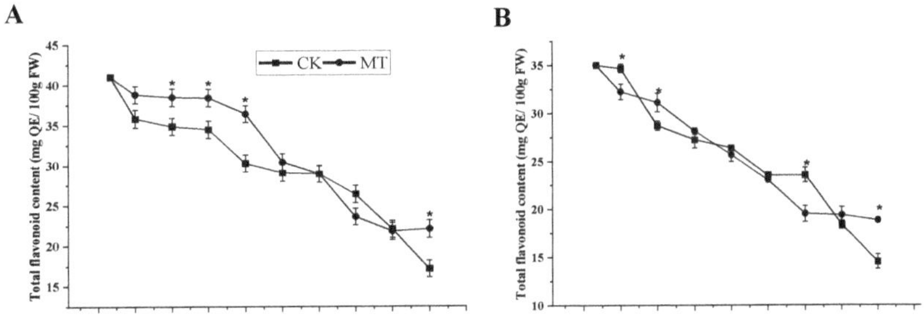

Figure 5. *Cont.*

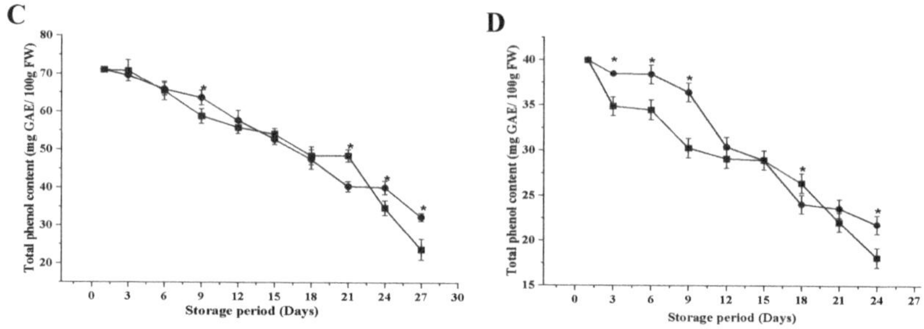

Figure 5. Total flavonoid content (**A**,**B**) and total phenol content (**C**,**D**) of 'Guiqi' (**A**,**C**) and 'Tainong 1' (**B**,**D**) fruits treated with 1000 μmol L^{-1} MT. Each value is the mean for three replicates, and vertical bars indicate the standard error (SE). Error bars with an asterisk (*) on the same storage period show a significant difference between the treatments ($p < 0.05$).

3.3. MDA Content

MDA, the main product of membrane lipid peroxidation, was measured in both cultivars during the 27 d for 'Guiqi' and 24 d for the 'Tainong 1' storage periods (Figure 6A, B). In Guiqi, the MDA content increased significantly up to 18 d of storage in both the control and MT-treated fruit and then slightly declined on 21 d of storage, then increased for the rest of the storage period. However, lower MDA content was observed in the MT treatment (1.60 ± 0.04 nmol g^{-1}) than in the control (1.89 ± 0.04 nmol g^{-1}) on 27 d. In 'Tainong 1', MDA slightly increased until 15 d, where a massive increase occurred until the final day of the storage period. However, MDA was lower in the MT-treated fruit (1.01 ± 0.02 nmol g^{-1}) than in the control fruit (1.06 ± 0.02 nmol g^{-1}) on 24 d. The difference between MT and the control in MDA content inhibition was more pronounced in 'Guiqi' than in 'Tainong 1' at the final stage of storage.

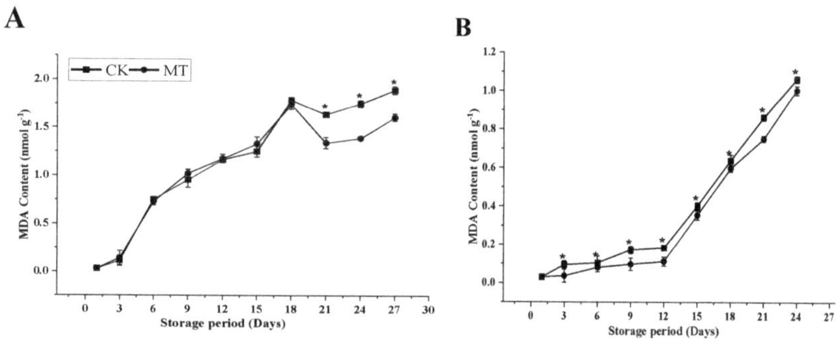

Figure 6. MDA content (**A**,**B**) of 'Guiqi' (**A**) and 'Tainong 1' (**B**) fruits treated with 1000 μmol L^{-1} MT. Each value is the mean for three replicates, and vertical bars indicate the standard error (SE). Error bars with an asterisk (*) on the same storage period show a significant difference between the treatments ($p < 0.05$).

3.4. *Effect of MT on Enzyme Activities*

3.4.1. PPO and PAL Activity

Figure 7A,B manifested a gradually increased PPO activity during storage, reaching a maximum level at the end of storage in the control and MT-treated fruits in both cultivars. The PPO activity in the control samples was significantly higher (0.35 ± 0.01 and 0.07 U min^{-1} g^{-1} FW for 'Guiqi' and 'Tainong 1', respectively) than in the MT-treated fruits (0.3 ± 0.01 and 0.05 U min^{-1} g^{-1} FW for 'Guiqi' and 'Tainong 1', respectively) at the end of the storage ($p < 0.05$). However, the difference between MT and control treatments in response to the PPO activity was more pronounced in 'Tainong 1' than in 'Guiqi'.

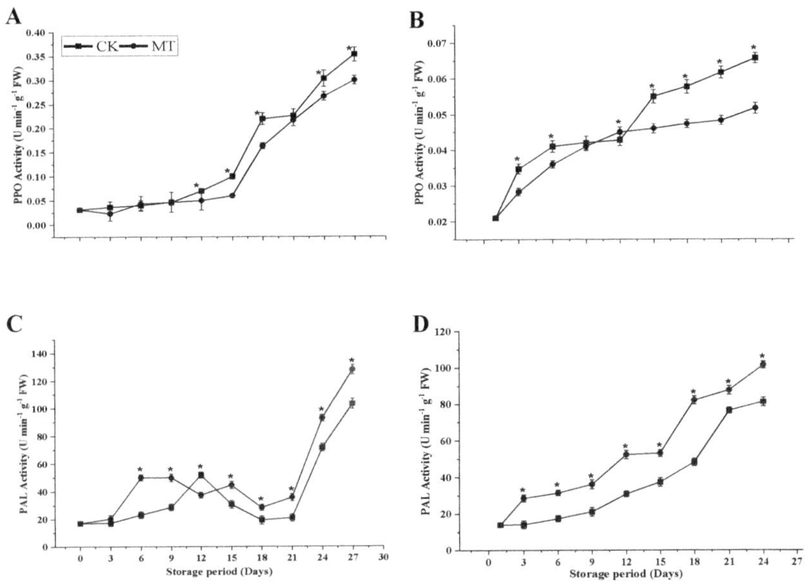

Figure 7. PPO activity (**A**,**B**) and PAL activity (**C**,**D**) of 'Guiqi' (**A**,**C**) and 'Tainong 1' (**B**,**D**) fruits treated with 1000 μmol L^{-1} MT. Each value is the mean for three replicates, and vertical bars indicate the standard error (SE). Error bars with an asterisk (*) on the same storage period show a significant difference between the treatments ($p < 0.05$).

As depicted in Figure 7C,D, the PAL enzyme activity in control fruits and fruits treated with MT increased within the first 12 d of storage. It then decreased until 21 d, and then massively increased until 27 d in 'Guiqi', while in 'Tainong 1', there was an increasing trend throughout the storage period. At the end of the storage period, the fruits treated with MT exhibited significantly higher PAL enzyme activity during storage ($p < 0.05$) than the control treatment, regardless of the cultivar (Figure 7C,D). However, higher PAL activity by MT was more pronounced in 'Tainong 1' than in 'Guiqi' during storage.

3.4.2. SOD and APX Activity

Figure 8A,B demonstrated that the SOD activity in both cultivars followed a similar trend during the storage period. MT treatment was 1.25 times higher than the control at 27 d in 'Guiqi', and 1.12 times higher at 24 d in 'Tainong 1'. However, the effect of MT on the SOD activity was more pronounced in 'Tainong 1' than in 'Guiqi' since MT ($p < 0.05$) did not significantly influence the treatment effect and storage period in 'Guiqi'.

As depicted in Figure 8C,D, there is a difference in the pattern of the APX activity between the two cultivars. The APX activity in 'Guiqi' increased in both treatments, where they reached the maximum peak on 12 d and then decreased until 27 d. After 12 d, the MT-treated fruits showed significantly higher APX activity than the control, in which the mean values for the MT treatment were twice as high as the control treatment in 'Guiqi'. On the contrary, the APX activity in 'Tainong 1' showed a continuous increase from 3 d to 24 d. MT treatment showed significantly higher APX activity (14.59 $\pm$ 0.13 U min^{-1} g^{-1} FW) than the control (13.82 $\pm$ 0.14 U min^{-1} g^{-1} FW) ($p < 0.05$). At the end of the storage period, the significantly higher APX between the MT treatment and the control was more pronounced in 'Guiqi' than in 'Tainong 1'.

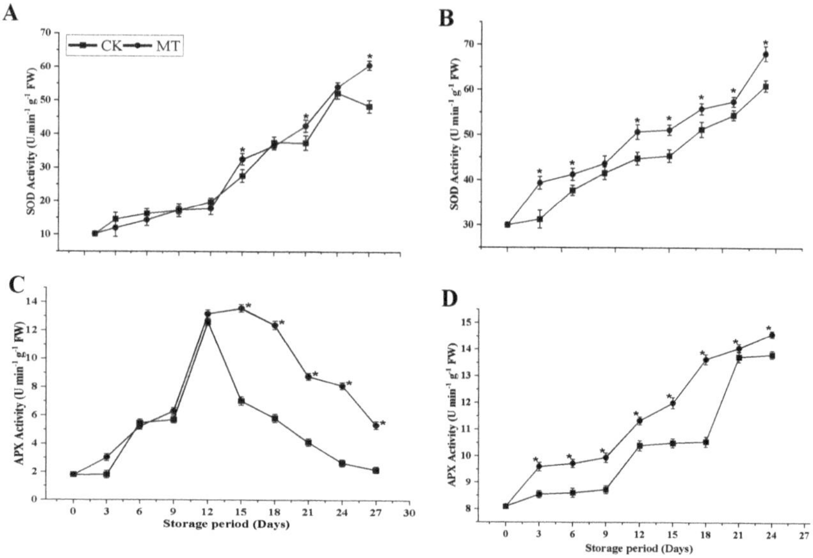

Figure 8. SOD activity (**A**,**B**) and APX activity (**C**,**D**) of 'Guiqi' (**A**,**C**) and 'Tainong 1' (**B**,**D**) fruits treated with 1000 μmol L^{-1} MT. Each value is the mean for three replicates, and vertical bars indicate the standard error (SE). Error bars with an asterisk (*) on the same storage period show a significant difference between the treatments ($p < 0.05$).

3.5. Effect of MT on Gene Relative Expression

3.5.1. PAL and PPO Genes

The expression of the *PAL* and *PPO* genes is depicted in Figure 9. The results showed that transcripts of these two genes in both the control fruits and the MT-treated fruits increased throughout the storage period. MT treatment upregulated the expression of the *PAL* gene (Figure 9A,B) and down-regulated the expression of the *PPO* gene (Figure 9C,D) during most stages of the storage. There was a statistically significant difference between the MT treatment and the control ($p < 0.05$) in the relative expression of both genes. The cultivars displayed slightly different gene expression profiles; 'Tainong 1' was more responsive to the *PPO* gene than 'Guiqi'. Nevertheless, MT displays a typical pattern of browning-related gene expression in both cultivars.

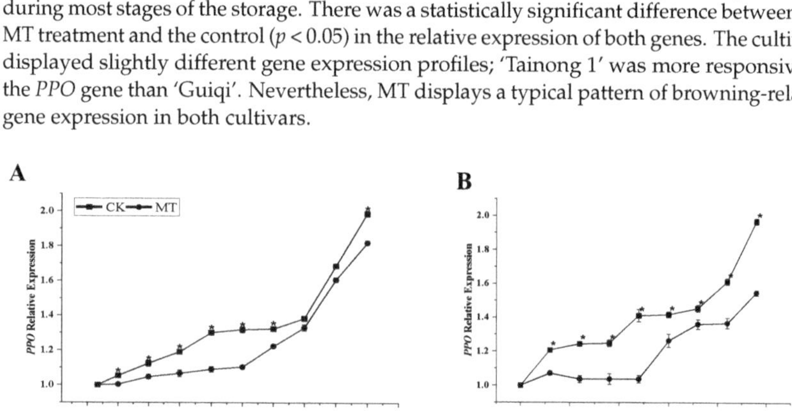

Figure 9. *Cont.*

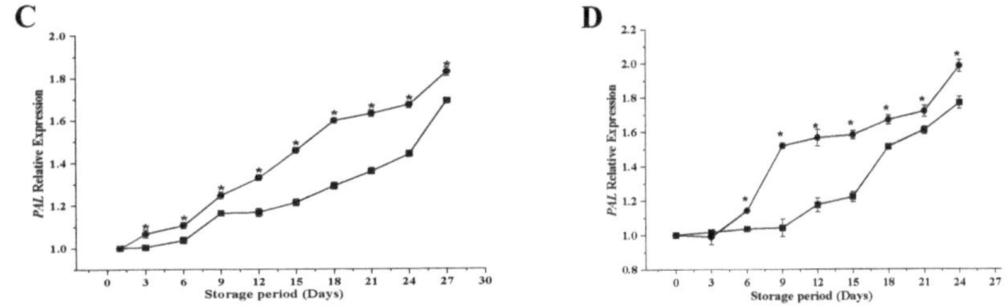

Figure 9. *PAL* relative gene expression (**A**,**B**) and *PPO* relative gene expression (**C**,**D**) of 'Guiqi' (**A**,**C**) and 'Tainong 1' (**B**,**D**) fruits treated with 1000 μmol L^{-1} MT. Each value is the mean for three replicates and vertical bars indicate the standard error (SE). Error bars with an asterisk (*) on the same storage period show a significant difference between the treatments ($p < 0.05$).

3.5.2. MnSOD and APX Genes

Figure 10 illustrated the effect of MT treatment on the relative gene expression level of antioxidant-related genes during cold storage. The results showed that MT treatment significantly upregulated the *MnSOD* genes (Figure 10A,B) and *APX* genes (Figure 10C,D) in both cultivars of mango during storage ($p < 0.05$) versus the control. Moreover, a slight difference in the gene expression profiles displayed by the two cultivars was observed but the difference was insignificant.

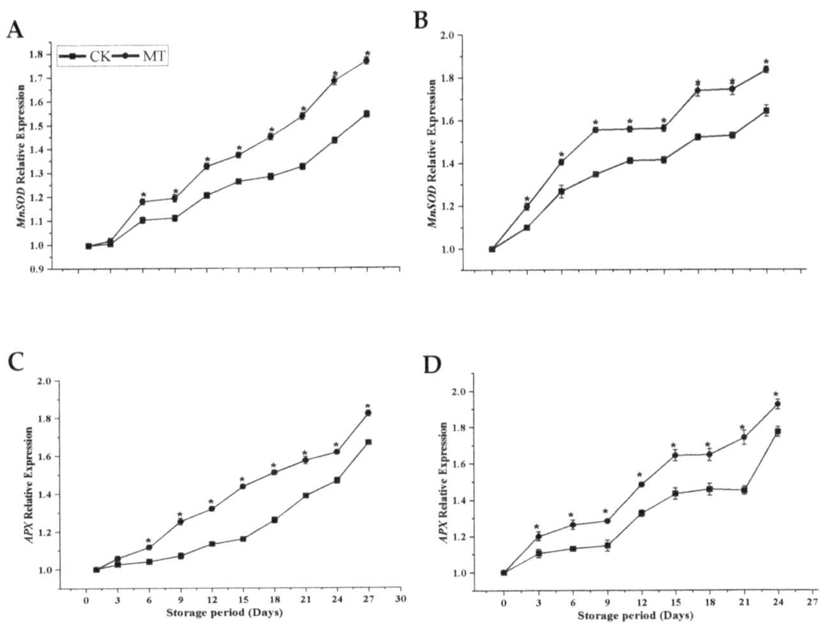

Figure 10. *MnSOD* relative gene expression (**A**,**B**) and *APX* relative gene expression (**C**,**D**) of 'Guiqi' (**A**,**C**) and 'Tainong 1' (**B**,**D**) fruits treated with 1000 μmol L^{-1} MT. Each value is the mean for three replicates, and vertical bars indicate the standard error (SE). Error bars with an asterisk (*) on the same storage period show a significant difference between the treatments ($p < 0.05$).

4. Discussion

Mango is a seasonal fruit; one of its primary constraints for continuous supply in the market is its climacteric nature. As a result, an attempt to minimize postharvest fruit deterioration, extend shelf life and maintain quality are critical in the supply chain of mango fruits. Concerning storage characteristics, our results showed that MT treatment delayed the increasing weight loss and slowed down the decreasing firmness of fruits compared to the control treatment in both cultivars (Figure 1A–D). Our results agree with previous research by Liu et al. and Bhardwaj et al. [15,16], confirming that MT treatment significantly delays fruit weight loss and firmness and the rate of respiration in mango fruits. Fruit weight loss and flesh firmness loss are two critical parameters that influence the storage of mango fruits due to their climacteric nature. Therefore, delaying fruit weight loss and firmness by MT treatment might result from the reduced activity of enzymes related to fruit softening and respiration, which was observed by Liu et al. [15].

These above results are supported by the fact that MT treatment significantly improves preservation via a mechanism involving the inhibition of respiration compared to the control in both cultivars (Figure 2C,D). However, the 'Guiqi' cultivar responded better to MT treatment concerning the delayed respiration rate than 'Tainong 1'. In agreement with our results, the respiration rate in peach and pear fruits was significantly inhibited by 0.1 Mm MT under storage conditions [11,25]. It was also confirmed by Bhardwaj et al. [16] that the efficiency of MT on quality preservation at chilling temperatures was associated with delayed ethylene production and the respiration rate of four mango cultivars. Michailidis et al. [27] assume that an active interplay may exist among respiration, cold stress, and MT in sweet cherries due to high displays of respiratory activity and low-temperature storage, which may induce an uncoupling of the respiratory chain, which gives rise to ROS. As a result, the inhibition of the respiration rate by MT suggests that MT can minimize the accumulation of ROS, thereby minimizing mango fruit softening and decay [28]. MT treatment significantly delayed fruit decay compared to control fruits in the current study regardless of the cultivar (Figure 2A,B). This result could be due to the decreased activities of browning-related (PPO) enzymes by MT treatment in both cultivars. In agreement with our results, Zhang et al. and Agham et al. [28,29] found that MT treatment decreases the rate of decay incidence in litchi and strawberry fruits, respectively, compared to the control treatment. The decay of postharvest fruits is usually accompanied by pathogen infection, which leads to fruit deterioration and spoilage. MT contains antioxidants with immune-modulatory and anti-inflammatory effects, indicating that it can inhibit bacterial, viral, and parasitic infections [7].

TSS and TA are critical parameters related to fruit ripening. Sugars represent a fundamental component of fruit's edible quality, predominantly conferring sweetness and significantly influencing consumer satisfaction. According to the present results, MT does not significantly affect the TSS and TA content (Figure 3A–D) in both cultivars studied. However, a continuous decline in TA is apparent and is considered an important marker of faster senescence. Because respiration utilizes organic acids as a substrate, this process tends to decrease TA [16]. The TSS:TA ratio is a strong indicator of ripeness while also influencing fruit taste and tends to increase in parallel to increasing TSS and decreasing TA significantly. The TSS:TA ratio was also not significantly lower in response to the MT treatment in 'Tainong 1', but it was higher in 'Guiqi' than in the control (Figure 3E,F), following a recent study [14]. MT's inability to influence important quality parameters like TSS, TA, and the TSS:TA ratio could be due to hydrolytic changes and the conversion of starch into simple sugars [30]. Han et al. [30] reported that there is a high possibility that MT might modulate sugar metabolism, which indicates that a higher TSS by MT treatment might reflect MT's effect on the loss of glucose and fructose, the main sugars found in many fruits. Moreover, Bhardwaj et al. and Miranda et al. [16,31] found that MT treatment significantly affects the TSS, TA, and TSS:TA ratio in three cultivars of mango fruits and sweet cherry, respectively during most stages of the storage period. However, more research is needed to understand the interaction between MT and sugar metabolism.

Oxidative damage can affect the quality of harvested fruit, which is a process of physiological deterioration [32]. This study determined the effects of MT treatment on the AsA content and total phenol and flavonoid content in two mango cultivars and how they influenced the postharvest quality of mango fruits and antioxidative activities. In this study, MT treatment significantly delayed the decreasing rate of AsA in both cultivars at the end of the storage period (Figure 4A,B). These results suggested that the exogenous application of MT could increase the ability of the mango fruits to resist oxidative damage and maintain fruit quality by activating the key enzyme activities and the increasing accumulation of antioxidants in the ascorbate-glutathione (AsA-GSH) cycle regardless of the cultivars under study. However, there was a slight variation in the cultivar response to MT treatment, but the difference was insignificant. The AsA-GSH cycle is an essential component in the elimination of ROS. APX is an essential enzyme in the AsA-GSH redox pathway, which plays a crucial role in promoting the regeneration of GSH and AsA, ensuring the normal functioning of metabolic pathways, and maintaining the balance of reduced substances in plants [33]. In our study, MT significantly increased APX activity compared to the control (Figure 8C,D), which corresponds to higher AsA content in both cultivars.

Mango fruit is rich in several other types of antioxidant phytochemicals, such as phenolic [34,35]. Therefore, the loss of astringency during mango fruit ripening is associated with decreased phenolic content [36]. In this study, a delayed decrease in the total phenol and flavonoid content was observed more in the MT-treated fruits (Figure 5A–D) than in the control, regardless of the cultivar, and the trend was similar in both cultivars during storage. Phenols play vital roles in maintaining the nutritional quality of fruits concerning bitterness, flavor, and stringency. Furthermore, phenolic compounds can protect membrane lipids from peroxidation by avoiding the occurrence and propagation of oxidative chain reactions [37]. Flavonoids, on the other hand, are potent antioxidants that play a crucial role in minimizing decay and disease in fruits. Both phenols and flavonoids are described as secondary metabolites that help in increasing the antioxidant characteristics of fruits via ROS scavenging activities [14]. Therefore, the higher content of total phenol and flavonoids by MT treatment at the end of the storage period indicates that MT plays a significant role in inhibiting postharvest fruit deterioration and minimizing qualitative loss in mango during storage. This result correlates with respiration inhibition by MT (Figure 2C,D) since an increase in respiration rate leads to an increase in the production of secondary metabolites.

To determine the physiological and molecular mechanism of the MT effect in two cultivars concerning fruit postharvest quality, an analysis of the enzyme activities related to browning and their relative gene expressions was determined during storage. MT treatment significantly delays the PPO enzyme activities compared to control fruits in the current study, regardless of the cultivar (Figure 7A,B). However, PPO activities in MT treatment were more pronounced in 'Tainong 1' than in 'Guiqi'. In agreement with our results, the delayed increase in browning-related enzymes such as PPO in litchi fruits upon MT treatment was consistent with the rise in total phenol, flavonoid, and anthocyanin contents, which delayed the pericarp browning and maintained quality [38]. Studies have shown that membrane damage triggers a loss of subcellular compartmentalization, resulting in contact between browning related-enzymes (PPO) and phenolic substrates, consequently leading to enzymatic browning in harvested horticultural crops. In the current study, delayed increases in PPO activities (Figure 7A,B) concurrent with higher levels of total phenol and flavonoid contents were shown in MT-treated mango fruit, indicating that MT might inhibit enzyme-catalyzed phenolic oxidation and delay fruit peel browning, which is an indicator of decay in mango. It was also observed in this study that the rate of PPO inhibition by MT treatment was cultivar dependent since 'Tainong 1' inhibited PPO activity more than 'Guiqi'.

The PAL enzyme, responsible for synthesizing polyphenols in plants, is located at a branch point between primary and secondary metabolism [39]. It enters secondary metabolism when the enzyme PAL catalyzes the elimination of ammonium, converting phenylalanine into cinnamic acid, a precursor of phenols such as flavonoids [39]. Our results showed a higher PAL activity in MT-treated fruits than in controls (Figure 7C,D) in both cultivars, suggesting that MT could minimize fruit decay by enhancing the defense-related secondary compounds that act against biotic and abiotic stress during storage. Furthermore, PAL has been demonstrated in the metabolic activity of many higher plants and is a crucial enzyme in synthesizing several defense-related secondary compounds like phenols and lignins [40]. In our study, the increase in the activity of PAL by MT treatment correlates to an increase in total phenol and flavonoid content, demonstrating that MT treatment effectively inhibited biotic and abiotic stresses, thereby minimizing fruit decay and elevating antioxidant processes during the cold storage of mango. However, the enhancement of the PAL enzyme was higher in 'Tainong 1' than in 'Guiqi' during the storage. In agreement with our results, the activity of the PAL enzyme was found to be increased by MT, which promotes the accumulation of total phenols and is beneficial to inhibiting fungal decay and prolonging the shelf-life of postharvest peach [13].

To shed more light on the activity of browning-related enzymes, we performed a qRT-PCR to determine the relative gene expression of the above enzymes. We found that MT treatment upregulated *PAL* genes and inhibited the *PPO* gene during storage (Figure 9A–D). The upregulation of the *PAL* gene and downregulation of the *PPO* gene correlate with their specific enzyme activities by MT treatment in this study.

ROS accumulation may damage lipids and form toxic products like MDA, indicating oxidative stress in plants. This phenomenon is known as lipid peroxidation and is triggered in part by lipoxygenase activity, resulting in cell membrane deterioration. Furthermore, our results showed that MT treatment minimized the MDA content (Figure 6A,B) at the end of the storage period compared to the control in both cultivars. However, the MT effect concerning the MDA content was more pronounced by 'Guiqi' than 'Tainong 1' at the end of the storage period. This result indicates that MT affects lipid peroxidation inhibition and minimizes oxidative stress in mango fruits during cold storage.

Plants have antioxidant systems which scavenge ROS and protect cells from ROS-induced injuries [40]. The enzymatic antioxidant system is also a primary way to control ROS production, which regulates the degree of lipid peroxidation [41]. SOD and APX are among the most important antioxidant enzymes for scavenging ROS. In our study, MT increased the activities of SOD and APX compared to the control-treated fruits (Figure 8A–D) in both cultivars, but the pattern of the MT effect was different between the two cultivars. This result indicates that MT could enhance the antioxidative processes to maintain fruit quality by improving ROS scavenging in mango fruits during cold storage. Exogenous MT treatment was also found to induce apple disease resistance by enhancing the activity of POD, SOD, and CAT and inducing the synthesis of endogenous MT [42].

Specific genes associated with antioxidant enzyme activities were unclear in mango and their roles in modulating fruit decay and maintaining quality. To understand the role of genes in the antioxidant system by MT treatment, the expression patterns of two genes were assessed in both mango cultivars during cold storage. The transcription levels of these genes (*MnSOD* and *APX*) in both cultivars were higher in MT-treated fruits than in control during storage (Figure 10A–D), but the magnitude of expression between the cultivars differs. Therefore, *MnSOD* and *APX* may be the significant genes regulating fruit's antioxidant system. The high expression of antioxidant-related genes was essential for maintaining the storage quality of fruit by delaying fruit decay and enhancing the antioxidant system of mango fruits. It was reported that the upregulated expression of *SOD*, *CAT*, and *POD* might contribute to prolonging the shelf life of kiwifruit [43], which agrees with our results in this study. Another study discovered that MT upregulated genes coding for *CAT*, *APX*, *Mn-SOD*, and *Cu/Zn-SOD* in two cultivars of sweet cherry led to improved fruit quality during cold storage [31]. The results suggested that MT treatment might mitigate the fruit decay of mango, possibly by activating antioxidant enzymes and related genes.

To simplify the understanding of the mechanism of MT's ability to delay mango fruit decay and enhanced the antioxidant system, the results of the experiment are summarize in the schematic diagram (Figure 11) including all the parameters assayed in this study. The diagram highlighted the pathway in which MT actions contributed to shelf life extension and maintaining quality in mango during cold storage.

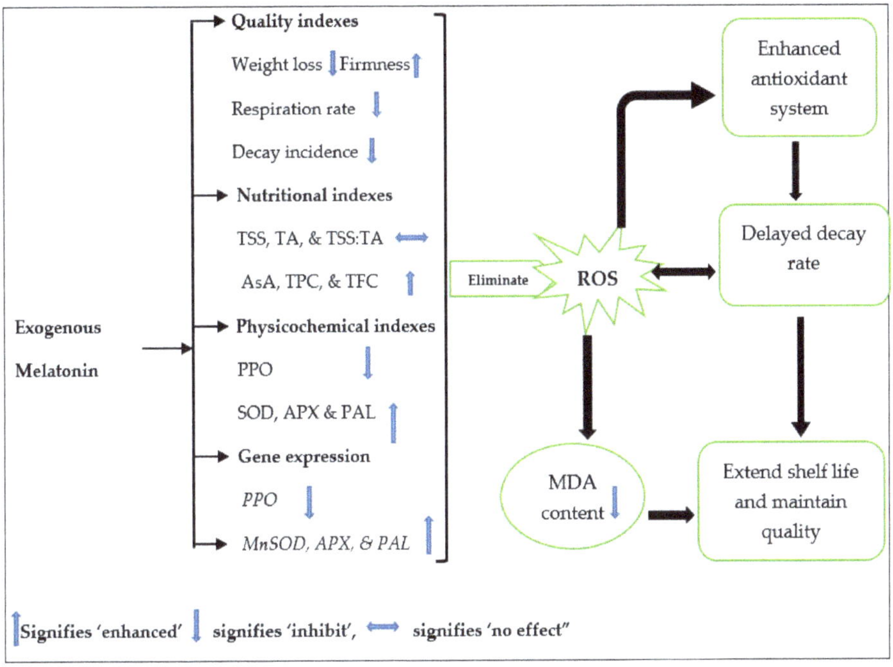

Figure 11. Schematic representation of melatonin (MT) antioxidant system enhancement. MT maintains the increase in weight loss, respiration rate, and decay and maintains the decrease in firmness. MT does not affect TSS, TA, and TS:TA. MT eliminates ROS directly and inhibits MDA content. It also increases the levels of PAL enzyme activity, enzymatic antioxidants (SOD, APX), and non-enzymatic antioxidants (AsA, TFC, TPC) and minimizes the activity of the PPO enzyme. MT enhanced the expression of genes related to the antioxidant system (*MnSOD* and *APX*) and the *PAL* gene and inhibited *PPO* relative gene expression. Abbreviations: TSS, total soluble solids; TA, titratable acidity; AsA, ascorbic acid content; TPC, total phenol content; TFC, total flavonoid content; PPO, polyphenol oxidase; SOD, superoxide dismutase; APX, ascorbate peroxidase; and PAL, phenylalanine ammonia-lyase.

5. Conclusions

In conclusion, it was observed that the storage behavior of the two cultivars differs slightly in response to MT treatment. 'Guiqi' lasted for a 27 d storage period while 'Tainong 1' lasted for 24 d. The study also discovered that MT maintained the quality of mango during storage by delaying weight loss, firmness, respiration rate, and decay incidence more than the control treatment in both mango cultivars. This result is evident with reduced PPO activity and enhanced PAL enzyme activity. However, MT did not affect TSS, TA, and the TSS:TA ratio in this study in both cultivars. Furthermore, MT was also found to delay the MDA content, while increasing SOD and APX enzyme activities, which correlates with maintaining a higher concentration of total phenols, flavonoid content, and AsA content during cold storage. The inhibition of the PPO enzyme and increase in the activities of the PAL, SOD, and APX enzymes by MT treatment correlates with the gene expressions. MT treatment was also cultivar dependent on specific parameters being studied. Therefore, this study could give a broader insight into the role of MT in the postharvest technology of mango fruits during storage.

Author Contributions: All authors contributed to the study's conception. Material preparation was performed by Q.L. Experimentation and data collection was performed by C.L. The design of the experiment, data analysis, and supervision was performed by W.Z., X.D. and X.P. The draft of the manuscript was written by A.N. All authors commented on previous versions of the manuscript. All authors have read and agreed to the published version of the manuscript.

Funding: This research is supported by Guizhou Provincial Science and Technology Projects [2019]4232 and the Hundred Levels Talent Training of Guizhou Province, grant number [2016]4038.

Institutional Review Board Statement: The study did not involve humans or animals.

Informed Consent Statement: The study did not involve humans.

Data Availability Statement: The data are available upon request from the corresponding author.

Conflicts of Interest: The authors declare no conflict of interest.

References

1. Tharanathan, R.; Yashoda, H.; Prabha, T. Mango(*Mangifera indica* L.), "The King of Fruits"—An Overview. *Food Rev. Int.* **2006**, *22*, 95–123. [CrossRef]
2. Zhang, Z.; Zhu, Q.; Hu, M.; Gao, Z.; An, F.; Li, M.; Jiang, Y. Low-temperature conditioning induces chilling tolerance in stored mango fruit. *Food Chem.* **2017**, *219*, 76–84. [CrossRef] [PubMed]
3. Cárdenas-Coronel, W.G.; Velez-de la Rocha, R.; Siller-Cepeda, J.H.; Osuna-Enciso, T.; Muy-Rangel, M.D.; Sañudo-Barajas, J.A. Changes in the composition of starch, pectins and hemicelluloses during the ripening stage of mango (Mangifera indica cv. Kent). *Rev. Chapingo. Ser. Hortic.* **2012**, *18*, 5–19.
4. Xu, T.; Chen, Y.; Kang, H. Melatonin Is a Potential Target for Improving Post-Harvest Preservation of Fruits and Vegetables. *Front. Plant Sci.* **2019**, *10*, 1388. [CrossRef] [PubMed]
5. Arnao, M.B.; Ruiz, J.H. Melatonin: A New Plant Hormone and/or a Plant Master Regulator? *Trends Plant Sci.* **2019**, *24*, 38–48. [CrossRef] [PubMed]
6. Hernández-Ruiz, J.; Arnao, M.B. Relationship of Melatonin and Salicylic Acid in Biotic/Abiotic Plant Stress Responses. *Agronomy* **2018**, *8*, 33. [CrossRef]
7. Debnath, B.; Islam, W.; Li, M.; Sun, Y.; Lu, X.; Mitra, S.; Hussain, M.; Liu, S.; Qiu, D. Melatonin Mediates Enhancement of Stress Tolerance in Plants. *Int. J. Mol. Sci.* **2019**, *20*, 1040. [CrossRef]
8. Shi, H.; Reiter, R.J.; Tan, D.-X.; Chan, Z. *Indole-3-acetic acid INDUCIBLE 17* positively modulates natural leaf senescence through melatonin-mediated pathway in *Arabidopsis*. *J. Pineal Res.* **2014**, *58*, 26–33. [CrossRef]
9. Bal, E. Physicochemical changes in 'Santa Rosa' plum fruit treated with melatonin during cold storage. *J. Food Meas. Charact.* **2019**, *13*, 1713–1720. [CrossRef]
10. Hu, W.; Yang, H.; Tie, W.; Yan, Y.; Ding, Z.; Liu, Y.; Wu, C.; Wang, J.; Reiter, R.J.; Tan, D.-X.; et al. Natural Variation in Banana Varieties Highlights the Role of Melatonin in Postharvest Ripening and Quality. *J. Agric. Food Chem.* **2017**, *65*, 9987–9994. [CrossRef] [PubMed]
11. Cao, S.; Song, C.; Shao, J.; Bian, K.; Chen, W.; Yang, Z. Exogenous Melatonin Treatment Increases Chilling Tolerance and Induces Defense Response in Harvested Peach Fruit during Cold Storage. *J. Agric. Food Chem.* **2016**, *64*, 5215–5222. [CrossRef] [PubMed]
12. Liu, C.; Zheng, H.; Sheng, K.; Liu, W.; Zheng, L. Effects of melatonin treatment on the postharvest quality of strawberry fruit. *Postharvest Biol. Technol.* **2018**, *139*, 47–55. [CrossRef]
13. Gao, H.; Lu, Z.; Yang, Y.; Wang, D.; Yang, T.; Cao, M.; Cao, W. Melatonin treatment reduces chilling injury in peach fruit through its regulation of membrane fatty acid contents and phenolic metabolism. *Food Chem.* **2018**, *245*, 659–666. [CrossRef] [PubMed]
14. Rastegar, S.; Khankahdani, H.H.; Rahimzadeh, M. Effects of melatonin treatment on the biochemical changes and antioxidant enzyme activity of mango fruit during storage. *Sci. Hortic.* **2020**, *259*, 108835. [CrossRef]
15. Liu, S.; Huang, H.; Huber, D.J.; Pan, Y.; Shi, X.; Zhang, Z. Delay of ripening and softening in 'Guifei' mango fruit by postharvest application of melatonin. *Postharvest Biol. Technol.* **2020**, *163*, 111136. [CrossRef]
16. Bhardwaj, R.; Pareek, S.; González-Aguilar, G.A.; Domínguez-Avila, J.A. Changes in the activity of proline-metabolising enzymes is associated with increased cultivar-dependent chilling tolerance in mangos, in response to pre-storage melatonin application. *Postharvest Biol. Technol.* **2021**, *182*, 111702. [CrossRef]
17. Cao, S.; Shao, J.; Shi, L.; Xu, L.; Shen, Z.; Chen, W.; Yang, Z. Melatonin increases chilling tolerance in postharvest peach fruit by alleviating oxidative damage. *Sci. Rep.* **2018**, *8*, 806. [CrossRef]
18. Wang, L.; Luo, Z.; Yang, M.; Li, D.; Qi, M.; Xu, Y.; Abdelshafy, A.M.; Ban, Z.; Wang, F.; Li, L. Role of exogenous melatonin in table grapes: First evidence on contribution to the phenolics-oriented response. *Food Chem.* **2020**, *329*, 127155. [CrossRef]

19. Li, T.; Wu, Q.; Zhu, H.; Zhou, Y.; Jiang, Y.; Gao, H.; Yun, Z. Comparative transcriptomic and metabolic analysis reveals the effect of melatonin on delaying anthracnose incidence upon postharvest banana fruit peel. *BMC Plant Biol.* **2019**, *19*, 289. [CrossRef]
20. Sharma, A.; Wang, J.; Xu, D.; Tao, S.; Chong, S.; Yan, D.; Li, Z.; Yuan, H.; Zheng, B. Melatonin regulates the functional components of photosynthesis, antioxidant system, gene expression, and metabolic pathways to induce drought resistance in grafted Carya cathayensis plants. *Sci. Total Environ.* **2020**, *713*, 136675. [CrossRef]
21. Quality, A.P.; Supervision, S. Analysis on the Quality and Aroma Components of Main Mango Fruits in Guizhou. *Chin. J. Trop. Crops* **2020**, *41*, 2305–2313. [CrossRef]
22. Ma, X.; Wu, H.; Liu, L.; Yao, Q.; Wang, S.; Zhan, R.; Xing, S.; Zhou, Y. Polyphenolic compounds and antioxidant properties in mango fruits. *Sci. Hortic.* **2011**, *129*, 102–107. [CrossRef]
23. Zhang, Z.; Gao, Z.; Li, M.; Hu, M.; Gao, H.; Yang, D.; Yang, B. Hot Water Treatment Maintains Normal Ripening and Cell Wall Metabolism in Mango (*Mangifera indica* L.) Fruit. *HortScience* **2012**, *47*, 1466–1471. [CrossRef]
24. Dong, J.; Kebbeh, M.; Yan, R.; Huan, C.; Jiang, T.; Zheng, X. Melatonin treatment delays ripening in mangoes associated with maintaining the membrane integrity of fruit exocarp during postharvest. *Plant Physiol. Biochem.* **2021**, *169*, 22–28. [CrossRef]
25. Daniels, A.J.; Poblete-Echeverría, C.; Opara, U.L.; Nieuwoudt, H.H. Measuring Internal Maturity Parameters Contactless on Intact Table Grape Bunches Using NIR Spectroscopy. *Front. Plant Sci.* **2019**, *10*, 1517. [CrossRef]
26. Wang, F.; Zhang, X.; Yang, Q.; Zhao, Q. Exogenous melatonin delays postharvest fruit senescence and maintains the quality of sweet cherries. *Food Chem.* **2019**, *301*, 125311. [CrossRef]
27. Zhang, Y.; Huber, D.J.; Hu, M.; Jiang, G.; Gao, Z.; Xu, X.; Jiang, Y.; Zhang, Z. Delay of Postharvest Browning in Litchi Fruit by Melatonin via the Enhancing of Antioxidative Processes and Oxidation Repair. *J. Agric. Food Chem.* **2018**, *66*, 7475–7484. [CrossRef]
28. Michailidis, M.; Tanou, G.; Sarrou, E.; Karagiannis, E.; Ganopoulos, I.; Martens, S.; Molassiotis, A. Pre- and Post-harvest Melatonin Application Boosted Phenolic Compounds Accumulation and Altered Respiratory Characters in Sweet Cherry Fruit. *Front. Nutr.* **2021**, *8*, 695061. [CrossRef]
29. Aghdam, M.S.; Fard, J.R. Melatonin treatment attenuates postharvest decay and maintains nutritional quality of strawberry fruits (Fragaria × anannasa cv. Selva) by enhancing GABA shunt activity. *Food Chem.* **2017**, *221*, 1650–1657. [CrossRef]
30. Han, Q.-H.; Huang, B.; Ding, C.-B.; Zhang, Z.-W.; Chen, Y.-E.; Hu, C.; Zhou, L.-J.; Huang, Y.; Liao, J.-Q.; Yuan, S.; et al. Effects of Melatonin on Anti-oxidative Systems and Photosystem II in Cold-Stressed Rice Seedlings. *Front. Plant Sci.* **2017**, *8*, 785. [CrossRef]
31. Miranda, S.; Vilches, P.; Suazo, M.; Pavez, L.; García, K.; Méndez, M.A.; González, M.; Meisel, L.A.; Defilippi, B.G.; del Pozo, T. Melatonin triggers metabolic and gene expression changes leading to improved quality traits of two sweet cherry cultivars during cold storage. *Food Chem.* **2020**, *319*, 126360. [CrossRef]
32. Hu, M.; Yang, D.; Huber, D.J.; Jiang, Y.; Li, M.; Gao, Z.; Zhang, Z. Reduction of postharvest anthracnose and enhancement of disease resistance in ripening mango fruit by nitric oxide treatment. *Postharvest Biol. Technol.* **2014**, *97*, 115–122. [CrossRef]
33. Sun, H.; Li, L.; Wang, X.; Wu, S.; Wang, X. Ascorbate–glutathione cycle of mitochondria in osmoprimed soybean cotyledons in response to imbibitional chilling injury. *J. Plant Physiol.* **2011**, *168*, 226–232. [CrossRef]
34. Hiwale, S. Mango (*Mangifera indica* L.). In *Sustainable Horticulture in Semiarid Dry Lands*; Springer: New Delhi, India, 2015; pp. 97–114. [CrossRef]
35. Mühlbauer, W.; Müller, J. *Mango (Mangifera indica L.)*; Woodhead Publishing Limited: Sawston, UK, 2020.
36. Trejo-Márquez, M.; Ramírez-Villatoro, G.; De La Rosa, N.C. Polyphenol oxidase and peroxidase activities in mangoes stored at chilling temperature. *Acta Hortic.* **2010**, 395–402. [CrossRef]
37. Pennycooke, J.; Cox, S.; Stushnoff, C. Relationship of cold acclimation, total phenolic content and antioxidant capacity with chilling tolerance in petunia (Petunia × hybrida). *Environ. Exp. Bot.* **2005**, *53*, 225–232. [CrossRef]
38. Wang, T.; Hu, M.; Yuan, D.; Yun, Z.; Gao, Z.; Su, Z.; Zhang, Z. Melatonin alleviates pericarp browning in litchi fruit by regulating membrane lipid and energy metabolisms. *Postharvest Biol. Technol.* **2019**, *160*, 111066. [CrossRef]
39. Aydaş, S.B.; Ozturk, S.; Aslım, B. Phenylalanine ammonia lyase (PAL) enzyme activity and antioxidant properties of some cyanobacteria isolates. *Food Chem.* **2013**, *136*, 164–169. [CrossRef]
40. Shewfelt, R.; Del Rosario, B. The Role of Lipid Peroxidation in Storage Disorders of Fresh Fruits and Vegetables. *HortScience* **2000**, *35*, 575–579. [CrossRef]
41. Gill, S.S.; Tuteja, N. Reactive oxygen species and antioxidant machinery in abiotic stress tolerance in crop plants. *Plant Physiol. Biochem.* **2010**, *48*, 909–930. [CrossRef]
42. Cao, J.J.; Yu, Z.C.; Zhang, Y.; Li, B.H.; Liang, W.X.; Wang, C.X. Control efficiency of exogenous melatonin against post-harvest apple grey mold and its influence on the activity of defensive enzymes. *Plant Physiol. J.* **2017**, *53*, 1760. [CrossRef]
43. Xia, H.; Ni, Z.; Hu, R.; Lin, L.; Deng, H.; Wang, J.; Tang, Y.; Sun, G.; Wang, X.; Li, H.; et al. Melatonin Alleviates Drought Stress by a Non-Enzymatic and Enzymatic Antioxidative System in Kiwifruit Seedlings. *Int. J. Mol. Sci.* **2020**, *21*, 852. [CrossRef] [PubMed]

foods MDPI

Article

A Synergistic Effect Based on the Combination of Melatonin with 1-Methylcyclopropene as a New Strategy to Increase Chilling Tolerance and General Quality in Zucchini Fruit

Jorge Medina-Santamarina, María Serrano, María Celeste Ruiz-Aracil, Mihaela Iasmina Madalina Ilea, Domingo Martínez-Romero and Fabián Guillén *

Postharvest Research Group of Fruit and Vegetables, Centro de Investigación e Innovación Agroalimentaria y Agroambiental (CIAGRO-UMH), University Miguel Hernández, Ctra. Beniel km. 3.2, Orihuela, 03312 Alicante, Spain

* Correspondence: fabian.guillen@umh.es; Tel.: +34-96-6749656

Citation: Medina-Santamarina, J.; Serrano, M.; Ruiz-Aracil, M.C.; Ilea, M.I.M.; Martínez-Romero, D.; Guillén, F. A Synergistic Effect Based on the Combination of Melatonin with 1-Methylcyclopropene as a New Strategy to Increase Chilling Tolerance and General Quality in Zucchini Fruit. *Foods* **2022**, *11*, 2784. https://doi.org/10.3390/foods11182784

Academic Editors: Maria Cefola and Bernardo Pace

Received: 3 August 2022
Accepted: 6 September 2022
Published: 9 September 2022

Abstract: Zucchini fruit are highly sensitive to low temperatures leading to significant peel depressions, increasing weight loss and making them impossible to be commercialized. In this study the effect on the reduction of chilling injury (CI) assaying different postharvest treatments to cv. Cronos was evaluated. We have compared the application of substances such as 1-methylcyclopropene (1-MCP) with the application of a natural origin compound as melatonin (MT), both with demonstrated activity against CI in different vegetal products. The effects of MT (1 mM) by dipping treatment of 1 h and 1-MCP (2400 ppb) have been evaluated on zucchini fruit during 15 days of storage at 4 °C plus 2 days at 20 °C. Treatments applied independently improved some fruit quality parameters in comparison with control fruit but were not able to manage CI even though they mitigated the impact on several parameters. However, when these two separated strategies were combined, zucchini cold tolerance increased with a synergic trend. This synergic effect affected in general all parameters but specially CI, being also the only lot in which zucchini fruit were most effectively preserved. This is the first evidence in which a clear positive effect on zucchini chilling tolerance has been obtained combining these two different strategies. In this sense, the combined effect of 1-MCP and MT could be a suitable tool to reach high quality standards and increasing shelf life under suboptimal temperatures.

Keywords: chilling injury; storage; melatonin; quality; 1-MCP

1. Introduction

Zucchini (*Cucurbita pepo* spp. *pepo*), is considered a non-climacteric fruit as many others unripen vegetal fruit. The commercial harvest stage is coincident with a high fruit metabolism which deteriorates rapidly quality. On the other hand, this fruit is very sensitive to cold storage being a major problem their marketability to European countries and worldwide at suboptimal temperatures. Common postharvest strategies have been demonstrated to be effective reducing chilling injury (CI) in zucchini, as physical treatments managing temperatures [1–3] or relative humidity during storage [4,5] and controlling atmospheres [6,7] also through a minimal packaging [8].

Recently, chemical treatments as nitric oxide [9] or widely used treatments as 1-methylciclopropene (1-MCP) [8,10] have demonstrated a protective role delaying zucchini chilling injury. 1-MCP effectively blocks ethylene action delaying zucchini senescence and different associated physiological changes as softening and CI [10]. Strategies based into plant hormones or natural elicitors as γ-aminobutyric acid (GABA) [11], glycine betaine [12], polyamines [13–15] and methyl jasmonate [16] have also reduced CI impact in zucchini fruit. In this sense, melatonin has shown a positive effect delaying CI when applied as a postharvest treatment in different non-climacteric species as strawberry and pomegranate [17,18].

With respect to climacteric fruit, MT increased CI tolerance in peach [19] and tomato fruit [20] Also showing effectiveness against CI after preharvest treatments on apricot trees [21]. In these research studies the CI tolerance was associated with a major energy status obtained through the GABA-shunt pathway activation which provides extra ATP. Plant tissues demand ATP especially under metabolic stress conditions. In this sense MT, methyl jasmonate, polyamines or GABA, have been proposed as responsible of GABA shunt pathway stimulation, providing extra ATP, and decreasing ROS accumulation [17,22]. For this reason, MT treated fruit could lead to maintenance membrane permeability and balanced antioxidant system under cold stress, increasing unsaturated/saturated fatty acids ratio in both non-climacteric, and climacteric fruit [17,23].

Several of the above-mentioned strategies have been shown an additional benefit when applied in combination with MCP describing a synergistic effect delaying senescence in comparison with these technologies when applied alone. In this sense, recently, a combination of hot air treatments with 1-MCP lead to delayed softening and quality maintenance of nectarines [24]. With respect to postharvest chemical treatments, 1-MCP in combination with chlorine dioxide postharvest treatment showed a synergistic inhibitory effect on chlorophyll degradation of green pepper fruit [25]. On the other hand, dipping in calcium chloride and then applying 1-MCP was a successful strategy to improve fresh-cut strawberries and watermelon quality during storage [26,27]. Other natural origin compounds as carvacrol [28] or elicitors as methyl salicylate [29] and melatonin [30] have increased shelf-life delaying ripening when these treatments were applied in combination with 1-MCP in red pitaya, tomato fruit, and apricot respectively. However, none of these studies have evaluated the chilling tolerance provided by a combined effect of the previous mentioned strategies. For this reason, the aim of this research has been to evaluate for the first time the effect of different strategies as melatonin and 1-MCP combined or alone over zucchini CI tolerance and quality during storage at suboptimal temperatures.

2. Materials and Methods

2.1. Plant Material and Postharvest Treatments

Zucchini fruit (*Cucurbita pepo* spp. *pepo*) commercial hybrid 'Cronos', were harvested from a commercial greenhouse located in Orihuela (Spain) and immediately transferred to the laboratory. Fruit of uniform size (20–22 cm) were randomly divided into 3 lots of 5 homogeneous fruit for each treatment and sampling date. Control fruits were submerged in distilled water containing 0.5% Tween 20 while MT treated fruit were dipped in a 1 mM MT solution (Sigma-Aldrich, Germany >98% M5250) for 60 min. In preliminary experiments, 9 zucchini fruit were selected per MT dose and different MT concentrations (0.1, 0.5 and 1 mM) and immersion times (from 10, 30, 60, 120 and 180 min) were evaluated. These treatments were assayed alone and in combination with 2400 ppb of 1-MCP for 48 h at 12 °C following Megías et al. [10] conditions for Cronos cultivar. 1 mM MT for 1 h and combined with 1-MCP showed the best effect on reducing CI symptoms. Thus, 3 replicates of 5 zucchini fruit for each treatment and sample time were selected to repeat the experiment at the optimal conditions observed. freshly prepared MT solutions (0 and 1 mM) with 0.5% Tween 20 were used to dip the different lots for 1 h at 20 °C. Then zucchini fruit were allowed to surface dry and then all the fruit were placed in 4 different 130 L hermetic containers. One lot with MT-treated and another lot with no immersed fruit were exposed to 2400 ppbL^{-1} of 1-MCP for 48 h at 12 °C. The other two lots previously immersed in MT solutions with 0 (Control) or 1 mM MT were treated with air and stored in the same conditions. After this period fruit were taken out from containers and placed under cold storage at 4 °C to induce CI following Megías et al. [10] conditions for 0, 3, 6, 9, 12 and 15 days + 2 days at 20 °C.

2.2. Postharvest Quality Parameters

Three replicates of 5 fruit were randomly selected from each treatment lot at 3 days interval during cold storage +2 days at 20 °C. Weight loss of individual zucchini fruit was calculated as percentage with respect to the weight on day 0. Firmness was determined individually as the force to achieve a 5% fruit diameter deformation in both sides and fruit firmness was expressed as N by using a Texture Analyzer (TX-XT2i, S Microsystems, Godalming, UK). Chilling injury was evaluated visually with a panel of 5 trained judges scoring superficial area affected by pitting damage and the pitting severity. Ratings were based on a 6-point hedonic scale, where the fruit surface affected was used to classify each fruit similarly to Megías et al. [31] with the following scale: 0 = no pitting, 1 = $\leq$5% pitting, 2 = 6–15% pitting, 3 = 16–25% pitting, 4 = 26–50% pitting, and 5 = $\geq$50% pitting. On the other hand, to assess the severity of pitting symptoms, the scale was 0 = no damage, 1 = very superficial damage, 2 = superficial damage, 3 = moderate damage, 4 = severe damage, 5 = very severe damage. The final CI index displayed in this manuscript was the average of both assessments.

CO_2 and ethylene production were determined by placing individually 6 randomly selected zucchini from each treatment in a 2.2 L plastic jar hermetically sealed with a rubber stopper for 30 min. After that, 1 mL gas sample per duplicate was taken from head space and carbon dioxide was quantified by using a Shimadzu TM 14A gas chromatograph (Kyoto, Japan) equipped with thermal conductivity detector and ethylene production was evaluated with a Hewlett-Packard^{TM} 5890A gas chromatograph. Chromatographic conditions were previously described [32]. Ethylene production and respiration rate were expressed as nL g^{-1} h^{-1} and mg of CO_2 kg^{-1} h^{-1}, respectively.

Malondialdehyde (MDA) content was assayed in the peel tissue of the zucchini samples following the method of Zhang et al. [33] with modifications. The tissue sample (1.0 g) was homogenized in 10 mL 10% trichloroacetic acid solution, then centrifuged at 10,000$\times$ g for 10 min. 2 mL of supernatant was added to a testing tube with 6 mL of 0.6% thiobarbituric acid per duplicate and mixed vigorously. Testing tubes were held at 95 °C for 20 min. Samples were cooled rapidly, tempered at room temperature, and evaluated in a spectrophotometer (1900 UV/Vis, Shimadzu, Kyoto, Japan) where absorbance was measured at 450, 532 and 600 nm. MDA content was calculated as described by Zhang et al. [33] and expressed as µmol kg^{-1}. Each assessment was repeated three times.

Electrolyte leakage (EL) was determined following Mao et al. [34] with some modifications. From each treatment, three replicates were measured, each consisting of 20 peel discs with 0.5 mm diameter obtained with a cork borer, from longitudinal 2 mm peel exocarp slices taken from opposite sides of each zucchini. After 3 rinses of 3 min each, discs were incubated in 50 mL of deionized water at room temperature with constant shaking for 30 min. Then electrical conductivity (EC) was measured (C1). Finally, samples were boiled at 100 °C for 15 min and measured to calculate total conductivity (C2). EL was expressed as percentage using the following formula: EL = (C1/C2) 100.

For chlorophyll measurement in peel tissue, six disks, each of 6.25 mm in diameter, were punched from same peel layers sliced for EL. Disks were weighed and placed immediately into 8 mL of 100% methanol. Pigments were allowed to be extracted in the dark at 30 °C for 24 h. Extract absorbance was measured using spectrophotometer (1900 UV/Vis, Shimadzu, Kyoto, Japan) at 652 and 665 nm [35].Two extractions were evaluated by replicate. Colour parameters (CIE a^* and CIE b^*) were individually measured on three points of the external (both sides) and internal longitudinal fruit perimeter by using a Minolta colorimeter (CRC200, Minolta Camera Co.; Kantō, Tokio, Japan) and colour was expressed as CIE *hue** (180 + $\tan^{-1} b^*/a^*$, if $a^* < 0$) according the CIELab coordinates.

Total soluble solids (TSS) were determined by duplicate in the juice obtained from the pulp of mix of 5 zucchini of each replicate per lot taken with a digital refractometer Atago PR-101 (Atago Co. Ltd.; Tokyo, Japan) at 20 °C, and expressed as percentage (g 100 g^{-1}). Also, for each replicate total acidity (TA) was determined by duplicate in the same juice by automatic titration with NaOH 0.1 N up to pH 8.1, using 1 mL of diluted juice in 25 mL distilled H_2O, and results were expressed as the percentage of malic acid (meq. malic acid = 0.067).

2.3. Statistical Analysis

All data in this paper are expressed as mean ± standard error (SE). Data were subjected to analysis of variance (ANOVA). Mean comparisons were carried out using a multiple range test (Tukey's HSD test) to find significant differences ($p < 0.05$). Different lowercase letters indicated a significant difference among treatments at the same sampling date. All analyses were performed using SPSS software package, version 22 (IBM Corp.; Armonk, NY, USA).

3. Results and Discussion

CI was evaluated in a previous experiment, carrying out a screen test with four different MT concentrations (0, 0.1, 0.5 and 1 mM) during 5 different immersion times (10, 30, 60, 120 and 180 min), to select the optimal MT treatment conditions (data not shown). Nine zucchini fruit were selected per MT treatment and individually evaluated after 7 d at 4 °C + 1 d at 20 °C. Visually, CI incidence was evaluated with 5 trained judges, and external quality was maintained specially for 1 mM MT dose assayed during 1 h immersion time. On the other hand, no additional benefits were observed by increasing the immersion time. Thus, immersions during 1 h with 1 mM MT were the conditions applied with or without 1-MCP in the present study.

3.1. Effect of Exogenous MT and 1-MCP on Weight Loss, Fruit Firmness and Cold Tolerance

Weight loss of zucchini fruit increased throughout cold storage regardless of the treatment applied. Zucchini weight losses were not significantly ($p \geq 0.05$) affected by MT dips and 1-MCP when applied alone. However, weight losses were significantly lower ($p < 0.05$) when combined treatments (1-MCP + MT) were evaluated during storage (Figure 1A).

In this sense the combined treatment (MT + 1-MCP) reduced weight loss (20.45%) after 6 days of cold storage plus an additional period at 20 °C as compared with the rest of the different lots evaluated. This trend was maintained until the end of the experiment.

Storage of zucchini at 4 °C plus 2 additional days at 20 °C resulted in a decrease in fruit firmness as expected (Figure 1B). However, fruit firmness levels remained higher in MT, 1-MCP and MT + 1-MCP treated fruit as compared to control fruit specially after 6 and 9 days of cold storage. On the other hand, MT and 1-MCP samples did not show significant differences ($p \geq 0.05$) as compared to control fruit at the end of the experiment showing a similar fruit firmness level. On the contrary MT + 1-MCP samples significantly ($p < 0.05$) maintained in general a higher fruit firmness along cold storage compared with the rest of the fruit evaluated.

Zucchini fruit are very sensitive to cold storage displaying CI after 3 days of cold storage in all fruit studied (Figure 1C). The CI index was in general significantly ($p < 0.05$) higher in control fruit as compared to treated fruit with MCP and MT during storage. Although MT and 1-MCP delayed CI symptoms even after 3 and 6 days of cold storage respectively MT + 1-MCP combined treatments showed the lowest CI incidence (42.66% lower as compared to control fruit) after 6 days of refrigerated storage. According to the observations (Figure 1C) only when 1-MCP was applied combined with MT, zucchini fruit still displaying an increased chilling tolerance after 9 days of cold storage showing additional benefits when both substances were applied together. The effect of MT and 1-MCP applied alone or as combined treatment on internal disorders can be clearly observed in the photographs performed (Figure 2).

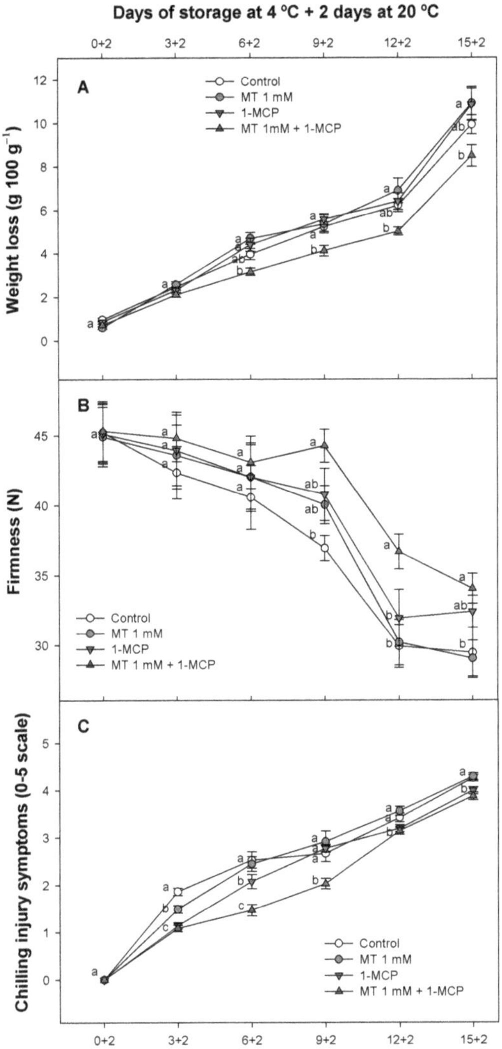

Figure 1. Evolution of weight losses (g 100 g^{-1}) (**A**) fruit flesh firmness (N) (**B**) and chilling injury (0–5 scale) (**C**) of 'Cronos' zucchini fruit treated with melatonin at 1 mM (MT) or distilled water (Control) with or without 1-MCP during cold storage plus 2 days at 20 °C. Data are the mean ± SE (n = 3). Different lowercase letters show significant differences ($p < 0.05$) among treatments for each sampling date.

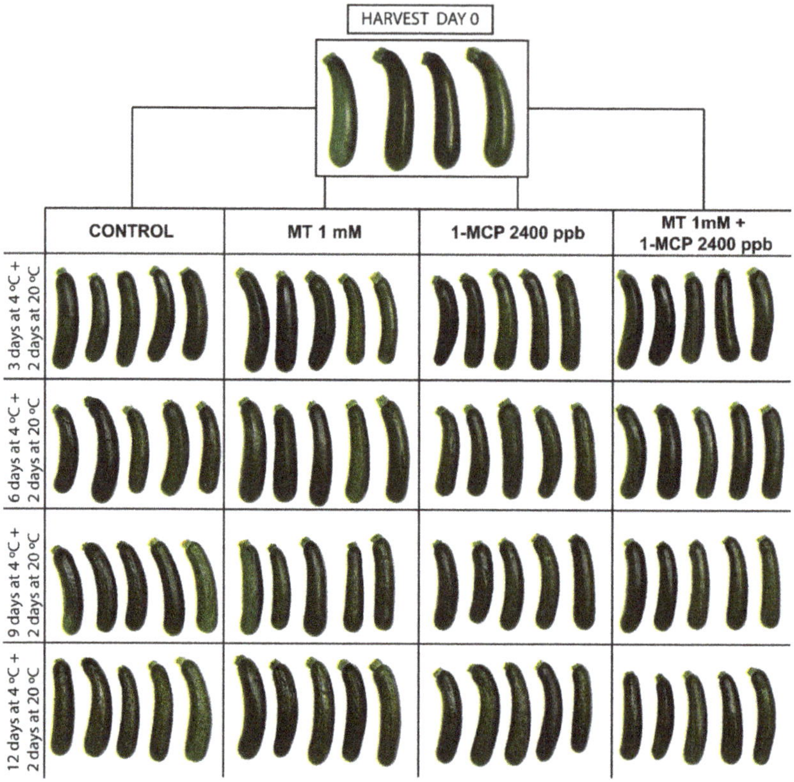

Figure 2. Photography displays the external visual aspect of 'Cronos' zucchini fruit treated with melatonin at 1 mM (MT) or distilled water (Control) with or without 1-MCP after 3, 6, 9 and 12 days of cold storage plus 2 days at 20 °C.

In general, in zucchini fruit, weight and fruit firmness decrease along storage specially when this fruit is stored at suboptimal temperatures mainly due to transpiration through a higher membrane permeability increased by pitting incidence. In this sense, weight loss affects cells turgor reducing fruit firmness in different fruit during storage [36,37]. For this reason, these two parameters use to be correlated between them and with CI incidence. Differences in weight loss between control and treatments were significant only when MT was applied in combination with 1-MCP. However, fruit firmness and CI were affected slightly by MT and 1-MCP alone but when were applied as a combined treatment a higher positive effect was exerted as compared to control fruit. For this reason, we inferred that 1-MCP combined with MT may have synergistic effect on these traits. 1-MCP can be effective controlling weight loss, fruit firmness and CI incidence in different non-climacteric and climacteric fruit [10,38–41]. In zucchini fruit, this trend could be cultivar dependent as it has been previously demonstrated. In fact, in Cronos cultivar, differences in weight loss were not significant when these parameters were analysed after 1-MCP treatment when stored at 4 °C [10]. These results were in consonance with our study, and we also observed an important effect for 1-MCP applied alone reducing CI. On the other hand, MT postharvest treatments have shown a strong effect delaying weight loss in some fruit [42] but a weak effect or even unaffected weight loss in different other fruit [43,44] as we observed when MT was applied alone in zucchini fruit. This slight effect also was observed on zucchini fruit firmness and CI with single MT applications. In previous studies, MT up-regulated cell wall structure-related genes [43,45]. MT also showed antioxidant activity that delayed membrane peroxidation and consequently caused a lower phenol oxidation through an effective inhibition of peroxidase (POD) and PPO activities [46,47]. Storing zucchini at suboptimal temperatures (below 7 °C) can lead to serious CI, characterized by an intense surface pitting, and sunken lesions on the skin surface, which could be caused by damage to the cell walls or cell membranes [2,12,48]. For this reason, and based in our results, we proposed that a combined effect of 1-MCP and MT delaying cell wall disassembly could allow MT antioxidant activity to control CI in a synergistic way when zucchini was treated with the combined treatment.

3.2. Effect of Exogenous MT and 1-MCP on Respiration Rate and Ethylene Production

Respiration rate in zucchini fruit for all treatments tended to increase during shelf life after cold storage. However, for 1-MCP and especially for MT + 1-MCP samples respiration just increased slightly during the beginning of the study (Figure 3A) showing significant differences between treatments ($p < 0.05$).

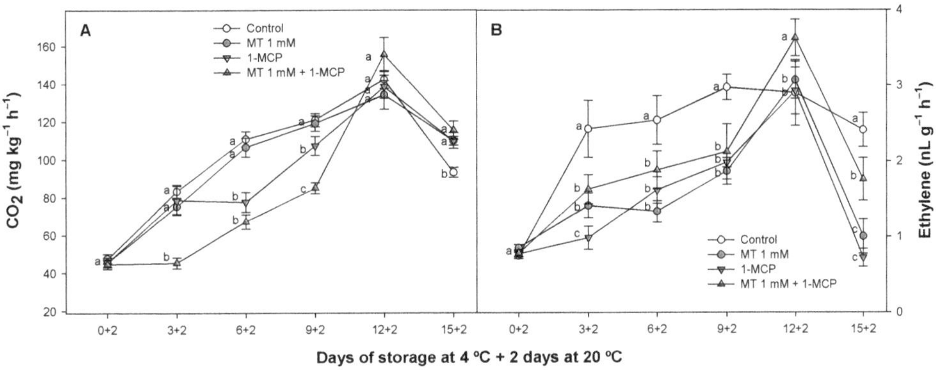

Figure 3. Respiration (mg CO_2 kg^{-1} h^{-1}) (**A**) and ethylene production rates (nL g^{-1} h^{-1}) (**B**) of 'Cronos' zucchini fruit treated with melatonin at 1 mM (MT) or distilled water (Control) with or without 1-MCP during cold storage and after cold storage plus 2 days at 20 °C. Data are the mean ± SE (n = 3). Different lowercase letters show significant differences ($p < 0.05$) among treatments for each sam-pling date.

At the end of storage CO_2 concentrations decreased for all zucchini fruit tested but all treatments applied containing 1-MCP delayed this pattern as compared with control and MT treated fruit. On the other hand, ethylene production in zucchini fruit was low during the experiment. Control fruit significantly increased ethylene concentration ($p < 0.05$) from the beginning of the study, but the different treatments applied delayed this pattern as compared with control fruit (Figure 3A).

Respiration is a key factor involved in weight loss process showing in this study a correlation between these two parameters. Megías et al. [10] found that 1-MCP was able to delay respiration process and ethylene production in different zucchini cultivars in consonance with our results. The reduction of respiration and ethylene production has been linked to an increased cold tolerance in zucchini fruit although the impact on fruit quality during cold storage depends on cultivar [10,49]. MT regulates γ-aminobutyric acid (GABA) content in non-climacteric and climacteric fruit [17,50], stimulating GABA-shunt pathway [51]. This increase in GABA provides the cell with an immediate energy substrate that it uses to recover from stress, increasing the net energy balance in plant cells and covering the energy needs of the plant [17]. In this study, MT-treated zucchini displayed lower medium CO_2 values but with no significant differences ($p \geq 0.05$) as compared with the rest of treatments applied, but 1-MCP treatment alone significantly ($p < 0.05$) delayed the respiration process. On the other hand, a synergistic effect delaying the respiration process was observed when MT and 1-MCP were applied as a combined treatment probably due to an additive effect between both treatments. This additional benefit applying the combined treatment was also observed on ethylene production though MT and 1-MCP reduced cold induced ethylene production when are applied alone (Figure 3B). In this sense all the treatment delayed ethylene production until peak also delaying the decrease of ethylene at the end of storage. MT has been described as a regulator of the expression of different genes reducing ethylene production [52] and 1-MCP is a potent inhibitor of ethylene action [53]. However, the combination of treatments did not reduce ethylene production with an additional effect as compared to these treatments when applied alone. For this reason, the synergistic and beneficial effect observed on cold tolerance of zucchini (Figure 1C; Figure 2) could be explained in relation to the stimulated antioxidant balance that MT exhibits when is applied as a postharvest treatment [52] as we will describe through the following parameters (MDA and total chlorophyll content) evaluated.

3.3. *Effect of Exogenous MT and 1-MCP on Membrane Permeability (MDA Content and EL)*

According to the results (Figure 4A) although all treatments showed a delay in the MDA accumulation, this parameter was significantly lower ($p < 0.05$) in the cases of 1-MCP and MT + 1-MCP compared to that of control fruit.

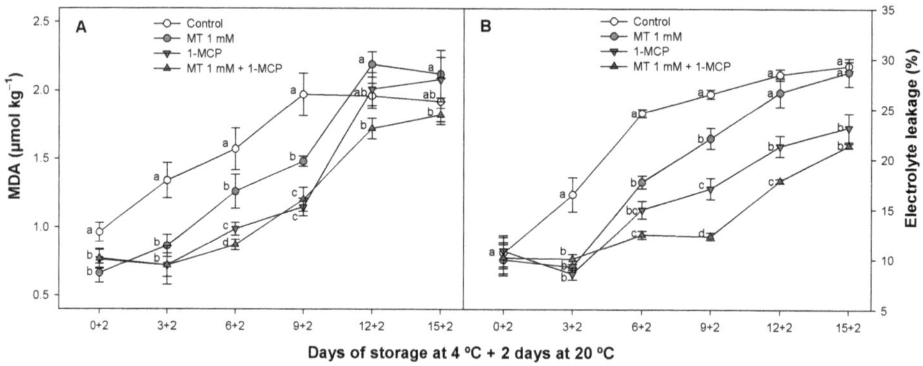

Figure 4. Evolution of malondialdehyde (MDA) content (µmol kg^{-1}) (**A**) an electron leakage (EL) (%) (**B**) of 'Cronos' zucchini fruit treated with melatonin at 1 mM (MT) or distilled water (Control) with or without 1-MCP during cold storage and after cold storage plus 2 days at 20 °C. Data are the mean ± SE (n = 3). Different lowercase letters show significant differences ($p < 0.05$) among treatments for each sam-pling date.

MDA in zucchini fruit treated with MT, 1-MCP or MT + 1-MCP was reduced by 19.7, 33.6 and 44.6% on the 6th day of storage respectively, as compared to control. This delay was observed only for MT + 1-MCP after 12 days of cold storage plus 2 days at 20 °C.

MT and 1-MCP alone or as a combined treatment (MT + 1-MCP) significantly ($p < 0.05$) delayed EL evolution (Figure 4B). However, a lower EL was observed especially when these compounds were applied as a combined treatment since the lowest EL was observed during storage when MT + 1-MCP were applied. On the contrary MT applied alone was the treatment with a weaker effect on EL. MDA reflects the lipid peroxidation of plasma membranes which directly affects the structural integrity of vegetal tissues [54]. Previous works have suggested that 1-MCP affects the activities of antioxidant enzymes [55,56] as well as MT treatments on different fruit [57,58] reducing MDA content and the impact on EL in consonance with our results. For this reason, the similar effect observed on these parameters after applying 1-MCP or MT alone, could be the reason why the reduction in MDA content and EL evolution was greater when both substances were applied together displaying an additive cold tolerance effect.

3.4. Effect of Exogenous MT and 1-MCP on Chlorophyll Content and External Colour

Chlorophyll content in treated and untreated samples showed a decreased pattern for all fruit tested as it was expected (Figure 5A).

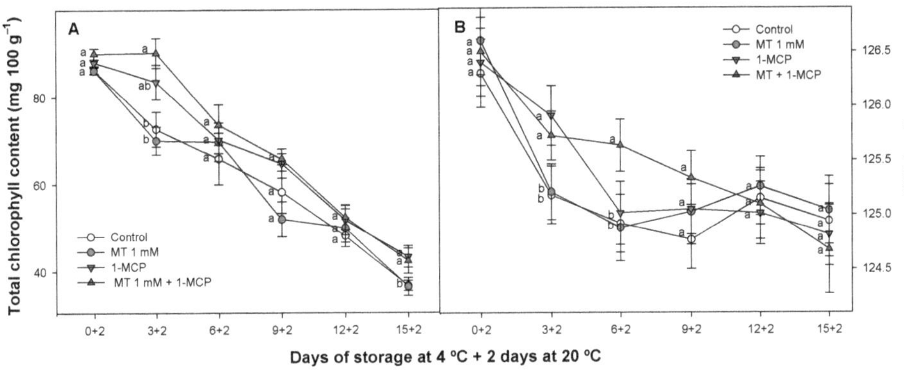

Figure 5. Evolution of total chlorophyll content (mg 100 g^{-1}) (**A**) and CIE *hue** (**B**) of 'Cronos' zucchini fruit treated with melatonin at 1 mM (MT) or distilled water (Control) with or without 1-MCP during cold storage and after cold storage plus 2 days at 20 °C. Data are the mean ± SE (n = 3). Different lowercase letters show significant differences ($p < 0.05$) among treatments for each sam-pling date.

Interestingly there was a clear effect provided by 1-MCP alone or combined with MT showing a positive effect on the maintenance of this parameter after 3 days of storage (83.44 ± 3.93 and 90.11 ± 3.41 mg 100 g^{-1} fw respectively). These values were significant higher ($p < 0.05$) than observed for MT and control fruit (70.00 ± 3.17 and 72.64 ± 4.01 mg 100 g^{-1} fw respectively). This positive effect was in general maintained along the whole experiment. On the other hand, when CIE *hue** slightly decreased during storage in all zucchini fruit tested, 1-MCP and MT + 1-MCP lots maintained significant differences ($p < 0.05$) as compared with the rest of treatments applied (Figure 5B) though these differences were reduced along the experiment. However, when MT and 1-MCP were applied as a combined treatment these differences were maintained delaying the evolution of this parameter for longer time than when these substances were applied alone.

Green colour in zucchini is determined by chlorophyll content and storage impact on pigment degradation is correlated with a reduced CIE *hue** in Cronos cultivar [59]. These authors also observed that zucchini fruit senescence is accompanied by decrease of chlorophyll pigments and CIE *hue** during storage in Cronos cultivar mainly due to loss of cell wall integrity, reducing firmness and contributing to chlorophyll pigment degradation [59]. In this sense 1-MCP treatments have been shown to maintain tissue firmness and chlorophyll content delaying senescence in climacteric and non-climacteric fruit [25,53]. On the other hand, chlorophyll pigments are also affected during postharvest as the main targets of ROS-linked damage since ROS detoxification systems decrease during plant senescence and other different stresses [60]. MT treatments have been shown to delay tissue degreening maintaining chlorophyll content in different other MT-treated vegetal products as broccoli, mango, or cucumber [61–63]. In our study, 1-MCP treatments maintained these parameters, but MT treatment did not affect both of them. For this reason, the synergistic effect observed when treatments were applied combined, could be due to a better antioxidant MT performance mediated by an improved cell homeostasis since 1-MCP when applied alone, also showed a better control over chlorophyll content, MDA and EL than observed for MT-treated fruit for all these parameters.

3.5. Effect of Exogenous MT and 1-MCP on TSS and TA

TSS in zucchini fruit is shown in Figure 6A exhibiting a decrease in all treatments.

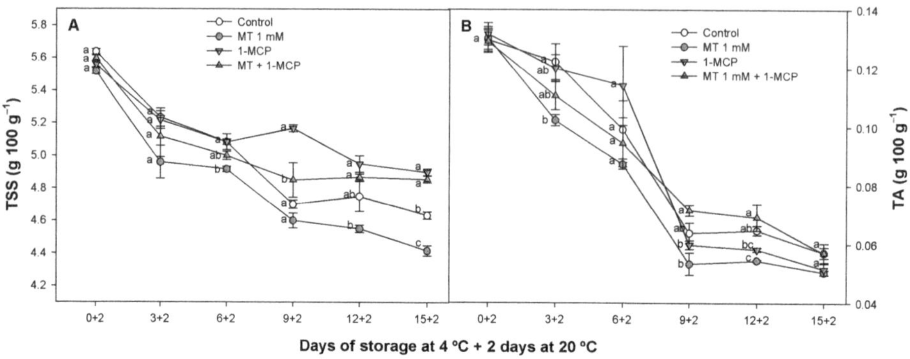

Figure 6. Evolution of total soluble solids (g 100 g^{-1}) (**A**) and titratable acitidity (g 100 g^{-1}) (**B**) of 'Cronos' zucchini fruit treated with melatonin at 1 mM (MT) or distilled water (Control) with or without 1-MCP during cold storage and after cold storage plus 2 days at 20 °C. Data are the mean ± SE (n = 3). Different lowercase letters show significant differences ($p < 0.05$) among treatments for each sam-pling date.

TSS content in 1-MCP and MT + 1-MCP groups was significantly higher ($p < 0.05$) as compared to control fruit after 9 days of 4 °C storage. On the other hand, MT fruit displayed the lowest values during the storage period decreasing to 4.41 ± 0.03 g 100 g^{-1} at the end of the storage.

Similarly to TSS, TA in all the groups evaluated decreased during storage conditions and when MT was applied alone TA levels were lower as compared to the rest of fruit groups studied (Figure 6B). 1-MCP alone retained initial TA levels during 6 days of storage but as compared to control fruit no significant differences ($p \geq 0.05$) were observed. However, fruit treated with MT + 1-MCP exhibited higher TA levels than the rest of treatments after 9 days of cold storage plus an additional period of 2 days at 20 °C.

TSS content is an important attribute which reflects the sugar concentration in cells which increases through the conversion of starch to sugar. However, in this study and in consonance with previous studies on zucchini fruit it seems that at 20 °C, the sugar consumption is faster than its accumulation leading to senescence [33,64]. On the other hand, is well documented the decreased TA level in zucchini and other different fruits by the use of organic acids as respiration substrates during ripening process [64,65]. Previous reports have revealed the effect of 1-MCP on delaying ripening processes as decreasing the respiration rate in zucchini [10] and maintaining TSS and TA content through this mechanism in different fruit species [53,65]. There are no previous studies of the effect of MT treatments on zucchini fruit though in a recent review MT was described as a substance capable of inducing cold tolerance by the accumulation of sugar and organic acids in different fruit species [66]. However, MT treatments did not increase TSS or TA in zucchini probably due to the similar respiration pattern in MT-treated fruit than observed for control fruit (Figure 3A). For this reason and based in our results, the combined treatment (MT + 1-MCP) though did not increase TSS to a higher concentration than that observed for 1-MCP alone, a synergistic effect maintaining higher TA levels was displayed when both treatments were applied together showing the highest levels of this parameter. We propose that this additional benefit could be due to a reduction in the ripening process and respiration caused by the 1-MCP but also to an increase in organic acids as have been observed in different fruit species treated with MT [66]. In this sense, a higher solute concentration is a positive factor for maintaining the protoplasm osmoregulation enhancing the cold tolerance. Higher levels in TSS and TA are directly related with a higher cell homeostasis and higher contents in sugars and organic acids as ascorbic or citric acid may also contribute to a reduced CI impact [66–68].

4. Conclusions

The present study confirmed that melatonin at 1 mM concentration as a postharvest dip treatment when combined with 1-MCP can extend the storage life of zucchini by reducing respiratory metabolism and maintaining fruit firmness reducing weight loss. The reduced metabolism observed by 1-MCP, combined with a melatonin treatment with an also displayed antioxidant effect observed maintaining chlorophyll content or reducing MDA accumulation were crucial factors on cell membrane integrity increasing zucchini cold tolerance. Also, an increased solute concentration observed in zucchini exposed to the combined treatment, could determine zucchini cold tolerance during storage at suboptimal temperatures. In this sense, results suggest that application of a combined treatment based on melatonin and 1-MCP could be a promising tool to increase storability of this fruit.

Author Contributions: J.M.-S.: Investigation, Methodology, Data curation, Formal analysis, Visualization, Writing-Original draft preparation, Software. M.S.: Investigation, methodology, Formal analysis, Validation, Visualization, Resources, Supervision. M.C.R.-A.: Investigation, Data curation, Formal analysis. M.I.M.I.: Investigation, Data curation. D.M.-R.: Investigation, Methodology, Data curation, Validation, Visualization. F.G.: Conceptualization, Investigation, Methodology, Formal analysis, Data curation, Validation, Visualization, Writing—Reviewing and Editing, Resources, Supervision. All authors have read and agreed to the published version of the manuscript.

Funding: This research was funded by Spanish Ministry of Science, Innovation and Universities and European Commission with FEDER funds through Project RTI2018-099664-B-100.

Institutional Review Board Statement: Not applicable.

Informed Consent Statement: Not applicable.

Data Availability Statement: Data are contained within the article.

Conflicts of Interest: The authors declare no conflict of interest.

References

1. Wang, C.Y. Combined Treatment of Heat Shock and Low Temperature Conditioning Reduces Postharvest Chilling Injury in Zucchini Squash. *Postharvest Biol. Technol.* **1994**, *4*, 65–73. [CrossRef]
2. Carvajal, F.; Palma, F.; Jamilena, M.; Garrido, D. Cell wall metabolism and chilling injury during postharvest cold storage in zucchini fruit. *Postharvest Biol. Technol.* **2015**, *108*, 68–77. [CrossRef]
3. Bokhary, S.U.F.; Wang, L.; Zheng, Y.; Jin, P. Pre-storage hot water treatment enhances chilling tolerance of zucchini (*Cucurbita pepo* L.) squash by regulating arginine metabolism. *Postharvest Biol. Technol.* **2020**, *166*, 1112–1129. [CrossRef]
4. Zuo, X.; Cao, S.; Jia, W.; Zhao, Z.; Jin, P.; Zheng, Y. Near-saturated relative humidity alleviates chilling injury in zucchini fruit through its regulation of antioxidant response and energy metabolism. *Food Chem.* **2021**, *351*, 129336. [CrossRef] [PubMed]
5. Zuo, X.; Cao, S.; Zhang, M.; Cheng, Z.; Cao, T.; Jin, P.; Zheng, Y. High relative humidity (HRH) storage alleviates chilling injury of zucchini fruit by promoting the accumulation of proline and ABA. *Postharvest Biol. Technol.* **2021**, *171*, 111344. [CrossRef]
6. Mencarelli, F. Effect of high CO_2 atmospheres on stored zucchini squash. *J. Am. Soc. Hortic. Sci.* **1987**, *112*, 985–988. [CrossRef]
7. Serrano, M.; Pretel, M.T.; Martinez-Madrid, M.C.; Romojaro, F.; Riquelme, F. CO_2 treatment of zucchini squash reduces chilling-induced physiological changes. *J. Agric. Food Chem.* **1998**, *46*, 2465–2468. [CrossRef]
8. Megías, Z.; Martínez, C.; Manzano, S.; García, A.; Rebolloso-Fuentes, M.M.; Garrido, D.; Valenzuela, J.L.; Jamilena, M. Individual shrink wrapping of zucchini fruit improves postharvest chilling tolerance associated with a reduction in ethylene production and oxidative stress metabolites. *PLoS ONE* **2015**, *10*, e0133058. [CrossRef]
9. Jiménez-Muñoz, R.; Palma, F.; Carvajal, F.; Castro-Cegrí, A.; Pulido, A.; Jamilena, M.; Romero-Puertas, M.C.; Garrido, D. Pre-storage nitric oxide treatment enhances chilling tolerance of zucchini fruit (*Cucurbita pepo* L.) by S-nitrosylation of proteins and modulation of the antioxidant response, Postharvest. *Biol. Technol.* **2021**, *171*, 111345. [CrossRef]
10. Megías, Z.; Martinez, C.; Manzano, S.; García, A.; Rebolloso-Fuentes, M.M.; Valenzuela, J.L.; Garrido, D.; Jamilena, M. Ethylene biosynthesis and signaling elements involved in chilling injury and other postharvest quality traits in the non-climacteric fruit of zucchini (*Cucurbita pepo*). *Postharvest Biol. Technol.* **2016**, *113*, 48–57. [CrossRef]
11. Palma, F.; Carvajal, F.; Jiménez-Muñoz, R.; Pulido, A.; Jamilena, M.; Garrido, D. Exogenous γ-aminobutyric acid treatment improves the cold tolerance of zucchini fruit during postharvest storage. *Plant Physiol. Biochem.* **2019**, *136*, 188–195. [CrossRef] [PubMed]
12. Yao, W.; Xu, T.; Farooq, U.; Jin, P.; Zheng, Y. Glycine betaine treatment alleviates chilling injury in zucchini fruit (*Cucurbita pepo* L.) by modulating antioxidant enzymes and membrane fatty acid metabolism. *Postharvest Biol. Technol.* **2018**, *144*, 20–28. [CrossRef]

13. Martínez-Téllez, M.Á.; Ramos-Clamont, M.G.; Gardea, A.A.; Vargas-Arispuro, I. Effect of infiltrated polyamines on polygalacturonase activity and chilling injury responses in zucchini squash (*Cucurbita pepo* L.). *Biochem. Biophys. Res. Commun.* **2002**, *295*, 98–101. [CrossRef]
14. Palma, F.; Carvajal, F.; Ramos, J.M.; Jamilena, M.; Garrido, D. Effect of putrescine application on maintenance of zucchini fruit quality during cold storage: Contribution of GABA shunt and other related nitrogen metabolites. *Postharvest Biol. Technol.* **2015**, *99*, 131–140. [CrossRef]
15. Palma, F.; Carvajal, F.; Jamilena, M.; Garrido, D. Putrescine treatment increases the antioxidant response and carbohydrate content in zucchini fruit stored at low temperature. *Postharvest Biol. Technol.* **2016**, *118*, 68–70. [CrossRef]
16. Wang, C.Y.; Buta, J.G. Methyl jasmonate improves quality of stored zucchini squash. *J. Food Qual.* **1999**, *22*, 663–670. [CrossRef]
17. Aghdam, M.S.; Fard, J.R. Melatonin treatment attenuates postharvest decay and maintains nutritional quality of strawberry fruits (Fragaria x anannasa cv. Selva) by enhancing GABA shunt activity. *Food Chem.* **2017**, *221*, 1650–1657. [CrossRef]
18. Molla, S.M.; Rastegar, S.; Omran, V.G.; Khademi, O. Ameliorative effect of melatonin against storage chilling injury in pomegranate husk and arils through promoting the antioxidant system. *Sci. Hortic.* **2022**, *295*, 110889. [CrossRef]
19. Cao, S.; Shao, J.; Shi, L.; Xu, L.; Shen, Z.; Chen, W.; Yang, Z. Melatonin increases chilling tolerance in postharvest peach fruit by alleviating oxidative damage. *Sci. Rep.* **2018**, *8*, 806. [CrossRef]
20. Jannatizadeh, A.; Aghdam, M.S.; Luo, Z.; Razavi, F. Impact of Exogenous Melatonin Application on Chilling Injury in Tomato Fruits During Cold Storage. *Food Bioprocess Tech.* **2019**, *12*, 741–750. [CrossRef]
21. Medina-Santamarina, J.; Zapata, P.J.; Valverde, J.M.; Valero, D.; Serrano, M.; Guillén, F. Melatonin treatment of apricot trees leads to maintenance of fruit quality attributes during storage at chilling and non-chilling temperatures. *Agronomy* **2021**, *11*, 917. [CrossRef]
22. Mohammadi, M.; Aelaei, M.; Saidi, M. Pre-harvest and pulse treatments of spermine, γ-and β-aminobutyric acid increased antioxidant activities and extended the vase life of gerbera cut flowers 'Stanza'. *Ornam. Hortic.* **2020**, *26*, 306–316. [CrossRef]
23. Gao, H.; Lu, Z.; Yang, Y.; Wang, D.; Yang, Y.; Cao, M.; Cao, W. Melatonin treatment reduces chilling injury in peach fruit through its regulation of membrane fatty acid contents and phenolic metabolism. *Food Chem.* **2018**, *245*, 659–666. [CrossRef]
24. Zhang, W.; Jiang, H.; Zhang, Y.; Cao, J.; Jiang, W. Synergistic effects of 1-MCP and hot air treatments on delaying softening and promoting anthocyanin biosynthesis in nectarines. *Postharvest Biol. Technol.* **2021**, *180*, 111598. [CrossRef]
25. Du, Y.; Jin, T.; Zhao, H.; Han, C.; Sun, F.; Chen, Q.; Yue, F.; Luo, Z.; Fu, M. Synergistic inhibitory effect of 1-methylcyclopropene (1-MCP) and chlorine dioxide (ClO2) treatment on chlorophyll degradation of green pepper fruit during storage. *Postharvest Biol. Technol.* **2021**, *171*, 111363. [CrossRef]
26. Aguayo, E.; Jansasithorn, R.; Kader, A.A. Combined effects of 1-methylcyclopropene, calcium chloride dip, and/or atmospheric modification on quality changes in fresh-cut strawberries. *Postharvest Biol. Technol.* **2006**, *40*, 269–278. [CrossRef]
27. Minh, N.P. Synergistic effect of calcium chloride and 1-Methylcyclopropene on storage of watermelon (*Citrullus lanatus*). *Plant Sci. Today.* **2021**, *8*, 118–122. [CrossRef]
28. Liu, R.; Gao, H.; Chen, H.; Fang, X.; Wu, W. Synergistic effect of 1-methylcyclopropene and carvacrol on preservation of red pitaya (*Hylocereus polyrhizus*). *Food Chem.* **2019**, *283*, 588–595. [CrossRef]
29. Min, D.; Li, F.; Zhang, X.; Shu, P.; Cui, X.; Dong, L.; Ren, C.; Meng, D.; Li, J. Effect of methyl salicylate in combination with 1-methylcyclopropene on postharvest quality and decay caused by Botrytis cinerea in tomato fruit. *J. Sci. Food Agric.* **2018**, *98*, 3815–3822. [CrossRef]
30. Guo, S.; Li, T.; Wu, C.; Fan, G.; Wang, H.; Shen, D. Melatonin and 1-methylcyclopropene treatments on delay senescence of apricots during postharvest cold storage by enhancing antioxidant system activity. *J. Food Process. Preserv.* **2021**, *45*, e15863. [CrossRef]
31. Megías, Z.; Martinez, C.; Manzano, S.; Barrera, A.; Rosales, R.; Valenzuela, J.L.; Garrido, D.; Jamilena, M. Cold-induced ethylene in relation to chilling injury and chilling sensitivity in the non-climacteric fruit of zucchini (*Cucurbita pepo* L.). *LWT Food Sci. Technol.* **2014**, *57*, 194–199. [CrossRef]
32. Martínez-Romero, D.; Serrano, M.; Carbonell, A.; Burgos, F. Riquelme, F.; Valero, D. Effects of postharvest putrescine treatment on extending shelf life and reducing mechanical damage in apricot. *J. Food Sci.* **2002**, *67*, 1706–1712. [CrossRef]
33. Zhang, M.; Liu, W.; Li, C.; Shao, T.; Jiang, X.; Zhao, H.; Ai, W. Postharvest hot water dipping and hot water forced convection treatments alleviate chilling injury for zucchini fruit during cold storage. *Sci. Hortic.* **2019**, *249*, 219–227. [CrossRef]
34. Mao, L.C.; Wang, G.Z.; Zhu, C.G.; Pang, H.Q. Involvement of phospholipase D and lipoxygenase in response to chilling stress in postharvest cucumber fruits. *Plant Sci.* **2007**, *172*, 400–405. [CrossRef]
35. Porra, R.J.; Thompson, W.A.; Kriedemann, P.E. Determination of accurate extinction coefficients and simultaneous equations for assaying chlorophylls a and b extracted with four different solvents: Verification of the concentration of chlorophyll standards by atomic absorption spectroscopy. *BBA Bioenergetics* **1989**, *975*, 384–394. [CrossRef]
36. Lin, T.T.; Pitt, R.E. Rheology of apple and potato tissue as affected by cell turgor pressure. *J. Texture Stud.* **1986**, *17*, 291–313. [CrossRef]
37. Heyes, J.A.; Sealey, D.F. Textural changes during nectarine (*Prunus persica*) development and ripening. *Sci. Hortic.* **1996**, *65*, 49–58. [CrossRef]
38. Lado, J.; Rodrigo, M.J.; Zacarías, L. Analysis of ethylene biosynthesis and perception during postharvest cold storage of Marsh and Star Ruby grapefruits. *Food Sci. Technol.* **2014**, *21*, 537–546. [CrossRef]

39. Tessmer, M.A.; Appezzato-da-Glória, B.; Kluge, R.A. Evaluation of storage temperatures to astringency 'Giombo' persimmon: Storage at 1 °C combined with 1-MCP is recommended to alleviate chilling injury. *Sci. Hortic.* **2019**, *257*, 108675. [CrossRef]
40. Zhang, W.; Zhao, H.; Jiang, H.; Xu, J.; Cao, J.; Jiang, W. Multiple 1-MCP treatment more effectively alleviated postharvest nectarine chilling injury than conventional one-time 1-MCP treatment by regulating ROS and energy metabolism. *Food Chem.* **2020**, *330*, 127256. [CrossRef]
41. Qian, C.; Ji, Z.; Zhu, Q.; Qi, X.; Li, Q.; Yin, J.; Liu, J.; Kan, J.; Zhang, M.; Jin, C.; et al. Effects of 1-MCP on proline, polyamine, and nitric oxide metabolism in postharvest peach fruit under chilling stress. *Hortic. Plant J.* **2021**, *7*, 188–196. [CrossRef]
42. Jayarajan, S.; Sharma, R.R. Melatonin: A blooming biomolecule for postharvest management of perishable fruits and vegetables. *Trends Food Sci. Technol.* **2021**, *116*, 318–328. [CrossRef]
43. Sun, Q.; Zhang, N.; Wang, J.; Zhang, H.; Li, D.; Shi, J.; Li, R.; Weeda, S.; Zhao, B.; Ren, S.; et al. Melatonin promotes ripening and improves quality of tomato fruit during postharvest life. *J. Exp. Bot.* **2015**, *66*, 657–668. [CrossRef]
44. Bal, E. Physicochemical changes in Santa Rosa plum fruit treated with melatonin during cold storage. *J. Food Meas.* **2019**, *13*, 1713–1720. [CrossRef]
45. Zhai, R.; Liu, J.L.; Liu, F.X.; Zhao, Y.X.; Liu, L.L.; Fang, C.; Wang, H.B.; Li, X.Y.; Wang, Z.G.; Ma, F.W.; et al. Melatonin limited ethylene production, softening and reduced physiology disorder in pear (*Pyrus communis* L.) fruit during senescence. *Postharvest Biol. Technol.* **2018**, *139*, 38–46. [CrossRef]
46. Dan, Y.; Zhang, S.; Zhong, H.; Yi, H.; Sainz, M.B. Novel compounds that enhance agrobacterium-mediated plant transformation by mitigating oxidative stress. *Plant Cell Rep.* **2015**, *34*, 291–309. [CrossRef] [PubMed]
47. Zhang, Y.; Huber, D.J.; Hu, M.; Jiang, G.; Gao, Z.; Xu, X.; Jiang, Y.; Zhang, Z. Delay of postharvest browning in Litchi fruit by melatonin via the enhancing of antioxidative processes and oxidation repair. *J. Agric. Food Chem.* **2018**, *66*, 7475–7484. [CrossRef]
48. Fernández-Trujillo, J.P.; Martínez, J.A. Ultrastructure of the onset of chilling injury in cucumber fruit. *J. Appl. Bot. Food Qual.* **2012**, *80*, 100–110.
49. Hakim, A.; Purvis, A.C.; Mullinix, B.G. Differences in chilling sensitivity of cucumber varieties depends on storage temperature and the physiological dysfunction evaluated. *Postharvest Biol. Technol.* **1999**, *17*, 97–104. [CrossRef]
50. Sharafi, Y.; Aghdam, M.S.; Luo, Z.; Jannatizadeh, A.; Razavi, F.; Fard, J.R.; Farmani, B. Melatonin treatment promotes endogenous melatonin accumulation and triggers GABA shunt pathway activity in tomato fruits during cold storage. *Sci. Hortic.* **2019**, *254*, 222–227. [CrossRef]
51. Shelp, B.J.; Bown, A.W.; McLean, M.D. Metabolism and functions of gamma-aminobutyric acid. *Trends Plant Sci.* **1999**, *4*, 1360–1385. [CrossRef]
52. Kou, X.; Feng, Y.; Yuan, S.; Zhao, X.; Wu, C.; Wang, C.; Xue, Z. Different regulatory mechanisms of plant hormones in the ripening of climacteric and non-climacteric fruits: A review. *Plant Mol. Biol.* **2021**, *107*, 477–497. [CrossRef] [PubMed]
53. Watkins, C.B. The use of 1-methylcyclopropene (1-MCP) on fruits and vegetables. *Biotechnol. Adv.* **2006**, *24*, 389–409. [CrossRef] [PubMed]
54. Bi, X.; Dai, Y.; Zhou, Z.; Xing, Y.; Che, Z. Combining natamycin and 1-methylcyclopropene with modified atmosphere packaging to evaluate plum (*Prunus salicina* cv. 'Cuihongli') quality. *Postharvest Biol. Technol.* **2022**, *183*, 111749. [CrossRef]
55. Cao, S.; Yang, Z.; Zheng, Y. Effect of 1-methylcyclopene on senescence and quality maintenance of green bell pepper fruit during storage at 20 °C. *Postharvest Biol. Technol.* **2012**, *70*, 1–6. [CrossRef]
56. Xu, F.; Liu, S. Control of postharvest quality in blueberry fruit by combined 1-methylcyclopropene (1-MCP) and UV-C irradiation. *Food Bioprocess Technol.* **2017**, *10*, 1695–1703. [CrossRef]
57. Arnao, M.B.; Hernández-Ruiz, J. Melatonin: A New Plant Hormone and/or a Plant Master Regulator? *Trends Plant Sci.* **2019**, *24*, 38–48. [CrossRef]
58. Wang, S.Y.; Shi, X.C.; Wang, R.; Wang, H.L.; Liu, F.; Laborda, P. Melatonin in fruit production and postharvest preservation: A review. *Food Chem.* **2020**, *320*, 126642. [CrossRef]
59. Blanco-Díaz, M.T.; Pérez-Vicente, A.; Font, R. Quality of Fresh Cut Zucchini as Affected by Cultivar, Maturity at Processing and Packaging. *Packag. Technol. Sci.* **2016**, *29*, 365–382. [CrossRef]
60. Khanna-Chopra, R. Leaf senescence and abiotic stresses share reactive oxygen species-mediated chloroplast degradation. *Protoplasma* **2012**, *249*, 81–469. [CrossRef]
61. Wei, L.; Liu, C.; Wang, J.; Younas, S.; Zheng, H.; Zheng, L. Melatonin immersion affects the quality of fresh-cut broccoli (*Brassica oleracea* L.) during cold storage: Focus on the antioxidant system. *J. Food Process. Preserv.* **2020**, *44*, e14691. [CrossRef]
62. Dong, J.; Kebbeh, M.; Yan, R.; Huan, C.; Jiang, T.; Zheng, X. Melatonin treatment delays ripening in mangoes associated with maintaining the membrane integrity of fruit exocarp during postharvest. *Plant Physiol. Biochem.* **2021**, *169*, 22–28. [CrossRef] [PubMed]
63. Madebo, M.P.; Luo, S.M.; Wang, L.; Zheng, Y.H.; Jin, P. Melatonin treatment induces chilling tolerance by regulating the contents of polyamine, γ-aminobutyric acid, and proline in cucumber fruit. *J. Integr. Agric.* **2021**, *20*, 3060–3074. [CrossRef]
64. Jafari, R.; Zandi, M.; Ganjloo, A. Effect of gelatin–alginate coating containing anise (*Pimpinella anisum* L.) essential oil on physicochemical and visual properties of zucchini (*Cucurbita pepo* L.) fruit during storage. *J. Food Process. Preserv.* **2022**, *46*, e16756. [CrossRef]
65. Valero, D.; Serrano, M. *Postharvest Biology and Technology for Preserving Fruit Quality*; CRC Press/Taylor & Francis: Boca Raton, FL, USA, 2010.

66. Zeng, W.; Mostafa, S.; Lu, Z.; Jin, B. Melatonin-Mediated Abiotic Stress Tolerance in Plants. *Front. Plant Sci.* **2022**, *13*, 847175. [CrossRef]
67. Boonyaritthongchai, P.; Supapvanich, S. Effects of methyl jasmonate on physicochemical qualities and internal browning of 'Queen' pineapple fruit during cold storage. *Hortic Environ Biotechnol.* **2017**, *58*, 479–487. [CrossRef]
68. Sangsoy, K.; Beckles, D.M.; Terdwongworakul, A.; Luengwilai, K. Discriminating pineapple batches for susceptibility to postharvest internal browning. *Sci. Hortic.* **2022**, *300*, 111069. [CrossRef]

foods

Article

Synergy of Nitric Oxide and 1-Methylcyclopropene Treatment in Prolong Ripening and Senescence of Peach Fruit

Xiaoqin Wu [1,2], Jiawei Yuan [1], Xiaoqing Wang [1], Mingliang Yu [3,*], Ruijuan Ma [3] and Zhifang Yu [2,*]

1 College of Biological and Food Engineering, Changshu Institute of Technology, Suzhou 215500, China; wuxiaoqin@cslg.edu.cn (X.W.); yuanjiawei5008@163.com (J.Y.); w1522079778@163.com (X.W.)
2 College of Food Science and Engineering, Nanjing Agricultural University, Nanjing 210095, China
3 Institute of Pomology, Jiangsu Academy of Agricultural Sciences/Jiangsu Key Laboratory for Horticulture Crop Genetic Improvement, Nanjing 210014, China; marj311@163.com
* Correspondence: mly1008@aliyun.com (M.Y.); yuzhifang@njau.edu.cn (Z.Y.); Tel.: +86-1395-169-2350 (Z.Y.)

Abstract: Peach is a putrescible fruit thus drastically restricting its postharvest storage life. In recent years, the application of 1-methylcyclopropene (1-MCP) and nitric oxide (NO) in postharvest fruit quality control has received considerable attention and investigative efforts due to the advantages of using relatively low concentrations and short-time treatment duration. In the present study, the effects of various 1-MCP and NO treatments on peach fruit (*Prunus persica* L. cv. Xiahui-8) stored at 25 °C were evaluated and compared. Results indicated that the combination treatment with both chemical agents (MN) was most effective in postponing peach ripening and preserving fruit quality, followed by 1-MCP and NO treatment alone. We also demonstrated that NO could delay fruit senescence mainly by stimulating antioxidant enzymes, while 1-MCP overly outperformed NO in the treatment of 'Xiahui-8' peach in slowing down respiration rate, inhibiting ethylene production, maintaining high firmness and reducing ROS content. NO treatment showed a greater influence on phenolic compounds than 1-MCP especially anthocyanins, flavanones and flavones according to LC/MS analysis. The phenolic change in MN group were highly associated to NO treatment. Through this study we provide informative physiological, biochemical and molecular evidence for the beneficial effects of the combined 1-MCP and NO treatment on peach fruit based on a functional synergy between these two chemical agents.

Keywords: fruit storage; antioxidant capacity; phenolic compounds; gene expression

Citation: Wu, X.; Yuan, J.; Wang, X.; Yu, M.; Ma, R.; Yu, Z. Synergy of Nitric Oxide and 1-Methylcyclopropene Treatment in Prolong Ripening and Senescence of Peach Fruit. *Foods* **2021**, *10*, 2956. https://doi.org/10.3390/foods10122956

Academic Editors: Maria Cefola and Bernardo Pace

Received: 27 October 2021
Accepted: 22 November 2021
Published: 1 December 2021

1. Introduction

Unlike other rosaceous fruit like apple or pear, peach (*Prunus persica* L.) is well known for the relatively short shelf-life of fruits due to high respiration rates, accelerated fruit ripening and fast flesh softening process that could significantly impede marketing and sales and lower commercial value. Hence, there is an urgent need to develop effective strategies for postharvest handling and storage in order to prolong shelf-life while maintaining consumer-desired fruit quality. Towards this goal, several previous studies investigated the use of 1-methylcyclopropene (1-MCP) and nitric oxide (NO) for postharvest treatment in peach fruits and demonstrated their high efficacy for delaying fruit ripening and senescence at relatively low concentrations and short treatment time duration [1–3].

The biological function of NO as a key signaling molecule in plant cells has long been recognized. For instance, Neill et al. in their investigation of the molecular events related to NO biosynthesis and functionality demonstrated that NO produced by plant cells can function as a critical signaling component in ABA-induced stomatal closure [4]. Fumigation treatments of climacteric and non-climacteric fruits with NO, later known as an ethylene antagonist, were found highly effective to considerably extend fruit postharvest life and delay senescence [5,6]. Likewise, another ethylene inhibitor 1-MCP has also been utilized in the postharvest treatment of fruits and vegetables due to its pronounced effects to dramatically delay ripening, lower ethylene production and respiratory rate and maintain desirable quality [1]. With these attractive properties, the utilization of either 1-MCP or NO in postharvest treatment of peach has been widely attempted [1,2,5,7,8]. However, to the best of our knowledge, no reports are available that describe the combined use of these two chemical agents and investigate if there are any synergistic effects as compared to single chemical treatment on peach ripening and senescence.

It is worth noting that the modes of action of these two chemical agents are strikingly different. 1-MCP is a competitive inhibitor of ethylene perception and is capable of interacting with ethylene receptor sites and thus preventing the ethylene-induced signaling that triggers ripening and senescence [9]. On the other hand, NO constitutes an important component in the endogenous signaling pathway in cellular metabolism and functions to modulate the physiological responses to phytohormones [10]. The fate of peach fruits upon treatment with both chemical agents remain unknown. Therefore, the objective of this work was to evaluate the effects of peach postharvest treatment with 1-MCP and NO individually or in combination and reveal the patterns of physiological response and gene expression associated with the treatments in order to explore better options for controlling ripening and decline in postharvest fruit quality.

2. Material and Methods

2.1. Peach Material and Treatment

Peach fruits (*Prunus persica* L. cv. Xiahui 8) were harvested from an orchard at Jiangsu Academy of Agricultural Sciences (JAAS) in Nanjing, Jiangsu, China. After 120 days post florescence, about 600 peaches with uniform size and without obvious defects or damages were picked and placed in a pre-cooled container, then transported to the lab immediately. The collected fruit were randomly divided into four groups and subjected to the following treatments: (1) CK or control group: 150 fruit were directly stored at 25 ± 2 °C with 85–90% humidity for 8 days; (2) N group with NO treatment: 150 fruit were placed in a sealed container and treated with 10 μL L^{-1} NO gas for 3 h [2]; (3) M group with 1-MCP treatment: 150 fruit were transferred to an enclosed container and treated with 10 μL L^{-1} 1-MCP (Sinopharm Chemical Reagent Beijing Co., Ltd., Beijing, China) for 12 h, and 1% (*w/v*) KOH solution was placed inside to prevent CO_2 accumulation [1]; (4) MN group with the combination of 1-MCP and NO treatments as described in published literature [11]: 150 peach fruit were first treated using the same condition as 1-MCP treatment for 12 h, and then subject to fumigation with 10 μL L^{-1} NO gas for 3 h. All fruits were stored at room temperature (25 ± 2 °C) with 80–90% relative humidity for 8 days. Samples were taken at 0, 2, 4, 6, and 8 days during storage and immediately used for measurement of respiration, ethylene production, firmness, total soluble solid (TSS), titratable acid (TA), H_2O_2, malondialdehyde (MDA) and O_2^- content. The rest of the fruits were peeled to remove skin and cut into pieces, frozen with liquid nitrogen and stored at −80 °C for further analysis. For each time point, 30 fruit samples were employed for each of three biological replicates, and only the mesocarp was used for analysis.

2.2. Respiratory Rate and Ethylene Production

For respiration and ethylene production, fifteen fruit were placed in three airtight containers equally for 1 h. CO_2 production rate was measured by a portable infrared CO_2 analyzer (PBI Dansensor CheckMate 3, Copenhagen, Denmark) and respiration rate was expressed as mg kg^{-1} h^{-1} of CO_2. Ethylene production was performed according to a method described by Huan [12], with minor modifications. One milliliter of the headspace gas was taken out form each jar and injected into a gas chromatograph (Agilent GC7890 A, Palo Alto, Santa Clara, CA, USA) equipped with an HP-AL/S column (30 m × 0.53 mm × 15 μm, Agilent, Palo Alto, Santa Clara, CA, USA) and a flame ionization detector (FID). The injector, oven and detector temperatures were 120, 100 and 200 °C, respectively. Ethylene production was expressed as μg kg^{-1} h^{-1}.

2.3. Firmness, MDA, H_2O_2 and O_2^- Detection

For fruit firmness, 10 fruit were used and evaluated by using a Fruit Hardness Tester (FHM-5, Tokyo, Japan). MDA, H_2O_2 and O_2^- were measured according to our previous report [13] and expressed as mmol per kilogram fresh weight (mmol kg^{-1} FW).

2.4. Enzymatic Assays

Activities of total superoxide dismutase (SOD) and catalase (CAT) were assayed as described in our previous report [14]. Peroxidase (POD) activity was measured according to the method of Zhang et al. [15] with minor modifications. Following steps were used for the assay: the reaction mixture was prepared by combining guaiacol (0.25%, 100 μL), crude enzyme extract (50 μL) and acetic acid buffer (100 mM, pH 5.4, 100 μL); the reaction was initiated by adding 50 μL of H_2O_2 (0.15%); and absorbance of the sample at 460 nm was measured. Polyphenol oxidase (PPO) activity was determined according to Yingsanga et al. [16]. Ascorbate peroxidase (APX) activity was assayed according to the method of Song et al. [17]. The absorbance changes of POD, PPO and APX reaction mixtures were measured using Microplate Reader (Tecan, Switzerland) for an assay duration of 6 min. One unit of these enzyme activities was defined as a change of 0.01 in absorbance per min and activities expressed as U per mg protein. PAL activity was assayed according to preciously published protocol [18]. Protein content in the extracts was determined by reading absorbance of the sample at 595 nm via the method of Bradford [19] using bovine serum albumin (BSA) as a standard.

2.5. RNA Isolation and Gene Expression Analysis

Sequence information on genes encoding POD and PAL was derived from Genome Database for Rosaceous (GDR; http://www.rosaceae.org/peach/genome (accessed on 16 October 2021)) and gene specific primers were designed using Primer 5.0 software and used for transcript sequencing. After screening of received sequencing data and discarding the redundant sequences, two *PpaPODs* and one *PpaPAL* were selected for further analysis. Primers for *PpaSOD*, *PpaCAT* and *PpaAPX* that were designed in previous research using the similar cultivar [20] were utilized herein. A translation elongation factor 2 (*PpaTEF 2*) was selected as s reference gene for its high expression stability [21]. All primers used for this study were showed in Supplementary Materials. Total RNA extraction, first-strand cDNA synthesis and real-time quantitative reverse transcription-polymerase chain reaction (qRT-PCR) analysis were performed according to our previous report [22].

2.6. LC/MS Analysis of Phenols

The obtained data manifested that respiration burst on D4 and D8, meanwhile ethylene production reached the peak on D8. We speculate that D4 and D8 is crucial time of peach fruit metabolism. Therefore, we choose peach materials of these two time point for further LC/MS analysis. The phenolic compounds extraction and LC/MS analysis were conducted followed by our previous report [22]. Briefly, approximately 10 g of peach tissue was ground with liquid nitrogen, then accurately weigh 5 g of ground sample and homogenized in 100 mL of 95% acidic (0.1 M HCl) methanol. After 4 h of extraction, the mixture was centrifuged at 10,000× *g* for 20 min. The supernatant was collected and evaporated to dryness. For LC/MS analysis, the residue was redissolved in 6 mL of methanol and filtered through a 0.22 μL membrane (Millipore) filter. LC/MS analysis system (G2-XS QTof, Waters) and liquid chromatography (UPLC) column (2.1 × 100 mm × 1.7 μm) was used in this study according to our previous research [22].

2.7. Statistical Analysis

The experiment was conducted in a completely randomized design. Figures were made with Origin Pro 2017 software program. Statistical analyses were performed with the SPSS 18.0 software using the Duncan's test with a significance level at $p < 0.05$. Pearson correlation test were performed with SPSS software.

3. Results and Discussion

Based on extensive characterization of the role in reducing aging of cut carnations, the use of 1-MCP as an ethylene antagonist in postponing the ripening of edible fruits and vegetables has been proposed previously [23]. Subsequently, numerous studies have shown that 1-MCP can extend postharvest life of a wide variety of food commodities [9]. NO was revealed to act as an important endogenous signaling molecule in many cellular metabolisms to modulate hormonal homeostasis during stress responses and plant developmental processes [10]. However, the detailed mechanisms of action through which NO affects fruit ripening and storage quality remain unclear. Other research reports have shown that free radical gas NO has anti-senescence properties similar to those of 1-MCP, which has been observed in tests with different fruits and vegetables [7]. This presumptive finding was somewhat confirmed in our investigations according to results of the following indices.

3.1. Firmness

Firmness is an important quality attribute of peaches that has the potential to enhance storage potential, improve resistance to decay organisms and mechanical injury and enhance market appeal and consumer preference [24]. Fruit softening is a consequence of the modifications and content changes of different cell wall polymers, which is a natural physiological process during ripening and senescence. In this work, firmness was excellently maintained by MN treatment followed by the 1-MCP treatment. Firmness value in NO treatment had no significant change since D4 to the end of storage as compared to that in CK (Figure 1C). Results showed that 1-MCP treatment alone or combined treatment can maintain high firmness of peach, which can delay fruit ripening by maintaining cell structure and improving resistance of decay organisms. NO application can inhibit flesh softening process at later storage time, but which effect was not superior to another two groups in this study.

3.2. Respiration and Ethylene Production

Fruit respiration converts storage compounds and sugars to energy via the generation of ATP to maintain normal metabolism. Respiration and ethylene production are critical indicators of peach ripening. As climacteric fruit, peach is characterized by an upsurge in the respiration rate coincided with a burst of ethylene production during ripening stage. In this research, the respiration rate showed a normal feature of climacteric fruit, which reached a respiratory peak at D4, thereafter decreased (Figure 1A). However, the respiration at D8 showed a high value, which might be induced by tissue damage in later stages of fruit senescence [25]. The ethylene release increased throughout the storage time, reaching a maximum level at D8, a trend consistent with previous studies [20]. 1-MCP or MN treatment significantly suppressed the ethylene release from D6 to the end of storage, but NO treatment had no distinctive effect on ethylene production throughout the storage duration (Figure 1B). Besides, the onset of ethylene productive peak was falling significantly behind respiratory climacteric peak, a phenomenon similar to what was reported in other research with peach [26]. MN treatment suppressed both respiratory and ethylene production rates to the lowest level, indicative of a better approach to postponing fruit ripening. Noticeably, 1-MCP treatment had a better effect in slowing down respiratory and ethylene production rates than NO treatment. We postulate that 1-MCP may be more efficient in delaying senescence of this cultivar. The results of combined 1-MCP and NO treatment observed herein were consistent with previous research with blueberry fruit [11], in which similar combination treatment significantly extended the postharvest life of one of the two compared blueberry cultivars. Several studies have also demonstrated that NO could inhibit CO_2 and ethylene production [7,27]. In particular, Zhu and coworkers theorized that NO is bound to 1-aminocyclopropane-1-carboxylate (ACC) oxidase and subsequently chelated to ACC to form an ACC-ACC oxidase-NO complex, thus decreasing enzymatic activity and reducing ethylene production [27]. NO at 10 μL/L was shown to exert excellent effect on a peach cultivar 'Feicheng' [27]. In spite of these reports, however, we found that treatments with NO at various concentrations on peach cultivar 'Xiahui-8' did not yield desirable results based on observation in several physiological indices. Furthermore, no statistical difference of ethylene release was observed between N and CK groups. These findings may reflect the distinct genotypic response of different cultivars to NO treatment with mechanisms thus far remained unknown.

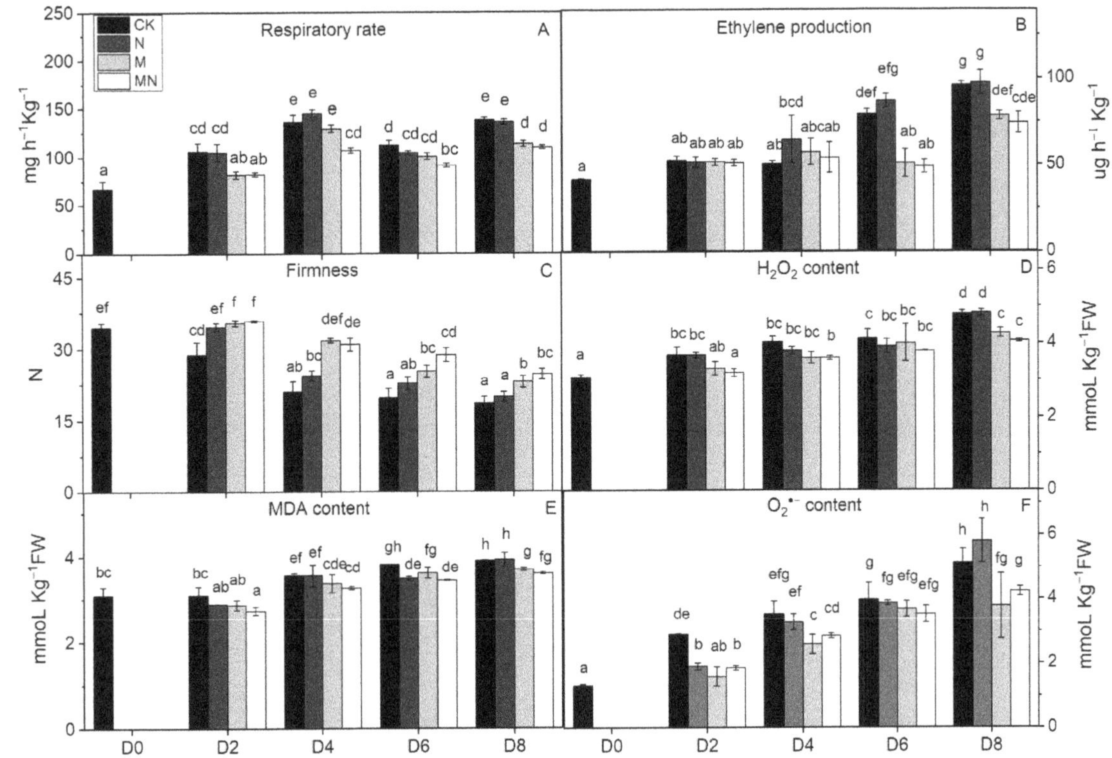

Figure 1. Physiological and biochemical indices of peach fruit including respiratory rate (**A**), ethylene production (**B**), firmness (**C**), H_2O_2 content (**D**), MDA content (**E**) and O_2^- content (**F**) used in this research. Each point represents means ± SE of three replicates. The lowercase letters indicate significant differences according to statistical analysis.

3.3. Reactive Oxygen Species (ROS) Production

Peach is putrescible fruit and can soften quickly at normal temperature, which makes it particularly vulnerable to internal and external stresses. When there is a serious imbalance in cell compartment between the production of ROS and antioxidant defense or ROS scavenging during peach ripening, the ROS increase will inevitably occur, leading to oxidative damage to many biological macromolecules, including proteins, DNA and lipids [14]. ROS can cause peroxidation of the membrane lipids resulting in cell membrane alterations and consequently the generation of MDA [7]. In this study, the tendencies of H_2O_2, MDA and O_2^- production were all similarly increasing gradually throughout the entire duration of storage (Figure 1), indicating that oxidative stress takes place during the natural course of ripening and senescence. The effect of 1-MCP or NO application alone on ROS reduction were mentioned in various researches such as apple [28], mango [29], winter jujube [30]. However, the comparison between these two treatments on peach fruit haven't not been reported yet. Our results showed that, as compared with control fruit, combination treatment MN or treatment with 1-MCP alone significantly reduced the production of ROS, thus delaying fruit senescence. For explaining the mechanism underlying this phenomenon, Lin [31] presumed that postharvest treatment could alleviate the damage action of ROS and the peroxidation process of membrane lipids, consequently retain the structure of pulp cellular membrane of fruit. Combined with the results we got, we assumed that the raised level of ROS and MDA content were highly related to the breakdown of cell structure (Figure 1C–F).

3.4. Enzymatic Activity

Effective reduction of ROS requires several antioxidant enzymes such as SOD, CAT, POD and APX. These enzymes act concomitantly with non-enzymatic antioxidants as a defense against excess ROS [32], consequently, inhibiting fruit quality deterioration. SOD is the first line of defense against ROS by catalyzing the dismutation of O_2^- to molecular oxygen and H_2O_2, and H_2O_2 is then scavenged by CAT, POD and APX [33]. In fruit, these antioxidant enzymes are well known for their roles in regulating the accumulation of ROS which can also act as signaling molecules in many biological processes, and recent studies showed that they are also involved in regulating fruit development [34] and ripening [35]. Furthermore, these antioxidant enzymes are readily activated by postharvest treatments such as hot water [20], brassinolide [36] and 1-MCP [37], which can effectively scavenge ROS to extend the shelf-life as well as improve fruit chilling tolerance [20]. In this study, similar trends of total SOD, APX, PPO and POD activities were observed for all treatments in contrast to CK (Figure 2): a discernable increase for the first four days (D0 to D4), followed by a slight decline (D6), and then a small increase at the end of observation period (D8). CAT activity in treated fruit was noticeably activated than untreated fruit (CK) with a further increased level at D8 during ripening (Figure 2B). Overall, MN treatment induced the highest enzymatic activities for all examined enzymes than those of the control during the entire storage with the exception of total SOD and CAT activities at D4, in which NO treatment generated higher levels. 1-MCP showed better effects for enhancing enzymatic activity than NO, suggesting the former is more effective in ROS elimination. Previous studies showed that exogenously applied NO increased the activity of total SOD, CAT and APX [38,39]. Our results consistent with pervious findings and showed enhanced enzymatic activities of these enzymes. However, the enzymatic activity of total SOD, CAT and APX as well as POD in N group most of the time were not superior to those in M group. Considering that a higher respiratory rate was observed in N group, the lower levels of activity of antioxidant enzymes could lead to increased accumulation of ROS. It is interesting that the general trends of total SOD, APX and POD activities are the same in all treated fruit, i.e., reaching the maximum level at D4 thereafter followed by a gradual decrease, and is similar with respiratory tendency. On the other hand, CAT activity continually increased till the end and did not show any peaks. Similar results for total SOD and CAT activities were also reported in previous studies with peach [40]. We speculate that high respiratory rate generates more ROS, which will in turn stimulate total SOD, APX and POD activities [20,41].

PAL is the first key enzyme in biosynthetic pathway of phenols in fruit and can be induced under various stress conditions [42]. PPO catalyzes the hydroxylation of monophenols that results in brown pigments. In this study, PPO and PAL activities manifested a similar changing trend in four groups, while maintaining a highest level at D4. MN outperformed individual NO or 1-MCP treatment in enhancing PPO and PAL activities, whereas untreated fruit (CK) possessed the lowest activity (Figure 2). It has previously been reported that exogenous NO can stimulate antioxidant enzymes such as PAL [43,44] and POD [43,45], which is in accordance with the results found in this study. However, unlike our results, 1-MCP application on strawberry [46], loquat [47] and nectarine [48] was found to inhibit activities of PAL and PPO and therefore was employed to prevent fruit browning. We speculate that the increased activities of PAL and PPO observed in our study were attributable to resistance response to biotic and abiotic stress processes [49,50].

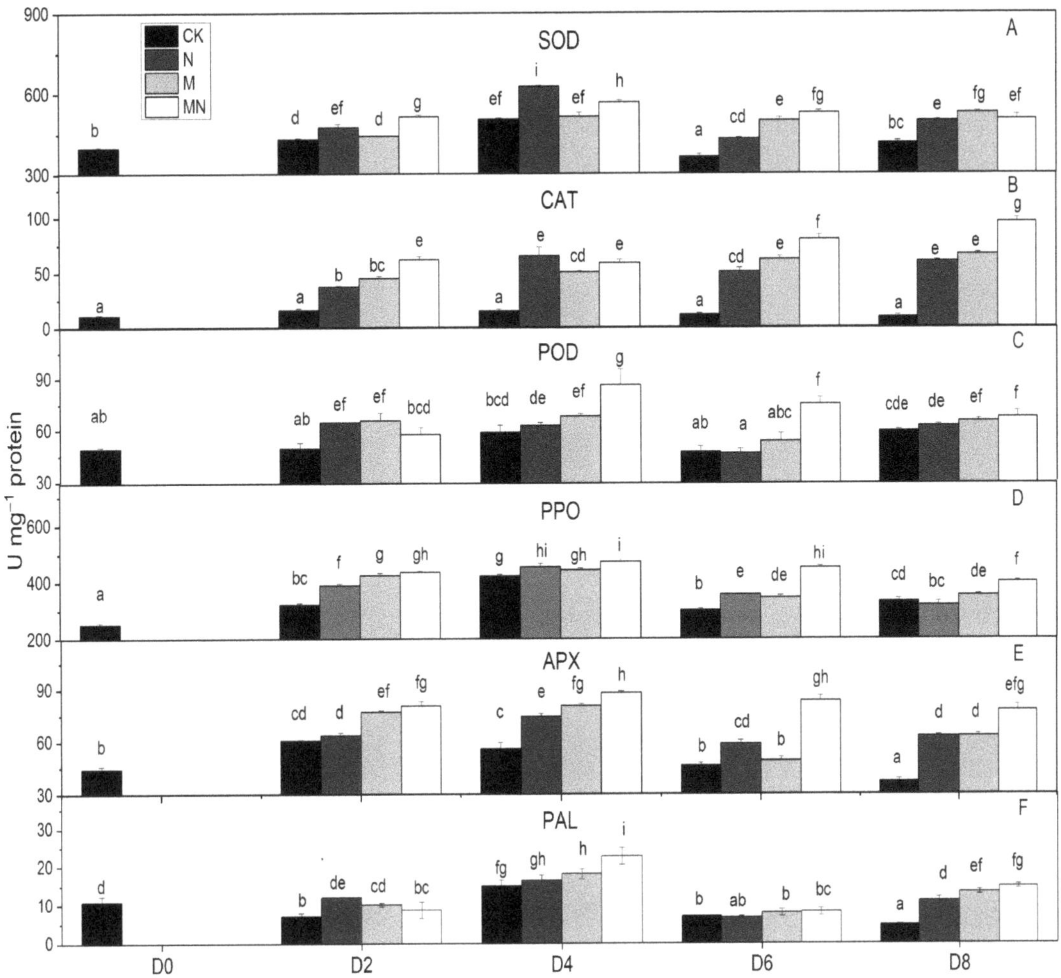

Figure 2. Enzymatic activity in peach fruit. SOD activity (**A**), CAT activity (**B**), POD activity (**C**), PPO activity (**D**), APX activity (**E**), PAL activity (**F**). Each point represents means ± SE of three replicates. The lowercase letters indicate significant differences according to statistical analysis.

3.5. Gene Expression Analysis

In order to check whether trends of enzymatic activity and related gene expression were similar or not in a quantitative way, we did Pearson correlation test, and the result are showed in Table 1. Through the values we found that almost all the genetic and enzymatic changes were inconsistent with each other except *PpaPAL*/PAL ($p < 0.01$). Similar results can be seen in series of published papers [3,51]. We postulate that these enzymes might be regulated by different, yet unidentified, gene members and factors. As showed in Figure 3, *PpaCAT* and *PpaAPX* exhibited a similar expression pattern across all treatments, while their levels of expression in CK were higher than that of treated fruit and at the same time followed a decreasing trend during storage time. Expression of *PpaSOD* declined at D2 then followed a continuous increase during the period from D4 to the end D8. No distinguishable patterns of change in the expression of these genes were observed amongst all three treatments.

Table 1. Pearson Correlation analysis of enzymatic activity and related gene expression.

	PpaSOD/SOD	*PpaCAT*/CAT	*PpaPOD*/POD	*PpaPOD-1*/POD	*PpaAPX*/APX	*PpaPAL*/PAL
Pearson correlation	−0.053	−0.533 **	0.112	−0.406	0.016	0.588 **

** $p < 0.01$.

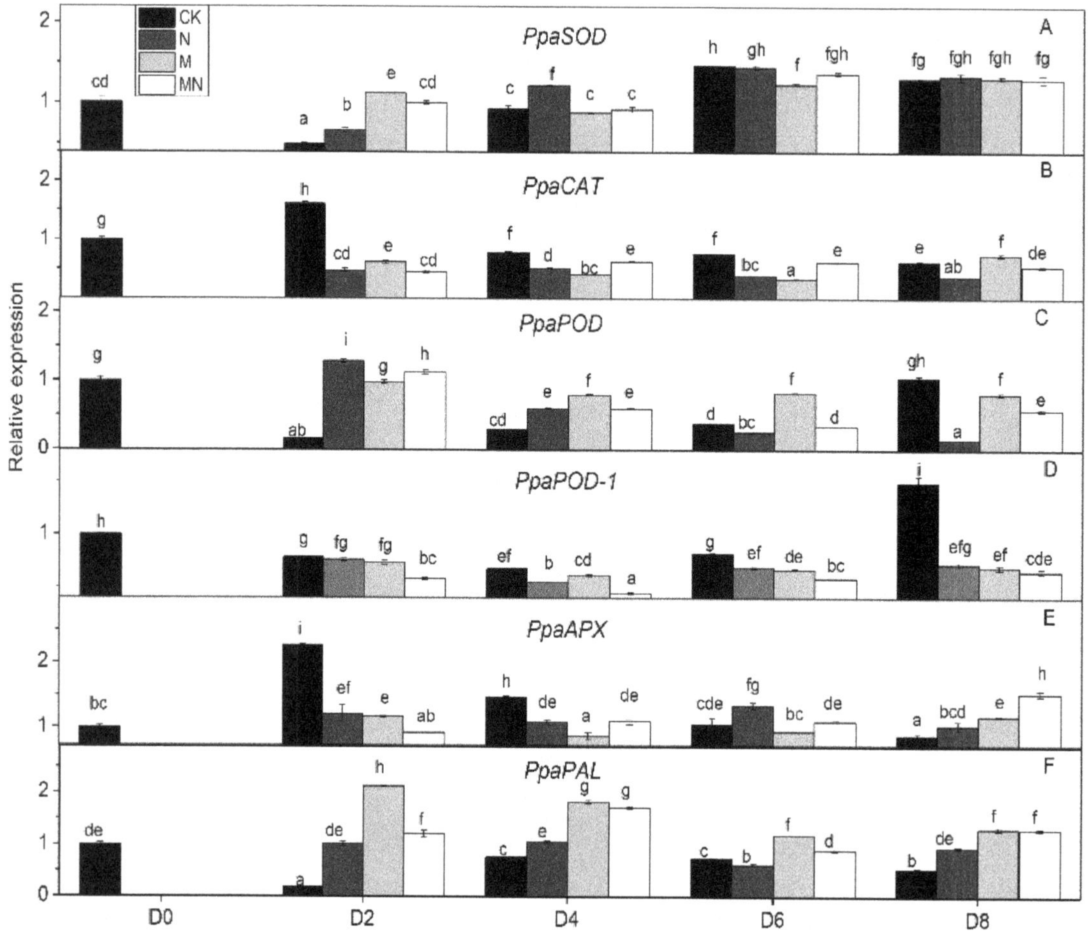

Figure 3. Gene expression profile in peach fruit including *PpaSOD* (**A**), *PpaCAT* (**B**), *PpaPOD* (**C**), *PpaPOD-1* (**D**), *PpaAPX* (**E**) and *PpaPAL* (**F**). Each point represents means ± SE of three replicates. The lowercase letters indicate significant differences according to statistical analysis.

The relative expression levels of *PpaPAL* in treated fruit were higher than that of untreated one (CK), and 1-MCP treatment or combined 1-MCP and NO treatment induced higher expression levels than NO treatment. Additionally, both PAL activity and *PpaPAL* expression exhibited a similar changing trend in response to postharvest treatments and storage process. Accordingly, the Pearson correlation value of *PpaPAL* and PAL was 0.588 ($p < 0.01$), which indicated that treatments tested herein might stimulate PAL activity by directly promoting the expression of *PpaPAL*.

Different trend patterns were observed between two POD genes: *PpaPOD* and *PpaPOD1* (Figure 3). The expression tendency in CK remained the same for these two genes, which decreased at first thereafter increased from D4 till the end D8. 1-MCP treatment stimulated *PpaPOD* expression from D2 to D6, thereafter maintained a stable expression level. The treatment methods showed a greater impact on the expression of *PpaPOD* than *PpaPOD1*, while *PpaPOD1* expression levels in variously treated fruit remained consistent with little changes, but all lower than that in CK group. In addition, similar dynamic changes in the levels of POD and *PpaPOD* expression indicated that POD enzyme might be regulated directly by expression activity of *PpaPOD*.

3.6. LC/MS Analysis of Phenols

20 phenolic compounds were successfully identified based on their retention times, MS data and the corresponding specific fragment, including anthocyanins, flavanones, flavanols, flavones, flavonols and phenolic acids. Representative mass spectrogram of galic acid at negative ionization mode were showed in Figure 4. We have already investigated the influence of 1-MCP on phenolics in our previous report [22], and here we emphasize the effect of NO treatment alone and the combined treatment. The relative amount of phenols were showed in Table 2. Four kinds of anthocyanins were successfully detected including Pigment A, peonidin 3-O-(6″-p-coumaroyl-glucoside), cyanidin 3-O-xylosyl-rutinoside and pelargonidin 3-O-rutinoside. 1-MCP elevated content of anthocyanins, which is benefit for color change in M group. However, NO treatment inhibited most of anthocyanin biosynthesis except peak 1. In addition, NO treatment here inhibits most of the phenolic compounds except peak 1,8,17,18. Intriguingly, the combined treatment showed the similar phenomenon with NO fumigation, but not the 1-MCP treatment. Previous researches such as NO treatment on strawberry [52] and Chinese winter jujube [53] showed that NO fumigation increased total phenolic content, but they did not provide more details about the specific increased or decreased phenolic compounds. In our study, we found that NO application exhibits the strong influence on phenolic biosynthesis, which effect even manifested in MN group. We deduced that NO might act as an internal signal and mediate secondary metabolism in plant cells [54]. However, the underlying mechanism of phenolic compounds and NO needs to be seen.

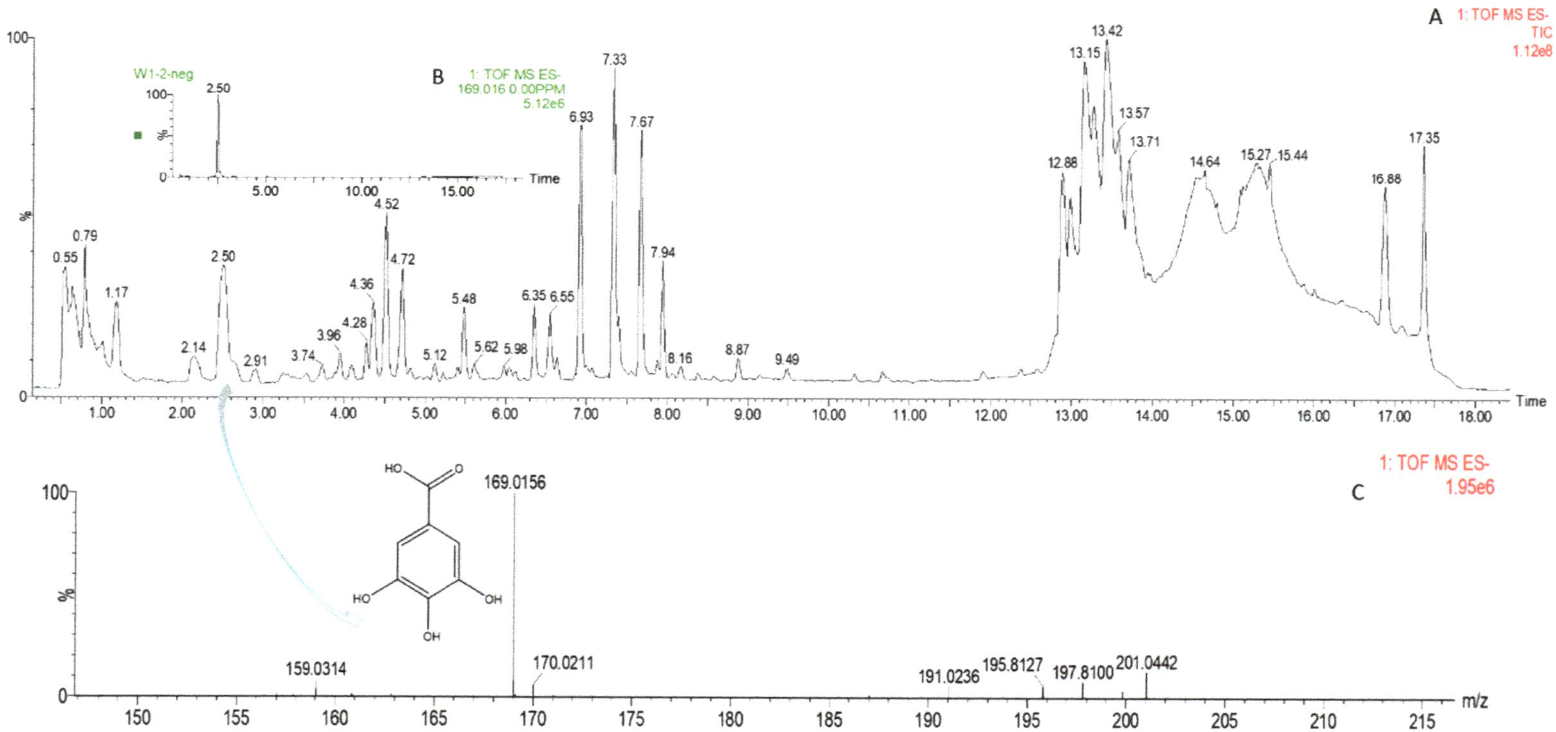

Figure 4. (**A**)Representative mass spectrogram of phenolic compounds from peach tissue extract from m/z 100 to 1000 (negative ionization mode); (**B**) mass spectrogram (MS) of galic acid at m/z 169.016; (**C**) MS/MS at 2.50 RM (negative ionization mode).

Table 2. Phenolic compounds analyzed by LC/MS.

Peak	Proposed Compounds	Catagory	RT (min)	Neutral Mass (Da)	[a] [M+H]/[b] [M-H] (m/z)	Mass Error (Ppm)	Formula	Fragment Number	Relative Amount of Phenolic Compounds [c]
1	Pigment A	Anthocyanins	4.09	609.1615	610.1688 [a]	0.1	C31H29O13	22	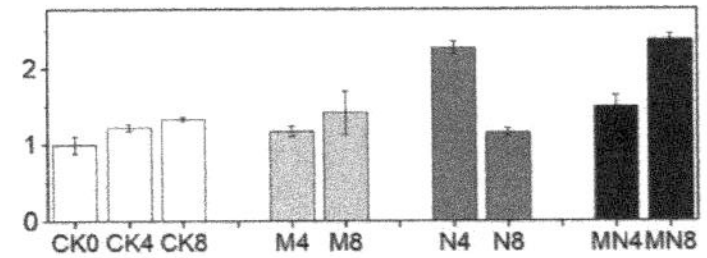
2	Peonidin 3-O-(6″-p-coumaroyl-glucoside)	Anthocyanins	4.82	609.1622	610.1695 [a]	−0.2	C31H29O13	37	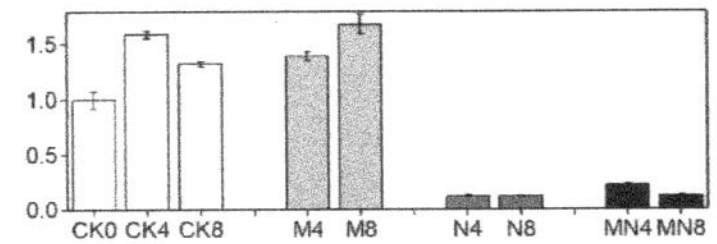
3	Cyanidin 3-O-xylosyl-rutinoside	Anthocyanins	9.64	727.2090	728.2163 [a]	0.5	C32H39O19	7	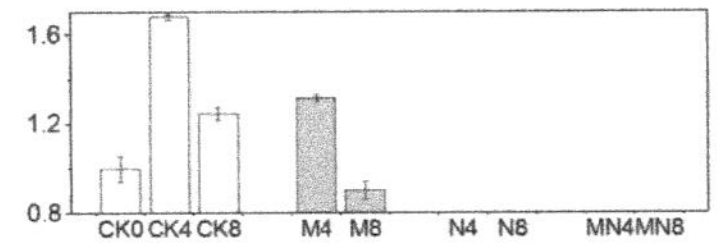
4	Pelargonidin 3-O-rutinoside	Anthocyanins	9.70	579.1731	580.1804 [a]	1.2	C27H31O14	6	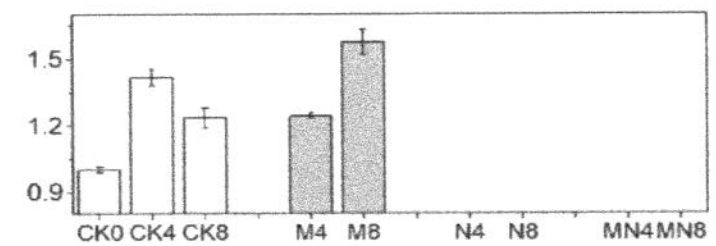
5	Naringenin 4′-O-glucuronide	Flavanones	3.94	448.1003	449.1076 [a]	0	C21H20O11	27	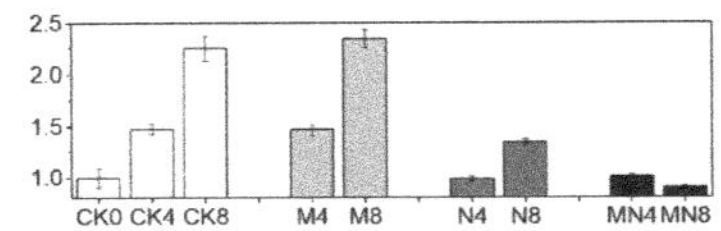
6	Poncirin	Flavanones	9.57	594.1942	595.2014 [a]	−0.3	C28H34O14	19	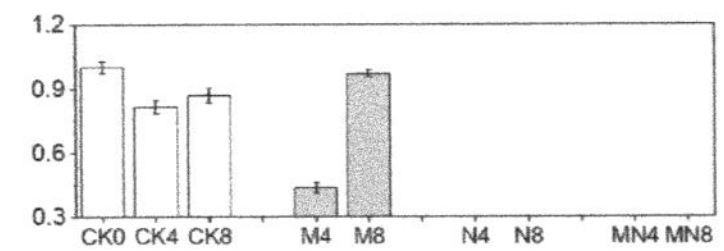

Table 2. *Cont.*

Peak	Proposed Compounds	Catagory	RT (min)	Neutral Mass (Da)	[a] [M+H]/[b] [M-H] (m/z)	Mass Error (Ppm)	Formula	Fragment Number	Relative Amount of Phenolic Compounds [c]
7	(+)-Gallocatechin	Flavanols	4.23	306.0725	307.0797 [a]	0.1	C15H14O7	6	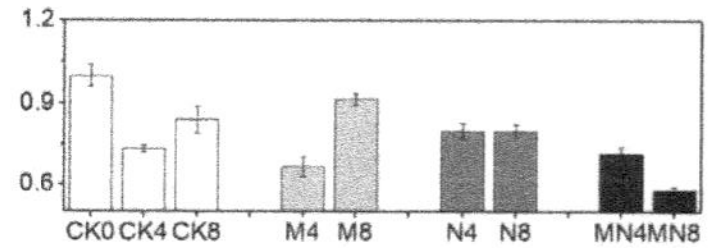
8	(+)-Catechin	Flavanols	3.46	290.0788	291.0861 [a]	−0.2	C15H14O6	3	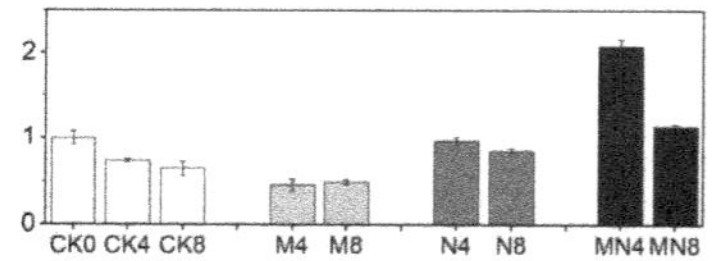
9	Luteolin 7-O-glucuronide	Flavones	0.92	462.0811	463.0883 [a]	1.4	C21H18O12	17	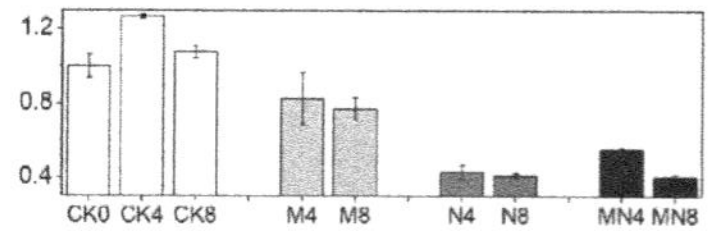
10	Kaempferol	Flavones	3.94	286.0479	287.0738 [a]	0	C15H10O6	5	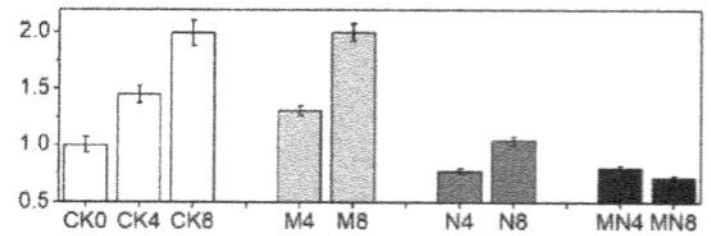
11	Kaempferol 3-O-galactoside	Flavones	3.95	448.1006	449.1079 [a]	0	C21H20O11	23	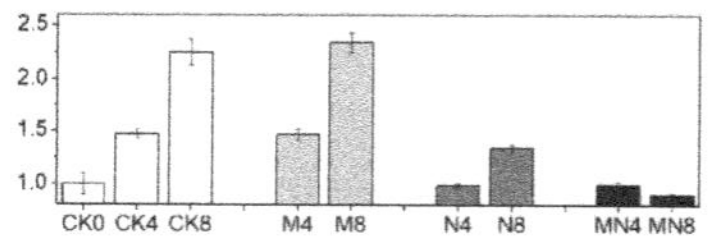
12	Apigenin 7-O-glucoside	Flavones	4.39	432.1066	433.1139 [a]	0.1	C21H20O10	17	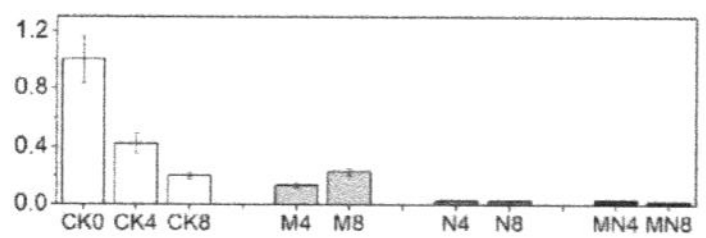

Table 2. *Cont.*

Peak	Proposed Compounds	Catagory	RT (min)	Neutral Mass (Da)	[a] [M+H]/[b] [M-H] (m/z)	Mass Error (Ppm)	Formula	Fragment Number	Relative Amount of Phenolic Compounds [c]
13	Luteolin 7-O-(2-apiosyl -6-malonyl)-glucoside	Flavones	9.48	666.1414	667.1486 [a]	−2.8	$C_{29}H_{30}O_{18}$	22	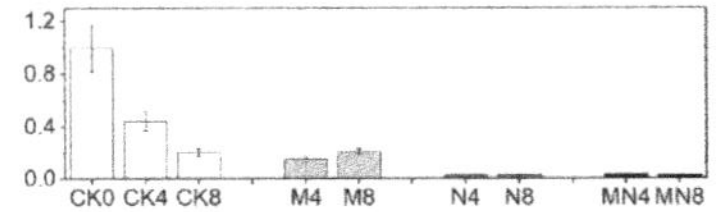
14	5,3′,4′-Trihydroxy-3-methoxy-6:7-methylenedioxy -flavone 4′-O-glucuronide	Flavonols	1.32	520.0844	521.0917 [a]	0.2	$C_{23}H_{20}O_{14}$	9	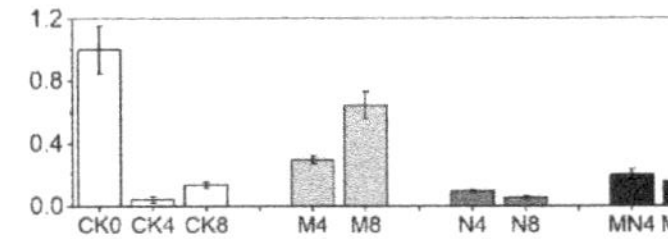
15	Isorhamnetin 3-O-glucoside 7-O-rhamnoside	Flavonols	4.84	624.1701	625.1774 [a]	1.2	$C_{28}H_{32}O_{16}$	40	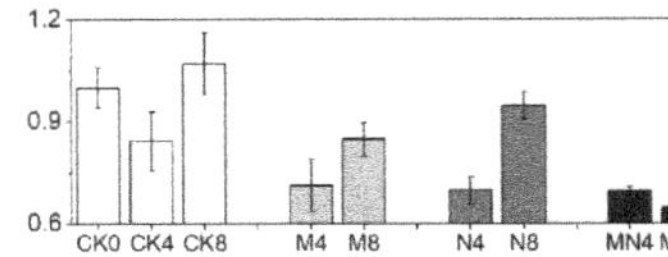
16	Gallic acid	Phenolic acids	2.50	170.0210	169.0156 [b]	0.1	$C_7H_6O_5$	2	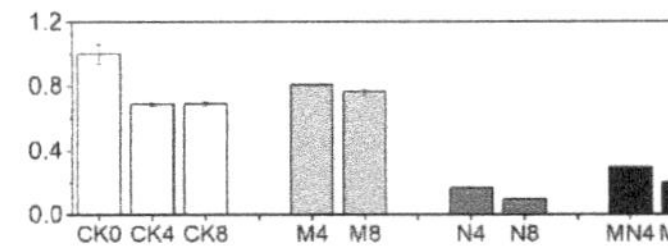
17	4-Hydroxybenzoic acid 4-O-glucoside	Phenolic acids	3.38	323.0734	299.0841 [b]	−0.2	$C_{13}H_{16}O_8$	10	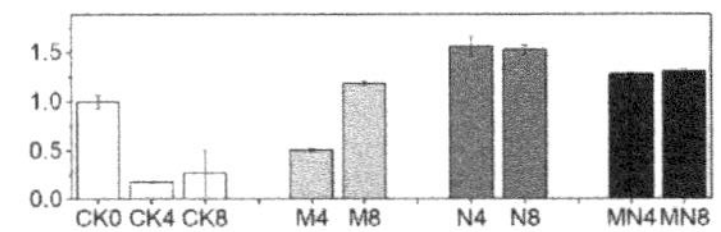
18	3-Caffeoylquinic acid	Phenolic acids	3.55	354.0945	353.1012 [b]	−0.6	$C_{16}H_{18}O_9$	8	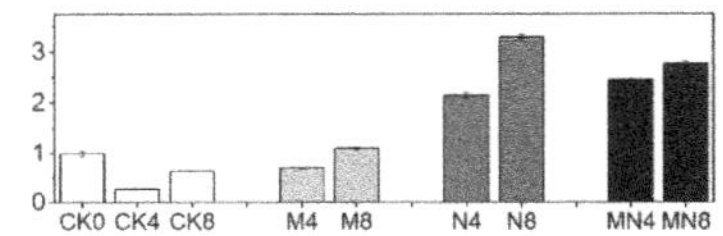

Table 2. *Cont.*

Peak	Proposed Compounds	Catagory	RT (min)	Neutral Mass (Da)	[a] [M+H]/[b] [M-H] (m/z)	Mass Error (Ppm)	Formula	Fragment Number	Relative Amount of Phenolic Compounds [c]
19	3-Feruloylquinic acid	Phenolic acids	4.36	368.1102	367.1101 [b]	−0.4	C17H20O9	13	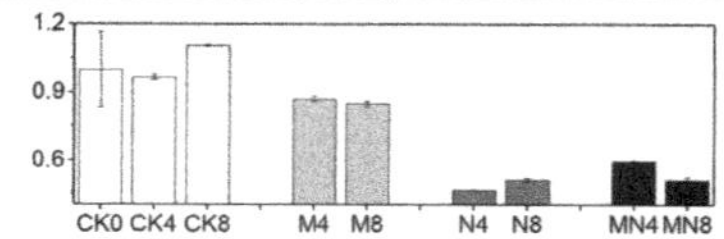
20	Caffeoyl glucose	Phenolic acids	7.44	342.0935	341.0570 [b]	1	C15H18O9	7	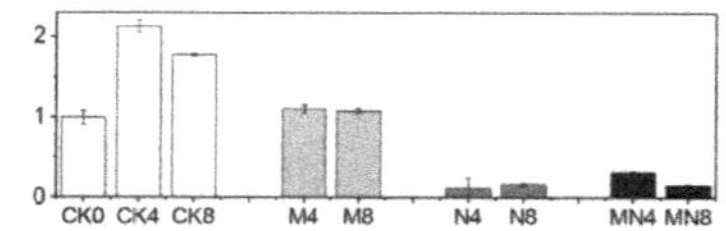

[a]: Positive ionization mode in LC/MS analysis; [b]: negative ionization mode in LC/MS analysis; [c]: Relative content of phenolics are expressed according to the peak value of each compound at D0, which are set to 1. Values are the mean ± SE from 3 replicates. CK0, CK4 and CK8 mean samples taken from day 0, day 4 and day 8 of CK group; M4 and M8 mean samples taken from day 4 and day 8 of M group; N4 and N8 mean samples taken from day 4 and day 8 of N group; MN4 and MN8 mean samples taken from day 4 and day 8 of MN group.

4. Conclusions

The combined 1-MCP and NO treatment showed the best effect on the improvement of postharvest fruit quality by maintaining good physical characteristics, decelerating fruit firmness, inhibiting ROS production, activating antioxidant enzymes and thus, postponing fruit ripening and senescence. NO application can extent peach shelf-life mainly by stimulating antioxidant enzymes. Moreover, NO application showed a greater effect on phenolic synthesis than 1-MCP. Regardless of the mode of action of NO and 1-MCP, for 'Xiahui-8′ peach, 1-MCP represents a more effective commercial option to inhibit senescence than NO treatment. Treatments with 1-MCP can enhance PAL and POD metabolism by activating via transcription upregulation the expression of *PpaPAL* and *PpaPOD* separately, while playing a lesser role in modulating the expression of *PpaCAT*, *PpaSOD* and *PpaAPX*. MN treatment manifested highest firmness, antioxidant enzymatic activities and lowest ROS content compared with 1-MCP or NO treatment alone. This study provides informative physiological, biochemical and molecular evidence for the benefits of using the combined 1-MCP and NO treatment on peach fruit due to a functional synergy between these two chemical agents.

Supplementary Materials: The following are available online at https://www.mdpi.com/article/10.3390/foods10122956/s1, Table S1: Primers used for quantification of mRNA levels by qRT-PCR.

Author Contributions: Writing—original draft preparation, X.W. (Xiaoqin Wu); data curation, J.Y. and X.W. (Xiaoqing Wang); resources, M.Y.; formal analysis, R.M.; supervision, Z.Y. All authors have read and agreed to the published version of the manuscript.

Funding: This research was funded by the National Natural Science Foundation of China (No. 32101860) and Natural Science Foundation of the Jiangsu Higher Education Institutions of China (No. 21KJD550001).

Institutional Review Board Statement: Not applicable.

Informed Consent Statement: Not applicable.

Data Availability Statement: Not applicable.

Acknowledgments: This study was supported in part by the National Natural Science Foundation of China (32101860) and Natural Science Foundation of the Jiangsu Higher Education Institutions of China (No. 21KJD550001). The authors also would like to show thanks to the anonymous reviewers and editors for their precious advices.

Conflicts of Interest: The authors declare no conflict of interest.

References

1. Jiang, L.; Zhang, L.; Shi, Y.; Lu, Z.X.; Yu, Z.F. Proteomic analysis of peach fruit during ripening upon post-harvest heat combined with 1-MCP treatment. *J. Proteom.* **2014**, *98*, 31–43. [CrossRef] [PubMed]
2. Kang, R.Y.; Zhang, L.; Jiang, L.; Yu, M.L.; Ma, R.J.; Yu, Z.F. Effect of postharvest nitric oxide treatment on the proteome of peach fruit during ripening. *Postharvest Biol. Technol.* **2016**, *112*, 277–289. [CrossRef]
3. Cai, H.F.; An, X.J.; Han, S.; Jiang, L.; Yu, M.L.; Ma, R.J.; Yu, Z.F. Effect of 1-MCP on the production of volatiles and biosynthesis-related gene expression in peach fruit during cold storage. *Postharvest Biol. Technol.* **2018**, *141*, 50–57. [CrossRef]
4. Neill, S.J.; Desikan, R.; Clarke, A.; Hancock, J.T. Nitric oxide is a novel component of abscisic acid signaling in stomatal guard cells. *Plant Physiol.* **2002**, *128*, 13–16. [CrossRef]
5. Han, S.; Cai, H.F.; An, X.J.; Huan, C.; Wu, X.Q.; Jiang, L.; Yu, M.L.; Ma, R.J.; Yu, Z.F. Effect of nitric oxide on sugar metabolism in peach fruit (cv. Xiahui 6) during cold storage. *Postharvest Biol. Technol.* **2018**, *142*, 72–80. [CrossRef]
6. Ren, Y.; Xue, Y.; Tian, D.; Zhang, L.; Xiao, G.; He, J. Improvement of Postharvest Anthracnose Resistance in Mango Fruit by Nitric Oxide and the Possible Mechanisms Involved. *J. Agric. Food Chem.* **2020**, *68*, 15460–15467. [CrossRef]
7. Flores, F.B.; Sanchez-Bel, P.; Valdenegro, M.; Romojaro, F.; Martinez-Madrid, M.C.; Egea, M.I. Effects of a pretreatment with nitric oxide on peach (*Prunus persica* L.) storage at room temperature. *Eur. Food Res. Technol.* **2008**, *227*, 1599–1611. [CrossRef]
8. Huan, C.; Jiang, L.; An, X.J.; Kang, R.Y.; Yu, M.L.; Ma, R.J.; Yu, Z.F. Potential role of glutathione peroxidase gene family in peach fruit ripening under combined postharvest treatment with heat and 1-MCP. *Postharvest Biol. Technol.* **2016**, *111*, 175–184. [CrossRef]

9. Al Ubeed, H.M.S.; Wills, R.B.H.; Bowyer, M.C.; Golding, J.B. Comparison of hydrogen sulphide with 1-methylcyclopropene (1-MCP) to inhibit senescence of the leafy vegetable, pak choy. *Postharvest Biol. Technol.* **2018**, *137*, 129–133. [CrossRef]
10. Arasimowicz, M.; Floryszak-Wieczorek, J. Nitric oxide as a bioactive signalling molecule in plant stress responses. *Plant Sci.* **2007**, *172*, 876–887. [CrossRef]
11. Grozeff, G.E.G.; Alegre, M.L.; Senn, M.E.; Chaves, A.R.; Simontacchi, M.; Bartoli, C.G. Combination of nitric oxide and 1-MCP on postharvest life of the blueberry (*Vaccinium* spp.) fruit. *Postharvest Biol. Technol.* **2017**, *133*, 72–80. [CrossRef]
12. Chen, H.; An, X.; Yu, M.; Li, J.; Ma, R.; Tu, M.; Yu, Z. Effect of combined heat and 1-MCP treatment on the quality and antioxidant level of peach fruit during storage. *Postharvest Biol. Technol.* **2018**, *145*, 193–202.
13. Wu, X.Q.; Mason, A.M.; Yu, M.L.; Ma, R.J.; Yu, Z.F. Quantitative proteomic analysis of pre- and post-harvest peach fruit ripening based on iTRAQ technique. *Acta Physiol. Plant* **2017**, *39*, 181. [CrossRef]
14. Wu, X.Q.; Jiang, L.; Yu, M.L.; An, X.J.; Ma, R.J.; Yu, Z.F. Proteomic analysis of changes in mitochondrial protein expression during peach fruit ripening and senescence. *J. Proteom.* **2016**, *147*, 197–211. [CrossRef]
15. Zhang, Z.Q.; Pang, X.Q.; Duan, X.W.; Ji, Z.L.; Jiang, Y.M. Role of peroxidase in anthocyanin degradation in litchi fruit pericarp. *Food Chem.* **2005**, *90*, 47–52. [CrossRef]
16. Yingsanga, P.; Srilaong, V.; Kanlayanarat, S.; Noichinda, S.; McGlasson, W.B. Relationship between browning and related enzymes (PAL, PPO and POD) in rambutan fruit (*Nephelium lappaceum* Linn.) cvs. Rongrien and See-Chompoo. *Postharvest Biol. Technol.* **2008**, *50*, 164–168. [CrossRef]
17. Song, H.W.; Yuan, W.M.; Jin, P.; Wang, W.; Wang, X.F.; Yang, L.M.; Zhang, Y.F. Effects of chitosan/nano-silica on postharvest quality and antioxidant capacity of loquat fruit during cold storage. *Postharvest Biol. Technol.* **2016**, *119*, 41–48. [CrossRef]
18. Assis, J.S.; Maldonado, R.; Munoz, T.; Escribano, M.I.; Merodio, C. Effect of high carbon dioxide concentration on PAL activity and phenolic contents in ripening cherimoya fruit. *Postharvest Biol. Technol.* **2001**, *23*, 33–39. [CrossRef]
19. Bradford, M.M. A rapid and sensitive method for the quantitation of microgram quantities of protein utilizing the principle of protein-dye binding. *Anal Biochem.* **1976**, *72*, 248–254. [CrossRef]
20. Huan, C.; Han, S.; Jiang, L.; An, X.J.; Yu, M.L.; Xu, Y.; Ma, R.J.; Yu, Z.F. Postharvest hot air and hot water treatments affect the antioxidant system in peach fruit during refrigerated storage. *Postharvest Biol. Technol.* **2017**, *126*, 1–14. [CrossRef]
21. Tong, Z.; Gao, Z.; Wang, F.; Zhou, J.; Zhang, Z. Selection of reliable reference genes for gene expression studies in peach using real-time PCR. *BMC Mol. Biol.* **2009**, *10*, 71. [CrossRef]
22. Wu, X.; An, X.; Yu, M.; Ma, R.; Yu, Z. 1-Methylcyclopropene Treatment on Phenolics and the Antioxidant System in Postharvest Peach Combined with the Liquid Chromatography/Mass Spectrometry Technique. *J. Agric. Food Chem.* **2018**, *66*, 6364–6372. [CrossRef] [PubMed]
23. Sisler, E.C.; Dupille, E.; Serek, M. Effect of 1-methylcyclopropene and methylenecyclopropane on ethylene binding and ethylene action on cut carnations. *Plant Growth Regul.* **1996**, *18*, 79–86. [CrossRef]
24. Crisosto, C.H.; Day, K.R.; Crisosto, G.M.; Garner, D. Quality attributes of white flesh peaches and nectarines grown under California conditions. *J. Amer. Pomol. Soc.* **2001**, *55*, 45–51.
25. Lu, G.H.; Li, C.J.; Lu, Z.H. Wound-Induced Respiration in Thin Slice of Chinese Jujube Fruit. *J. Plant Physiol.* **1993**, *141*, 115–119. [CrossRef]
26. Dal Cin, V.; Rizzini, F.M.; Botton, A.; Tonutti, P. The ethylene biosynthetic and signal transduction pathways are differently affected by 1-MCP in apple and peach fruit. *Postharvest Biol. Technol.* **2006**, *42*, 125–133. [CrossRef]
27. Zhu, S.H.; Liu, M.C.; Zhou, J. Inhibition by nitric oxide of ethylene biosynthesis and lipoxygenase activity in peach fruit during storage. *Postharvest Biol. Technol.* **2006**, *42*, 41–48. [CrossRef]
28. Sabban-Amin, R.; Feygenberg, O.; Belausov, E.; Pesis, E. Low oxygen and 1-MCP pretreatments delay superficial scald development by reducing reactive oxygen species (ROS) accumulation in stored 'Granny Smith' apples. *Postharvest Biol. Technol.* **2011**, *62*, 295–304. [CrossRef]
29. Xu, X.; Lei, H.; Ma, X.; Lai, T.; Song, H.; Shi, X.; Li, J. Antifungal activity of 1-methylcyclopropene (1-MCP) against anthracnose (*Colletotrichum gloeosporioides*) in postharvest mango fruit and its possible mechanisms of action. *Int. J. Food Microbiol.* **2017**, *241*, 1–6. [CrossRef]
30. Zhao, Y.; Zhu, X.; Hou, Y.; Wang, X.; Li, X. Postharvest nitric oxide treatment delays the senescence of winter jujube (*Zizyphus jujuba* Mill. cv. *Dongzao*) fruit during cold storage by regulating reactive oxygen species metabolism. *Sci. Hortic.* **2020**, *261*, 109009. [CrossRef]
31. Lin, Y.; Chen, G.; Lin, H.; Lin, M.; Wang, H.; Lin, Y. Chitosan postharvest treatment suppresses the pulp breakdown development of longan fruit through regulating ROS metabolism—ScienceDirect. *Int. J. Biol. Macromol.* **2020**, *165*, 601–608. [CrossRef] [PubMed]
32. Mittler, R. Oxidative stress, antioxidants and stress tolerance. *Trends Plant Sci.* **2002**, *7*, 405–410. [CrossRef]
33. Zhang, D.; Xu, X.; Zhang, Z.; Jiang, G.; Feng, L.; Duan, X.; Jiang, Y. 6-Benzylaminopurine improves the quality of harvested litchi fruit. *Postharvest Biol. Technol.* **2018**, *143*, 137–142. [CrossRef]
34. Hu, H.; Liu, Y.; Shi, G.L.; Liu, Y.P.; Wu, R.J.; Yang, A.Z.; Wang, Y.M.; Hua, B.G.; Wang, Y.N. Proteomic analysis of peach endocarp and mesocarp during early fruit development. *Physiol. Plant* **2011**, *142*, 390–406. [CrossRef] [PubMed]
35. Pandey, V.P.; Singh, S.; Jaiswal, N.; Awasthi, M.; Pandey, B.; Dwivedi, U.N. Papaya fruit ripening: ROS metabolism, gene cloning, characterization and molecular docking of peroxidase. *J. Mol. Catal. B Enzym.* **2013**, *98*, 98–105. [CrossRef]

36. Tang, R.X.; Yong-Hong, G.E.; Can-Ying, L.I.; Sun, R.H. Effect of Brassinolide Treatment on the Active Oxygen Metabolism of Postharvest 'Okubao' Peach. *Storage Process* **2016**, *16*, 5–9.
37. Shi, T.; Li, Z.; Zhang, Z.; Zhang, C.; Gao, Z. Effect of 1-methylcyclopropene (1- MCP) treatment on antioxidant enzymes of postharvest Japanese apricot. *Afr. J. Biotechnol.* **2013**, *12*, 689–694.
38. Clark, D.; Durner, J.; Navarre, D.A.; Klessig, D.F. Nitric oxide inhibition of tobacco catalase and ascorbate peroxidase. *Mol. Plant-Microbe Interact.* **2000**, *13*, 1380–1384. [CrossRef]
39. Zhu, S.H.; Sun, L.; Liu, M.C.; Zhou, J. Effect of nitric oxide on reactive oxygen species and antioxidant enzymes in kiwifruit during storage. *J. Sci. Food Agric.* **2008**, *88*, 2324–2331. [CrossRef]
40. Huan, C.; Jiang, L.; An, X.J.; Yu, M.L.; Xu, Y.; Ma, R.J.; Yu, Z.F. Potential role of reactive oxygen species and antioxidant genes in the regulation of peach fruit development and ripening. *Plant Physiol. Biochem.* **2016**, *104*, 294–303. [CrossRef]
41. Yang, N.; Wang, C.; Chen, X.; Yishen, L.I.; Zhang, X. Effect of Drought Stress on Antioxidant System and Reactive Oxygen in Chorispora bungeana Plantlets in vitro. *Acta Bot. Boreali Occident. Sin.* **2014**, *34*, 2483–2490.
42. Dixon, R.A.; Paiva, N.L. Stress-Induced Phenylpropanoid Metabolism. *Plant Cell* **1995**, *7*, 1085–1097. [CrossRef]
43. Zheng, X.L.; Hu, B.; Song, L.J.; Pan, J.; Liu, M.M. Changes in quality and defense resistance of kiwifruit in response to nitric oxide treatment during storage at room temperature. *Sci. Hortic.* **2017**, *222*, 187–192. [CrossRef]
44. Kovacik, J.; Klejdus, B.; Backor, M. Nitric oxide signals ROS scavenger-mediated enhancement of PAL activity in nitrogen-deficient Matricaria chamomilla roots: Side effects of scavengers. *Free Radic. Biol. Med.* **2009**, *46*, 1686–1693. [CrossRef]
45. Lai, T.F.; Wang, Y.Y.; Li, B.Q.; Qin, G.Z.; Tian, S.P. Defense responses of tomato fruit to exogenous nitric oxide during postharvest storage. *Postharvest Biol. Technol.* **2011**, *62*, 127–132. [CrossRef]
46. Shao, X.F.; Wang, H.F.; Xu, F.; Cheng, S. Effects and possible mechanisms of tea tree oil vapor treatment on the main disease in postharvest strawberry fruit. *Postharvest Biol. Technol.* **2013**, *77*, 94–101. [CrossRef]
47. Cao, S.F.; Zheng, Y.H.; Wang, K.T.; Rui, H.J.; Shang, H.T.; Tang, S.S. The effects of 1-methylcyclopropene on chilling and cell wall metabolism in loquat fruit. *J. Hortic. Sci. Biotechnol.* **2010**, *85*, 147–153. [CrossRef]
48. Ozkaya, O.; Yildirim, D.; Dundar, O.; Tukel, S.S. Effects of 1-methylcyclopropene (1-MCP) and modified atmosphere packaging on postharvest storage quality of nectarine fruit. *Sci. Hortic.* **2016**, *198*, 454–461. [CrossRef]
49. Constabel, C.P.; Bergey, D.R.; Ryan, C.A. Systemin activates synthesis of wound-inducible tomato leaf polyphenol oxidase via the octadecanoid defense signaling pathway. *Proc. Natl. Acad. Sci. USA* **1995**, *92*, 407–411. [CrossRef] [PubMed]
50. Sanchez-Ballesta, M.T.; Lafuente, M.T.; Zacarias, L.; Granell, A. Involvement of phenylalanine ammonia-lyase in the response of Fortune mandarin fruits to cold temperature. *Physiol. Plant* **2000**, *108*, 382–389. [CrossRef]
51. Liu, X.; Cui, X.; Ji, D.; Zhang, Z.; Tian, S. Luteolin-induced activation of the phenylpropanoid metabolic pathway contributes to quality maintenance and disease resistance of sweet cherry. *Food Chem.* **2020**, *342*, 128309. [CrossRef] [PubMed]
52. Huang, Y.; Peng, W.; Zhang, Y.; Yuanyuan, L.I.; Wang, L.; Shan, T.; Jin, P.; Zheng, Y. Effects of Nitric Oxide Treatment on Quality and Phenolic Metabolism in Strawberry Fruit. *J. Nucl. Agric. Sci.* **2016**, *30*, 1959–1966.
53. Zhou, S.J. Effects of nitric oxide fumigation on phenolic metabolism of postharvest Chinese winter jujube (*Zizyphus jujuba* Mill. cv. *Dongzao*) in relation to fruit quality. *LWT—Food Sci. Technol.* **2009**, *42*, 1009–1014. [CrossRef]
54. Dong, J.; Ming, Z.; Lu, L.; Sun, U.; Xu, M. Nitric oxide fumigation stimulates flavonoid and phenolic accumulation and enhances antioxidant activity of mushroom. *Food Chem.* **2012**, *135*, 1220–1225. [CrossRef] [PubMed]

foods

MDPI

Article

Application of the Hurdle Technology Concept to the Fresh Za'atar (*Origanum syriacum*) Preservation

Samer Mudalal [1,*], Doaa Kanan [1], Ola Anabtawi [1], Alma Irshaid [1], Mohammed Sabbah [1], Munqez Shtaya [2], Faisal Shraim [2] and Gianluigi Mauriello [3]

1 Department of Nutrition and Food Technology, Faculty of Agriculture and Veterinary Medicine, An-Najah National University, Nablus P.O. Box 707, Palestine
2 Department of Plant Production and Protection, Faculty of Agriculture and Veterinary Medicine, An-Najah National University, Nablus P.O. Box 707, Palestine
3 Department of Agricultural Science, University of Naples Federico II, 80055 Portici, Italy
* Correspondence: samer.mudalal@najah.edu

Abstract: Oregano (*Origanum syriacum*) is popularly called *za'atar* in the Middle East region. It is widely used in the Mediterranean diet as an aromatic herb. This study aimed to evaluate the preservation effect of natural additives, vacuum packaging, and refrigeration on the quality traits of fresh oregano. In total, 132 fresh oregano samples were formulated and split into 4 groups (n = 33) labeled group A (100% fresh oregano leaves, Control), group B (fresh oregano 63.2%, 15% fresh onion, 20% oil, 1.8% salt), group C (fresh oregano 61.91%, 15% fresh *Allium cepa*, 20% oil, 1.8% salt, 1.29% sumac), and group D (fresh oregano 59.2%, 15% fresh *Allium cepa*, 20% corn oil, 1.8% salt, 4% lactic acid, ultimate pH 4.4). Different quality traits such as color index (L*a*b*), microbiological analysis (total aerobic, anaerobic, and psychrotrophic bacteria and yeasts and molds), and sensory features (taste, flavor, appearance, saltiness, and overall acceptance) were assessed during the storage period (42 days) for all groups. Our study showed that the addition of lactic acid (group D) exhibited a strong preservation effect against aerobic and anaerobic bacteria. In this context, group D had significantly lower aerobic and anaerobic bacterial counts (5.12 vs. 6.7, 6, and 6.7 log (cfu/g); $p < 0.05$) and (4.75 vs. 6.6, 6.1, 6.77 (cfu/g); $p < 0.05$) than group A, B, and C; respectively. Group D exhibited significantly ($p < 0.05$) lower psychrotrophic bacterial count (3.6 log (cfu/g)) during the whole period of storage compared with control. Group B had a lower redness index (a*) (−3.3 vs. −1.8, −1.65, −1.23; $p < 0.05$) than groups A, C, and D; respectively. In conclusion, our study showed that there is a possibility of improving the preservation of oregano (*Origanum syriacum*) by using lactic acid and sumac combined with vacuum packaging under refrigeration conditions.

Keywords: oregano; vacuum packaging; *Allium cepa*; *Rhus coriaria*; refrigeration

Citation: Mudalal, S.; Kanan, D.; Anabtawi, O.; Irshaid, A.; Sabbah, M.; Shtaya, M.; Shraim, F.; Mauriello, G. Application of the Hurdle Technology Concept to the Fresh Za'atar (*Origanum syriacum*) Preservation. *Foods* **2022**, *11*, 3002. https://doi.org/10.3390/foods11193002

Academic Editors: Maria Cefola and Bernardo Pace

Received: 9 August 2022
Accepted: 14 September 2022
Published: 27 September 2022

1. Introduction

Origanum species are widely grown and cultivated in the Mediterranean basin. Out of the 43 species, 35 species occur solely in the East Mediterranean [1,2]. *Za'atar* is one the most common names for *Origanum syriacum* [3], and it is frequently used for different purposes and in different forms in Mediterranean cuisine, due to its fragrance, flavor, and properties that it adds to food, such as meat, vegetables, and baked goods [4].

Oregano has been used in the past to formulate a traditional medicine for several diseases. In this context, the traditional medicine containing oregano was dedicated to treating respiratory diseases such as whooping cough, bronchitis, and asthma [5]. The most common functional ingredients in oregano leaves are thymol and carvacrol [6]. *Origanum syriacum* leaves are considered a rich source of vitamins such as folic acid, beta carotene, and vitamins E, K, A, and C [7,8]. Its leaves are rich in phenolic compounds such as caffeic acid, rosmarinic acid, eriodyctiol, luteolin, naringenin, and apigenin [9–11]. Spain, France, Italy,

Switzerland, Bulgaria, Portugal, and Greece among other European countries cultivate and harvest wild thyme and oregano [12].

The oregano crop (*Oregano syriacum*) is a major component of the agricultural economy in Palestine; it is also called Palestine's green gold. Palestinians have known different types of oregano, especially wild oregano, and used it in their traditional cuisine. A few years ago, Palestinians realized how much its cultivation is due to its economic feasibility [13].

Thymol is one of the components of the essential oil of oregano that has antiseptic and antifungal properties [14,15]. Moreover, oregano has other volatile oils such as carvacrol, geraniol, and borneol that have antimicrobial properties. In addition, thyme contains linalool, apigenin, eugenol, and rosmarinic acid, which have antioxidant, anti-inflammatory and antiviral properties [10,11,14–16]. The studies revealed that thyme and oregano contained a significant amount of phenolic compounds such as zea-xanthin, which acts as a bronchodilator; apigenin, a muscle relaxer; and lutein, which supports brain and vision development, as well aso luteolin and thymosin [5,16]. The diversity of geography and climate in Palestine contributes to farm-wide varieties of *Origanum syriacum* species including wild and cultivated [17]. Several researchers investigated the ability to preserve *Origanum syriacum* by vacuum packaging [18] or by solar and freeze drying [19]. Mudalal and Abu-Khalaf [20] found a significant difference in some quality traits between freeze- and solar-dried *Origanum syriacum* using an electronic nose.

The market for *Origanum syriacum* has different channels for rural families and small farmers who can sell to local trades or small stores. In addition, some companies have private labels. There are numerous challenges that limit the exporting of *Origanum syriacum* including restricted and complicated access to fertile land (in particular in area C: political classification by the Israeli occupation); lack of quality control on agricultural inputs and practices; limited water for irrigation; and weak cold chain (for fresh and frozen products), which result in high waste during transit. On the other hand, dried *Origanum syriacum* exports have rapidly increased because of high demand in the region.

Oregano is only marketed as fresh in winter and spring, as it exhibits an unacceptable bitter taste in the remaining seasons. To overcome shortages of production due to off-seasons, our local small industries employ several conventional preservation techniques such as solar-drying, freezing, and packaging in plastic bottles. Similar techniques are usually used in other Mediterranean regions. However, the current conventional preservation techniques have adverse effects on the nutritional value and sensory characteristics (mainly color, taste, and flavor). So far, current conventional preservation techniques are not able to maintain the freshness of oregano, which is a very important issue for the production of local oregano-based bakery products. In the Mediterranean region and particularly in Palestine, fresh oregano leaves are usually mixed with oil, fresh onion, sumac, and salt then stuffed in flour dough in the form of a thin sheet, followed by baking. Our research aimed to study the preservation effects of refrigeration storage, vacuum packaging, and natural ingredients on the quality traits of fresh oregano.

2. Materials and Methods

2.1. Collection and Preparation of Samples

A small oregano field was cultivated in a village near Tulkarem city (Northern Palestine) and was used to collect about 5 kg of oregano stems. The selected areas in the field for sample collection had no weeds or plant diseases. The leaves were removed from the stems by hand. The final net weight of separated leaves after the removal of straw, stems, gravel, and any type of physical impurities was about 2.5 kg. Ultimately, the oregano leaves were subjected to a cleaning process to remove any types of soils and dust using running tap water. The cleaning process was stopped when the output water became clear. The wetted leaves were left to dry at room temperature for about one hour over paper towels to remove excess water due to the washing process, and the drying process did not remove any part of the native water in the fresh leaves. The whole quantities of leaves were mixed thoroughly to obtain a homogenous mixture. The total quantity of oregano was split into four batches to represent four treatments. In each group, there were 33 packs. Each pack contains a net weight of about 60 g of the product. Accordingly, the total number of samples for all treatments was 132.

In each group, different natural ingredients were added to oregano leaves, and then the ingredients were mixed and packaged under vacuum (−95% ambient pressure for 10 s, Figure 1). Transparent vacuum plastic packs made of coextruded multilayer flexible film with dimensions of 200×300 mm. The groups were as follows:

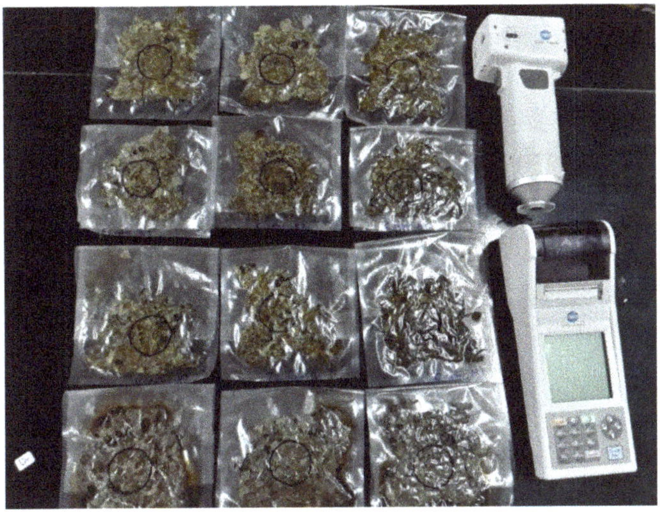

Figure 1. Oregano mix as the finished product was prepared for color index (L*a*b*) measurement.

Group A: (Control group): (100% Fresh oregano leaves)

Group B: (73.2% Fresh oregano, 15% Fresh *Allium cepa*, 20% corn oil, 1.8% salt)

Group C: (61.9% Fresh oregano, 15% Fresh *Allium cepa*, 20% corn oil, 1.8% salt, 1.29% sumac "*Rhus coriaria*")

Group D: (59.2% Fresh oregano, 15% Fresh *Allium cepa*, 20% corn oil, 1.8% salt, 4% lactic Acid, ultimate pH 4.4).

The percentages of ingredients were calculated by weight.

The samples were stored under refrigeration conditions (2–4 °C), and the relative humidity was about 40–60%.

2.2. Chemical Analysis

About 100 g of fresh oregano leaves was dedicated for proximate chemical analysis (moisture, fat, protein, fiber, and ash content). The proximate chemical composition of fresh oregano was assessed for each sample using official methods of AOAC [21]. Moisture content was determined by loss on drying using an air oven. About 5 g of the sample was accurately weighed and then dried at 105 °C for 16 h. The moisture content was calculated based on weight differences. For ash, about 2 g of fresh leaves of oregano was accurately weighed in the crucibles and incinerated in the muffle furnace for 5 h at 550 °C. After incineration, the weight of ash was recorded, and then ash content was calculated. The fat content was determined using an ANKOM XT15 extractor. About 1.5 g of fresh leaves of oregano was placed in the filter bags, then the fat extracted by petroleum ether in the extraction vessel. The fat content was calculated based on weight differences (before and after extraction). Fiber content was measured using an ANKOM 200 Fiber Analyzer. About 0.5 g of dried oregano leaves was placed in the filter bag. After that, the filter bags were placed in a fiber analyzer using the suspender tray. The samples were first digested using a heated acid solution (90 °C), followed by water washing and digestion using a heated alkali solution (sodium hydroxide). The remaining undigested matter was considered as total crude fiber.

2.3. Color Measurement

The CIE system (Commission Internationale de l'Eclairage) was used to measure color parameters according to standard values that are internationally used.

In each group, six different areas were highlighted with black circles over each pack. Color coordinates were measured in triplicates for each area, and the mean was determined. A reflectance colorimeter (Minolta Chroma Meter CR-400) with an illuminant source was used to measure the color index (L*a*b*). A reference white ceramic tile (Y = 93.9, x = 0.3130 and y = 0.3190) was used to calibrate the colorimeter before each measurement.

2.4. pH Measurement

Ten samples were selected from each group for pH measuring. About 2.5 g of fresh oregano was added to 25 mL of distilled water and then homogenized with ultra-turrax. The pH meter (ISFET, Model 98 # IQ150, IQ Scientific Instruments, San Diego, CA, USA) was calibrated at pH 4.0 and 7.0 before measuring the samples.

2.5. Microbiological Analysis

2.5.1. Aerobic, Anaerobic, and Psychrotrophic Bacterial Count

Total aerobic, anaerobic, and psychrotrophic bacteria counts were estimated in 12 replicates during the period of the study (42 days). 10 g of oregano samples was aseptically added with 90 mL ringer solution. Seven dilutions (10^{-1} to 10^{-7}) were used to count bacteria, mold, and yeast. Plate Count Agar (PCA) was used as a microbiological growth medium to estimate the total viable bacterial count. The plates were incubated for 48–72 h at 37 °C. Similar incubation conditions were used to count anaerobic bacteria, but the plates were kept in an anaerobic jar. Plate count agar (PCA) was also used to count psychrotrophic bacteria, and the plates were stored in a refrigerator for 7 days. The plates containing 25–250 colonies were considered for counting.

2.5.2. Yeast and Mold

From each group, four different samples were selected to determine yeast and mold counts. Potato dextrose agar (PDA) was used as a culture medium. Serial dilutions (10^{-1} to 10^{-5}) were used to obtain proper colony counts. The plates were incubated for 4–5 days at room temperature. The plates containing 25–250 colonies were considered for counting.

2.6. Sensory Analysis

Three packs of oregano samples were selected from each group to prepare bread stuffed with oregano mix as normally prepared in local bakeries that are already present in Palestine. The dough for the bread was made from wheat flour, dried yeast, salt, and a little sugar to activate the yeast. The dough was rolled into a thin layer, then stuffed with oregano mix and baked at 250 °C until the formation of a gold crust. Bread with oregano (*qraas za'atar*) was cut into small similar pieces. These pieces were subjected to sensory analysis by 30 panelists (trained) to evaluate 5 descriptors, taste, flavor, appearance, saltiness, and overall acceptance. The panelists were trained before they started the sensory analysis by explaining the scale hedonic test (in Arabic). The panelists expressed the intensity of each attribute with a mark on a 9-point scale (9 = Like Extremely; 1 = Dislike Extremely). The samples were coded in randomized blocks and presented to the panelists on plastic plates.

2.7. Statistical Analysis

Each quality trait (proximate analysis, color traits (L*a*b*), pH, sensory traits, and microbiological counts) was measured in replicates (4–12 replicates depending on the trait) during the storage period (42 days). The pooled effect of treatments during the storage period on quality traits (chemical, physical, and microbiological properties) was evaluated by ANOVA. The model investigated the main effects of treatments as well as the interaction effect (between treatments and the time of storage) using the general linear model (GLM) in SPSS version 18.0 (SPSS Inc., Chicago, IL, USA) for the main quality traits of fresh oregano. The means were separated using Tukey's range test, with $p \leq 0.05$ considered significant.

3. Results

3.1. Chemical Compositions of Fresh Oregano

Table 1 shows the proximate chemical analysis of fresh oregano including moisture, ash, fat, and fiber constituents.

Table 1. Proximate chemical composition of fresh oregano leaves.

	Composition	Mean ± SD (g/100 g)
1	Moisture	66.9 ± 0.42
2	Fiber	20.51 ± 1.32
3	Proteins	0.71 ± 0.05
4	Ash	5.50 ± 0.22
5	Fat	6.33 ± 0.22

3.2. pH Measurement Analysis

pH was measured during the refrigerated storage period (Figure 2) to understand the effect of microbial growth on pH. The initial pH for group A was significantly higher than the other groups, which represented the natural pH of oregano leaves (there were no additives). Group D exhibited the lowest significant value of all groups. There was no significant difference between groups B and C. The addition of sumac led to a reduction of about 0.3 units of pH, while the addition of lactic acid to group D caused a reduction in pH of about 2.8 units in comparison with control group A. In general, there were no significant changes in pH during the storage period in groups A, B, and C.

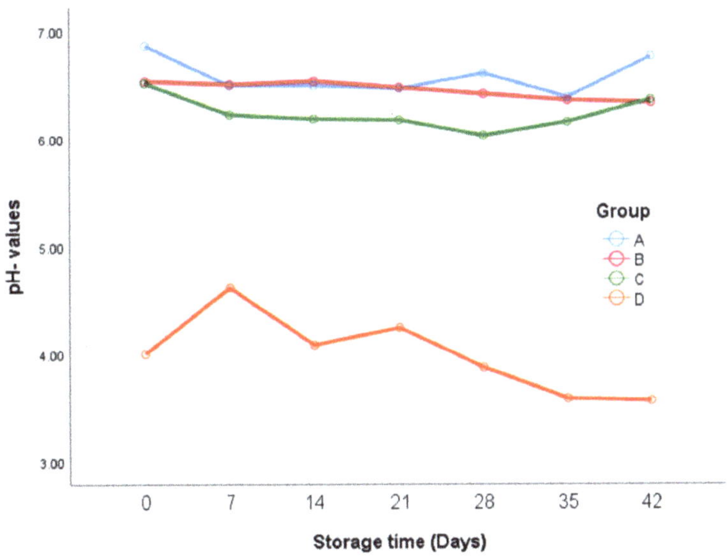

Figure 2. pH of fresh oregano at refrigerator temperature 4 °C during storage period (42 days). Group A: (Control group): (100% Fresh oregano leaves), Group B: (73.2% Fresh oregano, 15% Fresh *Allium cepa*, 20% corn oil, 1.8% salt), Group C: (61.9% Fresh oregano, 15% Fresh *Allium cepa*, 20% corn oil, 1.8% salt, 1.29% sumac "*Rhus coriaria*"), Group D: (59.2% Fresh oregano, 15% Fresh *Allium cepa*, 20% corn oil, 1.8% salt, 4% lactic Acid, ultimate pH 4.4.

3.3. Microbiological Analysis

Total plate count was used as an indicator of bacterial aerobic populations in different groups. It is not a measure of the entire bacterial population but rather a generic test for organisms that grow aerobically at mesophilic temperatures (25 to 40 °C). Figure 3 shows the growth of total bacteria during the storage period (42 days). There was an interaction between the effects of treatments and the effects of storage time. Groups A and C exhibited significantly ($p < 0.05$) higher aerobic counts than groups B and D during the entire period of storage.

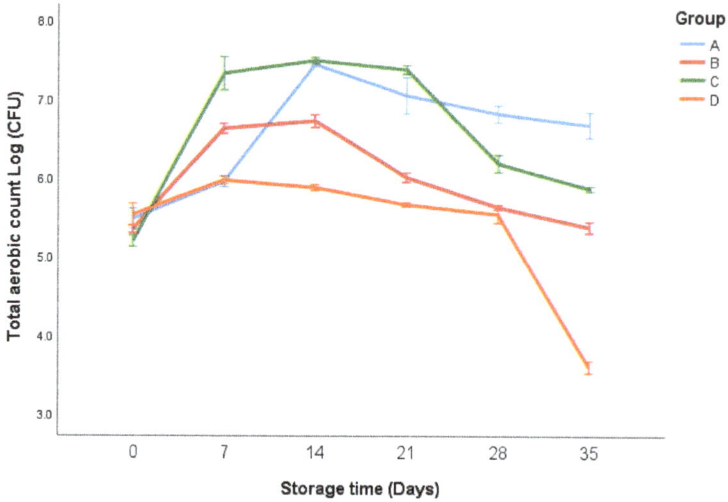

A Mean log (cfu/g)	B Mean log (cfu/g)	C Mean log (cfu/g)	D Mean log (cfu/g)	p-Value
6.7 [a]	6.0 [b]	6.7 [a]	5.12 [c]	<0.05

Figure 3. Total aerobic bacterial count of fresh oregano at refrigerator temperature 4 °C during storage period (35 days). The significant differences ($p < 0.05$) between means are indicated in different letters in the same row. The effect of treatments was pooled to consider the whole storage period and the means were separated by Duncan test.

3.4. Anaerobic Bacteria Analysis

The anaerobic bacterial count during the storage period (35 days) at refrigerator temperature is shown in Figure 4. There was a significant interaction between treatments and storage time. There was no significant difference in total anaerobic count between groups A and C during the whole storage period. Group D exhibited significantly lower anaerobic counts than the other groups. In general, the results of aerobic and anaerobic counts for all groups were quite similar. It was obvious that group D exhibited the lowest anaerobic bacteria counts during the whole period of storage.

The growth of psychrotrophic bacteria in fresh oregano leaves at refrigerator temperature 4 °C is shown in Figure 5. It was clear that group D exhibited significantly lower psychrotrophic bacterial count during the whole period of storage than the other groups. There were moderately significant differences between groups A and B (4.5 vs. 4.9, $p < 0.05$), respectively. Group C exhibited a significantly higher psychrotrophic bacterial count than group A (5.2 vs. 4.5, $p < 0.05$), respectively.

The growth of yeast and mold in fresh oregano leaves at a refrigerator temperature 4 °C is shown in Figure 6. In general, there were no significant differences in yeast and mold counts during the whole period of storage for all groups.

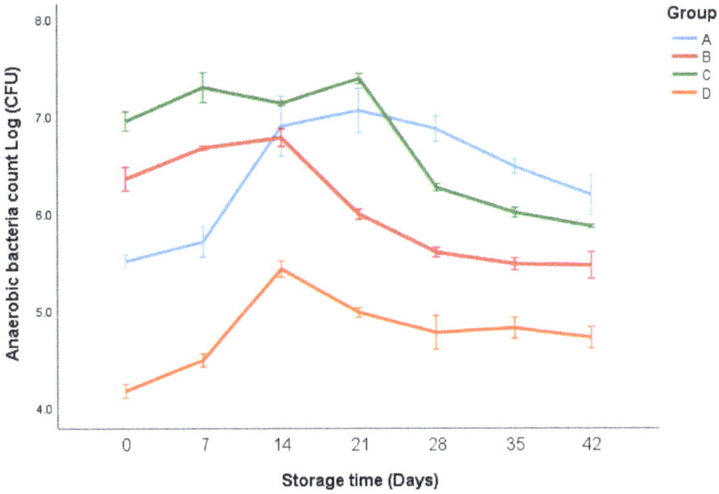

A Mean log (cfu/g)	B Mean log (cfu/g)	C Mean log (cfu/g)	D Mean log (cfu/g)	p-Value
6.6 [a]	6.1 [b]	6.77 [a]	4.75 [c]	<0.05

Figure 4. Anaerobic bacterial count of fresh oregano at refrigerator temperature 4 °C. The significant differences ($p < 0.05$) between means are indicated in different letters in the same row. The effect of treatments was pooled to consider the whole storage period and the means were separated by Duncan test.

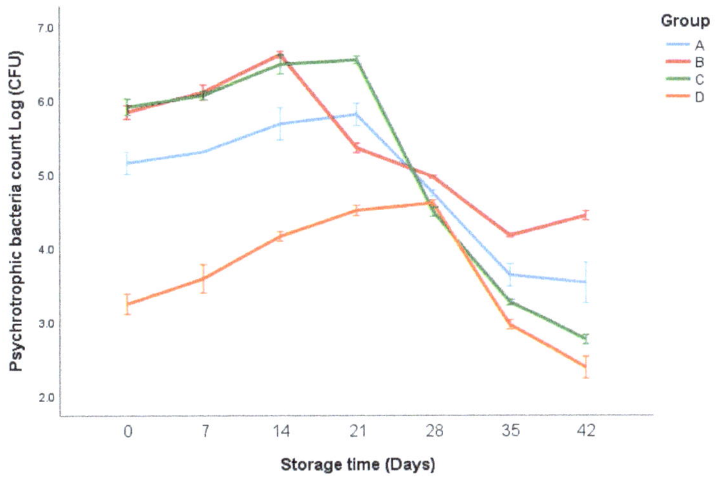

A Mean log (cfu/g)	B Mean log (cfu/g)	C Mean log (cfu/g)	D Mean log (cfu/g)	p-Value
4.5 [b]	4.9 [ab]	5.2 [a]	3.6 [c]	<0.05

Figure 5. Psychrotrophic bacterial count of fresh oregano samples at refrigerator temperature 4 °C. The significant differences ($p < 0.05$) between means are indicated in different letters in the same row. The effect of treatments was pooled to consider the whole storage period and the means were separated by Duncan test.

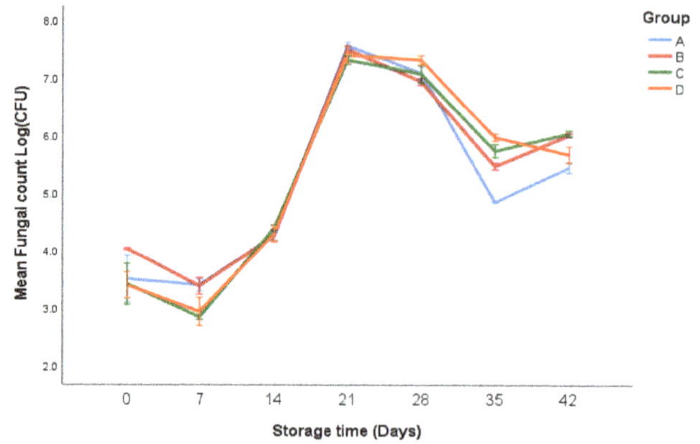

A Mean log (cfu/g)	B Mean log (cfu/g)	C Mean log (cfu/g)	D Mean log (cfu/g)	*p*-Value
5.3	5.5	5.5	5.6	0.51

Figure 6. Mold and yeast growth of fresh oregano at refrigerator temperature 4 °C. The significant differences ($p < 0.05$) between means are indicated in different letters in the same row. The effect of treatments was pooled to consider the whole storage period and the means were separated by Duncan test.

3.5. Color Index (L*a*b*)

The effects of treatments on a*-values during storage are shown in Figure 7. There were no significant differences between groups A, C, and D. During the whole period of storage, the a*-values of Group B changed much less than those of the other groups. Overall, group B exhibited significantly lower a-values than the other groups.

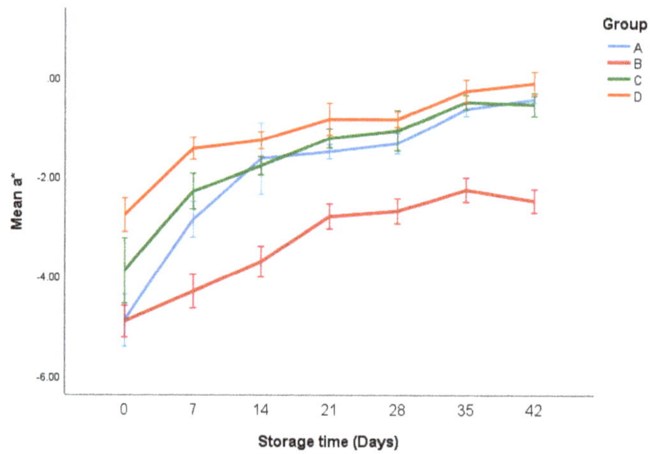

A a*-Value Mean	B a*-Value Mean	C a*-Value Mean	D a*-Value Mean	*p*-Value
-1.8 [a]	-3.3 [b]	-1.65 [a]	-1.23 [a]	<0.05

Figure 7. a*-values of fresh oregano at refrigerated temperature 4 °C during storage period (42 days). The significant differences ($p < 0.05$) between means are indicated in different letters in the same row. The effect of treatments was pooled to consider the whole storage period and the means were separated by Duncan test.

The effects of treatments on b*-values and L*-values are shown in Figures 8 and 9. It was found that there were no significant differences in b*-values between groups A, B, and C. Group D had significantly lower b*-values (9.9 vs. 12.1, 12.1, and 12.0, $p < 0.05$) than groups A, B, and C. It was observed that there was a sharp increase in b*-values in the first week in all groups.

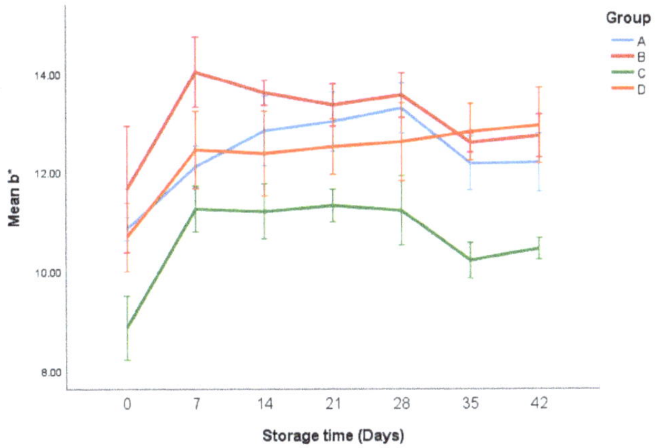

A b*-Values Mean	B b*-Values Mean	C b*-Values Mean	D b*-Values Mean	p-Value
12.1 [a]	12.1 [a]	12.0 [a]	9.9 [b]	<0.05

Figure 8. b*-values of fresh oregano at refrigerated temperature 4 °C during storage period (42 days). The significant differences ($p < 0.05$) between means are indicated in different letters in the same row. The effect of treatments was pooled to consider the whole storage period and the means were separated by Duncan test.

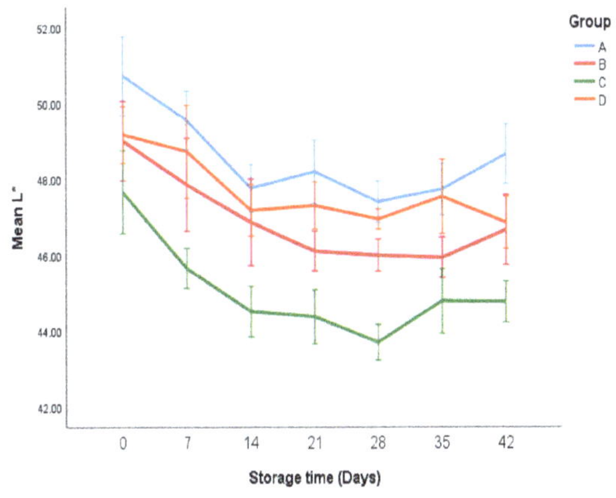

A L*-Values Mean	B L*-Values Mean	C L*-Values Mean	D L*-Values Mean	p-Value
48.8 [a]	46.27 [b]	44.52 [c]	47.7 [a]	<0.05

Figure 9. L*-values of fresh oregano at refrigerated temperature 4 °C during storage period (42 days). The significant differences ($p < 0.05$) between means are indicated in different letters in the same row. The effect of treatments was pooled to consider the whole storage period and the means were separated by Duncan test.

It was found that there were no significant differences in L*-values between group A and D. Group C exhibited significantly lower L*-values than the other groups. There was a drop in L*-values in the first two weeks of storage in all groups.

3.6. Sensory Analysis

The results of the sensory analysis of fresh oregano samples are reported in Figures 9–13. Breaded oregano (*qrass za'atar*) samples were obtained using the same traditional dough as that used by the local baker. The two fresh oregano samples in groups C and D proved to have similar characteristics (taste, flavor, saltiness, and appearance) to fresh oregano leaves that were produced from the traditional bakery (group B). Our study showed that there were no significant differences in taste between groups B, C, and D, except in group B, the taste was different on the last day of the storage period (Figure 10).

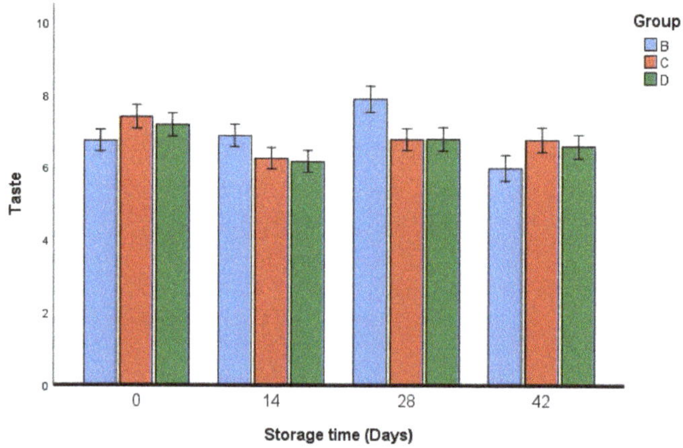

Figure 10. Taste analysis of fresh oregano during storage period (42 days). Group B: (73.2% Fresh oregano, 15% Fresh *Allium cepa*, 20% corn oil, 1.8% salt), Group C: (61.9% Fresh oregano, 15% Fresh *Allium cepa*, 20% corn oil, 1.8% salt, 1.29% sumac "*Rhus coriaria*"), Group D: (59.2% Fresh oregano, 15% Fresh *Allium cepa*, 20% corn oil, 1.8% salt, 4% lactic Acid, ultimate pH 4.4.

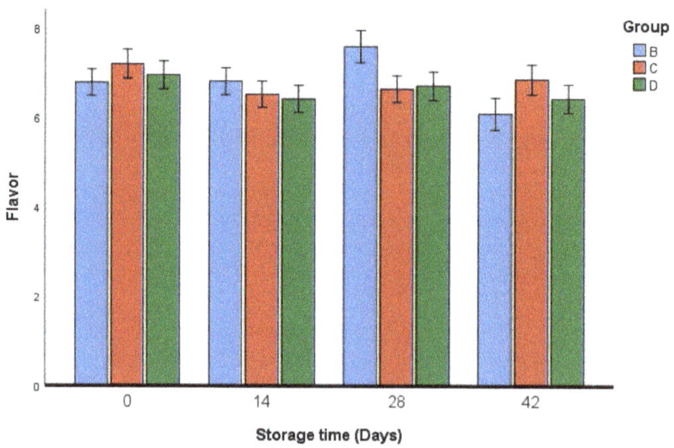

Figure 11. Flavor sensory analysis of fresh oregano during storage period (42 days). Group B: (73.2% Fresh oregano, 15% Fresh *Allium cepa*, 20% corn oil, 1.8% salt), Group C: (61.9% Fresh oregano, 15% Fresh *Allium cepa*, 20% corn oil, 1.8% salt, 1.29% sumac "*Rhus coriaria*"), Group D: (59.2% Fresh oregano, 15% Fresh *Allium cepa*, 20% corn oil, 1.8% salt, 4% lactic Acid, ultimate pH 4.4.

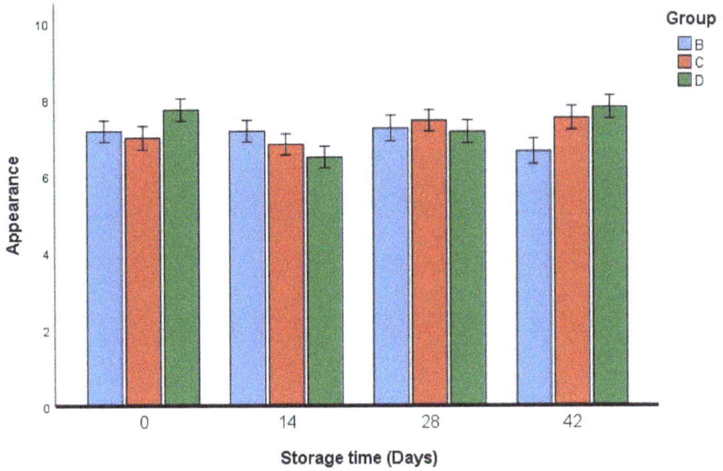

Figure 12. Appearance analysis of fresh oregano during storage period (42 days). Group B: (73.2% Fresh oregano, 15% Fresh *Allium cepa*, 20% corn oil, 1.8% salt), Group C: (61.9% Fresh oregano, 15% Fresh *Allium cepa*, 20% corn oil, 1.8% salt, 1.29% sumac "*Rhus coriaria*"), Group D: (59.2% Fresh oregano, 15% Fresh *Allium cepa*, 20% corn oil, 1.8% salt, 4% lactic Acid, ultimate pH 4.4.

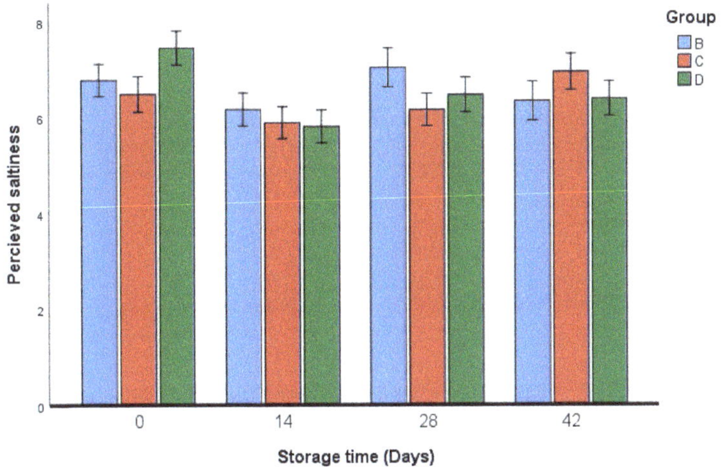

Figure 13. Saltiness sensory analysis of fresh oregano during storage period (42 days). Group B: (73.2% Fresh oregano, 15% Fresh *Allium cepa*, 20% corn oil, 1.8% salt), Group C: (61.9% Fresh oregano, 15% Fresh *Allium cepa*, 20% corn oil, 1.8% salt, 1.29% sumac "*Rhus coriaria*"), Group D: (59.2% Fresh oregano, 15% Fresh *Allium cepa*, 20% corn oil, 1.8% salt, 4% lactic Acid, ultimate pH 4.4.

However, in all of the studied recipes, there were significant differences in the flavor and appearance of the fresh oregano pastries produced using different ingredients (Figures 11 and 12).

All samples were judged to be satisfactory for perceived saltiness. Group D (which had lactic acid) showed significant differences in perceived saltiness only on day 42 (Figure 13). According to the evaluation of the overall acceptance (Figure 14), the oregano bread (*qrass za'atar*) produced using the different oregano mixes was the most appreciated by panelists. However, some complaints during the testing of group C were recorded by some of the tasters, who indicated a bitter taste.

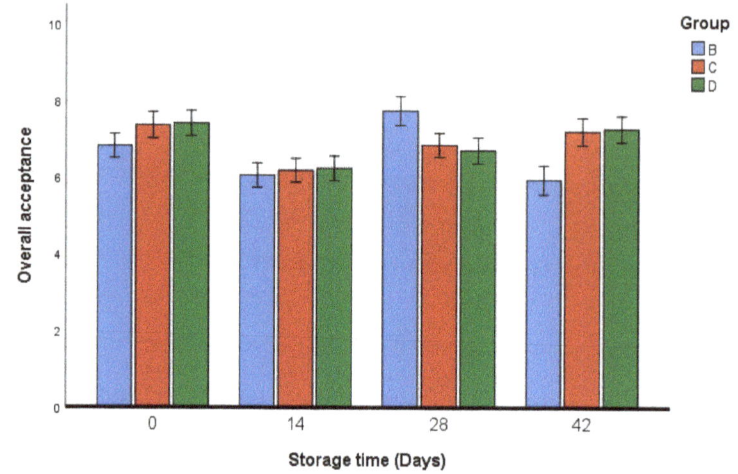

Figure 14. Overall acceptance sensory analysis of fresh oregano during storage period (42 days). Group B: (73.2% Fresh oregano, 15% Fresh *Allium cepa*, 20% corn oil, 1.8% salt), Group C: (61.9% Fresh oregano, 15% Fresh *Allium cepa*, 20% corn oil, 1.8% salt, 1.29% sumac "*Rhus coriaria*"), Group D: (59.2% Fresh oregano, 15% Fresh *Allium cepa*, 20% corn oil, 1.8% salt, 4% lactic Acid, ultimate pH 4.4.

4. Discussion

The proximate chemical compositions of local species of fresh oregano have not been well reported. The composition is usually affected by different factors such as cultivars, seasons, soil, and farming conditions [17]. This aromatic plant is characterized by a high density of nutrients such as minerals, vitamins, and phytonutrients [22]. Wide variations in the chemical compositions have the potential to contribute to the nutritional and health needs of consumers [22]. The results of proximate analysis are in agreement with previous studies conducted on thyme [17,23].

With respect to pH during the storage period, there was a significant decrease in pH in some groups, in particular group D. This may be due to the growth of lactic acid bacteria. In general, there were similarities in the patterns of changes in pH between groups stored at room conditions to those observed in a previous study [18]. The growth of lactic acid in group D also explains the gap in pH compared with the other groups. Cabello-Olmo et al. (2020) found a drop in pH in some vacuum-packed foods [24].

The pH of group A was the highest of all groups (Figure 2), which can be explained by the absence of any additives as in the control group. On the other hand, the lowest growth of aerobic bacteria was significantly observed in group D against the other groups (5.12 vs. 6.7, 6, and 6.7 logs, respectively). Group B exhibited a lower bacterial count than groups A and C; this may be attributed to the effect of onion addition. It was reported that allicin in onions had antifungal and antibacterial activity [25]. However, it is important to note that although group C contained onions, the addition of sumac may have contributed to the increase in the bacterial count. The sumac that was used in this experiment was not sterilized. The bacterial count in this study was quite lower than that in a similar previous study that was carried out at room temperature [18]. The growth of bacteria was lower at low temperatures, which meant the organism was not able to supply the maintenance requirement of the growth rate-limiting nutrient because of loss of affinity for that substrate [25].

Overall, the total anaerobic counts were relatively low across all groups (A, B, C, and D: 6.6, 6.1, 6.7, and 4.7 logs, respectively). This may be related to the combined effect of natural additives and refrigerated conditions. As mentioned earlier, the lower anaerobic count and sharp pH drop in group D could be associated with lactic acid. Our findings were in alignment with our previous study [18], which was carried out in similar experimental conditions except different storage temperature (4 vs. 25 °C). The sole difference between the current experiment and the previous one was in the absolute microbial count, where in the current study, the microbial counts were lower than the previous study, which can be attributed to differences in storage temperature. In this context, Russell et al. [26] found that low temperatures reduced the metabolic processes of microorganisms, which in turn led to a decrease in bacterial growth. In another study, it was found that cold environments disturbed homeostasis, which led to a drop in the growth rate of microorganisms [27]. Generally, psychrotrophic bacterial counts during the whole period of storage for all groups were relatively not high (the highest count was 5.2 logs in group C). This result may be attributed to the combined effect of natural additives, vacuum packaging, and cold environment, which led to improving the reduction performance in the growth of bacteria. Additionally, similar to the anaerobic bacterial growth, the slower rate in the growth of psychrotrophic bacteria in group D during the storage period may be attributed to the very low ultimate pH resulting from the addition of lactic acid.

In respect to fungal growth, our findings may be related to storage temperature, which is one of the most influential factors on yeast and mold growth [28]. Fungi can live in a relatively large range of temperatures, but growth and metabolism rates change at different temperatures even when other conditions, e.g., nutrient and water activity, are constant [29]. Yeast and mold growth were found to be lower at 10 °C compared with 15, 20, 25, and 30 °C [30].

In general, there was a gradual increase in a*-values during the whole period of storage in all groups. It was found that onion (*Allium Cepa* L.) reduced browning reactions [25], which may explain why group B had the lowest change in a*-value, while groups C and D contained sumac and lactic acid in addition to onions, which may have counteracted the effect of the onion.

The increase in b*-values in the first week of storage can be attributed to the degradation of chlorophylls and carotenoids during storage due to oxidative reactions of phenolic compounds by polyphenol oxidase, which produces quinones to various polymerized products [31,32]. The degradation of chlorophyll is usually very high at low pH, which may explain why group D exhibited the lowest b*-value [32].

L*-values in group C were significantly higher than in the other groups. This can be attributed to the addition of sumac, which contains a high level of natural pigments that may have contributed to darkening the color of oregano leaves.

There were no significant differences in overall sensory traits between groups, but perceived saltiness was higher in groups C and D, which may have been due to the addition of sumac and lactic acid.

5. Conclusions

Our study showed that it is possible to formulate fresh oregano recipes that are stable for a reasonable storage period. The most efficient addition for preserving various quality characteristics during storage was lactic acid. It was discovered that the stability of the oregano blend was much enhanced by the cold storage. Due to its limited storage stability, this product has not yet been fully utilized in the export market. Nevertheless, the results of this study may help to increase storage stability for such products.

Author Contributions: Conceptualization, S.M.; methodology, S.M., D.K. and G.M.; software, S.M.; validation, S.M. and G.M.; formal analysis, S.M., D.K. and G.M.; investigation, S.M., D.K. and G.M.; resources, S.M., D.K., O.A., A.I., M.S. (Mohammed Sabbah), M.S. (Munqez Shtaya), F.S. and G.M.; data curation, S.M., D.K. and G.M.; writing—original draft preparation, S.M., D.K., O.A., A.I., M.S. (Mohammed Sabbah), M.S. (Munqez Shtaya), F.S. and G.M.; writing—review and editing, S.M., D.K., O.A., A.I., M.S. (Mohammed Sabbah), M.S. (Munqez Shtaya), F.S. and G.M.; visualization, S.M., D.K. and G.M.; supervision, S.M., D.K. and G.M.; project administration, S.M.; funding acquisition, S.M. All authors have read and agreed to the published version of the manuscript.

Funding: This research received no external funding.

Data Availability Statement: Data is contained within the article.

Conflicts of Interest: The authors declare no conflict of interest.

References

1. Figuérédo, G.; Cabassu, P.; Chalchat, J.-C.; Pasquier, B. Studies of Mediterranean Oregano Populations. VIII—Chemical Composition of Essential Oils of Oreganos of Various Origins. *Flavour Fragr. J.* **2005**, *21*, 134–139. [CrossRef]
2. Alwafa, R.; Mudalal, S.; Mauriello, G. *Origanum syriacum* L. (Za'atar), from Raw to Go: A Review. *Plants* **2021**, *10*, 1001. [CrossRef]
3. Sulaiman, N.; Pieroni, A.; Sõukand, R.; Polesny, Z. Food Behavior in Emergency Time: Wild Plant Use for Human Nutrition during the Conflict in Syria. *Foods* **2022**, *11*, 177. [CrossRef]
4. Zhan, X.; Tan, Y.; Lv, Y.; Fang, J.; Zhou, Y.; Gao, X.; Zhu, H.; Shi, C. The Antimicrobial and Antibiofilm Activity of Oregano Essential Oil against *Enterococcus faecalis* and Its Application in Chicken Breast. *Foods* **2022**, *11*, 2296. [CrossRef] [PubMed]
5. Plati, F.; Papi, R.; Paraskevopoulou, A. Characterization of Oregano Essential Oil (*Origanum vulgare* L. subsp. *hirtum*) Particles Produced by the Novel Nano Spray Drying Technique. *Foods* **2021**, *10*, 2923. [CrossRef] [PubMed]
6. Xu, H.; Delling, M.; Jun, J.C.; Clapham, D.E. Oregano, Thyme and Clove-Derived Flavors and Skin Sensitizers Activate Specific TRP Channels. *Nat. Neurosci.* **2006**, *9*, 628–635. [CrossRef] [PubMed]
7. Jaworska, D.; Rosiak, E.; Kostyra, E.; Jaszczyk, K.; Wroniszewska, M.; Przybylski, W. Effect of Herbal Addition on the Microbiological, Oxidative Stability and Sensory Quality of Minced Poultry Meat. *Foods* **2021**, *10*, 1537. [CrossRef]
8. Hashim, A.M.; Alharbi, B.M.; Abdulmajeed, A.M.; Elkelish, A.; Hozzein, W.N.; Hassan, H.M. Oxidative Stress Responses of Some Endemic Plants to High Altitudes by Intensifying Antioxidants and Secondary Metabolites Content. *Plants* **2020**, *9*, 869. [CrossRef] [PubMed]
9. Daouk, R.K.; Dagher, S.M.; Sattout, E.J. Antifungal Activity of the Essential Oil of *Origanum syriacum* L. *J. Food Prot.* **1995**, *58*, 1147–1149. [CrossRef] [PubMed]
10. Exarchou, V.; Nenadis, N.; Tsimidou, M.; Gerothanassis, I.P.; Troganis, A.; Boskou, D. Antioxidant Activities and Phenolic Composition of Extracts from Greek Oregano, Greek Sage, and Summer Savory. *J. Agric. Food Chem.* **2002**, *50*, 5294–5299. [CrossRef]
11. Al-Kalaldeh, J.Z.; Abu-Dahab, R.; Afifi, F.U. Volatile Oil Composition and Antiproliferative Activity of Laurus Nobilis, Origanum Syriacum, Origanum Vulgare, and Salvia Triloba against Human Breast Adenocarcinoma Cells. *Nutr. Res.* **2010**, *30*, 271–278. [CrossRef] [PubMed]
12. Kregiel, D.; Pawlikowska, E.; Antolak, H. Urtica Spp.: Ordinary Plants with Extraordinary Properties. *Molecules* **2018**, *23*, 1664. [CrossRef] [PubMed]
13. Ali-Shtayeh, M.S.; Jamous, R.M.; Abu-Zaitoun, S.Y.; Akkawi, R.J.; Kalbouneh, S.R.; Dudai, N.; Bernstein, N. Secondary Treated Effluent Irrigation Did Not Impact Chemical Composition, and Enzyme Inhibition Activities of Essential Oils from Origanum Syriacum Var. Syriacum. *Ind. Crops Prod.* **2018**, *111*, 775–786. [CrossRef]
14. Mantzourani, I.; Daoutidou, M.; Dasenaki, M.; Nikolaou, A.; Alexopoulos, A.; Terpou, A.; Thomaidis, N.; Plessas, S. Plant Extract and Essential Oil Application against Food-Borne Pathogens in Raw Pork Meat. *Foods* **2022**, *11*, 861. [CrossRef]
15. Dorman, H.J.D.; Bachmayer, O.; Kosar, M.; Hiltunen, R. Antioxidant Properties of Aqueous Extracts from Selected Lamiaceae Species Grown in Turkey. *J. Agric. Food Chem.* **2004**, *52*, 762–770. [CrossRef] [PubMed]
16. Hazzit, M.; Baaliouamer, A.; Faleiro, M.L.; Miguel, M.G. Composition of the Essential Oils of Thymus and Origanum Species from Algeria and Their Antioxidant and Antimicrobial Activities. *J. Agric. Food Chem.* **2006**, *54*, 6314–6321. [CrossRef] [PubMed]
17. Sadowska, U.; Kopeć, A.; Kourimska, L.; Zarubova, L.; Kloucek, P. The Effect of Drying Methods on the Concentration of Compounds in Sage and Thyme. *J. Food Process. Preserv.* **2017**, *41*, e13286. [CrossRef]
18. Mudalal, S.; Kanan, D.; Abu Qaoud, H.; Mauriello, G. Effect of Vacuum Packaging and Natural Ingredients on the Physical and Microbiological Properties of Fresh Oregano (*Origanum syriacum*) Products. *J. Food Nutr. Res.* **2020**, *8*, 244–251. [CrossRef]
19. Alwafa, R.A.; Mudalal, S.; Shraim, F.; Mauriello, G. Comparison between Quality Traits of Solar-Dried and Freeze-Dried *Origanum syriacum* L. (Za'atar). *Plants* **2022**, *11*, 1110. [CrossRef]
20. Mudalal, S.; Abu-Khalaf, N. Electronic Nose to Differentiate between Several Drying Techniques for Origanum Syriacum Leaves. *Food Res.* **2021**, *5*, 260–265. [CrossRef]
21. AOAC. *Association of Official Analytical Chemists*, 15th ed.; AOAC: Washington, DC, USA, 1990; pp. 931–948.
22. Hels, O.; Larsen, T.; Christensen, L.P.; Kidmose, U.; Hassan, N.; Thilsted, S.H. Contents of Iron, Calcium, Zinc and β-Carotene in Commonly Consumed Vegetables in Bangladesh. *J. Food. Compost. Anal.* **2004**, *17*, 587–595. [CrossRef]
23. Rahimmalek, M.; Goli, S.A.H. Evaluation of Six Drying Treatments with Respect to Essential Oil Yield, Composition and Color Characteristics of Thymys Daenensis Subsp. Daenensis. Celak Leaves. *Ind. Crops. Prod.* **2013**, *42*, 613–619. [CrossRef]
24. Cabello-Olmo, M.; Oneca, M.; Torre, P.; Díaz, J.V.; Encio, I.J.; Barajas, M.; Araña, M. Influence of Storage Temperature and Packaging on Bacteria and Yeast Viability in a Plant-Based Fermented Food. *Foods* **2020**, *9*, 302. [CrossRef]
25. Yuniarti, T.; Sukarno, S.; Yuliana, N.D.; Budijanto, S. Inhibition of Enzymatic Browning by Onion (*Allium cepa* L.): Investigation on Inhibitory Mechanism and Identification of Active Compounds. *Curr. Res. Nutr. Food Sci.* **2018**, *6*, 770–780. [CrossRef]
26. Nedwell, D.B. Effect of Low Temperature on Microbial Growth: Lowered Affinity for Substrates Limits Growth at Low Temperature. *FEMS Microbiol. Ecol.* **1999**, *30*, 101–111. [CrossRef] [PubMed]

27. De Sarrau, B.; Clavel, T.; Clerté, C.; Carlin, F.; Giniès, C.; Nguyen-The, C. Influence of Anaerobiosis and Low Temperature on Bacillus Cereus Growth, Metabolism, and Membrane Properties. *Appl. Environ. Microbiol.* **2012**, *78*, 1715–1723. [CrossRef]
28. Ayerst, G. The Effects of Moisture and Temperature on Growth and Spore Germination in Some Fungi. *J. Stored Prod. Res.* **1969**, *5*, 127–141. [CrossRef]
29. Dales, R.E.; Schweitzer, I.; Bartlett, S.; Raizenne, M.; Burnett, R. Indoor Air Quality and Health: Reproducibility of Respiratory Symptoms and Reported Home Dampness and Molds Using a Self-Administered Questionnaire. *Indoor Air* **1994**, *4*, 2–7. [CrossRef]
30. Adan, O.G. *On the Fungal Defacement of Interior Finishes*, Technische Universiteit Eindhoven: Eindhoven, The Netherlands, 1994.
31. Pogson, B.J.; Morris, S.C. Consequences of Cool Storage of Broccoli on Physiological and Biochemical Changes and Subsequent Senescence at 20 °C. *J. Am. Soc. Hortic. Sci.* **1997**, *122*, 553–558. [CrossRef]
32. Manolopoulou, E.; Varzakas, T. Effect of Temperature in Color Changes of Green Vegetables. *Curr. Res. Nutr. Food Sci.* **2016**, *4*, 10–17. [CrossRef]

foods

MDPI

Article

Application of an Eco-Friendly Antifungal Active Package to Extend the Shelf Life of Fresh Red Raspberry (*Rubus idaeus* L. cv. 'Kweli')

Tiago M. Vieira, Vítor D. Alves * and Margarida Moldão Martins

LEAF—Linking Landscape, Environment, Agriculture and Food, Associated Laboratory TERRA, Instituto Superior de Agronomia, Universidade de Lisboa, Tapada da Ajuda, 1349-017 Lisbon, Portugal; tdmvieira@gmail.com (T.M.V.); mmoldao@isa.ulisboa.pt (M.M.M.)

* Correspondence: vitoralves@isa.ulisboa.pt; Tel.: +351-21-365-3546

Abstract: The main objective of this study was to extend the shelf life of fresh red raspberry (*Rubus idaeus*. L. cv. 'Kweli') by using active film-pads inside commercial compostable packages. The pads were produced with chitosan (Ch) with the incorporation of green tea (GTE) and rosemary (RSME) ethanolic extracts as natural antifungal agents. Pads were placed on the bottom of commercial fruit trays underneath the fruits, and the trays were heat-sealed with a polyacid lactic (PLA) film. Preservation studies were carried out over 14 days of storage at refrigeration temperature (4 °C). Raspberry samples were periodically analyzed throughout storage, in terms of quality attributes (fungal decay, weight loss, firmness, surface color, pH, total soluble solids), total phenolic content and antioxidant activity. Gas composition inside the packages was also analyzed over time. From the packaging systems tested, the ones with active film-pads Ch + GTE and Ch + RSME were highly effective in reducing fungal growth and decay of raspberry during storage, showing only around 13% and 5% of spoiled fruits after 14 days, respectively, in contrast with the packages without pads (around 80% of spoiled fruits detected). In addition, fruits preserved using packages with Ch + RSME active film-pads showed lower mass loss (5.6%), decreased firmness (3.7%) and reduced antioxidant activity (around 9% and 15% for DPPH and FRAP methods, respectively). This sustainable packaging presents a potential strategy for the preservation of raspberries and other highly perishable small fruits.

Keywords: antifungal active packaging; raspberry; chitosan; green tea extract; rosemary extract

Citation: Vieira, T.M.; Alves, V.D.; Moldão Martins, M. Application of an Eco-Friendly Antifungal Active Package to Extend the Shelf Life of Fresh Red Raspberry (*Rubus idaeus* L. cv. 'Kweli'). *Foods* **2022**, *11*, 1805. https://doi.org/10.3390/foods11121805

Academic Editors: Maria Cefola and Bernardo Pace

Received: 16 May 2022
Accepted: 17 June 2022
Published: 19 June 2022

1. Introduction

Raspberry commercialization has recently undergone significant changes due to increasing customer quality requirements, health and lifestyle (sustainable consumption) concerns, and the demand throughout the year [1]. Such changes require producers and traders' access to new varieties and to develop new strategies to reduce softening of the fruits and improve shelf life. Innovation in packaging technology can play a key role in improving the post-harvest quality. Because raspberries provide an important source of nutraceutical compounds (high phenolic compounds and anthocyanin contents) that are beneficial for human health [2], their marketing must integrate the themes of sustainable production and distribution processes [1,3]. The supply chain must consider packaging management, and the use of biodegradable materials derived from renewable sources is one key factor to increase sustainability of the post-harvest supply chain.

Raspberry fruits stored at 0–0.5 °C and 90–95% relative humidity (RH) can be maintained in an ambient atmosphere for 5–7 days. High CO_2 treatments and modified atmospheres (15–20% CO_2 and 5–10% O_2) (MAP) have also been studied [1] for improving the berries' shelf life. MAP is an active or passive dynamic process that involves the use of plastic films that limit gas diffusion from inside to outside and vice versa, leading to CO_2

enrichment and O_2 reduction inside the package [4]. Active MAP results from a rapid process of gas replacement establishing the desired gas mixture inside a package. On the other hand, passive MAP is generated by the natural process of fruit respiration and film permeability, and the gas composition eventually reaches a steady-state value, which should be suitable for the packed product [5]. The gaseous composition at steady-state depends on a series of factors, such as mass of the packed product, storage temperature, fruit respiration rate, cultivar and ripening stage. Moreover, the exchange of gases between the atmosphere inside the package and the exterior is affected by gas concentration differences, the exposed surface and permeability of the selected film [1]. Several films with various permeability values for water vapor, CO_2 and O_2 for fruit and vegetable packaging are commercially available [1,6], but the gas permeability values of most plastic films are too low to allow gas exchange and permit slow respiration [7]. During shipping, handling or retail display, a passive MAP storage system can cause O_2 depletion and CO_2 accumulation to levels outside the optimal range for the product, due to an inadequate permeability of the package film [8]. Raspberries should not be exposed to CO_2 levels greater than 20%, which can cause discoloration, softening and the development of off-flavors [1].

In the soft fruit sector, packaging is changing, and the reduction in packaging weight and the use of sustainable materials is becoming essential to respond to the increased environmental concerns of retailers and customers. Alternative biodegradable and sustainable packaging materials to the traditional non-biodegradable oil-based polymers are starting to be used in packaging for fresh and fresh-cut produce [1]. Polyester poly (lactic acid) (PLA) is one of the most economically competitive alternatives to traditional oil-based polymers, such as polyethylene terephthalate (PET) and high- and low-density polyethylene [9,10]. It can be produced from renewable sources, such as sugar beet and corn starch, or other renewable biomass products and wastes [11]. Several studies have shown the potential for this alternative material to be used for post-harvest storage of fruits [9,12,13], but data on the use of these materials to store highly perishable raspberry fruits under passive MAP conditions are limited [1].

Active packaging is another innovative approach that has emerged to maintain quality and extend the shelf life of fruits [14]. This packaging system is designed to interact with foods by releasing active components with biological properties (e.g., antioxidant and antimicrobial capacity), which may be obtained from natural sources, such as plants, that are in line with the increased demand by consumers for food products with natural additives over synthetic ones [15]. Other active packaging systems include the use of exudate absorbent pads placed underneath the product, in order to avoid product deterioration due its contact with moisture. These pads have been used in packages of meat, fish and small soft fruits such as fresh strawberries [16]. Although biodegradable polymers (e.g., cellulose and its derivatives, polyvinyl alcohol and starch) have been used to produce such pads [17], there is still a need to test novel materials for this application. Chitosan, a polysaccharide produced from various fungi and the shell waste of shrimps, crabs and bones of cuttlefish and squids, is an example. It is non-toxic, with an allergen-free nature (approved by the FDA and European Union) and possesses antimicrobial and antioxidant activities [18–20]. In addition, it is an effective carrier of antioxidant and antimicrobial agents, namely those of aromatic plant extracts, such as green tea and rosemary [20–22]. Rosemary plant is a good source of bioactive compounds, including flavonoids, phenolics, diterpenoids and triterpenes [23,24]. Extracts from rosemary have shown strong inhibitory activity against various bacteria strains [25], and high antifungal activity [26,27]. Furthermore, the green tea plant, rich in polyphenols and catechins, is also associated with health benefits and antimicrobial and antioxidant properties [28]. Previous research studies proved that the application of a chitosan film or coating incorporating natural antioxidant and antimicrobial agents improves the storability of several perishable foods [29–34].

Recently, novel chitosan active film-pads containing antifungal plant extracts (from green tea and rosemary) were developed. They have shown simultaneously great performance in terms of antimicrobial capacity against *P. expansum* and water absorption capacity, without compromising their mechanical resistance under high moisture conditions [35]. As such, chitosan may be used to produce dense hydrophilic films with added active plant extracts with interesting water absorption, showing potential to be used as fruit pads.

Fresh berries are highly susceptible to spoilage during storage due to microbial growth. To the best of our knowledge, there are no reported studies regarding the application of chitosan films enriched with aromatic plant extracts as active pads in the preservation of soft fruits. As such, this study is focused on the application of packaging with increased sustainability, composed of chitosan active film-pads with antifungal plant extracts (green tea and rosemary) inside a commercial tray sealed with a compostable polymer (PLA) film, aiming at the extension of fresh red raspberry fruits' shelf life.

2. Materials and Methods

2.1. Materials

To produce active film-pads, we used chitosan powder (Golden/Shell Biochemical Co., Ltd, Zhejiang, China), lactic acid (Panreac Quimica SAU, Barcelona, Spain), glycerol (Fisher Scientific, Loughborough, UK), dried leaves of commercial green tea (*Camellia sinensis* (L.) Kuntze) from Azores Island (Gorreana, São Miguel, Portugal) and rosemary (*Rosmarinus officinalis* L.), and ethanol (96% vol.) (Valente e Ribeiro. Lda, Lisbon, Portugal). For fruit preservation studies, we used red raspberries (*Rubus idaeus* L., cv. 'Kweli') acquired from a local farm in the Beja district, Portugal (37.65423, −8.756739), and a polylactic acid (PLA) compostable film (VGB 4 Vegware, London, UK) (O_2 permeability: 2.97×10^{-16} m^2 s^{-1} and CO_2 permeability: 1.22×10^{-15} m^2 s^{-1}). In the analytical methods, we used analytical grade sodium hydroxide (Merck Life Science S.L.U., Algés, Portugal), methanol, HCl, gallic acid, 1,1–diphenyl-2-picrylhydrazyl, Trolox, 2,4,6-Tris(2-pyridyl)-s-triazine (TPTZ) and $FeCl_3 \cdot 6H_20$ from Panreac Quimica SAU, Barcelona, Spain.

2.2. Development of the Active Film-Pads

Chitosan powder was dissolved in a lactic acid solution (1% v/v) at a concentration of 2% w/v. Glycerol was added as a plasticizer at a concentration of 25% w/w (chitosan basis). The resulting solution was transferred to flat plastic containers (0.4 mL cm^{-2}) and then dried at a temperature of 45 °C and 30% relative humidity for 24 h. After drying, the chitosan-based film obtained was cut into rectangular pads (7 cm × 5 cm).

Ethanolic extracts were obtained according to Vieira et al. [35]. Two different plants were used: green tea (*Camellia sinensis* (L.) Kuntze) and rosemary (*Rosmarinus officinalis* L.), both purchased from a local market in Lisbon (Portugal). The commercial dried plant leaves were stored at 5 °C in the dark until use. For each extract, the plant material was mixed with ethanol (96% vol.) in a 1:10 ratio (w/v). After vigorous stirring for 30 min protected from light at room temperature, the mixture was filtered using a Whatman No. 1 filter paper and a vacuum pump (Yuchengtech, AP-9925, Beijing, China). To avoid degradation, the ethanolic extracts of green tea (GTE) and rosemary (RSME) were stored in the dark at 4 °C for further use.

The active film-pads were then prepared by immersion of the rectangular pads into GTE and RSME ethanolic extracts in sealed Petri dishes, overnight at 4 °C, followed by drying at room temperature protected from light. The average pad thickness was around 215 µm, measured using a digital micrometer (APB-3D Mitutoyo, Kanagawa, Japan) [35].

2.3. Raspberry Fruits Preservation Using Packages with Active Antifungal Film-Pads

The fresh red raspberries (*Rubus idaeus* L., cv. 'Kweli') were acquired at commercial maturity. Fruits were selected to be free from visible defects and injuries, to avoid interference from natural infections before any packaging treatment.

The antifungal active film-pads were placed on the bottom of commercial packages for red raspberries. Packages were trays, which were wrapped with a polylactic acid (PLA) compostable film and heat-sealed, under normal atmospheric conditions, to promote passive MAP conditions [1]. A total of 80 sealed trays were stored at 4 °C and 95% of relative humidity divided into four groups: trays without film-pads (Ctr); trays with chitosan film-pads but without extracts incorporated (Ch); trays with chitosan film-pads containing green tea extract (Ch + GTE); trays with chitosan film-pads containing rosemary extract (Ch + RSME).

The fruits were analyzed over time (at 0, 3, 7, 14 days of storage at 4 °C), in terms of surface color, weight loss, firmness, pH, total soluble solids, volatile acidity, total phenolic content and antioxidant activity. The composition of the atmosphere inside fruit packages was also measured over time. Five independent packages were analyzed for each group on each day.

2.4. *Analytical Methods*

2.4.1. Composition of the Atmosphere inside the Packages

The concentration of carbon dioxide (CO_2) and oxygen (O_2) inside packages was measured using a gas analyzer (Checkmate 9900, PBI Dansensor, Ringsted, Denmark). An adhesive silicon septum was glued to the sampling point of packages to prevent gas leakage during analysis. The needle of the gas analyzer was inserted through the septum and results are expressed in percentages of O_2 and CO_2. Five packages per group were analyzed on each test day.

2.4.2. Fungal Decay Incidence

Fungal decay was visually inspected after 3, 7 and 14 days of storage at 4 °C. Raspberry fruit showing surface mycelia development was considered decayed. Fungal decay incidence was quantified as the percentage of total fruits that showed surface mycelia development.

2.4.3. Weight Loss, Firmness and Surface Color

The weight loss (% from the original weight [36]) from each sealed tray was measured using an electronic balance (TC-403, Denver Instrument Company, Vernon Hills, IL, USA). The results are expressed as an average of five replicates per group for each test day.

A Texturometer (TA-XT2, Stable Micro System, Surrey, UK) with a 5 kg load cell and equipped with a flat probe (37 mm diameter) was used to evaluate firmness in whole fresh raspberries. A number of 15 fruits from each package (three packages, randomly chosen from the total five packages used per group and test day) was assayed. Each fruit was positioned under the probe and compressed to 80% deformation at a speed of 2 mm s^{-1}. All fruit samples were tempered at least for 6 h at 25 °C before measurements.

A Konica Minolta CTR-300 colorimeter (Minolta, Williams Drive Ramsey, NJ, USA) was used to measure L^* (lightness), a^* (red-green) and b^* (yellow-blue) color parameters of raspberries. It was calibrated with a standard white plate, which was provided by the manufacturer. The color was measured on the non-moldy fruits from each package (three packages, randomly chosen from the total five packages used per group and test day). Measurements were taken on the side of a slightly flattened whole fruit [36].

2.4.4. Total Soluble Solids Content, pH and Volatile Acidity

The total soluble solids (TSS), pH and volatile acidity were measured in a fruit pulp produced by trituration of 15 fruits selected per tray (three packages, randomly chosen from each group per test day), using an Ultra-Turrax blender (IKA T18 basic Ultra-Turrax, Staufen, Germany).

The determination of TSS was performed using the clear juice obtained by pulp filtration with a double layer of gauze. The TSS was measured with a refractometer (Atago, Fisher Scientific, Ga., Bellevue, WA, USA). The pH values were measured using by a pH meter (Russel, Moder RL) with the electrode being directly immersed in the pulp [37].

The measurement of volatile acidity was carried as described in ISO 6632-1981 [38] with minor modifications. A mixture of the fruit pulp (10.0 g) and deionized water (10 mL) was steam-distilled using an automatic distiller (Gerhardt, Königswinter, Germany). The distillate was collected (250 mL), and the volatile compounds were neutralized with 0.1 M sodium hydroxide (Eka Pellets™, Amsterdam, The Netherlands) in the presence of 1% phenolphthalein (Himedialabs, Esdoornlaan, DB Maarn, The Netherlands) as an indicator. The consumption of 0.1 M sodium hydroxide represented the volatile acidity, being expressed in mass of acetic acid per mass of raspberries (mg acetic acid 100 g^{-1}).

2.4.5. Total Phenolic Content and Antioxidant Activity

For the quantification of total phenolic content and the antioxidant activity of packed raspberry fruits, extracts were obtained using a modified method from Kopjar et al. [39]. Five grams of raspberry pulp was extracted with 15 mL of methanol/HCl (99/1, *v*/*v*). The mixture of pulp and solvent was well mixed using a magnetic stirrer in the dark at 10 °C for 1 h. Then, the mixture was centrifuged at 14,881× *g* for 10 min at 4 °C and the supernatant was recovered, obtaining a final volume of 50 mL. The extraction was repeated three times for each group and test day. The supernatant was maintained at 4 °C in the dark before analytical determinations.

Total Phenolic Content (TPC)

TPC was analyzed by measuring the absorbance of the prepared extracts at a wavelength of 280 nm (UNICAM, UV/Vis spectrometer-UV4, Waltham, MA, USA), as described in other works in the literature [40,41]. Gallic acid (St. Louis, MO, EUA) was used as standard, and a calibration curve was performed measuring the absorbance of different gallic acid aqueous solutions (from 0 mg L^{-1} to 50 mg L^{-1}). The TPC was expressed as gallic acid equivalents (GAE) per mass of raspberry fresh weight (mg GAE 100 g^{-1}).

Antioxidant Activity (AOA)

1,1–diphenyl-2-picrylhydrazyl (DPPH) Assay

The DPPH method was used as described by Huang et al. [42], with minor modifications. A mass of 24 mg of DPPH was dissolved in 100 mL of methanol to make a stock solution. Before using, this stock solution was stored in the freezer for more than 2 h. A working solution was prepared by mixing 10 mL of stock solution with 45 mL of methanol to achieve an absorbance of less than 1.1 at 515 nm. An aliquot of 0.1 mL of extract was added to 4.9 mL of DPPH working solution and the absorbance was measured after 40 min (*Abs sample*). The analysis was carried out in triplicate. The blank consisted of 4.9 mL of DPPH working solution with 0.1 mL of methanol (*Abs blank*). The RSA (radical scavenging activity) was calculated by Equation (1).

$$RSA(\%) = \frac{Abs\ blank - Abs\ sample}{Abs\ blank} \times 100 \tag{1}$$

Trolox was used as standard, and a calibration curve was performed correlating RSA values with different Trolox concentrations (from 100–2000 µM). The results were expressed as Trolox Equivalent Antioxidant Capacity (TEAC), as mmol Trolox per mass of fresh raspberry (mmol Trolox 100 g^{-1}).

FRAP Assay

The FRAP test was performed using a modified version of that used by Suárez et al. [43]. A volume of 25 mL of acetate buffer 0.3 M was mixed with 2.5 mL of 2,4,6-Tris(2-pyridyl)-s-triazine (TPTZ) 0.01 M and 2.5 mL of Iron(III) chloride hexahydrate ($FeCl_3{\cdot}6H_2O$) solution 0.02 M to produce the working FRAP mixture. An aliquot of 90 µL of extract was added to a mixture of 270 µL of deionized water and 2.7 mL of the working FRAP solution. The absorbance was measured at 595 nm after reaction in a water bath (37 °C) for 40 min. The analysis was carried out in triplicate. Trolox was used as standard, and a calibration curve was performed correlating the absorbance at 595 nm with different Trolox concentrations (from 100–2000 µM). The results were expressed as Trolox Equivalent Antioxidant Capacity (TEAC), as mmol Trolox per mass of fresh raspberry (mmol Trolox 100 g^{-1}).

2.5. Statistical Analysis

After checking the normal distribution, ANOVA and Tukey tests were used to assess if there were significant differences between the data's average values, for a significance level of 0.05. Data analysis was performed using the software STATISTICA TM version 8.0 (StatSoft Inc., Tulsa, OK, USA).

3. Results and Discussion

3.1. Composition of the Atmosphere inside the Packages

The O_2 and CO_2 contents detected inside the packages for raspberries are reported in Figure 1.

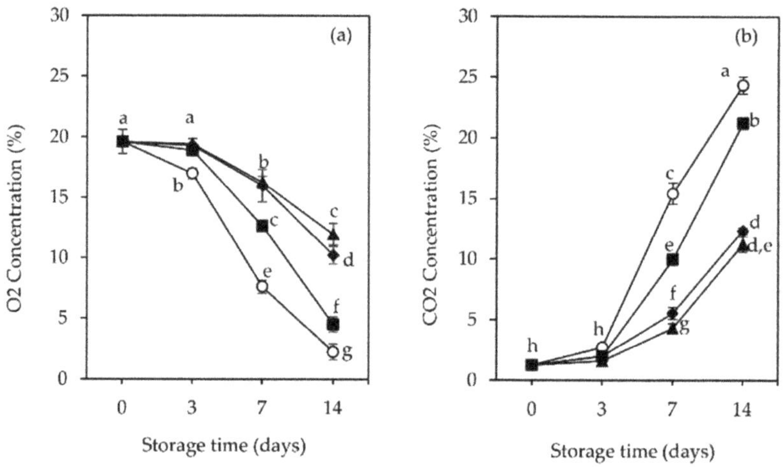

Figure 1. Changes in headspace gas composition: (**a**) O_2 and (**b**) CO_2 concentrations (%), of raspberry (*Rubus idaeus* L. cv. 'Kweli') fruit packages (no film-pad_Ctr, ○; with film-pad containing green tea extract_Ch + GTE, ♦, rosemary extract_Ch + RSME ▲, only chitosan ■), stored for 14 days at 4 °C under passive MAP conditions. Vertical bars indicate standard deviation of five replicates. Means with equal letters are not statistically different by Tukey's test with 5% significance level.

The initial atmosphere gas composition (19% O_2 and 1.25% CO_2) changed in the packages across all the packaging treatments applied. The exchange area (310.5 cm^2) through the film packages was constant, so the evolution of the atmosphere inside the trays was passively created by the respiration rate of the fruits, the metabolic activity of fungal development and the permeability of the film to O_2 and CO_2 [44], which are affected by temperature [1]. Nevertheless, significant differences ($p < 0.05$) were observed when comparing raspberry fruits in the presence of different film-pads.

The O_2 content in control and in Ch samples (both treatments without the extracts incorporated) were significantly much lower ($p < 0.05$) (Figure 1a) during storage, achieving minimal levels around to 2.24% and 4.52%, respectively, at the end of 14 days of storage, compared to the levels reached by the active film-pad samples Ch + GTE (10.91%) and Ch + RSME (11.95%). Given that the raspberries are non-climacteric, the drop in O_2 content may be related to O_2 consumption due to metabolic activity of fungal development [45]. Indeed, the control and Ch samples presented higher fungal incidence and decay compared to Ch + GTE and Ch + RSME samples (Section 3.2), which were significantly ($p < 0.05$) much lower. In these latter groups, possible interactions between antifungal agents from GTE and RSME loaded on chitosan and the fruit surface existed, along with pads' high water absorption capacity [35], which could absorb and inhibit fungal droplet moisture.

CO_2 contents were low in the first 3 days of refrigerated storage, without significant differences ($p > 0.05$) between all package groups (Figure 1b). After the third day, the content increased quickly, and at seventh day, it increased with significant differences ($p < 0.05$) between package groups. At the end of 14 days of storage, the CO_2 content in control samples packages reached the highest values (24.36%), followed by those with Ch pads (21.24%), which were significantly ($p < 0.05$) different, compared to the lowest CO_2 levels reached with Ch + GTE (12.35%) and Ch + RSME pads (11.21%). At this point, the high CO_2 atmosphere and minimal O_2 levels reached in package samples without any antifungal active film-pads inside became potentially harmful to fruit quality. According to Dennis et al. [46], the induction of anaerobic respiration can cause multiple undesirable changes in the fruits, including the development of off-flavors [47]. On the other hand, the O_2 and CO_2 levels in package samples containing Ch + GTE and Ch + RSME pads were around 10% and 10–20%, which are recommended as desirable to preserve fresh raspberry quality [48]. Indeed, the fruit quality in the presence of the antifungal active film-pads enhanced for 14 days of storage at low temperature. In contrast, fruits in package samples that did not contain the active film-pads became not marketable by visual analysis (Section 3.2).

3.2. Fungal Decay of the Raspberry Fruits

Raspberry fruit is highly perishable, due to high susceptibility to mechanical injury, water loss, high metabolic activity and mold and rot growing. To determine the effectiveness of the antimicrobial active film-pads for enhancing the fruit quality, fungal decay as the primary determinant of quality was observed during the storage time and statistically compared to the control samples (Figure 2).

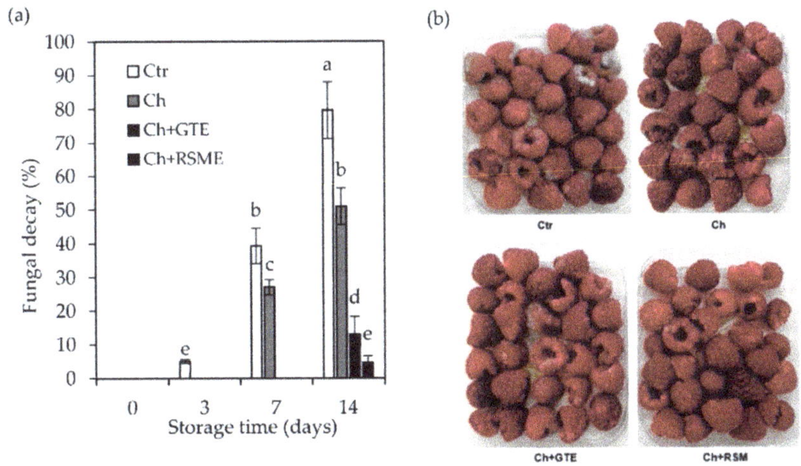

Figure 2. Evolution of fungal decay (**a**) and appearance at 14 days of storage (**b**) of packaged raspberry (*Rubus idaeus* L. cv. 'Kweli') fruits (no film-pad _Ctr; with film-pad containing green tea extract_Ch + GTE, rosemary extract_Ch + RSME and only chitosan_Ch) stored for 14 days at 4 °C under passive MAP conditions. Means with equal letters are not statistically different by Tukey's test with 5% significance level.

Fungal decay started quickly on the control samples, with 5% of fruits displaying signs of infection at day 3 of storage, and 39.4% displaying signs of infection at 7 days of storage at 4 °C. At the end of storage, we recorded 79.5% damage by mold spoilage in fruits in control samples, which was significantly higher ($p < 0.05$) than that of the other package groups (packages with film-pads) (Figure 2a). At this point, most of the raspberries in control samples showed the development of fungus *Botrytis cinerea* 'gray mold' mycelium and slightly brown spots (Figure 2b). The water vapor transmission rate of the PLA film used to wrap the trays is low regarding the transpiration rate of fresh raspberry fruits [49]. Therefore, high relative humidity conditions prevailed in the packages, causing condensation of water vapor and enhancing the conditions for fungal growth (in the absence of antimicrobial agents), resulting in fruit spoilage. In the case of Ch packages, the first signs of decay started to appear at day 7, showing 12.35% less infection than control packages, and at the end of 14 days, differences were significantly much higher ($p < 0.05$), showing 28.54% less decay. Chitosan was previously reported to have an antifungal effect when applied on raspberries [50], as well as on other soft fruits such as strawberries under cold storage [51,52]. Nevertheless, when fruits were packaged with the antifungal active film-pads (Ch + GTE and Ch + RSME), the fungal incidence and decay were observed only at the end of 14 days of storage, and differences were significantly much higher ($p < 0.05$) than control samples, with 66.54% and 74.92% less fungal decay, respectively. The Ch + RSME film-pad significantly ($p < 0.05$) reduced decay in raspberry fruits to less than 5% of spoiled fruits, compared to 13% with Ch + GTE active film-pad and 51% with Ch film-pad. The release of phenolic compounds from active film-pads to raspberries upon direct contact between them limits fungal growth. Furthermore, as some of the compounds of the extracts are volatile, their release and accumulation in the internal package atmosphere may also have played an important role in the antifungal activity. The suppressed spoilage decay over 14 days with Ch + GTE and Ch + RSME pads compared to control samples may have important economic implications, as the shelf life of raspberry fruits can be extended in the fresh market. Other reports have shown that chitosan antifungal activity can be enhanced by the addition of bioactive compounds such as lemon essential oil when applied to cold storage of soft fruits [53,54].

3.3. Weight Loss, Firmness, and Surface Color

As a result of their high water content (80–95%) and thin skin, raspberries present a high tendency to decrease in mass due to water loss. The effect of the active film-pads on raspberry fruits' weight loss is depicted in Figure 3a.

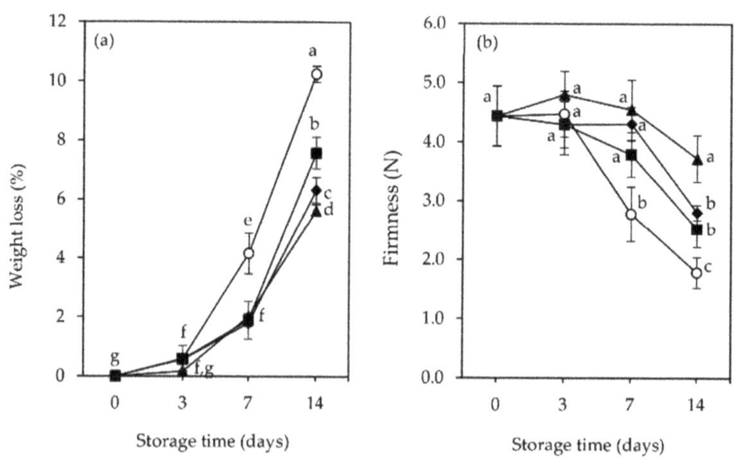

Figure 3. Changes in weight loss (**a**) and firmness (**b**) of packaged raspberry (*Rubus idaeus* L. cv. 'Kweli') fruits (no film-pad _Ctr, ○; with film-pad containing green tea extract_Ch + GTE, ♦, rosemary extract_Ch + RSME ▲, only chitosan ■), stored for 14 days at 4 °C under passive MAP conditions. Vertical bars indicate standard deviation of five replicates. Means with equal letters are not statistically different by Tukey's test with 5% significance level.

The data indicate that the % loss across all treatments increased with increasing storage period. No significant differences ($p > 0.05$) between the different film-pads tested were observed during the 7 days of storage. On the other hand, the weight loss in control samples sharply increased after the third day of storage, distancing from all the film-pad samples during storage and reaching the highest weight loss (10.24%), followed by Ch film-pad (7.57%) at the end of 14 days of storage. In contrast, the fruits packaged with the active film-pads Ch + GTE and Ch + RSME had the lowest weight loss (6.31% and 5.61%, respectively), which was significantly different from the Ch samples. Thus, the study indicates that both active film-pads Ch + GTE and Ch + RSME, by repressing fungal growth inside raspberry packages under cold storage (Section 3.2), retarded fruit tissue deterioration, which simultaneously contributed to reduced fruit water loss, along with changes in the respiration process (Section 3.1), the main metabolic process that is linked to moisture loss [50].

Firmness maintenance is one of the most important physical attributes for the quality of raspberries [50]. Significant differences ($p < 0.05$) between the film-pad samples and the control samples were observed in the loss of firmness during storage (Figure 3b). The firmness of control samples rapidly decreases after day 3, with a significant ($p < 0.05$) loss in firmness of 61% at day 14 in relation to the initial fruit firmness at day 0. The fruits packaged with Ch and Ch + GTE film-pads remained firmer over 7 days of storage, but after this period, both treatments were not effective, and fruits suffered an average loss in firmness at day 14 of 44.4% and 38.8%, respectively. Contrarily, when Ch + RSME was applied inside packages, the fruits remained much firmer during the storage period, with a lower loss in firmness (16.7%), significantly ($p < 0.05$) lower compared to that of fruits of the other packages at the end of 14 days. The differences in fruit firmness between type of packages may be related to moisture loss [55]. Indeed, raspberries packaged with Ch + RSME pads showed the lowest average weight loss rate during the storage period (Figure 3a), contributing to the maintenance of fruits' tissue integrity.

The chromatic characteristics of raspberries packaged with the film-pads as the function of storage time are shown in Table 1.

Table 1. Changes in color values (L^* and a^*/b^*) of packaged raspberry (*Rubus idaeus* L. cv. 'Kweli') fruits (no film-pad _Ctr; with film-pad containing green tea extract_Ch + GTE, rosemary extract_Ch + RSME and only chitosan_Ch) stored for 14 days at 4 °C under passive MAP conditions.

Type of Package	Storage Time (Days)			
	0	3	7	14
		L^* values		
Ctr	27.37 ± 0.47 [abA]	27.47 ± 0.31 [aA]	26.54 ± 0.29 [bB]	26.28 ± 0.19 [bcAB]
Ch	27.37 ± 0.47 [aA]	27.44 ± 0.24 [aA]	26.89 ± 0.24 [aAB]	26.56 ± 0.18 [aAB]
Ch-GTE	27.37 ± 0.47 [aA]	27.36 ± 0.41 [aA]	27.23 ± 0.28 [aA]	26.75 ± 0.22 [aA]
Ch-RSME	27.37 ± 0.47 [aA]	27.56 ± 0.28 [aA]	27.24 ± 0.24 [aA]	27.10 ± 0.29 [aA]
		a^*/b^* values		
Ctr	2.17 ± 0.04 [aA]	2.24 ± 0.04 [aA]	2.13 ± 0.08 [abA]	2.01 ± 0.07 [bB]
Ch	2.17 ± 0.04 [aA]	2.21 ± 0.06 [aA]	2.16 ± 0.05 [abA]	2.05 ± 0.06 [bB]
Ch-GTE	2.17 ± 0.04 [aA]	2.19 ± 0.05 [aA]	2.22 ± 0.08 [aA]	2.20 ± 0.07 [aA]
Ch-RSME	2.17 ± 0.04 [aA]	2.23 ± 0.014 [aA]	2.20 ± 0.05 [aA]	2.24 ± 0.02 [aA]

Data are means ± SD of three replicates. Means in each row with the same letters (a–c) and means in each column with the same letters (A,B) are not significantly different ($p \leq 0.05$), according to Tukey's test.

The CIELab color space was used, measuring three coordinates: L^*, the lightness, as well as a^* and b^*, which are the green-red and blue-yellow color components, respectively. The a^*/b^* ratio has been used as a color index in various types of fresh fruits [56]. The raspberry fruits before treatments presented a bright red color, and the initial L^* and a^*/b^* ratio values were 27.37 and 2.17, respectively. Furthermore, the external color derived from anthocyanins is related to consumers' perception of quality and is an important parameter for visual ripeness and freshness assessments of raspberry [57]. Statistical analysis showed that the interaction of package type and storage time was significant for L^* and a^*/b^* ratio parameters. L^* values were constant with no significant differences ($p > 0.05$) for all package types during the first three days of storage. After that, differences gradually became more significant ($p < 0.05$), as raspberry fruits in control packages and in packages with Ch pads lost more of their luminosity (showed greater decreases in L^*) than berries in Ch + GTE and Ch + RSME packages, with L^* values of 26.56 and 26.28, respectively, at the end of storage. A decrease in the L^* value reflected the darkening of fruits by anthocyanin accumulation and indicated that the ripening process had occurred in the fruits [57]. Contrarily to Briano et al. [1], we found no relation between the CO_2 accumulation inside packages in control samples and its effect on the maintenance of L^* value. Furthermore, the raspberries in Ch + RSME and Ch + GTE packages showed the highest L^* value (27.10 and 26.75) at the end of storage. According to Perdones et al. [54] and Duran et al. [58], high moisture loss can be linked to the darkness on fruits. Indeed, raspberries from Ch + RSME and Ch + GTE packages showed the lowest weight loss during storage (Figure 3a). During storage, the color of fruits in control and Ch packages became less reddish on the seventh day, suggesting the degradation of anthocyanins during the ripening process and browning reactions that are typical of fruit senescence [57], and this trend was significantly ($p < 0.05$) more evident at the end of 14 days of storage. Indeed, fruits in control in Ch packages showed the lowest a^*/b^* ratio values of 2.01 and 2.05, respectively, typified by a red-yellow color and a lower aesthetic appeal enhanced by fungal spoilage. In contrast, Ch + RSME and Ch + GTE samples showed better color retention, with a more vivid red color, given the higher a^*/b^* ratio values of 2.24 and 2.20, respectively, at the end of 14 days of storage, which were not significantly ($p > 0.05$) different from the initial value observed at day 0.

3.4. Total Soluble Content, pH and Volatile Acidity

Total soluble solids (TSS) content is an important parameter that affects fruit quality and consumer acceptability [59], and high values are required for good berry flavor [60]. The results obtained from the TSS analysis are presented in Table 2. TSS content in raspberry fruits decreased gradually during storage time, and significant differences were observed when comparing different package types. The TSS values of fruits packaged with Ch + RSME pads slightly decreased from 9.53% at time 0 to 9.35% at day 3. The values remained at 9.33 through 4 days followed by a decline to 8.53% at the end of storage. Regarding the TSS values of fruits in Ch + GTE and Ch packages, although presenting the same decreasing behavior of those packaged with Ch + RSME pads during storage time, the values were significantly lower ($p < 0.05$), reaching values of 7.95% and 7.58% for packages with Ch + GTE and Ch pads, respectively, at the end of storage. In addition, the TSS content in control samples dropped sharply during storage and was significantly different ($p < 0.05$) from the values of fruits packaged with Ch + RSME and Ch + GTE pads, achieving the lowest value of 7.43% at 14 days of storage. This decrease in TSS can be explained by the fact that the hydrolysis of sucrose to maintain physiological activity was more rapid [59]. Indeed, the results showed that fruits from the control packages had enhanced fungal growth and decay (Section 3.2) and presented the highest respiration rate (Section 3.1), as well as the highest weight loss by transpiration (Section 3.3), during the storage period, in comparison with the fruits packaged with active film-pads (particularly with the Ch + RSME).

Changes in pH values of raspberries after packaged with different film-pads during cold storage are also shown in Table 2. This parameter, which is related to the fruit senescence [61], increased significantly during storage due to the utilization of organic acids during respiration [62]. The pH difference between day 7 and day 14 was not statistically significant ($p > 0.05$) for all package types. However, pH values of fruits of control and Ch packages were significantly ($p < 0.05$) higher than those of fruits packaged with Ch + GTE and Ch + RSME pads at day 7 and day 14, indicating that chitosan combined with green tea and rosemary extracts can delay changes in the pH of raspberry fruits.

Volatile acidity (VA) is another important parameter related to sensory appreciation. High VA values indicate the accumulation of acetic acid, which gives an off-flavor, such as a vinegar or acidic flavor, as result from redox reactions of ethanol and acetaldehyde [63], which are both by-products of the fermentative metabolism in over-ripe and senescent fruits [64]. The VA values of raspberry fruits increased significantly ($p < 0.05$) during the refrigerated storage for all package types (Table 2). Indeed, the VA values in control and Ch packages rapidly increased after day 3 and day 7, achieving the highest values of 0.415 and 0.235, respectively, at the end of storage. In contrast, the VA values for fruits packaged with Ch + GTE and Ch + RSME pads had a slower increase after day 3 and day 7, achieving the lowest values of 0.150 and 0.105, respectively.

Table 2. Changes in TSS, pH and volatile acidity values of packaged raspberry (*Rubus idaeus* L. cv. 'Kweli') fruits (no film-pad _Ctr; with film-pad containing green tea extract_Ch + GTE, rosemary extract_Ch + RSME and only chitosan_Ch) stored for 14 days at 4 °C under passive MAP conditions.

Treatments	Storage Time (Days)			
	0	3	7	14
		TSS values (%)		
Ctr	9.53 ± 0.11 ^{aA}	8.90 ± 0.24 ^{bB}	8.48 ± 0.16 ^{bC}	7.43 ± 0.15 ^{cC}
Ch	9.53 ± 0.11 ^{aA}	8.78 ± 0.14 ^{bB}	8.76 ± 0.13b ^{BC}	7.58 ± 0.16 ^{cC}
Ch-GTE	9.53 ± 0.11 ^{aA}	9.00 ± 0.14 ^{bB}	8.87 ± 0.12 ^{bB}	7.95 ± 0.12 ^{cB}
Ch-RSME	9.53 ± 0.11 ^{aA}	9.35 ± 0.07 ^{aA}	9.33 ± 0.06 ^{aA}	8.53 ± 0.06 ^{bA}
		pH values		
Ctr	2.87 ± 0.01 ^{cA}	2.96 ± 0.01 ^{bA}	3.13 ± 0.02 ^{aA}	3.17 ± 0.02 ^{aA}
Ch	2.87 ± 0.01 ^{cA}	2.94 ± 0.02 ^{bA}	3.14 ± 0.02 ^{aA}	3.16 ± 0.04 ^{aA}
Ch-GTE	2.87 ± 0.01 ^{cA}	2.92 ± 0.03 ^{bA}	3.05 ± 0.01 ^{aB}	3.10 ± 0.02 ^{aB}
Ch-RSME	2.87 ± 0.01 ^{bA}	2.88 ± 0.01 ^{bB}	3.05 ± 0.01 ^{aB}	3.07 ± 0.01 ^{aB}
		Volatile acidity values (mg acetic acid 100 g^{-1})		
Ctr	0.059 ± 0.002 ^{dA}	0.074 ± 0.037 ^{cA}	0.150 ± 0.042 ^{bA}	0.415 ± 0.021 ^{aA}
Ch	0.059 ± 0.002 ^{cA}	0.068 ± 0.025 ^{cA}	0.120 ± 0.014 ^{bAB}	0.235 ± 0.024 ^{aB}
Ch-GTE	0.059 ± 0.002 ^{bA}	0.065 ± 0.014 ^{bA}	0.105 ± 0.021 ^{aBC}	0.150 ± 0.019 ^{aC}
Ch-RSME	0.059 ± 0.002 ^{bA}	0.059 ± 0.015 ^{bA}	0.075 ± 0.021 ^{aC}	0.105 ± 0.021 ^{aD}

Data are means ± SD of three replicates. Means in each row with the same letters (a–c) and means in each column with the same letters (A–D) are not significantly different ($p \leq 0.05$), according to Tukey's test.

3.5. Total Phenolic Content and Antioxidant Activity

The initial total phenolic content (TPC) of raspberry fruits was 127 mg GAE 100 g^{-1}, and the respective antioxidant activity (AOA) was 41.4 mmol Trolox 100 g^{-1} with the DPPH method and 2.0 mmol Trolox 100 g^{-1} with the FRAP method. Figure 4 represents the changes in TPC and AOA of raspberries over the storage time at 4 °C, packaged with the different film-pads.

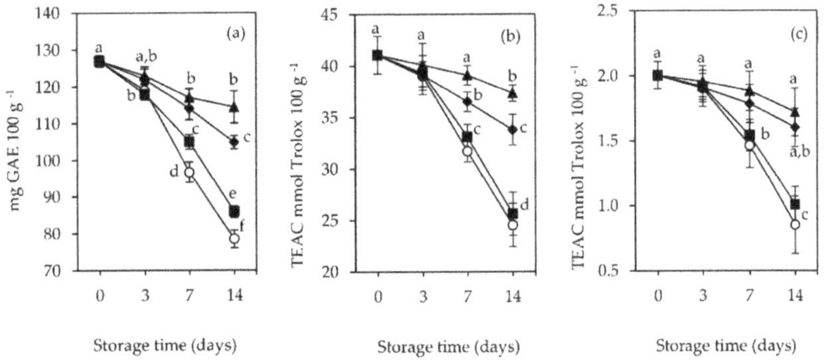

Figure 4. Total phenolic content (**a**) and antioxidant activity (AOA) [by DPPH (**b**) and FRAP (**c**) methods] of packaged raspberry (*Rubus idaeus* L. cv. 'Kweli') fruits (no film-pad_Ctr ○; with film-pad containing green tea extract_Ch + GTE ♦, rosemary extract_Ch + RSME ▲, only chitosan ■), stored for 14 days at 4 °C under passive MAP conditions. Vertical bars indicate standard deviation of three replicates. TEAC: Trolox equivalent antioxidant capacity; GAE: gallic acid equivalents. Means with equal letters are not statistically different by Tukey's test with 5% significance level.

As shown in Figure 4a, the TPC of raspberry fruits varied over time for all types of packages used. The maximum retention of TPC was found for fruits packaged with Ch + RSME pads (90.1%), followed by those of Ch + GTE packages (82.5%), at the end of the 14th day. The TPC of fruits of control and Ch packages showed a sharp decline after the third day of storage and the highest decrease (of 61.8% and 67.7%, respectively) at the end of storage. At 7 and 14 days of storage, the retention of TPC of fruits packaged with Ch + RSME and Ch + GTE pads was significantly ($p < 0.05$) different compared to TPC of fruits from control and Ch packages. Yang et al. [65] found a link between TPC values and the amount of several compounds, such as catechin and epicatechin, as well as phenolic acids (e.g., gallic and caffeic acid), reported to be present in raspberries. Ponder and Halmman [66] also reported the presence of myricetin, luteolin and kaempferol in these fruits. The rapid decline in the TPC values in fruits from control and Ch packages may be attributed to its higher respiration rate, resulting in the breakdown of total phenols [67], which may be related to the high fungal incidence, beginning on the third day for control packages and on the seventh day for Ch packages (Section 3.2), which progressively increased thereafter, causing decay in the majority of fruits (79.5% and 51%, respectively) at the end of storage, given the absence of the antifungal GTE and RSME in both cases.

The results of this work reveal a considerable variation in the AOA of raspberry across the different packaging types. In the DPPH and FRAP assays, a similar trend was observed for fruits from all the package types (Figure 4b,c). The highest AOA obtained with DPPH and FRAP methods at the end of the storage period was recorded for fruits from Ch + RSME packages (37.3 and 1.7 TEAC mmol Trolox 100 g^{-1}, respectively), followed by fruits from Ch + GTE packages (33.8 and 1.6 TEAC mmol Trolox 100 g^{-1}). On the other hand, the lowest AOA was observed for fruits from control (24.5 and 0.9 TEAC mmol Trolox 100 g^{-1}, respectively) and Ch packages (25.6 and 1.01 TEAC mmol Trolox 100 g^{-1}, respectively). As such, the antioxidant potential was found to be comparatively higher for the fruits packaged with the active film-pads. From those, the Ch + RSME film-pad was more effective in retaining the fruits' AOA during storage. The presence of the active film-pads inside raspberry fruit packages, by inhibiting fungal growth and spoilage (Section 3.2), seems to enhance the effect of packaging conditions applied, slowing down the metabolism in fruits during storage (Sections 3.1 and 3.3), thus minimizing the metabolism consumption of phenolics and flavonoids [68]. Under storage, secondary metabolites such as phenolics accumulate, causing an increase in the antioxidant levels [69].

Significant positive correlations ($p < 0.05$) were observed between TPC and AOA by DPPH and FRAP methods (Figure 5).

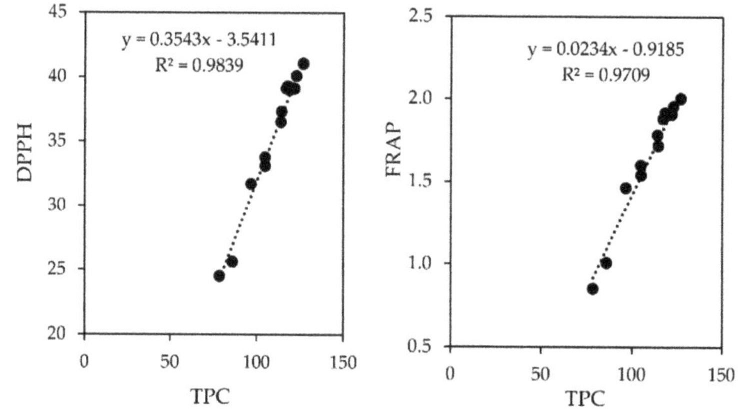

Figure 5. Correlations between AOA measured by DPPH and FRAP methods (TEAC mmol Trolox 100 g^{-1}) with TPC (mg GAE 100 g^{-1}). For both correlations, all data of the study (all sample types and all days of storage) were used.

The strong correlations observed indicate that the AOA of raspberries is mainly due to their composition in phenolic compounds. The different TEAC values obtained for the two methods are attributed to the different types of reactions taking place. The DPPH approach relies on radical scavenging, whereas the in the FRAP method, there is a reduction of Fe^{3+} [70]. These findings are consistent with those previously published [71], indicating that the FRAP and DPPH procedures are suitable for assessing antioxidant activity in raspberry fruit extracts with suitable reproducibility.

4. Conclusions

In this work, packaging strategies composed of trays sealed with compostable polylactic acid films and containing active film-pads made of chitosan with the incorporation of green tea (Ch + GTE) and rosemary ethanolic extracts (Ch + RSME) were tested in post-harvest preservation of raspberry fruits. The antifungal active film-pads tested in this work successfully decreased raspberry fruit fungal incidence, maintaining overall quality, as reflected in the fruits' physicochemical properties during storage, thereby extending the fruits' shelf life up to 14 days. These results can be attributed to the active compounds extracted from green tea and rosemary, acting as natural antifungal agents. At the end of storage (day 14), raspberry fruits packaged with both active film-pads maintained the most important qualitative and nutraceutical traits, close to those at harvest time, in comparison with fruits packaged without film-pads. Still, fruits packaged with Ch + RSME pads could be recommended, taking into account all the physical, chemical and microbial parameters studied. This green active packaging strategy may be suitable for other soft fruits to increase their marketability and as an effective substitute for non-biodegradable polymeric materials, applicable for products with clean and natural labels.

Author Contributions: Conceptualization, T.M.V., M.M.M. and V.D.A.; methodology, T.M.V., V.D.A. and M.M.M.; formal analysis, T.M.V.; resources, T.M.V., V.D.A. and M.M.M.; writing—original draft preparation, T.M.V.; writing—review and editing, T.M.V., V.D.A. and M.M.M.; supervision, M.M.M. and V.D.A.; project administration, M.M.M. and V.D.A.; funding acquisition, V.D.A. and M.M.M. All authors have read and agreed to the published version of the manuscript.

Funding: This research was funded by national funds through FCT—Fundação para a Ciência e a Tecnologia, I.P., under the project UIDB/04129/2020 of LEAF-Linking Landscape, Environment, Agriculture and Food, Research Unit. The scholarship of the first author was funded by the University of Lisbon (2015) and Instituto Superior de Agronomia (ISA), Portugal.

Institutional Review Board Statement: Not applicable.

Informed Consent Statement: Not applicable.

Data Availability Statement: Data is contained within the article.

Conflicts of Interest: The authors declare no conflict of interest. The funders had no role in the design of the study; in the collection, analyses, or interpretation of data; in the writing of the manuscript, or in the decision to publish the results.

References

1. Briano, R.; Giuggioli, N.R.; Girgenti, V.; Peano, C. Biodegradable and Compostable Film and Modified Atmosphere Packaging in Postharvest Supply Chain of Raspberry Fruits (cv. Grandeur). *J. Food Process. Preserv.* **2015**, *39*, 2061–2073. [CrossRef]
2. Hassanpour, H. Effect of Aloe vera gel coating on antioxidant capacity, antioxidant enzyme activities and decay in raspberry fruit. *LWT* **2015**, *60*, 495–501. [CrossRef]
3. Girgenti, V.; Peano, C.; Bounous, M.; Baudino, C. A life cycle assessment of non-renewable energy use and greenhouse gas emissions associated with blueberry and raspberry production in northern Italy. *Sci. Total Environ.* **2013**, *458–460*, 414–418. [CrossRef] [PubMed]
4. Belay, Z.A.; Caleb, O.J.; Opara, U.L. Influence of initial gas modification on physicochemical quality attributes and molecular changes in fresh and fresh-cut fruit during modified atmosphere packaging. *Food Packag. Shelf* **2019**, *21*, 100359. [CrossRef]
5. Siddiqui, M.W.; Rahman, M.S.; Wani, A.A. *Innovative Packaging of Fruits and Vegetables: Strategies for Safety and Quality Maintenance*, 1st ed.; Apple Academic Press: Palm Bay, FL, USA, 2018. [CrossRef]

6. Linke, M.; Geyer, M. Condensation dynamics in plastic film packaging of fruit and vegetables. *J. Food Eng.* **2013**, *116*, 144–154. [CrossRef]
7. Diego, A.C.; Aníbal, O.H. Modified Atmosphere Packaging: Design and Optimization Strategies for Fresh Produce. In *Postharvest Handling*; Kahramanoglu, I., Ed.; IntechOpen: London, UK, 2017. [CrossRef]
8. Jeong, M.; An, D.S.; Ahn, G.-H.; Lee, D.S. Master packaging system for sweet persimmon applicable to produce supply chains. *Postharvest Biol. Technol.* **2013**, *86*, 141–146. [CrossRef]
9. Murphy, S.H.; Marsh, J.J.; Kelly, C.A.; Leeke, G.A.; Jenkins, M.J. CO_2 assisted blending of poly(lactic acid) and poly(ε-caprolactone). *Eur. Polym. J.* **2017**, *88*, 34–43. [CrossRef]
10. Milovanovic, S.; Hollermann, G.; Errenst, C.; Pajnik, J.; Frerich, S.; Kroll, S.; Rezwan, K.; Ivanovic, J. Supercritical CO_2 impregnation of PLA/PCL films with natural substances for bacterial growth control in food packaging. *Food Res. Int.* **2018**, *107*, 486–495. [CrossRef]
11. Nofar, M.; Sacligil, D.; Carreau, P.J.; Kamal, M.R.; Heuzey, M.-C. Poly (lactic acid) blends: Processing, properties and applications. *Int. J. Biol. Macromol.* **2019**, *125*, 307–360. [CrossRef]
12. Peelman, N.; Ragaert, P.; De Meulenaer, B.; Adons, D.; Peeters, R.; Cardon, L.; Van Impe, F.; Devlieghere, F. Application of bioplastics for food packaging. *Trends Food Sci. Technol.* **2013**, *32*, 128–141. [CrossRef]
13. Lorite, G.S.; Rocha, J.M.; Miilumäki, N.; Saavalainen, P.; Selkälä, T.; Morales-Cid, G.; Gonçalves, M.P.; Pongrácz, E.; Rocha, C.M.R.; Toth, G. Evaluation of physicochemical/microbial properties and life cycle assessment (LCA) of PLA-based nanocomposite active packaging. *LWT* **2017**, *75*, 305–315. [CrossRef]
14. Jung, J.; Zhao, Y. Chapter 18—Antimicrobial Packaging for Fresh and Minimally Processed Fruits and Vegetables. In *Antimicrobial Food Packaging*; Barros-Velázquez, J., Ed.; Academic Press: San Diego, CA, USA, 2016; pp. 243–256. [CrossRef]
15. Sanches-Silva, A.; Costa, D.; Albuquerque, T.G.; Buonocore, G.G.; Ramos, F.; Castilho, M.C.; Machado, A.V.; Costa, H.S. Trends in the use of natural antioxidants in active food packaging: A review. *Food Addit. Contam. A* **2014**, *31*, 374–395. [CrossRef] [PubMed]
16. Bovi, G.G.; Caleb, O.J.; Klaus, E.; Tintchev, F.; Rauh, C.; Mahajan, P.V. Moisture absorption kinetics of FruitPad for packaging of fresh strawberry. *J. Food Eng.* **2018**, *223*, 248–254. [CrossRef]
17. Gaikwad, K.K.; Singh, S.; Ajji, A. Moisture absorbers for food packaging applications. *Environ. Chem. Lett.* **2019**, *17*, 609–628. [CrossRef]
18. USFDA. GRAS claim notification for chitosan. In *GRN000073. US Food and Drug Administration, Center for Food Safety and Applied Nutrition*; Office of Premarket Approval: Naperville, IL, USA, 2001.
19. EU. Establishing a list of permitted health claims made on foods, other than those referring to the reduction of disease risk and to children's development and health. *OJEU* **2012**, *136*, 1–40.
20. Nawaz, T.; Fatima, M.; Shah, S.Z.H.; Afzal, M. Coating effect of rosemary extract combined with chitosan on storage quality of mori (*Cirrhinus mrigala*). *J. Food Process. Preserv.* **2020**, *44*, e14833. [CrossRef]
21. Abdollahi, M.; Rezaei, M.; Farzi, G. Improvement of active chitosan film properties with rosemary essential oil for food packaging. *Int. J. Food Sci. Tech.* **2012**, *47*, 847–853. [CrossRef]
22. Siripatrawan, U. 14-Active food packaging from chitosan incorporated with plant polyphenols. In *Novel Approaches of Nanotechnology in Food*; Grumezescu, A.M., Ed.; Academic Press: Cambridge, MA, USA, 2016; pp. 465–507. [CrossRef]
23. Piñeros-Hernandez, D.; Medina-Jaramillo, C.; López-Córdoba, A.; Goyanes, S. Edible cassava starch films carrying rosemary antioxidant extracts for potential use as active food packaging. *Food Hydrocoll.* **2017**, *63*, 488–495. [CrossRef]
24. Andrade, M.A.; Ribeiro-Santos, R.; Costa Bonito, M.C.; Saraiva, M.; Sanches-Silva, A. Characterization of rosemary and thyme extracts for incorporation into a whey protein based film. *LWT* **2018**, *92*, 497–508. [CrossRef]
25. Mohsenabadi, N.; Rajaei, A.; Tabatabaei, M.; Mohsenifar, A. Physical and antimicrobial properties of starch-carboxy methyl cellulose film containing rosemary essential oils encapsulated in chitosan nanogel. *Int. J. Biol. Macromol.* **2018**, *112*, 148–155. [CrossRef]
26. Hendel, N.; Larous, L.; Belbey, L. Antioxidant activity of rosemary (*Rosmarinus officinalis* L.) and its in vitro inhibitory effect on Penicillium digitatum. *Int. Food Res. J.* **2016**, *23*, 1725–1732.
27. Moghtader, M.; Salari, H.; Farahmand, A. Evaluation of the antifungal effects of rosemary oil and comparison with synthetic borneol and fungicide on the growth of Aspergillus flavus. *J. Ecol. Nat.* **2011**, *3*, 210–214. [CrossRef]
28. Carrizo, D.; Taborda, G.; Nerín, C.; Bosetti, O. Extension of shelf life of two fatty foods using a new antioxidant multilayer packaging containing green tea extract. *Innov. Food Sci. Emerg. Technol.* **2016**, *33*, 534–541. [CrossRef]
29. Zivanovic, S.; Chi, S.; Draughon, A.F. Antimicrobial Activity of Chitosan Films Enriched with Essential Oils. *J. Food Sci.* **2005**, *70*, M45–M51. [CrossRef]
30. Ponce, A.G.; Roura, S.I.; del Valle, C.E.; Moreira, M.R. Antimicrobial and antioxidant activities of edible coatings enriched with natural plant extracts: In vitro and in vivo studies. *Postharvest Biol. Technol.* **2008**, *49*, 294–300. [CrossRef]
31. Duan, J.; Cherian, G.; Zhao, Y. Quality enhancement in fresh and frozen lingcod (*Ophiodon elongates*) fillets by employment of fish oil incorporated chitosan coatings. *Food Chem.* **2010**, *119*, 524–532. [CrossRef]
32. Alvarez, M.V.; Ponce, A.G.; Moreira, M.D.R. Antimicrobial efficiency of chitosan coating enriched with bioactive compounds to improve the safety of fresh cut broccoli. *LWT* **2013**, *50*, 78–87. [CrossRef]
33. Lekjing, S. A chitosan-based coating with or without clove oil extends the shelf life of cooked pork sausages in refrigerated storage. *Meat Sci.* **2016**, *111*, 192–197. [CrossRef]

34. Venkatachalam, K.; Lekjing, S. A chitosan-based edible film with clove essential oil and nisin for improving the quality and shelf life of pork patties in cold storage. *RSC Adv.* **2020**, *10*, 17777–17786. [CrossRef]
35. Vieira, T.M.; Brito, L.; Moldão-Martins, M.; Alves, V.D. Characterization of antifungal chitosan active film-pads based on chitosan with incorporation of green tea and rosemary extracts. *Membranes* **2022**. *under submission*.
36. Giuggioli, N.R.; Briano, R.; Baudino, C.; Peano, C. Effects of packaging and storage conditions on quality and volatile compounds of raspberry fruits. *CyTA-J. Food* **2015**, *13*, 512–521. [CrossRef]
37. Vieira, T.M.; Moldão-Martins, M.; Alves, V.D. Composite Coatings of Chitosan and Alginate Emulsions with Olive Oil to Enhance Postharvest Quality and Shelf Life of Fresh Figs (*Ficus carica* L. cv. 'Pingo De Mel'). *Foods* **2021**, *10*, 718. [CrossRef] [PubMed]
38. *ISO 6632:1981*; Fruits, Vegetables, and Derived Products—Determination of Volatile Acidity. International Organization for Standardization: Geneva, Switzerland, 1981. Available online: https://www.iso.org/obp/ui/#iso:std:13049:en (accessed on 16 June 2022).
39. Kopjar, M.; Orsolic, M.; Pilizota, V. Anthocyanins, Phenols, and Antioxidant Activity of Sour Cherry Puree Extracts and their Stability During Storage. *Int. J. Food Prop.* **2014**, *17*, 1393–1405. [CrossRef]
40. Gonçalves, J.C.; Coelho, M.T.; da Graça Diogo, M.; Alves, V.D.; Bronze, M.R.; Coimbra, M.A.; Martins, V.M.; Moldão-Martins, M. In vitro Shoot Cultures of Pterospartum tridentatum as an Alternative to Wild Plants as a Source of Bioactive Compounds. *Nat. Prod. Commun.* **2018**, *13*, 1934578X1801300415. [CrossRef]
41. Ricci, A.; Teslic, N.; Petropolus, V.-I.; Parpinello, G.P.; Versari, A. Fast Analysis of Total Polyphenol Content and Antioxidant Activity in Wines and Oenological Tannins Using a Flow Injection System with Tandem Diode Array and Electrochemical Detections. *Food Anal. Methods* **2019**, *12*, 347–354. [CrossRef]
42. Huang, D.; Ou, B.; Prior, R.L. The Chemistry behind Antioxidant Capacity Assays. *J. Agric. Food Chem.* **2005**, *53*, 1841–1856. [CrossRef]
43. Suárez, B.; Álvarez, Á.L.; García, Y.D.; Barrio, G.D.; Lobo, A.P.; Parra, F. Phenolic profiles, antioxidant activity and in vitro antiviral properties of apple pomace. *Food Chem.* **2010**, *120*, 339–342. [CrossRef]
44. Beaudry, R.M. Effect of O_2 and CO_2 partial pressure on selected phenomena affecting fruit and vegetable quality. *Postharvest Biol. Technol.* **1999**, *15*, 293–303. [CrossRef]
45. Nell, M.; Mammerler, R.; Steinkellner, S. Oxygen consumption-based evaluation of fungal activity. *Mycol. Res.* **2006**, *110*, 760–764. [CrossRef]
46. Dennis, W.J.; Arthur, C.C.; Ahmad, S.; Peter, D.P.; Randolph, M.B. Modified-atmosphere Packaging of 'Heritage' Red Raspberry Fruit: Respiratory Response to Reduced Oxygen, Enhanced Carbon Dioxide, and Temperature. *J. Am. Soc. Hortic. Sci.* **1994**, *119*, 540–545. [CrossRef]
47. Argenta, L.C.; Fan, X.; Mattheis, J.P. Responses of 'Fuji' apples to short and long duration exposure to elevated CO_2 concentration. *Postharvest Biol. Technol.* **2002**, *24*, 13–24. [CrossRef]
48. Adobatia, A.; Uboldib, E.; Franzettia, L.; Limbo, S. Shelf Life Extension of Raspberry: Passive and Active Modified Atmosphere Inside Master Bag Solutions. *Chem. Eng. Tans.* **2015**, *44*, 337–342. [CrossRef]
49. Halász, K.; Hosakun, Y.; Csóka, L. Reducing Water Vapor Permeability of Poly(lactic acid) Film and Bottle through Layer-by-Layer Deposition of Green-Processed Cellulose Nanocrystals and Chitosan. *Int. J. Polym. Sci.* **2015**, *2015*, 954290. [CrossRef]
50. Tezotto-Uliana, J.V.; Fargoni, G.P.; Geerdink, G.M.; Kluge, R.A. Chitosan applications pre- or postharvest prolong raspberry shelf-life quality. *Postharvest Biol. Technol.* **2014**, *91*, 72–77. [CrossRef]
51. El Ghaouth, A.; Arul, J.; Ponnampalam, R.; Boulet, M. Chitosan Coating Effect on Storability and Quality of Fresh Strawberries. *J. Food Sci.* **1991**, *56*, 1618–1620. [CrossRef]
52. Vargas, M.; Albors, A.; Chiralt, A.; González-Martínez, C. Quality of cold-stored strawberries as affected by chitosan–oleic acid edible coatings. *Postharvest Biol. Technol.* **2006**, *41*, 164–171. [CrossRef]
53. Vu, K.D.; Hollingsworth, R.G.; Leroux, E.; Salmieri, S.; Lacroix, M. Development of edible bioactive coating based on modified chitosan for increasing the shelf life of strawberries. *Food Res. Int.* **2011**, *44*, 198–203. [CrossRef]
54. Perdones, A.; Sánchez-González, L.; Chiralt, A.; Vargas, M. Effect of chitosan–lemon essential oil coatings on storage-keeping quality of strawberry. *Postharvest Biol. Technol.* **2012**, *70*, 32–41. [CrossRef]
55. Paniagua, A.C.; East, A.R.; Hindmarsh, J.P.; Heyes, J.A. Moisture loss is the major cause of firmness change during postharvest storage of blueberry. *Postharvest Biol. Technol.* **2013**, *79*, 13–19. [CrossRef]
56. Pathare, P.B.; Opara, U.L.; Al-Said, F.A.-J. Colour Measurement and Analysis in Fresh and Processed Foods: A Review. *Food Bioproc. Technol.* **2013**, *6*, 36–60. [CrossRef]
57. Krüger, E.; Dietrich, H.; Schöpplein, E.; Rasim, S.; Kürbel, P. Cultivar, storage conditions and ripening effects on physical and chemical qualities of red raspberry fruit. *Postharvest Biol. Technol.* **2011**, *60*, 31–37. [CrossRef]
58. Duran, M.; Aday, M.S.; Zorba, N.N.D.; Temizkan, R.; Büyükcan, M.B.; Caner, C. Potential of antimicrobial active packaging 'containing natamycin, nisin, pomegranate and grape seed extract in chitosan coating' to extend shelf life of fresh strawberry. *Food Bioprod. Process.* **2016**, *98*, 354–363. [CrossRef]
59. Aday, M.S.; Caner, C. Understanding the effects of various edible coatings on the storability of fresh cherry. *Packag. Technol. Sci.* **2010**, *23*, 441–456. [CrossRef]
60. Kader, A.A. Quality and its maintenance in relation to the postharvest physiology of strawberry. In *The Strawberry into the 21st.*; Dale, A., Luby, J.J., Eds.; Timber Press: Portland, OR, USA, 1991.

61. Han, C.; Zhao, Y.; Leonard, S.W.; Traber, M.G. Edible coatings to improve storability and enhance nutritional value of fresh and frozen strawberries (*Fragaria* × *ananassa*) and raspberries (*Rubus ideaus*). *Postharvest Biol. Technol.* **2004**, *33*, 67–78. [CrossRef]
62. Martínez-Ferrer, M.; Harper, C.; Pérez-Muntoz, F.; Chaparro, M. Modified Atmosphere Packaging of Minimally Processed Mango and Pineapple Fruits. *J. Food Sci.* **2002**, *67*, 3365–3371. [CrossRef]
63. Zoecklein, B.W.; Fugelsang, K.C.; Gump, B.H.; Nury, F.S. Volatile Acidity. In *Production Wine Analysis*; Zoecklein, B.W., Fugelsang, K.C., Gump, B.H., Nury, F.S., Eds.; Springer US: Boston, MA, USA, 1990; pp. 98–113. [CrossRef]
64. Wani, A.A.; Singh, P.; Gul, K.; Wani, M.H.; Langowski, H.C. Sweet cherry (*Prunus avium*): Critical factors affecting the composition and shelf life. *Food Packag. Shelf Life* **2014**, *1*, 86–99. [CrossRef]
65. Yang, J.; Cui, J.; Chen, J.; Yao, J.; Hao, Y.; Fan, Y.; Liu, Y. Evaluation of physicochemical properties in three raspberries (*Rubus idaeus*) at five ripening stages in northern China. *Sci. Hortic.* **2020**, *263*, 109146. [CrossRef]
66. Ponder, A.; Hallmann, E. The effects of organic and conventional farm management and harvest time on the polyphenol content in different raspberry cultivars. *Food Chem.* **2019**, *301*, 125295. [CrossRef]
67. Ali, A.; Muhammad, M.T.M.; Sijam, K.; Siddiqui, Y. Effect of chitosan coatings on the physicochemical characteristics of Eksotika II papaya (*Carica papaya* L.) fruit during cold storage. *Food Chem.* **2011**, *124*, 620–626. [CrossRef]
68. Gonzalez-Aguilar, G.A.; Villa-Rodriguez, J.A.; Ayala-Zavala, J.F.; Yahia, E.M. Improvement of the antioxidant status of tropical fruits as a secondary response to some postharvest treatments. *Trends Food Sci. Technol.* **2010**, *21*, 475–482. [CrossRef]
69. Frusciante, L.; Carli, P.; Ercolano, M.R.; Pernice, R.; Di Matteo, A.; Fogliano, V.; Pellegrini, N. Antioxidant nutritional quality of tomato. *Mol. Nutr. Food Res.* **2007**, *51*, 609–617. [CrossRef] [PubMed]
70. Drogoudi, P.; Pantelidis, G.E.; Goulas, V.; Manganaris, G.A.; Ziogas, V.; Manganaris, A. The appraisal of qualitative parameters and antioxidant contents during postharvest peach fruit ripening underlines the genotype significance. *Postharvest Biol. Technol.* **2016**, *115*, 142–150. [CrossRef]
71. Obeng, E.; Kpodo, F.M.; Tettey, C.O.; Essuman, E.K.; Adzinyo, O.A. Antioxidant, total phenols and proximate constituents of four tropical leafy vegetables. *Sci. Afr.* **2020**, *7*, e00227. [CrossRef]

foods

MDPI

Article

Quality Attributes of Refrigerated Barhi Dates Coated with Edible Chitosan Containing Natural Functional Ingredients

Kashif Ghafoor *, Fahad Y. Al-Juhaimi, Elfadil E. Babiker, Isam A. Mohamed Ahmed, Syed Ali Shahzad and Omer N. Alsawmahi

Department of Food Science and Nutrition, College of Food and Agricultural Sciences, King Saud University, P.O. Box 2460, Riyadh 11451, Saudi Arabia; faljuhaimi@ksu.edu.sa (F.Y.A.-J.); ebabiker.c@ksu.edu.sa (E.E.B.); iali@ksu.edu.sa (I.A.M.A.); syedalishahzad@gmail.com (S.A.S.); omernasser2009@hotmail.com (O.N.A.)

* Correspondence: kghafoor@ksu.edu.sa; Tel.: +966-11-4691951

Abstract: Edible chitosan coatings with natural functional ingredients were used to preserve quality attributes of fresh Barhi date fruit. Fruits were coated with chitosan and/or 1 and 2% olive cake extract (OCE) or orange peel extract (OPE). Both coated and uncoated fruits were stored at 4 °C for 4 weeks. A slight decrease in the pH and increase in acidity with storage was observed. However, when chitosan was mixed with OCE or OPE, an increase in pH was observed with a concomitant decrease in acidity. The phenolic content of the samples was decreased with time. However, coating the date with OCE or OPE significantly ($p \leq 0.05$) increased the total phenolic with a concomitant increase in radical scavenging activity. The textural properties, particularly hardness, were better preserved in case of coated dates. The sensory evaluation data showed non-significant changes in the acceptability of the Barhi dates throughout the storage period. Chitosan-coating significantly ($p \leq 0.05$) inhibited mold growth over time. Scanning electron microscope (SEM) imaging showed difference among different coatings. According to principal component analysis (PCA), OCE and OPE were found to have protective effects on fruit quality.

Keywords: Barhi dates; *Phoenix dactylifera*; chitosan; orange peel; olive cake; coating; quality attributes; scanning electron microscopy; surface structure

Citation: Ghafoor, K.; Al-Juhaimi, F.Y.; Babiker, E.E.; Mohamed Ahmed, I.A.; Shahzad, S.A.; Alsawmahi, O.N. Quality Attributes of Refrigerated Barhi Dates Coated with Edible Chitosan Containing Natural Functional Ingredients. *Foods* **2022**, *11*, 1584. https://doi.org/10.3390/foods11111584

Academic Editors: Maria Cefola and Bernardo Pace

Received: 6 May 2022
Accepted: 25 May 2022
Published: 28 May 2022

1. Introduction

Date (*Phoenix dactylifera*) palm is one of the essential subsistence crops and a major fruiting tree in the Middle East and some other countries [1]. This tree bears a climacteric fruit which has certain characteristic physical stages while maturing due to the production of ethylene. Date fruit ripening is associated with four maturation stages that are generally recognized using Arabic terms. An immature fruit is having a hard texture and green color and is considered at Kimri stage of it its maturity. A hard, yellow, and edible Barhi date is considered to be in the Khalal stage; a soft, brown, and semi-ripe stage reflects the Rutab stage of dates and a final soft, dark brown, and fully ripe date fruit is considered to reflect the Tamar stage [2]. These different phases of date fruit maturation are also characterized by their specific textural, sensory, and chemical characteristics [3,4]. Date fruits are consumed in different stages of maturity, such as Khalal, Rutab, and Tamer stages, due to certain organoleptic characteristics and consumer preferences [5]. Barhi is a seasonal date and is cherished when at its Khalal stage (sweet taste and yellow color). Barhi dates have a short span at this stage after harvest (August–October in Saudi Arabia) and change to a Rutab stage quickly (approximately 1 week) if not preserved properly. The Khalal Barhi has high importance from a market and consumer perspectives due to its high demand, characteristic taste, and associated health benefits. These quality attributes of Barhi dates at a Khalal stage may be due to bioactive compounds and certain flavor components. Due to these reasons, it is important to study and establish techniques that can increase the shelf-life of

Barhi at the Khalal stage and preserve its characteristics for a longer time [6]. Edible film coating is a developing and frequently used technique in which a thin film layer of edible materials is applied on fruit and vegetable surfaces. Coating can aid in the control of gas exchange and transfer of moisture, thereby modifying the internal atmosphere, maintaining quality, and prolonging the postharvest shelf life of fruit and vegetables [7,8]. There are different types of materials for such coatings, including biodegradable polymers such as polysaccharides, lipids, and proteins [9]. Chitosan is a polysaccharide material, derived from nature, that is non-toxic, biodegradable, biocompatible, and antimicrobial, and has good film-forming properties; thus, it is often used as coating material [10]. Generally, natural products extracted from different types of plant by-products contain phenolic substances with beneficial bioactivities, such as antimicrobial and antioxidant activity, and hence, there is ongoing research on the use of these phytochemicals into chitosan films and coating them on fresh agro-produce [11]. The current study was carried out to investigate the effect of edible chitosan coatings with natural functional ingredients obtained from orange peel and olive cake on the quality attributes of fresh Barhi date fruit. Evaluation of different physicochemical attributes of Barhi dates (coated or uncoated) along with characterization of surface structure using electron microscopy were also carried out.

2. Materials and Methods

2.1. Materials

Fresh Barhi dates were obtained from local dealers in Riyadh and Qassim provinces during August–October 2021, and various trials were immediately carried out to date samples. The fruit samples were identified by experts in the College of Food and Agriculture for the appropriate maturity at Khalal stage and then properly labeled. Fruits were inspected and any damaged fruits were removed from the spikelet and only good quality fruits on the spikelet were selected for further treatment and experiment. Fresh olives were obtained from the local market and the oil was extracted using mechanical oil extractor (SUS 304, SUS Machines, Shanghai, China) using the cold pressing technique. The remaining cake material was dried and ground to powder form (moisture 11.32 ± 1.43%). Orange peels were manually removed from the fresh oranges obtained from the local farm. Orange peels were dried, and then ground to make the powder form (moisture 7.67 ± 1.85%). Both olive cake and orange peel powders were stored at 4 °C before further use. All chemicals used were of reagent grade.

2.2. By-Product Extracts' Preparation

A known weight of olives cake and orange peel powders (50 g) was extracted in distilled water at 70 °C for 30 min followed by cooling at room temperature and filtration to remove impurities. The filtrate was dried under vacuum at 50 °C using a rotary evaporator and the extract was saved in sterilized bags and stored at −20 °C before use for physico-chemical analysis. The extraction process was repeated for the filtered residue and substantial quantities of both olive cakes extract (OCE) and orange peel extract (OPE) were obtained. The extraction process was carried out at a low temperature to preserve the phytochemical compounds in the extract [12,13].

2.3. Preparation of Coating Solution and Application

A stock solution of chitosan was prepared by dissolution of chitosan powder (2% w/v) 1% acetic acid along with 1% glycerol as a plasticizer. The OCE and OPE were added to the chitosan solution at the ratios of 1% and 2.0%, mixed and homogenized in a blender (Acapulco 30564, Palson Co., Kunshan, China) for 5 min until a smooth solution was obtained [14]. Fresh Barhi date fruit was washed and allowed to dry at room temperature. Once completely dried, each spikelet of date fruit was dipped in various coating solutions for 5 min. Spikelets were carefully removed from the coating solution and kept on a sieved stand to allow drying at room temperature for 2–3 h. A small plastic fan (CK2215, Clickon,

Liwan, China) was also used to allow quicker drying of the coated dates. Afterward, dates were removed from the spikelet and grouped into six batches:

- Batch A: Uncoated as a control;
- Batch B: Coated with chitosan (2%) solution;
- Batch C: Coated with chitosan (2%) and OCE (1%) solution;
- Batch D: Coated with chitosan (2%) and OCE (2%) solution;
- Batch E: Coated with chitosan (2%) and OPE (1%) solution;
- Batch F: Coated with chitosan (2%) and OPE (2%) solution.

Each batch was then packed in a polyethylene plastic container with 5–6 holes in the lid and stored at 4 ° C for different periods (0, 7, 14, 21, and 28 days). Each batch was prepared in triplicates.

2.4. Physicochemical Characterization

The moisture content was estimated using the oven drying method [15]. Water activity was measured using a water activity meter (Aqualab CX3-TE, Labo-Scientifica, Parma, Italy). After equilibration, the water activity value was recorded. The total soluble solids (TSS) contents of the samples, either uncoated or coated (°Brix) were determined a 20 °C using a digital refractometer (DR 6000, A. Kruss Optronic GmbH, Hamburg, Germany), and the °Brix value was calculated using the dilution factor. The pH determination included homogenization of 5 g of sample with 50 mL of deionized distilled water followed by pH measurements using a Corning 240 pH meter (Corning Scientific Products, New York, NY, USA). A potentiometric titration method [15] using NaOH solution (0.1 N) with pH 8.1 was applied to measure the titratable acidity of the slurry made from date samples.

2.5. Microstructure Determination Using Scanning Electron Microscopy (SEM)

In order to carry out the SEM study, small cuttings of date fruit surfaces (coated and uncoated) were obtained and freeze dried for 3 days. The surface morphology of coated and uncoated date skin samples was studied and scanned using a field-emission scanning electron microscope (JSM-7600, Jeol Ltd., Tokyo, Japan) at a resolution of 1000 [16]. It was also equipped with energy dispersive spectroscopy or EDS. Previously dried samples were coated with platinum for 35 s and the total layer thickness of the sample was 25 mm to avoid sample charging under the electron beam. The imaging was done using secondary electrons. The working distance was kept 4.5 mm and an accelerating voltage (15 kV) was used during this study.

2.6. Total Phenolics Determination

The total phenolic content of the samples was determined by the method described by Singleton and Rossi [17] using a Folin–Ciocalteu as the main reagent and gallic acid as a standard, and the results were expressed in mg/100 g gallic acid equivalent (GAE).

2.7. Antioxidant Activity

The method of Lee et al. [18] was used to evaluate the anti-DPPH radical activities of date fruit samples. One mL of the extract was diluted in methanol, then added mixed with a 2 mL DPPH solution. Methanol was used as a control and the absorbance of the mixtures was read with a spectrophotometer at 517 nm. The DPPH inhibition was estimated using the formula:

$$\text{DPPH inhibition (\%)} = \left(\frac{A_{\text{control517}} - A_{\text{sample517}}}{A_{\text{control517}}}\right) \times 100$$

where A is the absorbance recorded and DPPH inhibition was expressed in percentage.

2.8. Color Measurement

The color evaluation of the samples was carried out using instrumental color method before and during storage. A Hunter Lab colorimeter (Model No. Miniscan® XE plus 4500 L, Hunter Associates Laboratory, Inc., Reston, VA, USA) was used for this evaluation as described previously [12].

2.9. Texture Measurement

The samples before and after storage were subjected to texture profile analysis using a texture analyzer (Model CT3, Brookfield, Middleboro, MA, USA). The hardness (kg), cohesiveness, and springiness (mm) were measured in triplicate using a two-cycle test.

2.10. Sensory Evaluation

Sensory analysis was carried out on the first day after coating and then after each week of storage. Twenty semi-trained panelists (male, age range 20–35 years old) were recruited for the evaluation of sensory attributes of date fruit samples using a five-point hedonic scale. The serving of the samples was random using coded and panelists evaluated the texture, color, taste, odor, and overall acceptability. The evaluation was based on the scale of 1–5 (1 = extremely dislike, 5 = extremely like). The sensory evaluation studies were carried out in in three sessions for each storage period (0, 7, 14, 21, and 28 days). The scores were for each sample and each session were calculated as means before subjecting to data analysis. A mean score in the range of 3 to 5 was considered acceptable.

2.11. Mold Counts

A pour plate method was applied for the total count of molds at 25 °C for 3–5 days using enumeration (Yeast Extract Glucose Chloramphenicol Agar (YGC), Merck, Darmstadt, Germany). Molds were counted aseptically which involved mixing 25 g of the samples with 225 mL sterile Ringer solution in a Stomacher blender for 1 min.

2.12. Principal Component Analysis (PCA)

Coating treatment and storage times were evaluated, for their effects and interrelations on the physicochemical properties of Barhi date samples, using principal component analysis techniques in MultiPlot software [19].

2.13. Statistical Analysis

The experiments were carried out using a completely randomized block design with six treatments (control, CH, 1% OCE + CH, 2% OCE + CH, 1% OPE + CH, and 2% OPE + CH) in triplicates, and the physicochemical attributes were evaluated on five storage periods (0, 7, 14, 21, and 28 days). The measurements for each quality attribute were carried out in triplicates. The entire blocks were triplicated independently. SAS software (SAS Institute, Inc., Cary, NC, USA) was used to analyze the data obtained using two-way analysis of variance (ANOVA) and Duncan's Multiple Range Tests. The treatments and storage times were considered as fixed effects and the replications of the experiments as random effects in model studies. The comparison of sensory evaluation scores was carried out between the treatments and storage times. A General Linear Model (GLM) and Duncan's Multiple Range Test were used for the comparison of means. The data from triplicate measurements were presented as means and standard deviation (SD). The statistical significances were defined at a probability value of ≤ 0.05.

3. Results and Discussion

3.1. Changes in Water Activity, Moisture, and Total Soluble Solids of Coated Fresh Barhi Dates

Table 1 shows changes in moisture, water activity, and total soluble solids of fresh Barhi dates coated with chitosan and/or olive cake or orange peel extracts during cold storage (4 °C). The moisture content of all samples remained above 50% irrespective of the storage period and coating treatment, although it decreased significantly ($p \leq 0.05$) with an increase in storage time. The coating treatments showed variable effects on the prevention of moisture loss from the date surface. According to Iqbal et al. [20], moisture loss was observed in date fruits during storage as they progressed from the Kimri to the Rutab stage. Coatings were more effective in preventing moisture loss from dates at 4 °C, and coatings experienced slightly less moisture loss during storage. Hoa et al. [21] found that hydrophobic coatings were more effective than hydrophilic coatings in slowing the weight loss of mango fruits during storage than hydrophilic coatings, which resulted in weight losses comparable to uncoated controls. The water activity of the samples stored at 4 °C was above 0.900 even after 28 days of storage. There were no insignificant differences among different treatments; however, with time, the water activity of the samples during the first two weeks was slightly high compared to those stored for 3 or 4 weeks. The reduction in water activity of the date samples with time could be due to a reduction in moisture content. Similarly, an increase in TSS was observed with storage time, which can be attributed to the loss in the samples' moisture content. TSS was varied between coating materials, with 1% OPE + CH coating providing the lowest percentage of TSS and the control sample providing a higher value than the other treatments at week 4. The increase in TSS with storage time implies that the fruits underwent anaerobic respiration, in which simple sugars are broken down into alcohol and acetaldehyde [22], and the value of TSS as a predictor of the transition from Khalal to Rutab is lost. Permeability in some coating materials, which maintain aerobic respiration, was expected. Moreover, the differences between the storage periods studied were large enough to be statistically significant. This increase could be due to the conversion of some insoluble compounds into soluble compounds (for example, protopectin to pectin) or to water loss from fruits. Thompson and Abboodi [23] found that lower moisture content had a positive effect on the TSS percentage. The current findings are consistent with those reported by Abd El-Moneim et al. [24], where it was reported that the soluble solid contents of Zaghloul date palm cv. were the lowest at zero time, and increased consistently with increasing cold storage period up to sixty days.

Table 1. Changes in water activity, moisture, and total soluble solids of fresh Barhi dates coated with chitosan and/or olive cake or orange peel extracts during cold storage (4 °C).

Treatment	Storage Period (Days)				
	0	7	14	21	28
			Moisture (%)		
Uncoated	67.72 ± 1.03 aB	67.17 ± 2.01 aA	63.11 ± 1.66 bA	62.36 ± 0.97 cA	61.99 ± 2.88 dB
CH	67.39 ± 2.11 aB	65.56 ± 0.88 bB	63.38 ± 2.03 cA	61.90 ± 1.05 dB	61.67 ± 1.67 dB
1% OCE + CH	66.81 ± 1.55 aB	66.25 ± 0.87 aB	63.17 ± 3.01 bA	62.64 ± 1.44 cA	61.35 ± 1.27 dB
2% OCE + CH	67.89 ± 0.99 aB	66.26 ± 1.01 bB	62.44 ± 1.44 cB	61.63 ± 2.07 dB	62.61 ± 2.04 cA
1% OPE + CH	69.07 ± 1.07 aA	65.26 ± 1.09 bC	62.57 ± 1.62 cB	62.31 ± 2.01 cA	62.01 ± 1.77 cA
2% OPE + CH	67.55 ± 1.23 aB	66.47 ± 2.01 bB	63.03 ± 1.67 cA	62.43 ± 3.07 cA	61.14 ± 2.15 dB
			Water activity (aw)		
Uncoated	0.940 ± 0.11	0.928 ± 0.08	0.915 ± 0.06	0.909 ± 0.01	0.905 ± 0.07
CH	0.946 ± 0.12	0.932 ± 0.03	0.911 ± 0.05	0.905 ± 0.04	0.904 ± 0.13
1% OCE + CH	0.934 ± 0.22	0.926 ± 0.04	0.913 ± 0.22	0.916 ± 0.02	0.906 ± 0.23
2% OCE + CH	0.943 ± 0.31	0.938 ± 0.12	0.912 ± 0.34	0.904 ± 0.13	0.901 ± 0.17
1% OPE + CH	0.935 ± 0.07	0.929 ± 0.21	0.911 ± 0.24	0.909 ± 0.16	0.900 ± 0.04
2% OPE + CH	0.940 ± 0.03	0.932 ± 0.09	0.917 ± 0.06	0.909 ± 0.13	0.903 ± 0.07
			Total soluble solids (%)		
Uncoated	25.40 ± 1.07 eA	27.20 ± 1.13 dA	29.20 ± 2.02 cA	33.20 ± 2.33 bA	35.80 ± 1.09 aA
CH	25.00 ± 0.98 eA	26.50 ± 1.09 dB	28.34 ± 1.04 cB	29.46 ± 1.23 bB	30.71 ± 2.01 aB
1% OCE + CH	22.20 ± 0.77 dC	22.58 ± 2.01 dE	26.50 ± 1.22 cC	28.45 ± 1.19 bC	29.66 ± 1.67 aC
2% OCE + CH	23.00 ± 0.68 cB	23.64 ± 0.99 cD	25.40 ± 0.88 bD	25.65 ± 2.05 bE	28.90 ± 1.45 aD
1% OPE + CH	21.80 ± 0.69 dC	21.86 ± 0.89 dE	28.76 ± 0.79 cB	29.33 ± 2.07 bB	30.58 ± 1.84 aB
2% OPE + CH	23.40 ± 0.59 eB	24.56 ± 1.05 dC	26.40 ± 0.97 cC	27.66 ± 1.16 bD	28.83 ± 1.58 aD

Values are means of triplicate samples (±SD). Means not sharing common lowercase letters in a row or capital letters in a column are significantly different at $p \leq 0.05$ as assessed by Duncan's Multiple Range Test. CH = Chitosan, OCE = Olive cake extract, OPE = Orange peel extract.

3.2. Changes in pH, Acidity, Total Phenolics, and Antioxidant Activity (DPPH Inhibition) of Coated Fresh Barhi Dates

Table 2 shows changes in pH, acidity, total phenolics, and antioxidant activity of fresh Barhi dates coated with chitosan and/or olive cake or orange peel extracts during cold storage (4 °C). Regardless of coating materials, the results showed that a slight decrease in the pH and increase in acidity was observed. However, when chitosan was mixed with olive cake or orange peel extracts, an increase in pH was observed with a concomitant decrease in acidity regardless of the storage time.

Table 2. Changes in pH, acidity, total phenolics, and antioxidant activity of fresh Barhi dates coated with chitosan and/or olive cake or orange peel extracts during cold storage (4 °C).

Treatment	Storage Period (Days)				
	0	7	14	21	28
			pH		
Uncoated	6.39 ± 0.21 a	6.12 ± 0.82 a	5.89 ± 0.72 ab	5.22 ± 0.41 bBC	4.23 ± 0.63 cB
CH	6.35 ± 0.08 a	6.22 ± 0.67 a	6.13 ± 0.66 a	5.78 ± 0.46 abB	5.12 ± 0.55 bA
1% OCE + CH	6.33 ± 0.11 a	6.39 ± 0.54 a	5.98 ± 0.47 a	5.24 ± 0..91 bAB	5.11 ± 0.71 bA
2% OCE + CH	6.34 ± 0.26 a	6.41 ± 0.44 a	6.12 ± 0.48 a	5.89 ± 0.28 abA	5.69 ± 0.88 bA
1% OPE + CH	6.17 ± 0.56 a	6.52 ± 0.35 a	6.21 ± 0.29 a	6.12 ± 0.51 aA	5.46 ± 0.38 bA
2% OPE + CH	6.34 ± 0.61 a	6.68 ± 0.49 a	6.25 ± 0.34 a	6.12 ± 0.43 aA	5.79 ± 0.39 aA
			Titratable acidity (% malic acid)		
Uncoated	0.096 ± 0.002 b	0.172 ± 0.02 abA	0.199 ± 0.012 aA	0.236 ± 0.031 aA	0.257 ± 0.013 aA
CH	0.109 ± 0.01 b	0.131 ± 0.004 bB	0.157 ± 0.013 bB	0.187 ± 0.007 aA	0.203 ± 0.021 aA
1% OCE + CH	0.110 ± 0.02 b	0.129 ± 0.012 bB	0.179 ± 0.006 aA	0.213 ± 0.021 aA	0.232 ± 0.022 aA
2% OCE + CH	0.111 ± 0.003 b	0.125 ± 0.031 bB	0.156 ± 0.003 aB	0.186 ± 0.002 aA	0.202 ± 0.011 aA
1% OPE + CH	0.104 ± 0.001 b	0.130 ± 0.001 aB	0.145 ± 0.005 aB	0.173 ± 0.0024 aA	0.188 ± 0.033 aA
2% OPE + CH	0.104 ± 0.004 b	0.116 ± 0.003 bB	0.114 ± 0.014 aB	0.136 ± 0.0021 aB	0.148 ± 0.002 aB
			Total phenolics (mg GAE/g)		
Uncoated	7.18 ± 0.31 aE	5.516 ± 0.12 bF	3.854 ± 0.19 cF	3.458 ± 0.51 cF	2.842 ± 0.11 dF
CH	8.37 ± 0.42 aD	6.705 ± 0.36 bE	5.050 ± 0.21 cE	4.968 ± 0.38 cE	4.125 ± 0.23 dE
1% OCE + CH	9.54 ± 0.45 aC	7.989 ± 0.42 bD	6.442 ± 0.23 cD	5.460 ± 0.46 dD	5.147 ± 0.31 dD
2% OCE + CH	10.19 ± 0.62 aB	8.550 ± 0.45 bC	7.897 ± 0.33 cC	6.987 ± 0.52 dC	6.789 ± 0.41 dC
1% OPE + CH	10.76 ± 0.57 aB	9.408 ± 0.61 bB	8.057 ± 0.37 cB	7.458 ± 0.43 cdB	7.244 ± 0.55 dB
2% OPE + CH	13.10 ± 0.72 aA	11.564 ± 0.39 bA	10.036 ± 0.61 cA	9.546 ± 0.37 cA	9.785 ± 0.67 cA
			DPPH inhibition (%)		
Uncoated	43.782 ± 1.22 bD	54.00 ± 0.88 aE	31.69 ± 0.71 cE	15.26 ± 0.95 eF	20.15 ± 0.48 dF
CH	42.760 ± 0.98 bE	54.60 ± 0.78 aE	29.90 ± 0.84 cF	24.25 ± 0.48 dE	22.15 ± 0.39 eE
1% OCE + CH	48.722 ± 2.01 bC	60.65 ± 0.77 aD	38.07 ± 0.87 cD	35.25 ± 0.69 dD	26.25 ± 0.55 eD
2% OCE + CH	78.250 ± 1.34 aA	71.72 ± 0.65 bC	66.25 ± 1.32 cB	55.24 ± 0.75 dC	48.25 ± 0.43 eC
1% OPE + CH	70.102 ± 1.83 bB	82.03 ± 1.03 aB	60.86 ± 1.56 cC	59.58 ± 1.11 dB	58.36 ± 0.57 eB
2% OPE + CH	77.683 ± 0.93 bA	89.61 ± 1.13 aA	65.33 ± 1.72 cA	62.58 ± 2.01 dA	60.45 ± 0.92 eA

Values are means of triplicate samples (±SD). Means not sharing lowercase letters in a row or capital letters in a column are significantly different at $p \leq 0.05$ as assessed by Duncan's Multiple Range Test. CH = Chitosan, OCE = Olive cake extract, OPE = Orange peel extract.

The study's data revealed that increasing the storage period increased the acidity of Barhi fruits at different rates for all studied treatments, but most of the studied coating treatments had lower decreasing rates compared to the control sample until the fifth week of storage. The findings are consistent with those reported by Abd El-Moneim et al. [24] on Zaghloul date fruits during the orange season. This was most likely since the film formed by materials altered the fruit's endogenous CO_2 and O_2 concentrations, causing ripening to be delayed. The effect of edible coating on acidity loss has been observed in chitosan- and alginate-coated peaches [25], as well as in avocado coated with methylcellulose [26]. Because organic acids are substrates for many reactions during aerobic respiration in plant cells, the effect of coating on acidity retention could be due to the lower respiration rate found in coated fruits.

As shown in Table 2, the phenolic contents of the control dates were lower than that of the coated samples. The rise in phenolic compounds was due to the presence of phenolic compounds in the extracts obtained from olive cake and orange peel, as these by-products are considered to be rich in phenolic compounds. During the first week of storage, the dates coated using OPE had higher total phenolics than other samples. Variations were observed during the storage of the samples; however, the phenolic compounds of Barhi date alone remained lower than in other samples. Some interesting results were obtained for the coated samples. It was observed that the phenolic compounds of date samples coated with OPE decreased slightly with storage although they were initially higher than the samples coated with OCE, which may be attributed to the stability of phenolic compounds in the coating materials, prepared using olive cake phytochemicals. However, as it has been established already that phenolic compounds are important phytochemicals having various health beneficial properties as well as the ability to prevent microbial spoilage, it seems that the use of olive cake and orange peel extracts can significantly increase the contents of these compounds in coated Barhi dates.

Consistent with the results of total phenolic compounds, the radical scavenging activity of uncoated dates was lower than those of chitosan, OCE + chitosan, and OPE + chitosan-coated dates. This is evidence that coating materials used in this study improve the functional properties of Barhi dates. The radical activity of all samples was decreased with the storage time, but coated fruits still exhibited higher radical scavenging activities as compared to control samples. The radical scavenging activity seems to be less affected by the storage temperature although low-temperature storage or refrigeration is recommended for all types of fresh fruits and vegetables. The above reported results were consistent and well correlated in terms of the bioactive compounds of dates and antioxidant activity results similar to other fruits such as grapes [13]. The declining trend in phenols and antioxidant compounds may be attributed to enzymatic oxidation (polyphenol oxidase and peroxidase) during storage [27]. Both OCE and OPE are expected to contribute significantly to the occurrence of various type of antioxidants, phytochemicals and bioactive compounds, which are capable of preserving oxidations reactions, reduce microbial growth, and enhance health benefits, as reported earlier [12,16]

3.3. Changes in Color and Texture of Coated Fresh Barhi Dates

The results of the color are presented in Table 3, where L* indicates whiteness or brightness/darkness, a* redness/greenness, and b* yellowness/blueness. In terms of L* values, it can be observed that the lightness or brightness of the date color decreased with the storage time for both coated and uncoated dates, which indicated that the dates were brighter at the start of the storage time, whereas this decreased with the progression of the storage time. The brightness of the samples on the first day showed insignificant differences; however, a naked eye observation revealed that the dates coated with a mixture of OCE and chitosan were brighter in appearance as compared to the other samples and coated dates. However, this visual difference was not detected during the instrumental measurement of color values.

Table 3. Changes in color and texture of fresh Barhi dates coated with chitosan and/or olive cake or orange peel extracts during cold storage (4 °C).

Treatment	Storage Time														
	0			7			14			21			28		
	L*	a*	b*	L*	a*	b*	L*	a*	b*	L*	a*	b*	L*	a*	b*
	Color														
Uncoated	58.76 ± 0.73 aA	1.89 ± 0.04 aA	38.99 ± 1.23 aA	46.01 ± 1.22 cC	1.20 ± 0.03 dA	26.75 ± 1.02 cE	51.14 ± 1.31 bD	0.41 ± 0.01 eB	31.60 ± 1.01 bE	42.44 ± 1.12 dE	1.65 ± 0.07 bA	25.75 ± 1.03 dD	39.19 ± 1.05 eF	1.52 ± 0.03 cA	24.87 ± 1.11 eB
CH	59.27 ± 1.02 aA	1.61 ± 0.12 bA	35.00 ± 1.09 aC	53.05 ± 0.89 bB	0.58 ± 0.05 cB	27.20 ± 1.23 cD	49.66 ± 1.09 cE	0.35 ± 0.04 dB	29.30 ± 0.79 bE	43.87 ± 0.89 dD	2.08 ± 0.14 aA	23.32 ± 0.99 dE	41.25 ± 1.22 eE	1.98 ± 0.12 aA	19.48 ± 0.89 eD
1% OCE + CH	57.54 ± 0.88 aB	0.96 ± 0.34 bB	35.74 ± 2.03 aC	55.88 ± 1.04 bA	1.46 ± 0.06 aA	32.96 ± .96 cC	54.73 ± 0.69 cB	0.53 ± 0.03 cB	34.55 ± 1.03 bB	49.45 ± 0.74 dC	1.47 ± 0.21 aA	28.36 ± 0.79 dC	45.29 ± 1.34 eC	1.65 ± 0.21 aA	22.48 ± 0.78 eC
2% OCE + CH	57.37 ± 0.79 aB	0.62 ± 0.24 cC	37.20 ± 0.89 aB	56.29 ± 0.79 bA	1.22 ± 0.12 aA	33.53 ± 0.55 cB	56.16 ± 067 bA	0.51 ± 0.02 dB	35.05 ± 0.78 bA	42.28 ± 1.08 cE	0.78 ± 0.05 bB	20.67 ± 0.67 dF	42.75 ± 1.52 cD	0.76 ± 0.08 bB	19.56 ± 0.66 eD
1% OPE + CH	59.26 ± 0.67 aA	1.35 ± 0.67 dA	38.56 ± 0.78 aA	55.95 ± 0.81 bA	1.98 ± 0.14 aA	34.37 ± 0.73 bA	53.92 ± 0.55 dC	1.11 ± 0.06 eA	32.77 ± 0.64 cCD	55.53 ± 1.04 bB	1.72 ± 0.08 bB	34.95 ± 0.59 bA	54.26 ± 1.65 cA	1.52 ± 0.05 cA	32.49 ± 1.02 cA
2% OPE + CH	58.05 ± 0.99aA	1.46 ± 0.46 bA	33.23 ± 0.48 aD	43.78 ± 0.69 cD	1.32 ± 0.08 bA	21.51 ± 0.66 dF	47.04 ± 0.63 dF	1.05 ± 0.02 cA	26.63 ± 0.73 cF	56.28 ± 1.07 bA	1.91 ± 0.06 aA	32.64 ± 0.78 bB	50.25 ± 0.98 cB	1.48 ± 0.06 bA	32.85 ± 0.77 bA
	Texture														
Treatment	Ha	Co	Sp.	Ha	Co	Sp.	Ha	Co	Spr	Ha	Co	Sp.	Ha	Co	Sp.
Uncoated	1020.8 ± 5.34 aF	0.83 ± 0.03 a	0.867 ± 0.12 a	262.17 ± 2.11 bF	0.82 ± 0.01 a	0.911 ± 0.07 a	244.33 ± 4.33 cC	0.74 ± 0.02 a	0.80 ± 0.11 b	55.50 ± 1.07 dF	0.77 ± 0.13 a	0.80 ± 0.04 b	39.50 ± 1.02 eE	0.73 ± 0.11 a	0.77 ± 0.13 b
CH	1069.0 ± 3.66 aE	0.85 ± 0.04 a	0.911 ± 0.08 a	649.17 ± 3.14 bB	0.82 ± 0.04 a	0.867 ± 0.05 a	282.33 ± 3.24 cB	0.77 ± 0.04 a	0.83 ± 0.03 a	78.00 ± 0.98 dA	0.82 ± 0.21 a	0.91 ± 0.12 a	47.33 ± 1.23 eC	0.72 ± 0.04 a	0.81 ± 0.09 a
1% OCE + CH	1166.3 ± 7.55 aD	0.84 ± 0.02 a	0.911 ± 0.04 a	662.83 ± 5.34 bA	0.84 ± 0.07 a	0.867 ± 0.03 a	353.17 ± 2.88 cA	0.81 ± 0.07 a	0.83 ± 0.04 a	57.83 ± 0.83 dE	0.79 ± 0.08 a	0.83 ± 0.06 a	46.50 ± 1.63 eD	0.75 ± 0.13 a	0.77 ± 0.12 b
2% OCE + CH	1344.8 ± 8.55 aA	0.84 ± 0.08 a	0.911 ± 0.06 a	540.67 ± 4.38 bC	0.84 ± 0.05 a	0.911 ± 0.05 a	244.83 ± 4.11 cC	0.77 ± 0.05 a	0.77 ± 0.12 b	63.67 ± 0.91 dD	0.72 ± 0.03 a	0.77 ± 0.05 b	51.67 ± 0.89 eA	0.75 ± 0.06 a	0.77 ± 0.21 b
1% OPE +CH	1238.6 ± 8.56 aB	0.85 ± 0.03 a	0.911 ± 0.07 a	450.67 ± 6.12 bD	0.78 ± 0.09 a	0.833 ± 0.06 a	244.17 ± 5.23 cC	0.74 ± 0.11 a	0.81 ± 0.02 a	67.83 ± 0.69 dB	0.69 ± 0.01 b	0.77 ± 0.04 b	47.50 ± 0.79 eC	0.68 ± 0.03 b	0.73 ± 0.05 b
2% OPE + CH	1227.6 ± 7.99 aC	0.87 ± 0.05 a	0.911 ± 0.02 a	425.00 ± 5.26 bE	0.79 ± 0.03 a	0.800 ± 0.09 a	239.33 ± 2.89 cD	0.72 ± 0.08 a	0.83 ± 0.06 a	65.67 ± 0.59 dC	0.76 ± 0.05 a	0.83 ± 0.13 a	48.83 ± 0.46 eB	0.73 ± 0.08 a	0.81 ± 0.16 a

Values are means of triplicate samples (±SD). Means not sharing common lowercase letters in a row or capital letters in a column are significantly different at $p \leq 0.05$ as assessed by Duncan's Multiple Range Test. Abbreviations: CH, Chitosan; OCE, Olive cake extract; OPE, Orange peel extract; Ha, hardness; Co, Cohesiveness; Sp., Springiness.

The a* values for the control sample were decreased with time and fluctuated in treated samples and the differences were varied between treatments. The b* values were decreased with the storage time, particularly on the 28th day of storage, showing a decrease in the yellowness of the dates. The treatment seemed to have invariable effects on the color values; for example, the brightness seemed to be better preserved in coated dates stored at 4 °C. Overall coating with OCE extract has positive effects on the color values of dates. The decrease in b* values indicates that the yellowing of the samples was significantly lower in the coated fruit when compared to the control sample. The lower change in b values was most likely due to the retention of fruit pigments, with Maskan [28] reporting that a lower b* value might be due to the hydrolysis of chlorophyll and carotenoid pigments, non-enzymatic Maillard browning, and formation of brown pigments. Coatings have been found to mitigate the effects of such interactions. It has been reported that changes in the brightness of fruits can be used to predict fruit browning [29]. Surface coating of the fruits with chitosan and/or OCE and OPE reduced the rate of loss in the brightness of the fruits by lowering the rate of loss in L values, and it was discovered that as the concentration of OCE or OPE increased, the rate of loss in L value decreased. The findings revealed that coating preserved the brightness of the fruits by giving them a shiny appearance. According to Eissa [30], chitosan coating reduces oxidative enzyme activity, such as polyphenol-oxidase, peroxidase, catalase, and laccase, which are associated with discoloration. As a result, the changes in color parameters of the control samples in this study were more pronounced than those of the coated samples. In a study by Jiang and Li [31], it was determined that chitosan coating inhibits the growth of some fungi and delays further decay of stored longan fruit. Similarly, chitosan coating appears to reduce the pH of mushrooms during storage, as reported by Eissa [30]. This is an indication that chitosan coating reduces pathogen development and accordingly could be partially useful in delaying discoloration and browning during storage.

The results of textural analyses of the date samples were shown in Table 3. The hardness, cohesiveness, and springiness of date fruits were decreased with storage time with a significant reduction observed in hardness. However, coating of the date samples alleviates the effect of storage on date texture. The loss of firmness is an important criterion that indicates the quality of the date during storage. The chitosan and/or OCE and OPE mixture coating improved the texture of the date significantly. The firmness of all samples decreased with storage, but chitosan and/or OCE and OPE-coated dates retained their firmness longer than the control sample. At the end of the third week, both the control and coated samples experienced rapid firmness loss, with the control samples experiencing the most rapid changes. According to Mannozzi et al. [32], who studied the effect of edible coatings on the quality of blueberry fruits during shelf-life, the higher firmness values of the coated date are likely due to the presence of the coating agent, which provides structural rigidity at the product's surface. Date softening during storage is determined by cell structure deterioration, cell wall composition, and intracellular materials, as reported for guava by Hong et al. [33]. The firmness of chitosan and/or OCE and OPE-treated dates may be retained due to a reduction in respiration and other maturation processes during storage as a result of covering the date's cuticle and lentils with the materials' coating. The observed firmness loss is consistent with the findings of Hong et al. [33], who investigated the effect of chitosan coating on guava.

3.4. Sensory Evaluation of Coated Fresh Barhi Dates Samples

Samples were coded anonymously and the fruit was evaluated for texture, color, taste, odor, astringency, and overall acceptability using a five-point hedonic scale, where one denotes "disliked extremely" and five reflects "liked extremely". The results of the sensory evaluation are presented in Table 4. The sensory evaluation data showed non-significant changes in acceptability and sensory properties of the Barhi dates due to different coating materials throughout the storage period. The overall acceptability of control or those coated with 2% OCE or 1% OPE and 2% chitosan were closer to each other. However, the sensory properties of all the samples seemed to get a low score but were not significant with the advent of storage time. Dates stored at 4 °C were fairly acceptable until the end of the storage period and showed fairly high acceptability compared to control samples, revealing a positive effect of coating materials on the color, texture, and sensory quality of Barhi dates. Abu-Shama [34] investigated the effect of edible coatings on the fruit quality of the Barhi date cultivar and concluded that all edible coating treatments studied had little to no effect on the organoleptic characteristics of Barhi date fruits, implying that these treatments could be used to extend the shelf life of Barhi date fruits.

3.5. Changes in Molds of Coated Fresh Barhi Dates

Figure 1 shows molds (cfu/g) of fresh Barhi dates coated with chitosan and/or olive cake or orange peel extracts during cold storage (4 °C). There was a significant increase in the fungal content of the control dates during storage. However, in chitosan-coated date molds increased during the first week, and thereafter, dropped significantly, but in other coating materials, they were significantly lower than in the control samples. Molds had a maximum value of 170 cfu/g on day 28 for the control samples, while coated date had a value of 10 cfu/g at the same storage time. The increase in simple sugars and decrease in moisture content of the fruits during date maturation create a better microenvironment for fungal growth [35]. In comparison to the control samples, the results showed that chitosan and/or OCE and OPE-containing coatings performed best at 4 °C in terms of keeping fungal numbers under control. Differences could be attributed to molds' inability to use sugars as a substrate [35]. Furthermore, it appeared that the preservation of simple sugars in coated samples inhibited fungal growth. According to Lasram et al. [36], all coatings were more effective in retarding fungal growth at 3 °C than at 25 °C because the lower temperature was suboptimal for fungal growth and metabolism.

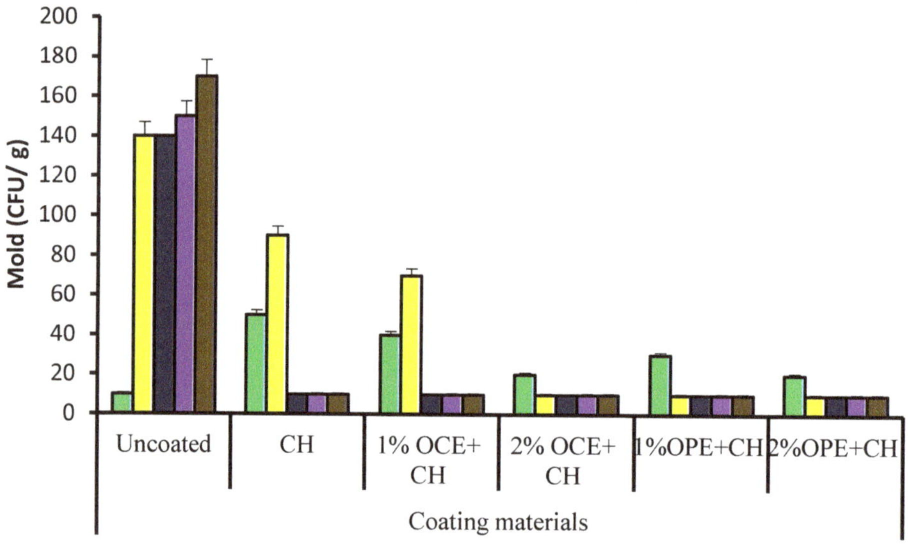

Figure 1. Mold (CFU/g) of fresh Barhi dates coated with chitosan and/or olive cake or orange peel extracts during cold storage (4 °C). CH, Chitosan; OCE, Olive cake extract; OPE, Orange peel extract. Columns from left to right (storage period, days), 0 (green), 7 (yellow), 14 (blue), 21 (purple), and 28 (brown).

Table 4. Sensory evaluation of fresh Barhi dates coated with chitosan and/or olive cake or orange peel extracts during cold storage (4 °C).

Treatment	Storage Period (Days)											
	0						7					
	Texture	Color	Taste	Odor	Astringency	Overall Accept.	Texture	Color	Taste	Odor	Astringency	Overall Accept.
Uncoated	4.17 ± 0.23	4.33 ± 0.32	4.33 ± 0.57	4.25 ± 0.67	4.33 ± 0.28	4.00 ± 0.23	4.25 ± 0.38	3.75 ± 0.22	3.25 ± 0.44	3.50 ± 0.77	3.75 ± 0.38	4.00 ± 0.83
CH	4.50 ± 0.31	4.50 ± 0.35	4.00 ± 0.56	4.50 ± 0.49	4.17 ± 0.47	3.83 ± 0.35	3.75 ± 0.29	3.50 ± 0.76	3.50 ± 0.22	3.75 ± 0.62	3.75 ± 0.42	4.33 ± 0.72
1% OCE + CH	4.17 ± 0.44	4.50 ± 0.62	4.05 ± 0.55	4.58 ± 0.44	4.00 ± 0.51	3.95 ± 0.67	3.75 ± 0.66	3.75 ± 0.61	4.00 ± 0.82	3.50 ± 0.56	3.50 ± 0.83	4.33 ± 0.76
2% OCE + CH	4.00 ± 0.54	4.00 ± 0.47	4.42 ± 0.49	4.50 ± 0.29	3.67 ± 0.47	4.17 ± 0.82	3.75 ± 0.27	4.00 ± 0.53	3.50 ± 0.33	3.25 ± 0.51	4.00 ± 0.91	4.33 ± 0.59
1% OPE + CH	4.00 ± 0.34	3.83 ± 0.56	4.17 ± 0.38	4.05 ± 0.78	3.17 ± 0.66	4.00 ± 0.54	3.50 ± 0.48	3.00 ± 0.19	3.75 ± 0.17	3.50 ± 0.41	3.50 ± 0.36	4.00 ± 0.62
2% OPE + CH	3.83 ± 0.45	3.83 ± 0.33	3.50 ± 0.28	4.00 ± 0.81	3.25 ± 0.43	3.72 ± 0.37	4.00 ± 0.53	3.25 ± 0.26	3.50 ± 0.28	3.75 ± 0.55	3.75 ± 0.47	4.00 ± 0.28
	14						21					
Treatment	Texture	Color	Taste	Odor	Astringency	Overall accept.	Texture	Color	Taste	Odor	Astringency	Overall accept.
Uncoated	3.67 ± 0.24	4.00 ± 0.47	4.00 ± 0.46	4.00 ± 0.83	3.67 ± 0.42	3.63 ± 0.16	3.15 ± 0.41	3.17 ± 0.29	3.14 ± 0.82	3.85 ± 0.22	3.42 ± 0.37	3.16 ± 0.46
CH	3.67 ± 0.66	3.33 ± 0.38	4.00 ± 0.34	3.67 ± 0.46	3.33 ± 0.65	3.63 ± 0.33	3.65 ± 0.72	3.33 ± 0.32	3.48 ± 0.33	3.45 ± 0.53	3.28 ± 0.46	3.48 ± 0.63
1% OCE + CH	3.67 ± 0.37	3.00 ± 0.39	3.67 ± 0.87	3.33 ± 0.36	3.00 ± 0.72	3.25 ± 0.27	3.85 ± 0.82	3.00 ± 0.41	3.58 ± 0.46	3.52 ± 0.61	2.78 ± 0.23	3.75 ± 0.27
2% OCE + CH	3.67 ± 0.57	3.00 ± 0.27	3.00 ± 0.77	3.33 ± 0.33	3.33 ± 0.39	3.13 ± 0.28	3.58 ± 0.88	3.00 ± 0.52	3.24 ± 0.37	3.48 ± 0.72	3.47 ± 0.29	3.65 ± 0.41
1% OPE + CH	4.33 ± 0.88	3.33 ± 0.81	3.33 ± 0.48	3.67 ± 0.87	3.00 ± 0.22	3.63 ± 0.38	4.18 ± 0.92	3.33 ± 0.84	3.45 ± 0.32	3.75 ± 0.54	3.17 ± 0.37	3.88 ± 0.46
2% OPE + CH	3.67 ± 0.89	3.33 ± 0.71	4.00 ± 0.43	3.33 ± 0.67	3.50 ± 0.87	3.63 ± 0.44	3.52 ± 0.26	3.33 ± 0.73	3.78 ± 0.49	3.17 ± 0.29	3.46 ± 0.27	3.63 ± 0.88
	28											
Treatment	Texture	Color	Taste	Odor	Astringency	Overall accept.						
Uncoated	3.25 ± 0.35	3.27 ± 0.43	3.23 ± 0.32	3.75 ± 0.32	3.39 ± 0.32	3.22 ± 0.26						
CH	3.57 ± 0.68	3.29 ± 0.22	3.46 ± 0.23	3.53 ± 0.33	3.32 ± 0.19	3.46 ± 0.37						
1% OCE + CH	3.61 ± 0.48	3.11 ± 0.31	3.61 ± 0.26	3.29 ± 0.53	2.81 ± 0.37	3.47 ± 0.51						
2% OCE + CH	3.57 ± 0.27	3.13 ± 0.32	3.29 ± 0.17	3.51 ± 0.52	3.39 ± 0.41	3.71 ± 0.65						
1% OPE + CH	3.27 ± 0.44	3.39 ± 0.54	3.25 ± 0.52	3.69 ± 0.42	3.27 ± 0.47	3.89 ± 0.33						
2% OPE + CH	3.21 ± 0.63	3.23 ± 0.53	3.68 ± 0.39	3.23 ± 0.43	3.52 ± 0.38	3.74 ± 0.66						

Values are means of triplicate samples (±SD). Means not sharing common lowercase letters in a row or capital letters in a column are significantly different at $p \leq 0.05$ as assessed by Duncan's Multiple Range Test. CH, Chitosan; OCE, Olive cake extract; OPE, Orange peel extract.

3.6. Surface Characteristics of Coated and Uncoated Date Fruit

The surface and cross-section microstructures of coated dates with chitosan (2%) alone, chitosan (2%) and OCE (1–2%), and chitosan (2%) and OPE (1–2%) were examined using a scanning electron microscope and compared with those of uncoated date fruit surfaces. The visualization of the structural characteristics of coated date fruit and that of fresh date fruit surface was carried out for the first time (to the best of our knowledge). Figure 2A shows the scanning electron microscope (SEM) images (resolution $\times 1000$) of fresh Barhi dates (without coating). The fruit surface was rough due to the natural structure of the date fruit surface and can be due to the cellulose and pectin network on the surface of the fruit. These natural openings may be helping in the transfer of gases and may also in some cases aid in the ripening process. Chitosan, which is normally used in the development of edible coatings, was coated alone (2% solution) and the SEM studies of chitosan-coated date fruit surface (Figure 2B) showed that the cracks that appeared naturally on the fruit surfaces were covered and the surface of the fruit became smooth. The coating material (chitosan 2%) was modified with OCE (1–2%) and the SEM studies (Figure 2C,D) showed that the fruit surface became very smooth, particularly when the OCE concentration was 2% in the coating solution. It appears that the OCE 1% chitosan solution might be less viscous than the OCE 2% chitosan solution, which might have completely covered the rough surface structures. SEM studies (Figure 2E,F) of citrus peel extract in edible chitosan coating showed a less smooth surface and some cracks were visible somewhat similar to cracks that appeared when fruits were coated with chitosan alone. Hence, the SEM studies showed that olive cake extract produced the best coating smoothness, and this can be attributed to the presence of certain fatty materials in the extracts that can also help in improving the surface color. Tran et al. [11] also studied the characteristics of the chitosan coating film after the incorporation of plant essential oils and observed that chitosan alone showed a smooth (observed by SEM) and homogeneous surface, whereas essential oils made the membrane less uniform, and the higher the content of essential oil, the less homogeneous the surface was. This might have been due to the addition of essential oils in the membrane matrix, which broke down the continuous structure of the polymer matrix [37]. However, our findings were not in agreement with those of Tran et al. [11], as both the addition of OCE and OPE caused smoothness of the chitosan coating.

3.7. Principal Component Analysis (PCA) of Physical Properties of Coated Fresh Barhi Dates

To assess the combined effects of treatments on the physical properties of Barhi date, PCA was conducted, and the results are shown in Figure 3. The results indicated a high contribution of the PC1 (65.14%) to the total variability of the plotted components (80.05%) followed by PC2 (14.91%). In the biplot, the cosine of the angle between the vectors of the traits indicated the correlations between them, in which acute, obtuse or straight, or straight angles indicate positive, negative, and no correlations, respectively [38]. Highly positive correlations were seen among color attributes (L, b, and a), moisture and water activity, and texture attributes (hardness, cohesiveness, and springiness), whereas these attributes were negatively correlated with TSS and redness (a). Three clusters of the treatments were seen based on their effect on the physical properties of Barhi dates. The first group (upper right of the graph, black circle symbol) is characterized by a high level of TSS, and this group is composed of the samples untreated (negative control), CH-coated (positive control), and 1 and 2% OCE-coated Barhi dates stored at 4 °C for longer time (21 and 28 days). The high TSS of these samples during prolonged storage indicates rapid decomposition of intact matters of the dates by enzymatic or microbial action and thereby releasing more soluble materials. The second group (right of the graph, red square symbol) is characterized by higher levels of redness (a) than other groups. This group was composed of negative and positive control samples stored at 4 °C for 7 and 14 days, CH-coated with 1% OCE stored at 4 °C for 7 and 14 days, and CH-coated with OPE (1 and 2%) stored at 4 °C for 7, 14, 21, and 28 days. The third group (left of the graph, blue tringle symbol) is characterized by higher levels of color attributes (L, b, and a), moisture, water activity, and texture attributes (hardness, cohesiveness, and springiness) than the other groups. This group was composed of the fresh samples (0 day) of positive and negative controls, and those treated with different concentrations of OCE and OPE indicating that storage at different temperatures and durations adversely affected these attributes. However, the effect of the storage time and temperature were less in the samples treated with OCE and OPE compared to positive and negative controls, suggesting the protective effects of OCE and OPE.

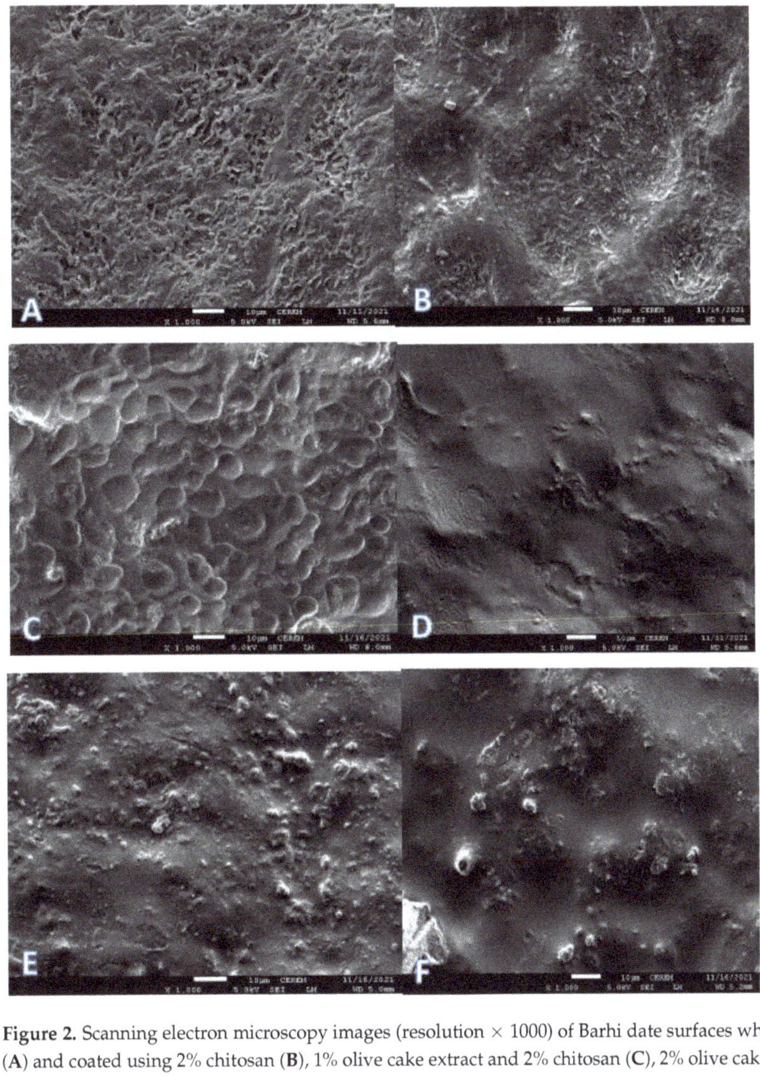

Figure 2. Scanning electron microscopy images (resolution × 1000) of Barhi date surfaces when fresh (**A**) and coated using 2% chitosan (**B**), 1% olive cake extract and 2% chitosan (**C**), 2% olive cake extract and 2% chitosan (**D**), 1% orange peel extract and 2% chitosan (**E**), and 2% orange peel extract and 2% chitosan (**F**).

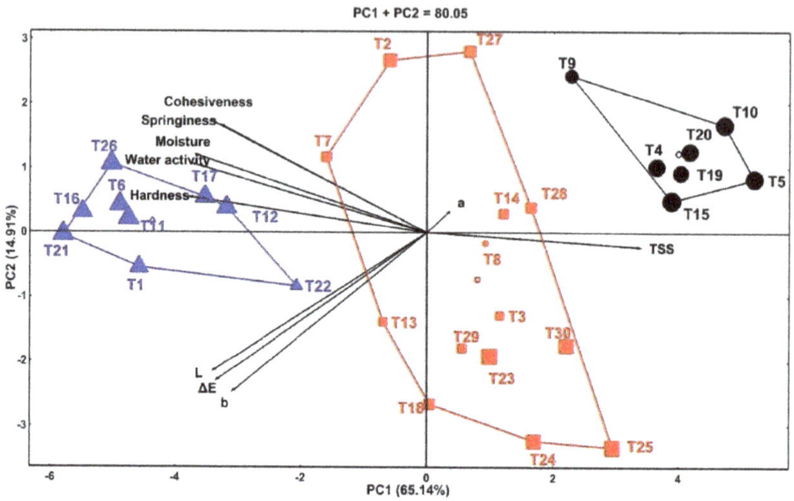

Figure 3. HJ biplot of Barhi dates coated chitosan fortified with 1 and 2 % orange peel extract (OPE) and olive cake extract (OCE). T1, T2, T3, T4, and T5 = negative control (uncoated dates); T6, T7, T8, T9, and T10 = positive control (chitosan-coated dates); T11, T12, T13, T14, and T15 = coated dates fortified with 1% olive cake extract; T16, T17, T18, T19, and T20 = coated dates fortified with 2% olive cake extract; T21, T22, T23, T24, and T25 = coated dates fortified with 1% orange peel extract; T26, T27, T28, T29, and T30 = coated dates fortified with 2% orange peel extract stored for 0, 7, 14, 21, and 28 days, respectively.

4. Conclusions

The current study found that all coating materials increased the shelf life of Barhi date fruits when compared to the control sample. However, when chitosan was combined with OCE and OPE, a pronounced effect was observed. Furthermore, all coating materials increased TSS, were more effective in preventing moisture and firmness loss during storage, did not affect the sensory characteristics of Barhi date fruits, and were extremely effective in preventing fungal growth. SEM image showed that the surface of coated and the uncoated date differed with OCE producing the best smooth coating. Based on these findings, it is possible to conclude that all coating materials tested may be useful in extending the shelf life and maintaining the quality of Barhi date fruits. The PCA analysis showed that the effect of the storage time and temperature were less in the samples treated with OCE and OPE compared to positive and negative controls suggesting the protective effects of OCE and OPE. Further studies demonstrating the effects of storing these fruits at room temperature may also provide valuable information for the preservation of Barhi dates.

Author Contributions: Conceptualization, methodology, resources, supervision, project administration, funding acquisition, writing—review and editing, K.G. validation, F.Y.A.-J., investigation, writing—original draft preparation, visualization, E.E.B., software and data curation, I.A.M.A.; formal analysis, S.A.S. and O.N.A. All authors have read and agreed to the published version of the manuscript.

Funding: This project was funded by the National Plan for Science, Technology and Innovation (MAARIFAH), King Abdulaziz City for Science and Technology, Kingdom of Saudi Arabia, award number (2-17-04-001-0031).

Institutional Review Board Statement: Not applicable.

Informed Consent Statement: Not applicable.

Data Availability Statement: The data presented in this study are available in this article.

Conflicts of Interest: The authors declare no conflict of interest.

References

1. Chao, C.T.; Krueger, R.R. The date palm (*Phoenix dactylifera* L.): Overview of biology, uses, and cultivation. *HortScience* **2007**, *42*, 1077–1082. [CrossRef]
2. Siddiq, M.; Greiby, I. Overview of date fruit production, postharvest handling, processing, and nutrition. In *Dates: Postharvest science, Processing Technology and Health Benefits*; Siddiq, M., Aleid, S.M., Kader, A.A., Eds.; Wiley: Sussex, UK, 2007; pp. 1–28.
3. Awad, M.A.; Al-Qurashi, A.D.; Mohamed, S.A. Biochemical changes in fruit of an early and a late date palm cultivar during development and ripening. *Int. J. Fruit Sci.* **2011**, *11*, 167–183. [CrossRef]
4. Mehyar, G.F.; El Assi, N.M.; Alsmairat, N.G.; Holley, R.A. Effect of edible coatings on fruit maturity and fungal growth on Berhi dates. *Int. J. Food Sci. Technol.* **2014**, *49*, 2409–2417. [CrossRef]
5. Al-Qurashi, A.D.; Awad, M.A. Quality characteristics of bisir "Barhee" dates during cold storage as affected by postharvest dipping in gibberellic acid, naphthaleneacetic acid and benzyladenine. *Fruits* **2011**, *66*, 343–352. [CrossRef]
6. Botes, A.; Zaid, A. The economic importance of date production and international trade. In *Date Palm Cultivation. FAO Plant Production and Protection Paper No. 156*; Zaid, A., Ed.; Food and Agriculture Organization of the United Nations: Rome, Italy, 2002; pp. 45–56.
7. Dubey, N.K.; Dubey, R. Edible films and coatings: An update on recent advances. In *Biopolymer-Based Formulations*; Pal, K., Banerjee, I., Sarkar, P., Kim, D., Deng, W.P., Dubey, N.W., Majumder, K., Eds.; Elsevier: Amsterdam, The Netherlands, 2020; pp. 675–695.
8. Otoni, C.G.; Avena-Bustillos, R.J.; Azeredo, H.M.; Lorevice, M.V.; Moura, M.R.; Mattoso, L.H.; McHugh, T.H. Recent advances on edible films based on fruits and vegetables—A review. *Comp. Rev. Food Sci. Food Saf.* **2017**, *16*, 1151–1169. [CrossRef] [PubMed]
9. Paul, S.K.; Sarkar, S.; Sethi, L.N.; Ghosh, S.K. Development of chitosan based optimized edible coating for tomato (*Solanum lycopersicum*) and its characterization. *J. Food Sci. Technol.* **2018**, *55*, 2446–2456. [CrossRef]
10. Zahoorullah, S.M.; Dakshayani, L.; Rani, A.S.; Venkateswerlu, G. Effect of chitosan coating on the postharvest quality of banana during storage. *Asian J. Biotechnol. Bioresource Technol.* **2017**, *1*, 1–10. [CrossRef]
11. Tran, V.T.; Kingwascharapong, T.; Tanaka, F.; Tanaka, F. Effect of edible coatings developed from chitosan incorporated with tea seed oil on Japanese pear. *Sci. Hortic.* **2021**, *288*, 110314. [CrossRef]
12. Adiamo, O.Q.; Ghafoor, K.; Al-Juhaimi, F.; Mohamed Ahmed, I.A.; Babiker, E.E. Effects of thermosonication and orange by-products extracts on quality attributes of carrot (*Daucus carota*) juice during storage. *Int. J. Food Sci. Technol.* **2017**, *52*, 2115–2125. [CrossRef]
13. Ghafoor, K.; Al-Juhaimi, F.Y.; Choi, Y.H. Supercritical fluid extraction of phenolic compounds and antioxidants from grape (*Vitis labrusca* B.) seeds. *Plant Food Hum. Nutr.* **2012**, *67*, 407–414. [CrossRef]
14. Alsaggaf, M.S.; Moussa, S.H.; Tayel, A.A. Application of fungal chitosan incorporated with pomegranate peel extract as edible coating for microbiological, chemical and sensorial quality enhancement of *Nile tilapia* fillets. *Int. J. Biologic. Macromol.* **2017**, *99*, 499–505. [CrossRef] [PubMed]
15. AOAC. *Official Method of Analysis*, 7th ed.; Association of Official Agricultural Chemists: Washington, DC, USA, 2000.
16. Khalifa, I.; Barakat, H.; El-Mansy, H.A.; Soliman, S.A. Preserving apple (*Malus domestica* var. *Anna*) fruit bioactive substances using olive wastes extract-chitosan film coating. *Inform. Process. Agric.* **2017**, *4*, 90–99. [CrossRef]
17. Singleton, V.L.; Rossi, J.A. Colorimetry of total phenolics with phosphomolybdic-phosphotungstic acid reagents. *Am. J. Enol. Vitic.* **1965**, *16*, 144–158.
18. Lee, S.K.; Mbwambo, Z.H.; Chung, H.; Luyengi, L.; Gamez, E.J.; Mehta, R.G.; Pezzuto, J.M. Evaluation of the antioxidant potential of natural products. *Comb. Chem. High Throughput Screen.* **1998**, *1*, 35–46. [CrossRef]
19. Vicente-Villardon, J.L. *MULTBIPLOT: A Package for Multivariate Analysis Using Biplots. Computer Software*; Departamento de Estadıstica, Universidad de Salamanca: Salamanca, Spain, 2010.
20. Iqbal, M.; Imaranullah, M.; Niamatullah, M. Physiochemical characteristics of date palm (*Phoenix dactylifera* L.) cultivars at various maturity stages under environmental conditions of dera Ismail Khan. *J. Agric. Res.* **2011**, *49*, 249–259.
21. Hoa, T.T.; Ducamp, M.N.; Lebrun, M.; Baldwin, E.A. Effect of different coating treatments on the quality of mango fruit. *J. Food Qual.* **2002**, *25*, 471–486. [CrossRef]
22. Al-Redhaiman, K.N. Modified atmosphere improves storageability, controls decay and maintains quality and antioxidant contents of brahi date fruits. *Food Agric. Environ.* **2004**, *2*, 25–33.
23. Thompson, K.A.; Abboodi, A.H. Modified Atmosphere Packaging. In Proceedings of the International Conference on Date Palm; King Saud University: Riyadh, Saudi Arabia, 2003; pp. 363–394.
24. Abd El-Moneim Eman, A.A.; EL-Gioushy, S.F.; Baiea, M.H.M. Effect of some Natural Coating Materials on Storability and Fruit Quality of Zaghloul Date Palm cv. under Cold Storage. *Middle East J. Agric. Res.* **2015**, *4*, 602–612.
25. Maftoonazad, N.; Ramaswamy, H.S.; Marcotte, M. Shelflife extension of peaches through sodium alginate and methyl cellulose edible coatings. *Int. J. Food Sci. Technol.* **2008**, *43*, 951–957. [CrossRef]
26. Maftoonazad, N.; Ramaswamy, H.S. Postharvest shelf life extension of avocado using methyl cellulose-based coatings. *LWT Food Sci. Technol.* **2005**, *38*, 617–624. [CrossRef]

27. Arendse, E.; Fawole, O.M.; Opara, U.L. Influence of storage temperature and duration on postharvest physico-chemical and mechanical properties of pomegranate fruit and arils. *CyTA J. Food* **2014**, *12*, 389–398. [CrossRef]
28. Maskan, M. Kinetics of color change of kiwifruits during hot air and microwave drying. *J. Food Eng.* **2000**, *48*, 169–175. [CrossRef]
29. Dadalt, G.; Apar, D.K.; Ozbek, B. Color change kinetics of okra undergoing microwave drying. *Dry. Technol.* **2007**, *25*, 925–936. [CrossRef]
30. Eissa, H.A. Effect of chitosan coating on shelf life and quality of fresh-cut mushroom. *J. Food Qual.* **2007**, *30*, 623–645. [CrossRef]
31. Jiang, Y.; Li, Y. Effects of chitosan coating on postharvest life and quality of longan fruit. *Food Chem.* **2001**, *73*, 139–143. [CrossRef]
32. Mannozzi, C.; Cecchini, J.P.; Tylewicz, U.; Siroli, L.; Patrignani, F.; Lanciotti, R.; Rocculiac, P.; Dalla, M.; Rosaac, M.; Romaniace, S. Study on the efficacy of edible coatings on quality of blueberry fruits during shelf-life. *LWT-Food Sci. Technol.* **2017**, *85*, 440–444. [CrossRef]
33. Hong, K.; Xie, J.; Zhang, L.; Sun, D.; Gong, D. Effects of chitosan coating on postharvest life and quality of guava (*Psidium guajava* L.) fruit during cold storage. *Sci. Hortic.* **2012**, *144*, 172–178. [CrossRef]
34. Abu-Shama, H.S.; Abou-Zaid, F.O.F.; El-Sayed, E.Z. Effect of using edible coatings on fruit quality of Barhi date cultivar. *Sci. Hortic.* **2020**, *265*, 109262. [CrossRef]
35. Ray, B.; Bhunia, A. *Fundamental Food Microbiology*, 4th ed.; CRC Taylor & Francis Group: Boca Raton, FL, USA, 2008; pp. 63–71.
36. Lasram, S.; Queslati, S.; Valero, A.; Marin, S.; Ghaorbel, A.; Sanchis, V. Water activity and temperature effects in fungal growth and ochratoxin A production by ochratoxigenic *Aspergillus carbonarius* isolated from Tunisian grapes. *J. Food Sci.* **2010**, *75*, M89–M97. [CrossRef]
37. Siracusa, V.; Romani, S.; Gigli, M.; Mannozzi, C.; Cecchini, J.P.; Tylewicz, U.; Lotti, N. Characterization of active edible films based on citral essential oil, alginate and pectin. *Materials* **2018**, *11*, 1980. [CrossRef]
38. Mutwali, N.I.A.; Mustafa, A.A.; Gorafi, Y.S.; Mohamed Ahmed, I.A. Effect of environment and genotypes on the physic chemical quality of the grains of newly developed wheat inbred lines. *Food Sci. Nutr.* **2016**, *4*, 508–520. [CrossRef] [PubMed]

foods

MDPI

Article

Efficacy of Pectin-Based Coating Added with a Lemon Byproduct Extract on Quality Preservation of Fresh-Cut Carrots

Valeria Imeneo [1], Amalia Piscopo [2], Olga Martín-Belloso [3] and Robert Soliva-Fortuny [3,*]

[1] Department of Food, Environmental and Nutritional Sciences (DeFENS), University of Milan, G. Celoria 2, 20133 Milan, Italy; valeria.imeneo@unimi.it

[2] Department of AGRARIA, University Mediterranea of Reggio Calabria, Vito, 89124 Reggio Calabria, Italy; amalia.piscopo@unirc.it

[3] Department of Food Technology, University of Lleida—Agrotecnio CERCA Center, Av. Alcalde Rovira Roure 191, 25198 Lleida, Spain; olga.martin@udl.cat

* Correspondence: robert.soliva@udl.cat; Tel.: +34-973-702678

Abstract: The effect of an edible pectin-based coating supplemented with a lemon byproduct extract on the quality attributes of fresh-cut carrots was studied. Color, hardness, microbial growth, respiratory activity, and antioxidant properties of fresh-cut carrots were studied during 14 days of storage at 4 °C. The application of a pectin-based coating containing a lemon byproduct extract preserved carrots' physiological parameters, reduced their physiological activity and, thus, delayed senescence. This aspect was also confirmed by the reduced O_2 consumption of the coated carrots due to the slowing down of the product's metabolic reactions. Moreover, coated carrots were characterized by limited changes in colour ($\Delta E < 3$) and white-blush development on both cortical tissue and vascular cylinder, and the presence of calcium chloride in the coating formulation helped to maintain carrots' hardness throughout storage. In addition, treatment with pectin-based coating and lemon byproduct extract improved microbiological stability of fresh-cut carrots, showing the lowest value of total bacterial count immediately after treatment (2.58 log CFU g^{-1}). This kind of treatment also resulted in a significant preservation of valuable compounds (17.22 mg GAE 100 g^{-1}) and antioxidant activity level (289.49 μM Trolox 100 g^{-1}), reducing the wounding stress induced by processing operations for at least ten days.

Keywords: pectin-based coating; quality attributes; fresh-cut carrot; respiratory activity; carotenoids and phenolic compounds; lemon byproduct; antioxidant

Citation: Imeneo, V.; Piscopo, A.; Martín-Belloso, O.; Soliva-Fortuny, R. Efficacy of Pectin-Based Coating Added with a Lemon Byproduct Extract on Quality Preservation of Fresh-Cut Carrots. *Foods* **2022**, *11*, 1314. https://doi.org/10.3390/foods11091314

Academic Editors: Maria Cefola and Bernardo Pace

Received: 8 April 2022
Accepted: 26 April 2022
Published: 30 April 2022

1. Introduction

Consumption of fresh fruit and vegetables is known to be beneficial to human health. In this context, carrot consumption is becoming ever more popular thanks to their nutritional value. Carrots represent a good source of bioactive compounds, namely, fibre and antioxidants such as carotenoids, phenolic compounds, and vitamin C [1,2]. However, like other fresh fruit and vegetables, they are highly perishable. During post-harvest handling and storage, significant losses of vitamins and other phytonutrients can occur, depending on the nutrient, genotype, physical damage, temperature and storage conditions [3]. The colour and appearance of minimally processed carrots are critical quality attributes. Their characteristic bright orange colour may be rapidly lost due to dehydration and the development of white blush on the surface, hence reducing their acceptability [4,5]. Maintaining the overall quality of minimally processed fruit and vegetables becomes even more difficult. Therefore, fresh-cut fruit and vegetables have become a challenging problem among food scientists and technologists, over the years [6].

In this respect, the application of techniques such as the dipping of minimally processed carrots in acid solutions demonstrated to be an efficient approach to preserve their quality, limit colour changing and microbial growth, reduce enzymatic activity and increase the preservation of total carotenoid content [5].

Moreover, polysaccharide edible coatings have proven to be a valid strategy to boost food appearance and extend the shelf-life of fresh-cut fruits and vegetables. Coatings can form a semi-permeable barrier to gasses and be potential carriers of additives to help in preserving or improving the quality of produce [6]. Polysaccharide-based edible coatings can help to prolong the shelf life of fresh fruits and vegetables by slowing down decay processes associated to water loss [7]. Among polysaccharides, pectins are valuable compounds for coating formulations thanks to their ability to create rigid and stable gels, enabling effective food applications [8–11]. Pectins are α-1,4 bonded D-galacturonic acid polymers characterized by the presence of hydrophobic groups, consisting of methyl ester, acetyl, and protein residues, that promote the absorption of organic lipid substances, hence contributing to their emulsifying capacity [12]. In particular, the hydrocolloidal and polyelectrolyte properties of pectins define their unique capacities, such as high water retention in colloidal systems and their stabilization, and aptitude to plasticize with glycerol [13]. According to their content of methyl esters or the degree of esterification (DE), pectins are divided into high methoxyl (DE > 50%) or low methoxyl (DE < 50%), showing a significant influence on solubility and gel-formation characteristics [14]. In coating formulations, low-methoxyl pectins are generally used, mainly because of their ability to develop strong gels or insoluble polymers when multivalent metal cations, such as calcium, are present [15–17]. Overall, edible coatings are applied on fruit and vegetables due to their ability to reduce respiration and senescence, retain moisture and slow down colour changes throughout storage [6,18]. Several authors have studied the incorporation of additives, namely, antioxidant, antimicrobial and antibrowning compounds or texture enhancers, into edible coating formulations, making them a valid option for decreasing physiological postharvest deterioration rate in minimally processed fruits and vegetables [19–22]. The application of plant byproducts and extracts is regarded as a way of incorporating these active compounds to fresh-cut fruit and vegetables while maintaining a label free from synthetic additives. In this regard, lemon byproducts, consisting of lemon peel, pulp and seeds, represent a natural source of minerals, organic acids, dietary fibre and phenolic compounds, such as phenolic acids and flavonoids (flavanones, flavonols, flavones), characterised by antioxidant, anti-inflammatory and antimicrobial properties [23–28]. Among these bioactive compounds, flavanones and flavones are the most plentiful flavonoids, followed by neohesperidin, naringin, rutin and apigenin. Other compounds identified in lemon peels are furocoumarins and coumarins. The highest quantity of phenolic acids is found in lemon seeds, the only waste characterised by the presence of gallic acid, protocatechic acid and p-cumaric acid, as well as obacunone, a compound belonging to the class of limonoids [29]. In view of the antioxidant and antimicrobial activities exhibited by these compounds, the goal of this study was to determine the effect of an edible pectin-based coating supplemented with a lemon byproduct extract on the quality attributes of fresh-cut carrots. This approach would contribute to the shelf-life extension of the ready-to-eat commodity, while valorising the potential of a byproduct generated from the citrus industry.

2. Materials and Methods

2.1. Chemical and Reagents

Food-grade low-methoxyl pectin (~30% esterified) (Sigma–Aldrich Chemic, Steinhein, Germany) was the carbohydrate biopolymer used to prepare the coating formulations. Glycerol (Merck, Whitehouse Station, NJ, USA) was used as plasticizer. Calcium chloride (Sigma–Aldrich Chemic, Steinhein, Germany) was added to promote pectin gelation by crosslinking. Folin-Ciocalteu reagent and ethanol were purchased from Scharlau S.L. (Barcelona, Spain); sodium carbonate was purchased from Fisher Scientific Scharlau Chemie (Loughborough, UK). DPPH (2,2-diphenyl-1-picrylhydrazyl), ABTS (2,2′-azino-bis acid (3- ethylbenzothiazolin-6-sulfonic acid), gallic acid and Trolox (6-hydroxy-2,5,7,8-tetramethylchroman-2-carboxylic acid) were bought from Sigma-Aldrich (St. Louis, MO,

USA). Sodium hydroxide was acquired from Acros Organics (New Jersey, NJ, USA). Methanol was acquired from J.T. Baker S.A. (Sowińskiego, Poland). Acetone was purchased from Fisher Chemical (Loughborough, UK). Phenolphthalein was bought from POCH S.A. (Sowińskiego, Poland). PCA, GCA and buffer peptone water were purchased from Biokar Diagnostics (Beauvais, France). Sodium hypochlorite was purchased from Productes Sant Mateu (Barcelona, Spain).

2.2. Lemon Extract (LE)

Lemon byproducts (*Citrus limon (L.) Osbeck)*, consisting of peels, pulp and seeds, obtained after the extraction of lemon juice and essential oils, were provided by Agrumaria Reggina company, situated in Reggio Calabria (Italy). The lemon byproduct extract (LE) used in this study was obtained by an aqueous extraction at 70 °C for 30 min, as reported and characterised by Imeneo et al. [30].

2.3. Preparation of the Dipping and Coating Solutions

Carrots (*Daucus carota* cv. Nantes) were purchased in a local supermarket in Lleida (Spain). The carrots were carried to the laboratory of the University of Lleida and immediately processed for this study. Whole carrots (length of 17 ± 3 cm) were sanitized in a 200 μL L^{-1} NaClO solution for 2 min. Sanitized carrots were rinsed with tap water and the excess of water was blotted away at room temperature. Carrots were peeled and afterwards cut lengthwise into two parts and then transversely in two semi-discs (diameter 28 ± 2, height 20 ± 2 mm).

A portion of the carrot pieces (W + LE) was dipped in an aqueous solution containing LE (1%, v/v). Other carrot pieces were coated using a pectin-based coating (PC), prepared in accordance with Oms-Oliu et al. [15] description. The coating solution was prepared by dissolving pectin (2%, w/v) powder in distilled water and heating at 70 °C while stirring, until the solution became clear. Glycerol (1.5%, w/v) was added as a plasticizer to the pectin solution and the carrot pieces were dipped into it for 2 min. Excess coating material was drained for 1 min. Subsequently, carrot pieces were dipped for 2 min into a calcium-chloride aqueous solution (2%, w/v) with the addition of LE (1%, v/v) for crosslinking pectin (PC + LE).

Treatments without the addition of LE (W and PC) were prepared as control references.

Then, 100 g of carrots were placed in manually closed, side-perforated polyethylene terephthalate trays (150 mL) and stored at 4 ± 1 °C in the dark until they were withdrawn for analyses after 1, 3, 7, 10 and 14 days of storage.

Two trays of each treatment condition were taken at each monitoring time to carry out repetition of analyses.

2.4. Physicochemical Properties

Titratable acidity (% of citric acid) and pH (pH meter Crison micropH 2000, Crison Instruments S.A., Alella, Barcelona, Spain) were evaluated according to AOAC methods [31,32].

Total soluble solids (°Brix) were determined through the measurement of the refraction index with a digital refractometer (PR-32, 3412-J01, Atago Company Ltd., Tokyo, Japan) at 25 °C.

Dry matter (%) of carrot pieces was quantified according to AOAC method [33], by calculating loss weight at 70 °C until constant values were reached.

- Colour

Colour values of cortical tissue (external side) and vascular cylinder (internal side) were measured with a colorimeter (Minolta Chroma Meter Model CR-400, Minolta Sensing Inc., Osaka, Japan). Ten readings were performed on each replicate. Total colour difference (ΔE), as reported by Thompson [34], and the whiteness index (WI), according to Piscopo et al. [5], were determined, considering CIELab coordinates in freshly cut carrot pieces and the values obtained at each day of monitoring according to the equations below:

$$\Delta E = \sqrt{\left[(L - L_0)^2 + (a - a_0)^2 + (b - b_0)^2\right]} \quad (1)$$

$$WI = 100 - \sqrt{[(100 - L^2) + a^2 + b^2]} \quad (2)$$

where L_0, a_0, and b_0 are the values measured on the 1st day, and L, a and b relate to data measured at each sampling time.

- Hardness

Hardness of cortical tissue and vascular cylinder of carrots was measured with a TA-XT2 texture analyser (Stable Micro Systems Ltd., Surrey, UK), fitted with a 4-mm-diameter cylinder steel probe, which went through the carrot surface for 10 mm at a constant speed of 5 mm·s^{-1} and automatic return. The probe movement was orthogonal to the carrot tissue. Hardness (N s^{-1}) was defined as the area under the curve between the graphic depicting force vs time [35]. Five randomly withdrawn carrot pieces were analysed per treatment and sampling time.

2.5. Microbiological Analysis

Total aerobic bacterial counts (TBC) were evaluated throughout storage. Two replicate packages were analysed per treatment and sampling time and two counts were obtained from each one. Under sterile conditions, 10 g of carrot pieces were homogenized for 3 min with 90 mL of 0.1% sterile peptone water using a Stomacher Lab Blender 400 (Seward Medical, London, UK). Serial dilutions of the homogenates were placed on plate count agar (PCA; Biokar Diagnostics. Beauvais, France) and incubated at 35 ± 1 °C for 48 h for TBC determination. The results were expressed as log CFU g^1 [36].

2.6. Respiratory Activity

A Micro-GC gas analyser (Model CP 2002, Chrompack International, Middelburg, The Netherlands) characterised by a thermal conductivity detector was applied to evaluate the respiratory activity of carrot pieces. A variation of the procedure reported by López-Gómez et al. [37] was used. At each monitoring time, 50 g of carrots were placed in airtight containers of 250 mL and stored at 4 °C for 3 h. Subsequently, 1.7 mL of gas was collected from the headspace with a syringe via a rubber septum. Carbon dioxide production ($RRCO_2$) and oxygen consumption (RRO_2) were reported as mg kg^{-1} h^{-1}, as described by Tappi et al. [38].

2.7. Extraction and Determination of Total Phenolic and Content and Antioxidant Activity

The extraction of phenolic compounds was carried out as reported by Formica-Oliveira et al. [39], with minor modifications. A portion of 5 g of carrot was homogenised with 20 mL of methanol with an Ultra-Turrax T25 (IKA®-Werke GmbH & Co., Staufen, Germany) for 2 min. Homogenates were centrifuged at 13,500× *g* and 4 °C for 20 min (Centrifuge AVANTITM J-25, Beckman Instruments Inc., Fullerton, CA, USA). The supernatants were collected and then filtered through Whatman no. 1 filter.

The total phenolic content (TPC) was determined following the Folin–Ciocalteu procedure tailored to 96-well microplates [2]. After an aliquot of 30 µL of methanolic extract was introduced into a microplate, 150 µL of 10% (*v*/*v*) Folin–Ciocalteu reagent and 120 µL of Na_2CO_3 7.5% (*w*/*v*) were added. After incubating for 90 min at room temperature in darkness, the absorbance at 765 nm was determined with a microplate reader (Thermo Scientific Multiskan GO, Vantaa, Finland). The results were expressed as mg of gallic acid equivalents per 100 g (mg 100 g^{-1}) on a fresh-weight basis. Phenolic compounds extraction was carried out twice per treatment repetition and fourfold spectrophotometrically determined.

Total carotenoids content (TCC) was determined spectrophotometrically (CECIL CE 2021; Cecil Instruments Ltd., Cambridge, UK) as described by González-Casado et al. [40], with slight modifications. An aliquot of fresh-cut carrots (2 g) was homogenised with 25 mL of acetone:ethanol (1:1, *v*/*v*) with an Ultra-Turrax T25 (IKA® WERKE, Germany). Sample was extracted in the dark, filtered through Whatman No. 4 filter paper, and washed with the acetone:ethanol solution until the residue was colourless. Samples were adjusted to 100 mL, and the absorbance was read at 470 nm versus a blank of acetone:ethanol. TCC was determined by the equation below:

$$\text{Total Carotenoids Content} = \frac{A_{470} \cdot V \cdot 10^4}{A_{1cm}^{1\%} \cdot G} \quad (3)$$

where A_{470} is the absorbance at 470 nm, V is the total volume of extract (mL), $A_{1cm}^{1\%}$ is the extinction coefficient of a mixture of carotenoids established as 2500 by Gross [41] and G is the sample weight (g). Total carotenoids were expressed as mg per 100 g of fresh weight (mg TCC 100 g^{-1}).

The antioxidant activity of fresh-cut carrots was determined by DPPH and ABTS assays, using a 96-well microplate reader (Thermo Scientific Multiskan GO, Vantaa, Finland).

The determination of free radical scavenging effect on 1,1-diphenyl-2-picrylhydrazyl (DPPH) radical was performed as reported by Ribas-Agustí et al. [35] with some modifications. An aliquot of 20 μL of methanolic extract (Section 2.7) was placed into a microplate and 280 μL of a 6×10^{-5} M methanolic solution of DPPH were added. The homogenate was kept in darkness for 30 min under continuous stirring and the absorption of the samples was measured at 515 nm against a blank of methanol without DPPH. A calibration curve was built with Trolox (from 6 to 30 μM). Results were expressed as μM Trolox equivalents per 100 g (μmol 100 g^{-1}) on a fresh-weight basis.

The ABTS (2,2′-azino-bis acid 3-ethylbenzothiazolin-6-sulfonic acid) assay was carried out as reported by Re at al. [42]. An aliquot of 40 μL of methanolic extract (Section 2.7) was placed into a microplate and 260 μL of ABTS ethanol solution were added. The homogenate was kept in darkness for 6 min under continuous stirring and the absorption of the samples was measured at 734 nm against a blank of ethanol without ABTS. A calibration curve was built with Trolox (from 30 to 120 μM). The results were expressed as μM Trolox equivalents per 100 g (μmol 100 g^{-1}) on a fresh-weight basis.

2.8. Statistical Analysis

All the experimental results were expressed as mean value ($n = 4$) ± standard deviation (mean ± SD). Significance of the results and statistical differences were analysed using SPSS Software (Version 15.0, SPSS Inc., Chicago, IL, USA). One-way analysis of variance (ANOVA) and several multiple comparisons, by Tukey's post-hoc test, were conducted to identify individual significant differences ($p < 0.05$). The Pearson's correlation test was performed to determine correlation coefficients (r) among polyphenolic compounds and antioxidant assays.

3. Results and Discussion

3.1. Physicochemical Properties

Physicochemical characterisation results of uncoated (W and W + LE) and coated (PC and PC + LE) fresh-cut carrots are reported in Table 1.

Table 1. Changes in acidity, pH, total soluble solids (Brix°) and dry matter of uncoated and coated fresh-cut carrots during 14 days of storage (4 °C).

	Storage Time (Day)				
	1	3	7	10	14
Acidity	Mean ± SD	Mean ± SD	Mean ± SD	Mean ± SD	Mean ± SD
W	0.10 ± 0.00 ^{ab,A}	0.07 ± 0.01 ^{C}	0.07 ± 0.01 ^{b,BC}	0.08 ± 0.01 ^{b,B}	0.10 ± 0.00 ^{a,A}
W + LE	0.10 ± 0.01 ^{ab,A}	0.07 ± 0.00 ^{BC}	0.11 ± 0.00 ^{a,A}	0.08 ± 0.00 ^{b,AB}	0.05 ± 0.02 ^{b,C}
PC	0.11 ± 0.00 ^{a,A}	0.08 ± 0.01 ^{B}	0.08 ± 0.00 ^{b,B}	0.12 ± 0.03 ^{a,A}	0.07 ±0.00 ^{ab,B}
PC + LE	0.09 ± 0.01 ^{b}	0.07 ± 0.01	0.09 ± 0.01 ^{a,b}	0.07 ± 0.01 ^{b}	0.09 ± 0.02 ^{a}
Significance	*	ns	**	**	**
pH					
W	6.34 ± 0.04 ^{BC}	6.32 ± 0.03 ^{C}	6.43 ± 0.01 ^{AB}	6.6 ± 0.03 ^{a,A}	6.49 ± 0.10 ^{a,B}
W + LE	6.27 ± 0.09	6.46 ± 0.10	6.54 ± 0.08	6.51 ± 0.25 ^{a}	6.43 ± 0.00 ^{a}
PC	6.31 ± 0.05 ^{A}	6.38 ± 0.01 ^{A}	6.40 ± 0.06 ^{A}	6.04 ± 0.26 ^{b,B}	6.21 ± 0.00 ^{b,AB}
PC + LE	6.31 ± 0.01 ^{C}	6.29 ± 0.16 ^{C}	6.52 ± 0.13 ^{B}	6.73 ± 0.06 ^{a,A}	6.24 ± 0.06 ^{b,C}
Significance	ns	ns	ns	**	**
Soluble solids content (°Brix)					
W	3.6 ± 0.14 ^{d}	3.25 ± 0.07 ^{c}	3.60 ± 0.57 ^{c}	3.70 ± 0.28 ^{c}	3.90 ± 0.99 ^{b}
W + LE	5.65 ± 0.21 ^{b,A}	5.20 ± 0.42 ^{b,AB}	5.00 ± 0.42 ^{b,B}	5.65 ± 0.07 ^{b,A}	5.60 ± 0.28 ^{a,A}
PC	5.35 ± 0.07 ^{c}	5.40 ± 0.14 ^{b}	4.90 ± 0.42 ^{b}	5.30 ± 0.28 ^{b}	5.35 ± 1.34 ^{a,b}
PC + LE	6.4 ± 0.14 ^{a,A}	6.05 ± 0.07 ^{a,C}	6.25 ± 0.07 ^{a,ABC}	6.35 ± 0.21 ^{a,AB}	6.10 ± 0.14 ^{a,BC}
Significance	**	**	**	**	**
Dry matter					
W	9.24 ± 0.18 ^{b}	9.94 ± 0.94 ^{a}	9.99 ± 1.15	9.60 ± 0.45 ^{ab}	9.50 ± 0.07
W + LE	10.47 ± 0.78 ^{a}	9.71 ± 0.21 ^{a}	10.48 ± 1.76	10.11 ± 0.32 ^{a}	9.19 ± 0.41
PC	10.96 ± 0.50 ^{a,A}	9.73 ± 0.10 ^{a,B}	9.18 ± 0.18 ^{B}	9.41 ± 0.03 ^{b,B}	9.31 ± 0.84 ^{B}
PC + LE	9.20 ± 0.36 ^{b}	8.68 ± 0.28 ^{b}	8.73 ± 0.68	9.55 ± 0.50 ^{ab}	9.22 ± 1.53
Significance	**	**	ns	*	ns

Values are the mean of four determinations ± SD (n = 4). Different superscript letters within a row or column indicate statistically significant differences between uncoated and coated fresh-cut carrots. Lower case letters denote differences among treatments for a set treatment time. Upper case letters denote differences among treatment times for a set treatment. Absence of letters indicate non statistically significant differences. ** Significance at $p < 0.01$; * Significance at $p < 0.05$; ns, not significant. W, carrots dipped in water; W + LE, carrots dipped in water + LE solution; PC, carrots coated with a pectin-based coating; PC + LE, carrots coated with a pectin-based coating with LE.

The application of a pectin-based coating containing lemon byproduct extract (PC + LE) on minimally processed carrots helped to preserve their physiological parameters, helping to the maintain higher total soluble solids and acidity values during refrigerated storage at 4 °C. This is suggestive of reduced physiological activity in the coated product [43]. Immediately after treatment, PC + LE carrots showed the lowest acidity values, compared to other fresh-cut carrots. The combination of the lemon byproduct extract and pectin-based coating favoured a relevant preservation of the acidity levels, showing no significant changes ($p > 0.05$) throughout storage. In addition, PC + LE carrots exhibited the highest soluble solids content (°Brix) from the first to the last day of storage, with a very slight decreasing trend over time. Physiological parameters, such as titratable acidity, pH and sugar content, are good indicators of fruit maturation and senescence. In this regard, the coatings successfully demonstrated their potential in slowing down the natural physiological behaviour of fruits and vegetables, retarding metabolic reactions and, thus, senescence, thanks to the different gas permeability of the coating and its influence on the vegetables' respiratory activity [6,44]. Moreover, thanks to its content in bioactive compounds, the simultaneous presence of LE also contributed to delaying changes induced by stressful conditions concomitant with minimal carrots processing, such as the metabolism of soluble sugars.

Colour changes were monitored over time by determining CIELab parameters (lightness, L *; green-red chromaticity, a *; and blue-yellow chromaticity, b *) and total colour difference (Table 2), on both cortical tissue and vascular cylinder of uncoated and coated fresh-cut carrots. Total colour difference (ΔE) values throughout storage were significantly ($p < 0.05$) reduced by the application of a pectin-based coating and a lemon byproduct extract.

Table 2. Total colour difference (ΔE) of uncoated and coated fresh-cut carrots throughout 14 days of storage (4 °C).

	ΔE									
	Storage Time (Day)									
	1		**3**		**7**		**10**		**14**	
	Cortical Tissue	Vascular Cylinder	Cortical Tissue	Vascular Cylinder	Cortical Tissue	Vascular Cylinder	Cortical Tissue	Vascular Cylinder	Cortical Tissue	Vascular Cylinder
W	3.63 ± 0.56 ^{a}	7.09 ± 0.53 ^{ab,B}	5.85 ± 1.24 ^{a}	11.31 ± 1.48 ^{a,A}	5.60 ± 1.27 ^{a}	9.67 ± 1.42 ^{a,A}	5.83 ± 2.04 ^{a}	9.84 ± 1.52 ^{A}	5.47 ± 0.94 ^{a}	9.62 ± 1.72 ^{a,A}
W + LE	3.01 ± 0.30 ^{ab}	6.49 ± 1.35 ^{a,B}	2.33 ± 0.93 ^{b}	13.87 ± 2.00 ^{a,A}	2.87 ± 0.04 ^{b}	9.01 ± 1.88 ^{a,AB}	3.53 ± 1.26 ^{a,b}	7.89 ± 2.35 ^{B}	2.33 ± 0.91 ^{b}	9.57 ± 1.67 ^{a,AB}
PC	2.20 ± 0.64 ^{bc}	3.17 ± 0.56 ^{ab}	2.07 ± 0.56 ^{b}	6.21 ± 0.70 ^{b}	1.42 ± 0.61 ^{b}	4.41 ± 1.26 ^{b}	1.66 ± 0.58 ^{b}	5.93 ± 2.58	1.63 ± 0.69 ^{b}	6.14 ± 2.06 ^{a,b}
PC + LE	1.63 ± 0.19 ^{c,B}	1.74 ± 0.66 ^{b,B}	3.06 ± 0.26 ^{b,A}	2.07 ± 0.71 ^{c,AB}	1.00 ± 0.28 ^{b,B}	3.02 ± 0.19 ^{b,AB}	1.22 ± 0.07 ^{b,B}	4.38 ± 1.78 ^{A}	2.65 ± 0.36 ^{b,A}	3.76 ± 0.28 ^{b,AB}
Sign.	**	*	**	**	**	**	**	ns	**	**

Values are reported as mean ± standard deviation (n = 10). Different superscript letters within a row or column indicate statistically significant differences between uncoated and coated fresh-cut carrots. Lower case letters denote differences among treatments for a set treatment time. Upper case letters denote differences among treatment times for a set treatment. Absence of letters indicate non statistically significant differences. Abbreviations: **; *; ns; Sign.; W; W + LE; PC; PC + LE (see Table 1).

As shown in Table 2, coated carrots were characterized by the lowest ΔE values on both cortical tissue and vascular cylinder immediately after treatment (day 1). Moreover, the combination of the pectin coating and lemon byproducts extract (PC + LE) limited colour changes throughout storage (ΔE < 3). It is remarkable that total-colour-difference values greater than 3 denote differences that are easily noticeable by the human eye [45]. In contrast, uncoated carrots underwent colour changes (ΔE > 3) over storage that are noticeable by the human eye in both cortical and vascular cylinder tissues. Interestingly, W + LE-treated carrots were characterised by significantly lower ΔE values ($p < 0.05$) compared to W-treated carrots, which can be attributed to the antibrowning effect of the lemon extract.

The coating application was related to a significant ($p < 0.05$) greater stability of lightness (L *) values and with higher a * and b * values throughout storage in comparison to uncoated carrots (data not reported). The higher L * values observed in W and W + LE carrots could be related to the biosynthesis of lignin in wounded carrot tissues by enzymes stimulated during minimal processing operations, such as phenylalanine ammonia lyase (PAL) [46]. As reported by several authors [47–50], PAL promotes surface discoloration by increasing levels of soluble phenolics required for lignin biosynthesis and its activity could increase due to wound-induced stimulation throughout storage.

In this study, the presence of the pectin-based coating and LE on carrots' surfaces appears to have contributed to the retention of colour over time on both the cortical tissue and vascular cylinder, by counteracting the enzymatic action stimulated by abiotic stress.

As shown in Figure 1, coating treatments significantly ($p < 0.05$) affected the development of white blush on fresh-cut carrots during storage. The WI values of coated carrots were significantly lower than those of the uncoated ones, throughout the evaluated storage period. In all the fresh-cut carrots, vascular cylinder surfaces exhibited higher WI values than cortical surfaces over time, which is consistent with the observed ΔE values (Table 2). PC and PC + LE carrots did not suffer an increase in WI throughout storage, recording constant values on both cortical tissue and vascular cylinder. Comparable findings were also noted by Vargas et al. [51] for carrot slices, who reported that a chitosan coating significantly reduced the development of white blush on carrots' surfaces. In contrast, an increase in WI values was observed in the vascular cylinder surfaces of uncoated carrots (W and W + LE) from the third day of storage onwards. Moreover, the addition of the lemon byproduct extract into the pectin-coating formulation did not show any influence on the development of white blush throughout storage.

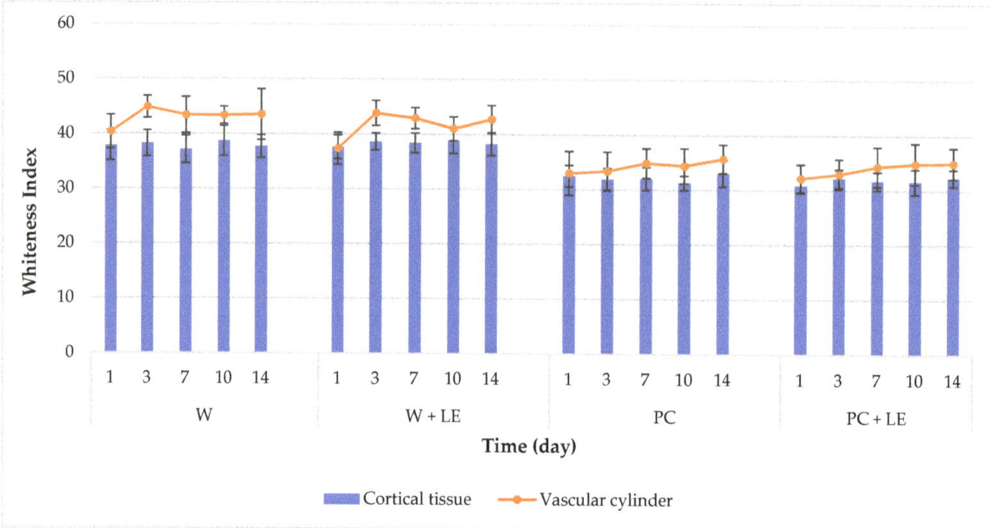

Figure 1. Changes in whiteness index (WI) values of uncoated and coated fresh-cut carrots on cortical tissue and vascular cylinder throughout 14 days of storage (4 °C). Abbreviations: W; W + LE; PC; PC + LE (see Table 1).

In this study, the increase in white-blush development was significantly ($p < 0.05$) retarded by the application of the pectin-based coating. Considering that the main reason for carrot white discoloration is surface dehydration, the edible coating delayed whitening by working as a surface moisturizer [52,53]. Hence, acting as a humectant thanks to its hydrophilic nature, pectin keeps the surface of peeled carrots moist to retard white discolouration (Figure 2).

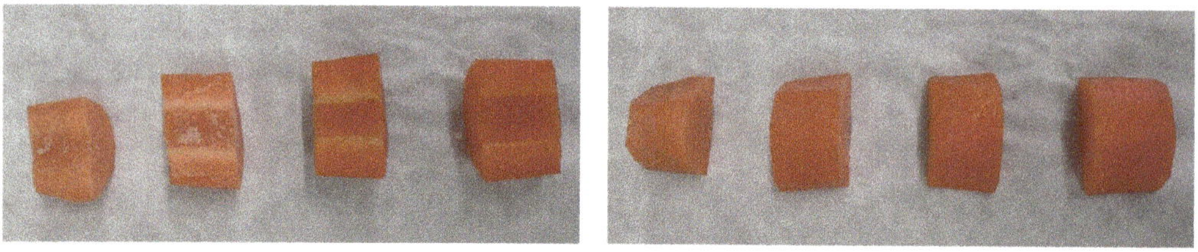

Figure 2. Representative photograph of white blush in vascular cylinder (**on the left**) and cortical tissue (**on the right**) of uncoated and coated fresh-cut carrots on the 14th day of storage at 4 °C. From left to right: W; W + LE; PC; PC + LE (abbreviations, see Table 1).

Regarding texture (Figure 3), the coated fresh-cut carrots showed similar ($p > 0.05$) hardness values in both cortical tissue and vascular cylinder throughout storage. This contrasts with uncoated carrots, whose texture values significantly ($p < 0.05$) differed, with vascular cylinders softer than cortical tissues. This difference between tissues, observed from day 1 in W and W + LE carrots, was consistent throughout storage.

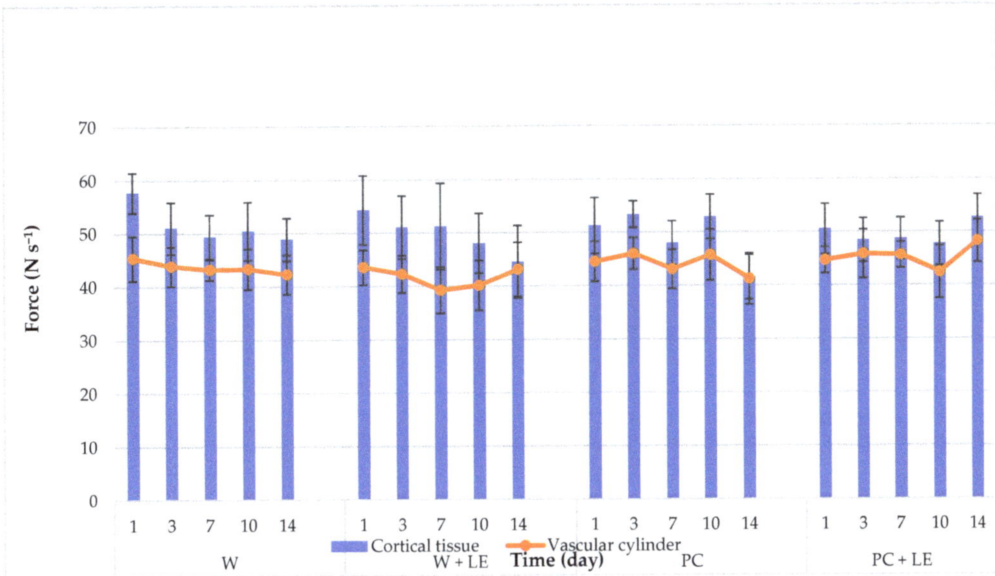

Figure 3. Hardness changes throughout storage of fresh-cut carrots (cortical and vascular cylinder tissues) as affected by pectin coating and the addition of a lemon byproduct extract. Abbreviations: W; W + LE; PC; PC + LE (see Table 1).

The use of calcium chloride in the coating formulation in combination with LE seems to be a determinant factor in maintaining the hardness of fresh-cut carrots. Various authors described the positive impact of incorporating calcium chloride into coating formulations on the retention of hardness in fresh-cut fruits [15,54–56]. However, the presence of LE helped, as well, to preserve the hardness of carrots, probably thanks to the extract's action in counteracting the activity of specific enzymes such as polygalacturonase. Hardness loss, in fact, may be related with the deterioration of compounds responsible for vegetable structural rigidity, primarily insoluble pectin and protopectin. In maturation, pectinesterase and polygalacturonase activities intensify, producing the solubilisation of pectin substances [57]. In addition, the observed decrease in hardness of the fresh-cut carrots may be linked not only to the action of pectinolytic enzymes, but also to the increased activity of glycolytic enzymes, which contribute to the hydrolysis of hemicellulose and other cell wall components and which might be activated as a defence mechanism in cases of microbiological attack and/or injury [58]. The structural integrity of carrot tissues could also be related to the highest soluble solids content observed for PC + LE carrots (Table 2). In fact, although soluble sugars are known to be a signal of the regulation of various mechanisms associated to growth, development and metabolic responses in plants, they also operate as metabolic resources and structural elements of cells [59].

3.2. Microbial Growth

The results of microbiological analyses are reported in Figure 4. Microbial counts on PC + LE significantly differed from those observed in carrots subjected to other treatments. PC + LE carrots showed the lowest value of TBC (2.58 ± 0.06 log CFU g^{-1}) immediately after treatment. Microbial counts for treatments without the incorporation of LE were significantly ($p < 0.01$) higher (3.66 ± 0.31 and 3.81 ± 0.44 log CFU g^{-1}, for W and PC carrots, respectively). After the first day of storage, PC + LE continued to exhibit the lowest TBC values ($p < 0.01$) for at least one week. Our results agree with those obtained by Amanatidou et al. [60], who studied the effect of an alginate-based coating added with 0.1% of citric acid on sliced carrots stored at 8 °C under modified atmosphere conditions.

The authors reported that the combination of 0.1% citric acid and 2% $CaCl_2$ in coating formulations significantly reduced the initial total microbiota for at least 1 log CFU g^{-1} during up to 8 days of storage.

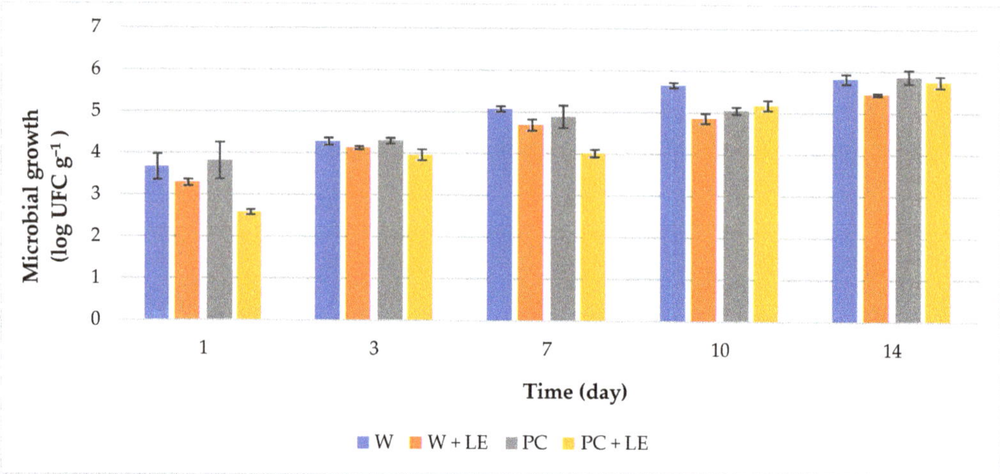

Figure 4. Growth of total bacterial load on uncoated and coated fresh-cut carrots throughout 14 days of storage (4 °C). Abbreviations: W; W + LE; PC; PC + LE (see Table 1).

The inclusion of natural antioxidants and antimicrobials to the pectin-based coating, by means of the incorporation of LE, decisively led to the shelf-life extension of fresh-cut carrots. This effect could be related to the presence of phenolic compounds dispersed within the pectin matrix of the coating, which allowed them to be progressively released to the carrot surface over storage [61]. As reported by Budiati et al. [62] and Ivasenko et al. [63], compounds such as apigenin and gallic acid, detected in lemon byproduct extracts, are characterised by an intense antibacterial activity, by limiting microbial adhesion and deactivatong bacterial enzymes and cell transport proteins. In addition, hardness decay may be, as well, related to the proliferation of pectolytic Pseudomonas [60]. In fact, the higher and more stable texture values observed for PC + LE (Figure 3) correlate well with lower total bacterial counts, which could be associated to reduced bacterial enzymatic activities.

3.3. Respiratory Activity

The effect of treatments on the respiratory activity of fresh-cut carrots, expressed as oxygen consumption (RRO_2) and carbon dioxide production ($RRCO_2$), is shown in Figure 5.

Significant differences ($p < 0.05$) in terms of O_2 consumption and CO_2 production were found between treated carrots immediately after treatment. Particularly, coated fresh-cut carrots (PC and PC + LE) showed the lowest values of RRO_2 (Figure 5a). Furthermore, an increasing trend was observed for RRO_2 values over storage, with the highest values observed for W-treated carrots. The reduced oxygen consumption in coated fresh-cut carrots is probably attributable to the slowing down of the product's metabolic reactions, which would result in a better preservation of physicochemical parameters. On the other hand, regarding $RRCO_2$ (Figure 5b), all treatments showed an initial increase in values followed by a decrease after the 7th day of storage.

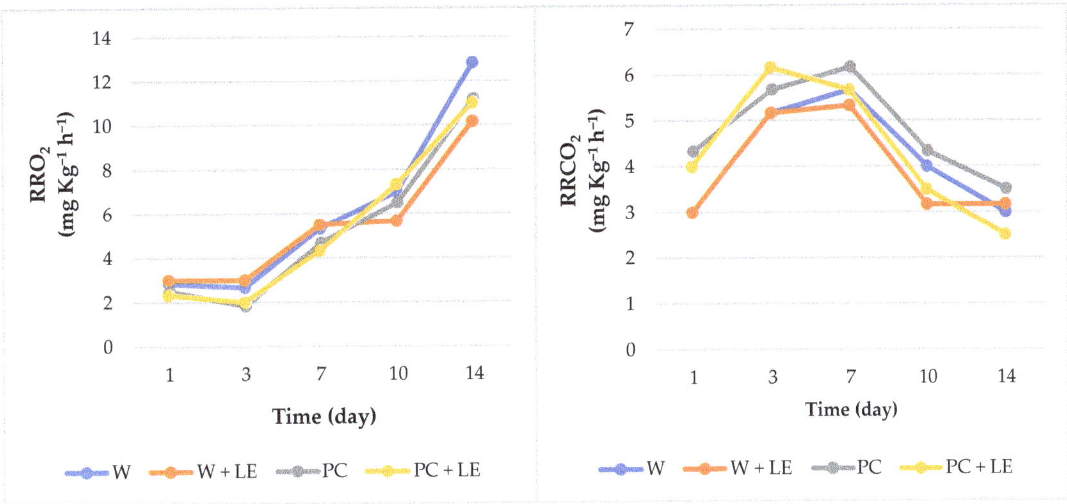

(**a**) (**b**)

Figure 5. (**a**) Oxygen consumption values (RRO_2) of uncoated and coated fresh-cut carrots throughout 14 days of storage at 4 °C; (**b**) carbon dioxide production values ($RRCO_2$) of uncoated and coated fresh-cut carrots throughout 14 days of storage at 4 °C. Abbreviations: W; W + LE; PC; PC + LE (see Table 1).

Variations in respiration rates furnish information of the general metabolic activity of carrot tissues as affected by postharvest and minimal processing conditions [37]. As reported by Tappi et al. [38], the inhibition of respiration relative to O_2 consumed did not always match a significant decrease in CO_2 production ($RRCO_2$). This contrasting behaviour could be caused by the stress generated during processing and coating operations, which promoted a slight increase in CO_2 production. However, the application of the lemon byproduct extract in the coating formulation does not appear to have had any significant impact on the respiration rate of fresh-cut carrots.

In addition, the coating itself could behave as a barrier to gas transport and prolong the commercial shelf-life of fresh products by modifying their internal atmosphere [51]. At the same time, the increase in the respiration process observed in coated carrots could be associated to modifications in the barrier properties of the coating matrix, which may be stimulated by the high values of carrots' water activity when stored in chilling conditions [64]. The changes in respiratory rates could also be linked to vegetable-tissue stress induced by minimal processing operations, such as trimming, peeling and cutting, or the application of the coating matrix, which generates stressful conditions [4]. However, the presence of the pectin-based coating encouraged a significant reduction in the respiration rate of fresh-cut carrots for at least one week of storage compared to the control W, most likely as a consequence of the interaction of calcium ions with pectin and plant-cell-wall components.

3.4. Bioactive Compounds Content and Antioxidant Activity

The total carotenoids content (TCC) of fresh-cut carrots, as affected by the coating treatments, is reported in Figure 6. The highest total carotenoid content was found in just-processed carrots subjected to treatments with the incorporation of LE, with values of 16.83 $\pm$ 0.67 and 13.29 $\pm$ 0.75 mg TCC 100 g^{-1} for W + LE and PC + LE, respectively. Among them, PC + LE showed constant values in TCC for one week of storage. This aspect confirms the simultaneous protective effect of the coating and the extract on the stability of carotenoids over time, a parameter that could be related to the minimal colour variation found in PBC + LE (Table 2). Changes in colour can be linked to dehydration and discolouration of the carrots' surfaces, together with the oxidation of carotenoids [51]. In fact, the main cause of carotenoid losses in vegetables is the oxidation of carotenoids, due to their highly unsaturated structure, which can occur by the spontaneous reaction of vegetable tissues in the presence of oxygen and other environmental factors such as light [65]. After the first week, a progressive decrease in TCC was observed until the last day of storage, except for W and PC, which showed an increment in total carotenoid content after the 10th day of storage.

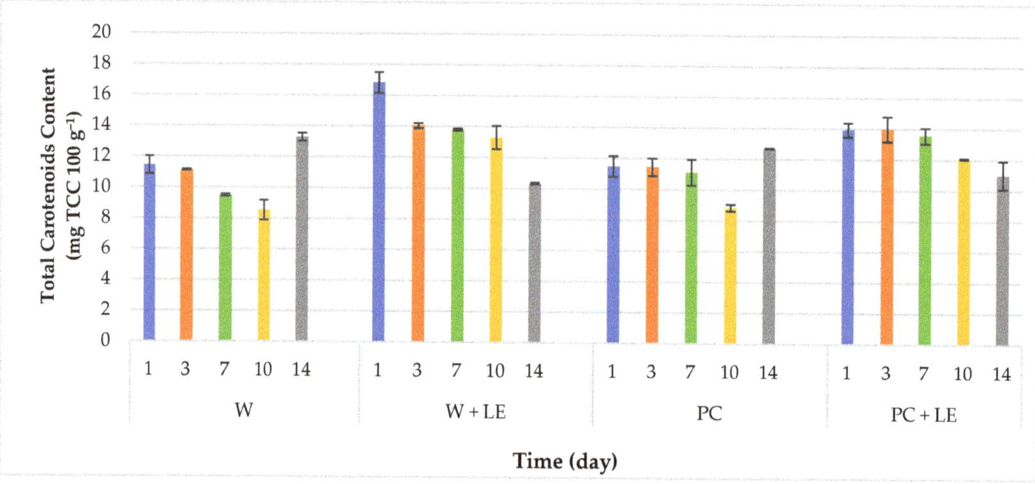

Figure 6. Total carotenoids content of uncoated and coated fresh-cut carrots throughout 14 days of storage (4 °C). Abbreviations: W; W + LE; PC; PC + LE (see Table 1).

Total phenolic content (TPC) and antioxidant activity values (Figure 7) exhibited a similar trend to that observed for TCC, especially in fresh-cut carrots with added LE. The incorporation of the lemon byproduct extract into the pectin-based coating formulation assured a greater stability of the bioactive composition of the carrots during storage. TPC values clearly denote a protecting effect of the coating against oxidative phenomena [61]. Hence, treatments incorporating the lemon byproduct extract (PC + LE and W + LE) allowed maintaining a higher total polyphenols content than control samples (W and PC), until the 10th day of storage. This contrasts with the values observed for W and PC carrots, which noticeably decreased during the first week of storage. However, on the 10th day, treatments without the incorporation of LE showed a significant increase in TPC as well as antioxidant activity values, as also observed by Ranjitha et al. [66].

This phenomenon may be ascribed to the onset of metabolic pathways leading to the production of phenolic compounds, as a part of the plant defence mechanisms [61]. An increase in TPC is one of the most widely studied events in response to wounding in several fresh-cut products, which has been further confirmed in the case of carrots. Namely, an increase in TPC and TCC might also be related to the wound-induced stimulation of the plant enzyme phenylalanine ammonia lyase (PAL), which transforms L-phenilalanine into trans-cinnamic acid, which acts as a precursor for different phenylpropanoids, such as lignin [66]. In this regard, several authors assign development of a superficial white blush to lignin synthesis in reply to cellular injuries, where lignin performs as a new barrier [51,67,68]. This is consistent with total-colour-difference (Table 2) and whiteness-index (Figure 1) results. However, the application of the edible coatings added with LE was effective in preserving the the TPC and antioxidant activity of coated carrots, which is related to the contents of different compounds, such as polyphenols and carotenoids. A high positive correlation between TCC and DPPH assays was found for W + LE carrots (r = 0.90) and PC + LE (r = 0.83), which agrees with the well-established relationship between the protective affect exerted by the lemon byproduct extract on carotenoids. As reported by de Oliveira et al. [43], the application of coatings could reduce the loss of phenolic compounds and preserve coated fruit and vegetables from oxidative damage and the accumulation of free radicals, thanks to their contribution in slowing down enzymatic activity and, thus, retarding changes in colour and other physicochemical parameters associated to senescence [69].

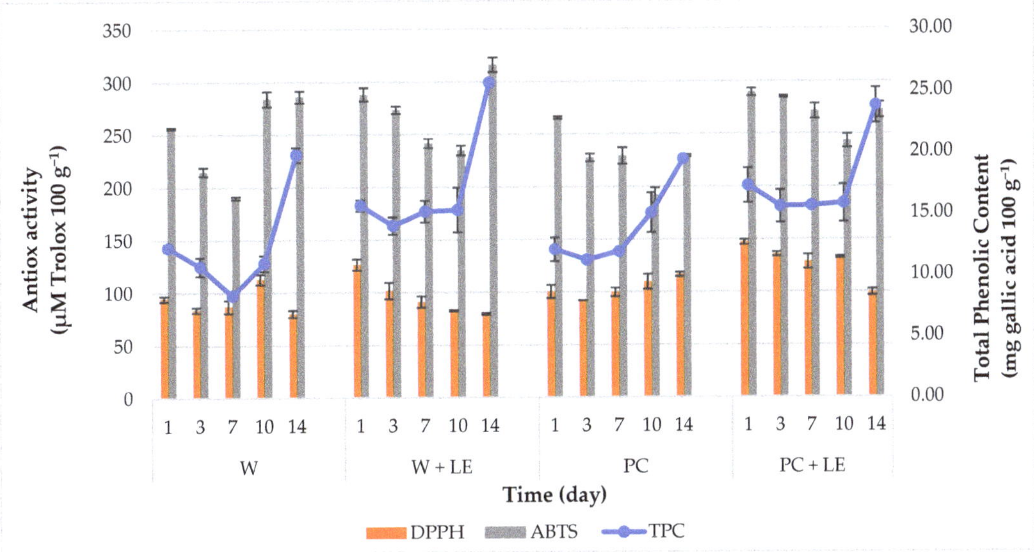

Figure 7. Total phenolic content (TPC) and expression of antioxidant activity (ABTS and DPPH assays) of uncoated and coated fresh-cut carrots throughout 14 days of storage (4 °C). Abbreviations: W; W + LE; PC; PC + LE (see Table 1).

4. Conclusions

The results of this study demonstrate that a pectin-based edible coating could be a simple and affordable technique for carrying bioactive compounds, such as those contained in the lemon byproduct extract, with considerable applicability in minimally processed carrots to improve the nutritional value and quality attributes of these fresh goods.

In fact, fresh-cut carrots treated with a pectin-based coating and lemon byproduct extract were characterized by a good preservation of physiological parameters and limited changes in colour ($\Delta E < 3$) and white-blush development on both cortical tissue and vascular cylinder throughout the storage period. The application of the pectin-based coating with the lemon byproduct extract ensured that carrots maintained stable structural integrity throughout the 14 days of storage at 4 °C, thanks to the reduction in enzymatic bacterial activity. This kind of treatment on minimally processed carrots also resulted in higher levels of carotenoids, phenolic compounds and antioxidant activity, as evident from higher ABTS and DPPH radical scavenging activity values.

This strategy represents an interesting option to decrease the rate of the physiological postharvest decay of fresh-cut carrots, slowing down the respiration process and senescence throughout storage and, at the same time, could be considered a valid approach to valorise a food industry byproduct. In this regard, edible coatings have proven to be promising carrier systems for bioactive ingredients that can improve food functional properties, representing the ideal choice for fruit and vegetables to maintain quality attributes for a longer shelf-life.

Author Contributions: Conceptualization, V.I., O.M.-B. and R.S.-F.; Investigation, V.I.; Methodology, V.I.; Supervision, A.P., O.M.-B. and R.S.-F.; Writing—original draft, V.I.; Writing—review and editing, O.M.-B. and R.S.-F. All authors have read and agreed to the published version of the manuscript.

Funding: This research received no external funding.

Informed Consent Statement: Not applicable.

Data Availability Statement: Data is contained within the article.

Acknowledgments: This work was supported by "European Commission, European Social Fund and the Calabria Region" and by University of Lleida.

Conflicts of Interest: The authors declare no conflict of interest.

References

1. Arscott, S.A.; Tanumihardjo, S.A. Carrots of many colors provide basic nutrition and bioavailable phytochemicals acting as a functional food. *Compr. Rev. Food Sci. Food Saf.* **2010**, *9*, 223–239. [CrossRef]
2. López-Gámez, G.; Elez-Martínez, P.; Martín-Belloso, O.; Soliva-Fortuny, R. Enhancing phenolic content in carrots by pulsed electric fields during post-treatment time: Effects on cell viability and quality attributes. *Innov. Food Sci. Emerg. Technol.* **2020**, *59*, 102252. [CrossRef]
3. Guerreiro, A.C.; Gago, C.M.L.; Miguel, M.G.C.; Faleiro, M.L.; Antunes, M.D.C. The influence of edible coatings enriched with citral and eugenol on the raspberry storage ability, nutritional and sensory quality. *Food Packag. Shelf Life* **2016**, *9*, 20–28. [CrossRef]
4. Mastromatteo, M.; Conte, A.; Del Nobile, M.A. Packaging strategies to prolong the shelf life of fresh carrots (*Daucus carota* L.). *Innov. Food Sci. Emerg. Technol.* **2012**, *13*, 215–220. [CrossRef]
5. Piscopo, A.; Zappia, A.; Princi, M.P.; De Bruno, A.; Araniti, F.; Lupini, A.; Abenavoli, M.R.; Poiana, M. Quality of shredded carrots minimally processed by different dipping solutions. *J. Food Sci. Technol.* **2019**, *56*, 2584–2593. [CrossRef]
6. Huber, K.C.; Embuscado, M. *Edible Films and Coatings for Food Applications*; Springer: New York, NY, USA, 2009. [CrossRef]
7. Bourtoom, T. Review article-edible films and coatings: Characteristics and properties. *Int. Food Res. J.* **2008**, *15*, 237–248.
8. Campos, C.A.; Gerschenson, L.N.; Flores, S.K. Development of Edible Films and Coatings with Antimicrobial Activity. *Food Bioproc. Technol.* **2011**, *4*, 849–875. [CrossRef]
9. Salmieri, S.; Lacroix, M. Physicochemical properties of alginate/polycaprolactone-based films containing essential oils. *J. Agric. Food Chem.* **2006**, *54*, 10205–10214. [CrossRef]
10. Guerreiro, A.C.; Gago, C.M.L.; Faleiro, M.L.; Miguel, M.G.C.; Antunes, M.D.C. The use of polysaccharide-based edible coatings enriched with essential oils to improve shelf-life of strawberries. *Postharvest Biol. Technol.* **2015**, *110*, 51–60. [CrossRef]
11. Thakur, B.R.; Singh, R.K.; Handa, A.K. Chemistry and Uses of Pectin—A Review. *Crit. Rev. Food Sci. Nutr.* **1997**, *37*, 47–73. [CrossRef]
12. Zhao, S.; Ren, W.; Gao, W.; Tian, G.; Zhao, C.; Bao, Y.; Cui, J.; Lian, Y.; Zheng, J. Effect of mesoscopic structure of citrus pectin on its emulsifying properties: Compactness is more important than size. *J. Colloid Interface Sci.* **2020**, *570*, 80–88. [CrossRef]
13. Baeva, M.; Panchev, I. Investigation of the retaining effect of a pectin-containing edible film upon the crumb ageing of dietetic sucrose-free sponge cake. *Food Chem.* **2005**, *92*, 343–348. [CrossRef]
14. Corbo, M.R.; Campaniello, D.; Speranza, B.; Bevilacqua, A.; Sinigaglia, M. Non-conventional tools to preserve and prolong the quality of minimally-processed fruits and vegetables. *Coatings* **2015**, *5*, 931–961. [CrossRef]
15. Oms-Oliu, G.; Soliva-Fortuny, R.; Martín-Belloso, O. Edible coatings with antibrowning agents to maintain sensory quality and antioxidant properties of fresh-cut pears. *Postharvest Biol. Technol.* **2008**, *50*, 87–94. [CrossRef]
16. Tumbarski, Y.; Petkova, X.; Todorova, M.; Ivanov, I.; Deseva, I.; Mihaylova, D.; Ibrahim, S.A. Effects of pectin-based edible coatings containing a bacteriocin of Bacillus methylotrophicus BM47 on the quality and storage life of fresh blackberries. *Ital. J. Food Sci.* **2020**, *32*, 420–427. [CrossRef]
17. Martínez-Hernández, G.B.; Amodio, M.L.; Colelli, G. Potential use of microwave treatment on fresh-cut carrots: Physical, chemical and microbiological aspects. *J. Sci. Food Agric.* **2016**, *96*, 2063–2072. [CrossRef]
18. Attar, R.F.; Sedaghat, N.; Pasban, A.; Yeganehzad, S.; Hesarinejad, M.A. Modeling the respiration rate of chitosan coated fresh in-hull pistachios (Pistacia vera L. cv. Badami) for modified atmosphere packaging design. *J. Food Meas. Charact.* **2022**, *16*, 1049–1061. [CrossRef]
19. Valencia-Chamorro, S.A.; Palou, L.; Delfio, M.A.; Pérez-Gago, M.B. Antimicrobial edible films and coatings for fresh and minimally processed fruits and vegetables: A review. *Crit. Rev. Food Sci. Nutr.* **2011**, *51*, 872–900. [CrossRef]
20. Song, Z.; Li, F.; Guan, H.; Xu, Y.; Fu, Q.; Li, D. Combination of nisin and ε-polylysine with chitosan coating inhibits the white blush of fresh-cut carrots. *Food Control* **2017**, *74*, 34–44. [CrossRef]
21. Arnon-Rips, H.; Porat, R.; Poverenov, E. Enhancement of agricultural produce quality and storability using citral-based edible coatings; the valuable effect of nano-emulsification in a solid-state delivery on fresh-cut melons model. *Food Chem.* **2019**, *277*, 205–212. [CrossRef]

22. Rangel-Marrón, M.; Mani-López, E.; Palou, E.; López-Malo, A. Effects of alginate-glycerol-citric acid concentrations on selected physical, mechanical, and barrier properties of papaya puree-based edible films and coatings, as evaluated by response surface methodology. *LWT* **2019**, *101*, 83–91. [CrossRef]
23. Nair, A.K.; Mukherjee, M.; Nag, S.; Pandimadevi, M. Antioxidant and antimicrobial activities of citrus lemon peels encapsulated in PVA. *Carpathian J. Food Sci. Technol.* **2019**, *11*, 111–126.
24. Mathew, B.B.; Shajie, D.; Wadhwa, N.; Murthy, N.K.; Murthy, T.K.; Rashmi, M.V. Comparative antioxidant efficacy of Citrus limonum pulp and peel—An in vitro study. *Drug Inven. Today* **2013**, *5*, 296–301. [CrossRef]
25. Al-Qassabi, J.S.A.; Weli, A.M.; Hossain, M.A. Comparison of total phenols content and antioxidant potential of peel extracts of local and imported lemons samples. *Sustain. Chem. Pharm.* **2018**, *8*, 71–75. [CrossRef]
26. O'Shea, N.; Arendt, E.K.; Gallagher, E. Dietary fibre and phytochemical characteristics of fruit and vegetable by-products and their recent applications as novel ingredients in food products. *Innov. Food Sci. Emerg. Technol.* **2012**, *16*, 1–10. [CrossRef]
27. Putnik, P.; Bursać Kovačević, D.; Režek Jambrak, A.; Barba, F.J.; Cravotto, G.; Binello, A.; Lorenzo, J.M.; Shpigelman, A. Innovative "green" and novel strategies for the extraction of bioactive added value compounds from citrus wastes—A review. *Molecules* **2017**, *22*, 680. [CrossRef]
28. Imeneo, V.; Romeo, R.; Gattuso, A.; De Bruno, A.; Piscopo, A. Functionalized Biscuits with Bioactive Ingredients Obtained by Citrus Lemon Pomace. *Foods* **2021**, *10*, 2460. [CrossRef]
29. Russo, M.; Bonaccorsi, I.; Torre, G.; Sarò, M.; Dugo, P.; Mondello, L. Underestimated sources of flavonoids, limonoids and dietary fibre: Availability in lemon's by-products. *J. Funct. Foods* **2014**, *9*, 18–26. [CrossRef]
30. Imeneo, V.; Romeo, R.; De Bruno, A.; Piscopo, A. Green-sustainable extraction techniques for the recovery of antioxidant compounds from "citrus Limon" by-products. *J. Environ. Sci. Health B* **2022**, 1–13. [CrossRef]
31. AOAC, Association of Official Analytical Chemists. Hydrogen-ion activity (pH) method. In *Method 14.022*, 13th ed.; Horwitz, W., Ed.; Association of Official Analytical Chemists: Washington, DC, USA, 1980; p. 213.
32. AOAC, Association of Official Analytical Chemists. Method 942.15. Acidity of fruit products. In *Official Methods of Analysis*, 17th ed.; Horwitz, W., Ed.; Association of Official Analytical Chemists: Washington, DC, USA, 2000; p. 213.
33. AOAC, Association of Official Analytical Chemists. *Determination of Water/Dry Matter (Moisture) in Animal Feed, Grain, and Forage (Plant Tissue); In Official Methods of Analysis*, 17th ed.; Horwitz, W., Ed.; Association of Official Analytical Chemists: Washington, DC, USA, 2000; p. 12.
34. Thompson, B. *Printing Materials Science and Technology*, 2nd ed.; Pira International: Surrey, UK, 2004.
35. Ribas-Agustí, A.; Martín-Belloso, O.; Soliva-Fortuny, R.; Elez-Martínez, P. Enhancing hydroxycinnamic acids and flavan-3-ol contents by pulsed electric fields without affecting quality attributes of apple. *Food Res. Int.* **2019**, *121*, 433–440. [CrossRef]
36. Fan, L.; Song, J. Microbial quality assessment methods for fresh-cut fruits and vegetables. *Stewart Postharvest Rev.* **2008**, *4*, 1–9. [CrossRef]
37. López-Gámez, G.; Elez-Martínez, P.; Martín-Belloso, O.; Soliva-Fortuny, R. Pulsed electric fields affect endogenous enzyme activities, respiration and biosynthesis of phenolic compounds in carrots. *Postharvest Biol. Technol.* **2020**, *168*, 111284. [CrossRef]
38. Tappi, S.; Berardinelli, A.; Ragni, L.; Dalla Rosa, M.; Guarnieri, A.; Rocculi, P. Atmospheric gas plasma treatment of fresh-cut apples. *Innov. Food Sci. Emerg. Technol.* **2014**, *21*, 114–122. [CrossRef]
39. Formica-Oliveira, A.C.; Martínez-Hernández, G.B.; Díaz-López, V.; Artés, F.; Artés-Hernández, F. Effects of UV-B and UV-C combination on phenolic compounds biosynthesis in fresh-cut carrots. *Postharvest Biol. Technol.* **2017**, *127*, 99–104. [CrossRef]
40. González-Casado, S.; Martín-Belloso, O.; Elez-Martínez, P.; Soliva-Fortuny, R. Enhancing the carotenoid content of tomato fruit with pulsed electric field treatments: Effects on respiratory activity and quality attributes. *Postharvest Biol. Technol.* **2018**, *137*, 113–118. [CrossRef]
41. Gross, J. *Pigments in Vegetables: Chlorophylls and Carotenoids*; Van Nostrand Reinhold: New York, NY, USA, 1991; 278p.
42. Re, R.; Pellegrini, N.; Proteggente, A.; Pannala, A.; Yang, M.; Rice-Evans, C. Antioxidant activity applying an improved ABTS radical cation decolorization assay. *Free Radic. Biol. Med.* **1999**, *26*, 1231–1237. [CrossRef]
43. de Oliveira, K.Á.R.; Fernandes, K.F.D.; de Souza, E.L. Current advances on the development and application of probiotic-loaded edible films and coatings for the bioprotection of fresh and minimally processed fruit and vegetables. *Foods* **2021**, *10*, 2207. [CrossRef]
44. Han, C.; Zhao, Y.; Leonard, S.W.; Traber, M.G. Edible coating to improve storability and enhance nutritional value of fresh and frozen strawberries (Fragaria x Ananassa) and raspberries (*Rubus ideaus*). *Postharvest Biol. Technol.* **2004**, *33*, 67–78. [CrossRef]
45. Fai, A.E.C.; Alves de Souza, M.R.; de Barros, S.T.; Bruno, N.V.; Ferreira, M.S.L.; Gonçalves, T.C.B.D.A.; de Andrade, É.C.B. Development and evaluation of biodegradable films and coatings obtained from fruit and vegetable residues applied to fresh-cut carrot (*Daucus carota* L.). *Postharvest Biol. Technol.* **2016**, *112*, 194–204. [CrossRef]
46. Howard, L.R.; Griffin, L.E. Lignin Formation and Surface Discoloration of Minimally Processed Carrot Sticks. *J. Food Sci.* **1993**, *58*, 1065–1067. [CrossRef]
47. Hager, T.J.; Howard, L.R. Processing effects on carrot phytonutrients. *HortScience* **2006**, *41*, 74–79. [CrossRef]
48. Howard, L.R.; Griffin, L.E.; Lee, Y. Steam treatment of minimally processed carrot sticks to control surface discoloration. *J. Food Sci.* **1994**, *59*, 356–358. [CrossRef]
49. Leja, M.; Mareczek, A.; Wojciechowska, R.; Stanisław, R. Phenolic metabolism in root slices of selected carrot cultivars. *Acta Physiol. Plant.* **1997**, *19*, 319–325. [CrossRef]

50. Pace, B.; Capotorto, I.; Cefola, M.; Minasi, P.; Montemurro, N.; Carbone, V. Evaluation of quality, phenolic and carotenoid composition of fresh-cut purple Polignano carrots stored in modified atmosphere. *J. Food Compost. Anal.* **2020**, *86*, 103363. [CrossRef]
51. Vargas, M.; Chiralt, A.; Albors, A.; González-Martínez, C. Effect of chitosan-based edible coatings applied by vacuum impregnation on quality preservation of fresh-cut carrot. *Postharvest Biol. Technol.* **2009**, *51*, 263–271. [CrossRef]
52. Mei, Y.; Zhao, Y.; Furr, H.C. Using Edible Coating to Enhance Nutritional and Sensory Qualities of Baby Carrots. *J. Food Sci.* **2002**, *67*, 1964–1968. [CrossRef]
53. Cisneros-Zevallos, L.; Saltveit, M.E.; Krochta, J.M. Hygroscopic Coatings Control Surface White Discoloration of Peeled (Minimally Processed) Carrots during Storage. *J. Food Sci.* **1997**, *62*, 363–366. [CrossRef]
54. Lee, J.Y.; Park, H.J.; Lee, C.Y.; Choi, W.Y. Extending shelf-life of minimally processed apples with edible coatings and antibrowning agents. *Lebensm.-Wiss. Technol.* **2003**, *36*, 323–329. [CrossRef]
55. Olivas, G.I.; Mattinson, D.S.; Barbosa-Canovas, G.V. Alginate coatings for reservation of minimally processed 'Gala' apples. *Postharvest Biol. Technol.* **2007**, *45*, 89–96. [CrossRef]
56. Rojas-Grau, M.A.; Tapia, M.S.; Martín-Belloso, O. Using polysaccharide-based edible coatings to maintain quality of fresh-cut Fuji apples. *Lebensm.-Wiss. Technol.* **2008**, *41*, 139–147. [CrossRef]
57. Ferrari, C.C.; Sarantópoulos, C.I.G.L.; Carmello-Guerreiro, S.M.; Hubinger, M.D. Effect of Osmotic Dehydration and Pectin Edible Coatings on Quality and Shelf Life of Fresh-Cut Melon. *Food Bioprocess Technol.* **2013**, *6*, 80–91. [CrossRef]
58. Shigematsu, E.; Dorta, C.; Santos, D.N.; Ferreira, K.A.; Góes-Favoni, S.P.; Oshiiwa, M.; Mauro, M.A. Edible coating with coconut water to preserve probiotic strains and sensory characteristics of minimally processed carrots. *Int. Food Res. J.* **2019**, *26*, 1285–1292.
59. Rosa, M.; Prado, C.; Podazza, G.; Interdonato, R.; González, J.A.; Hilal, M.; Prado, F.E. Soluble sugars: Metabolism, sensing and abiotic stress: A complex network in the life of plants. *Plant Signal. Behav.* **2009**, *4*, 388–393. [CrossRef] [PubMed]
60. Amanatidou, A.; Slump, R.A.; Gorris, L.G.M.; Smid, E.J. High oxygen and high carbon dioxide modified atmospheres for shelf-life extension of minimally processed carrots. *J. Food Sci.* **2000**, *65*, 61–66. [CrossRef]
61. Ben-Fadhel, Y.; Maherani, B.; Manus, J.; Salmieri, S.; Lacroix, M. Physicochemical and microbiological characterization of pectin-based gelled emulsions coating applied on pre-cut carrots. *Food Hydrocoll.* **2020**, *101*, 105573. [CrossRef]
62. Budiati, T.; Suryaningsih, W.; Yudistira, H.; Azhar, S.W. Antimicrobial activity of jengkol and petai peel extract to inhibit listeria monocytogenes. *IOP Conf. Ser. Earth Environ. Sci.* **2021**, *672*, 012046. [CrossRef]
63. Ivasenko, S.; Orazbayeva, P.; Skalicka–wozniak, K.; Ludwiczuk, A.; Marchenko, A.; Ishmuratova, M.; Poleszak, E.; Korona-Glowniak, I.; Akhmetova, S.; Karilkhan, I.; et al. Antimicrobial activity of ultrasonic extracts of two chemotypes of thymus serpyllum l. Of central kazakhstan and their polyphenolic profiles. *Maced. J. Med. Sci.* **2021**, *9*, 61–67. [CrossRef]
64. Leceta, I.; Molinaro, S.; Guerrero, P.; Kerry, J.P.; De la Caba, K. Quality attributes of map packaged ready-to-eat baby carrots by using chitosan-based coatings. *Postharvest Biol. Technol.* **2015**, *100*, 142–150. [CrossRef]
65. Odriozola-Serrano, I.; Soliva-Fortuny, R.; Hernández-Jover, T.; Martín-Belloso, O. Carotenoid and phenolic profile of tomato juices processed by high intensity pulsed electric fields compared with conventional thermal treatments. *Food Chem.* **2009**, *112*, 258–266. [CrossRef]
66. Ranjitha, K.; Sudhakar Rao, D.V.; Shivashankara, K.S.; Oberoi, H.S.; Roy, T.K.; Bharathamma, H. Shelf-life extension and quality retention in fresh-cut carrots coated with pectin. *Innov. Food Sci. Emerg. Technol.* **2017**, *42*, 91–100. [CrossRef]
67. Fonseca, S.C.; Oliveira, F.A.R.; Brecht, J.K. Modelling respiration rate of fresh fruits and vegetables for modified atmosphere packages: A review. *J. Food Eng.* **2002**, *52*, 99–119. [CrossRef]
68. Izumi, H.; Watada, A.E. Calcium Treatments Affect Storage Quality of Shredded Carrots. *J. Food Sci.* **1994**, *59*, 106–109. [CrossRef]
69. Muley, A.B.; Singhal, R.S. Extension of postharvest shelf life of strawberries (*Fragaria ananassa*) using a coating of chitosan-whey protein isolate conjugate. *Food Chem.* **2020**, *329*, 127213. [CrossRef] [PubMed]

foods

Article

Profiles of Volatile and Phenolic Compounds as Markers of Ripening Stage in Candonga Strawberries

Rosaria Cozzolino [1,*], Bernardo Pace [2,*], Michela Palumbo [2,3], Carmine Laurino [1], Gianluca Picariello [1], Francesco Siano [1], Beatrice De Giulio [1], Sergio Pelosi [2] and Maria Cefola [2]

1 Institute of Food Science, National Research Council (CNR), Via Roma 64, 83100 Avellino, Italy; carmine.laurino@isa.cnr.it (C.L.); gianluca.picariello@isa.cnr.it (G.P.); francesco.siano@isa.cnr.it (F.S.); beatrice.degiulio@isa.cnr.it (B.D.G.)

2 Institute of Sciences of Food Production, National Research Council of Italy (CNR), c/o CS-DAT, Via M. Protano, 71121 Foggia, Italy; michela.palumbo@ispa.cnr.it (M.P.); sergio.pelosi@ispa.cnr.it (S.P.); maria.cefola@ispa.cnr.it (M.C.)

3 Department of Agriculture, Food, Natural Resources and Engineering, University of Foggia, Via Napoli 25, 71122 Foggia, Italy

* Correspondence: rosaria.cozzolino@isa.cnr.it (R.C.); bernardo.pace@ispa.cnr.it (B.P.)

Abstract: Volatile compounds, quality traits (total phenols and antioxidant capacity) and High-performance liquid chromatography (HPLC)-isolated polyphenols of strawberries, variety Sabrosa, commercially referred to as "Candonga", harvested at three different times (H1, H2 and H3) and at two different ripening stages, namely half-red (Half-red-H1, Half-red-H2 and Half-red-H3) and red (Red-H1, Red-H2 and Red-H3) were evaluated. Dominant anthocyanins, namely cyanidin-3-*O*-glucoside, pelargonidin-3-*O*-glucoside and pelargonidin-3-*O*-rutinoside, as well as *p*-coumaryl hexoside increased during harvesting, differently from flavonoids, such as quercetin-3-*O*-glucoside, kaempferol-3-*O*-glucoronide and quercetin 3-*O*-glucoronide, that declined. Samples clustered in different quadrants of the principal component analysis (PCA) performed on volatiles, quality traits and phenolic compounds, highlighting that only the red samples were directly correlated to volatile components, as volatiles clearly increased both in number and amount during ripening. In particular, volatiles with a positive impact on the consumers' acceptance, including butyl butyrate, ethyl hexanoate, hexyl acetate, nonanal, terpenes and lactones, were positively associated with the Red-H1 and Red-H2 strawberries, while volatiles with negative coefficients related to consumer liking, including isopropyl butyrate, isoamyl butyrate and mesifurane directly correlated with the Red-H3 samples. Accordingly, strawberries harvested at Red-H1 and Red-H2 ripening stages could be preferred by the consumers compared to the Red-H3 fruit. Altogether, these results could help to individuate quality traits as putative markers of the ripening stage, and optimize the process of post-harvesting ripening to preserve or improve the desirable aromatic characteristics of strawberries.

Keywords: *Fragaria* × *ananassa* Duch.; Sabrosa; ripening stage; headspace solid phase microextraction (HS SPME GC/MS); HPLC-MS/MS principal component analysis

Citation: Cozzolino, R.; Pace, B.; Palumbo, M.; Laurino, C.; Picariello, G.; Siano, F.; De Giulio, B.; Pelosi, S.; Cefola, M. Profiles of Volatile and Phenolic Compounds as Markers of Ripening Stage in Candonga Strawberries. *Foods* **2021**, *10*, 3102. https://doi.org/10.3390/foods10123102

Academic Editor: Jinhe Bai

Received: 12 November 2021
Accepted: 9 December 2021
Published: 14 December 2021

1. Introduction

Strawberry (*Fragaria* × *ananassa* Duch.) belongs to the Rosaceae family, and is one of the most commonly consumed berry fruit and cash crops worldwide, with more than 2000 varieties [1]. Because of the high content of ellagic acid (EA) and its precursors, strawberries are considered functional foods [1].

The functional traits of the strawberry originate from to the combination of vitamin C and other antioxidant components, primarily flavonoids, anthocyanins and EA. In spite of their relative low content (0.001–0.01% of fresh fruit weight), the volatile organic compounds (VOCs) are the main responsible for the strawberries' flavor, which is a crucial factor in determining the consumer's preference and the sensory quality of the fruit [2–4].

Strawberry fruit can be considered a typical example of a complex fruit aroma, since several hundred VOCs concur to determine the flavor of fresh strawberries [4].

Qualitative and quantitative profiles of strawberry VOCs stem partially from genetic traits and growth-dependent activation of specific metabolic pathways, so that they show specific patterns and distinctive volatile components depending on the cultivar and degree of ripening [5]. Thus, aroma can be a fingerprint to distinguish among varieties and stage of fruit development [3].

The content of VOCs increases during the maturation in climacteric fruits, and a similar effect could be expected in non-climacteric fruit as well, such as the strawberry [3].

Harvesting fruit before it is fully ripe is a common practice for many fruits in the supply chain, as it allows to complete ripening in post-harvest storage [6].

Clearly, the maturity stage affects the consumer liking of these fruits, including strawberries, as a consequence of the changes affecting the volatile profile [7].

Monitoring the VOCs pattern during the maturation process could offer specific markers for establishing the optimal harvest time, ensuring a standardized aroma to consumers as well as maximizing quality and phytosanitary characteristics, thus contributing to minimizing post-harvest losses [5]. Volatile esters are the compounds most associated with strawberry fruit ripening, although VOCs belonging to different classes could be considered markers of maturity, varying with the cultivar [3].

Among the phenolic compounds, anthocyanins are responsible for the bright red color of the strawberries. They are the most abundant phenolic compounds in most of the 27 cultivars analyzed by Aaby et al. [8], and are known to possess health benefits. The anthocyanins also vary with cultivars, and steadily increase during ripening, as assessed by analyzing the number of strawberry cultivars [8].

The present study aimed at evaluating for the first time the changes in fruit quality traits, content of VOCs and phenolic compounds in strawberry samples of the variety Sabrosa, commercially referred to as "Candonga". Strawberries were collected at two different ripening stages (half-red and red) in three different harvesting times. The goal was to gain information about the modification at a metabolite level occurring during strawberry maturation and the contributing individuate putative molecular markers, in order to optimize the selection of the timing of harvest.

2. Materials and Methods

2.1. Plant Material

"Candonga" strawberries (*Fragaria* × *ananassa* Duch. var. Sabrosa) were harvested by Apofruit (Scanzano Jonico, Italy) on 21 May (first harvest time, H1), 27 May (second harvest time, H2) and 1 June (third harvest time, H3) at two different ripening stages, namely half-red (in ripening phase, fully expanded and 50% red, indicated as Half-red-H1, Half-red-H2 and Half-red-H3) and red (in ripening phase, fully expanded and 100% red, indicated as Red-H1, Red-H2 and Red-H3), according to visual criteria (Figure 1).

Half-red and red strawberries showed soluble solids at harvest of about 8.8 ± 0.7 and 9.8 ± 0.2 °Brix, respectively. At each harvest time, fruit berries were packed into PET trays (Cartonpack spa, Rutigliano, Ba) (about 500 g for each tray) and transported in refrigerated conditions (4 ± 1 °C) to the Consiglio Nazionale delle Ricerche (CNR) laboratories, and were analyzed within 3 h since harvesting. Here, strawberries were visually inspected and selected to choose fruits free of physical or biological damage. Respiration rate titratable acidity, total soluble solids, pH, color parameters, antioxidant activity, total phenols, HPLC-separated individual phenolic compounds and VOCs were evaluated at each harvest time and for each ripening stage.

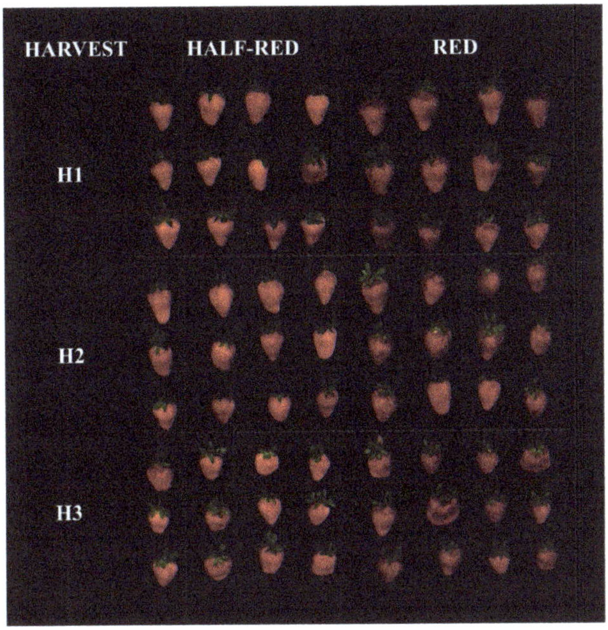

Figure 1. "Candonga" strawberries (var. Sabrosa) harvested at two different ripening stages, namely half-red (in ripening phase, fully expanded and 50% red, indicated as Half-red-H1, Half-red-H2 and Half-red-H3) and red (in ripening phase, fully expanded and 100% red, indicated as Red-H1, Red-H2 and Red-H3).

2.2. Reagents and Chemicals

Chemicals, standards and reagents were from Sigma-Aldrich (St. Louis, MO, USA). Folin–Ciocalteu's phenol reagent was purchased from Merck (Darmstadt, Germany). Ultrapure water (resistivity at 25 °C of 18 MΩ cm) was from a Millipore Milli-Q purification system (Millipore Corp., Bedford, MA, USA), while helium at a purity of 99.999% (Rivoira, Milan, Italy) was used as GC carrier gas. The HS-SPME fibers and the glass vials were purchased from Supelco (Bellofonte, PA, USA); the capillary GC-MS column High Polarity (HP)-Innowax (30 m × 0.25 mm × 0.5 μm) was acquired from Agilent J&W (Agilent Technologies Inc., Santa Clara, CA, USA).

2.3. Respiration Rate

The respiration rate of strawberries was measured at 8 °C using a closed system, according to the method reported by Kader [9]. In particular, for each ripening stage and replicate (n = 3), about 500 g of sample were put into 3.6 L sealed plastic jar (one jar for each replicate) where CO_2 was allowed to accumulate up to 0.1% of the standard concentration of the CO_2. The time taken to get to this value was detected by measuring the CO_2 amounts at regular intervals of time. The CO_2 analysis was conducted by injecting 1 mL of gas sample from the headspace of the plastic jars through a rubber septum into a gas chromatograph (p200 micro GC-Agilent, Santa Clara, CA, USA) fitted with dual columns and a thermal conductivity detector. CO_2 was analyzed with a retention time of 16 s and a total run time of 120 s using a 10 m porous polymer (PPU) column (Agilent, Santa Clara, CA, USA) at a constant temperature of 70 °C. Respiration rate was reported as mL CO_2/kg h.

2.4. Total Soluble Solids, Titratable Acidity and pH

For each replicate and at each ripening stage, about 100 g of strawberries was homogenized to obtain the fruit juice which was used to assay total soluble solids (TSS), titratable acidity (TA) and pH. The TSS content was determined using a digital refractometer (DBR35-XS Instruments, Carpi, Italy) and results were expressed in °Brix.

The pH of the fruit juice was determined using a pH meter (PH-Burette 24-Crison Instrument, Barcelona, Spain) and the titratable acidity (% citric acid) was measured by titration using 0.1 M NaOH to the final pH 8.1, revealed with phenolphthalein as the indicator.

2.5. Antioxidant Activity and Total Phenolic Content

The analyses of the antioxidant activity (AA) and the total phenolic content (TPC) of strawberry samples were carried out on samples extracted as follows: for each replicate and at each ripening stage, 5 g of strawberries (chopped into small pieces) was homogenized in 20 mL methanol/water solution (80:20 *v*/*v*) for 2 min, using a homogenizer (T-25 digital ULTRA-TURRAX®-IKA, Staufen, Germany) and then centrifuged (Prism C2500-R, Labnet, Edison, NJ, USA) at 15,000 rpm for 5 min at 4 °C. The extracts were collected and stored at −20 °C before the analysis.

The AA was measured on the methanol extract using the DPPH (1,1-diphenyl-2-picrylhydrazyl) assay as described by Cefola et al. [10]. The absorbance was measured at 515 nm after 40 min in the dark, using a spectrophotometer (UV-1800, Shimadzu, Kyoto, Japan). The results were expressed as mg of Trolox per 100 g of fw (fresh weight) using a Trolox calibration curve (82–625 μM; $R^2 = 0.999$).

The TPC was determined according to Fadda et al. [11]. In detail, 100 μL of each extract was mixed into 1.58 mL of water, 100 μL of Folin–Ciocalteu's reagent and 300 μL of sodium carbonate solution (200 g/L). The resulting absorbance was measured at 765 nm after 2 h in the dark and the results were expressed as mg of gallic acid equivalents (GAE) per 100 g of fw. The calibration curve of gallic acid was prepared with five points, from 50 to 500 μg/mL, with $R^2 = 0.998$.

2.6. Analysis of Polyphenols Compounds

2.6.1. Reversed Phase-High Performance Liquid Chromatographic-Diode Array Detector (RP-HPLC-DAD) Semi-Quantitative Determination of Polyphenols

Strawberry methanolic (80%, *v*/*v*) extracts, prepared as described above (Par. 2.5), were ten-fold diluted with aqueous 0.1% (*v*/*v*) trifluoroacetic acid (TFA) and 100 μL of the diluted sample was separated using a modular HP 1100 chromatographer (Agilent Technologies, Paolo Alto, CA, USA) equipped with a 250 × 2.0 mm i.d. C18 reversed-phase column, 4 mm particle diameter (Jupiter Phenomenex, Torrance, CA, USA) held at 37 °C in a thermostatic oven. HPLC runs were performed at a constant flow rate of 0.2 mL/min applying the following gradient of solvent B: isocratic elution at 5% B for 5 min, 5–60% linear gradient of B for 5–65 min and 60–100% B at 65–70 min. Eluent A and B were 0.1% TFA in HPLC-grade water and 0.1% TFA in acetonitrile, respectively. Samples were run in triplicate and monitored at wavelengths λ = 520, 360, 320 and 280 nm using a diode array detector (DAD), also acquiring a UV-Vis spectrum every second in the 200–700 nm range. A home-prepared multicomponent standard solution containing 15 (poly)phenols among which were gallic acid, *p*-coumaric acid, quercetin 3-*O*-glucoside, rutin, EA and quercetin and kaempferol aglycones (all from Sigma-Aldrich, St. Louis, MI, USA), was used to confirm or exclude the assignment of some phenolic compounds. Flavonoids were semi-quantified by plotting the area of peaks integrated at 360 nm on an external calibration curve built with standard rutin at a known concentration in the 0.10–5.00 μg/mL range ($R^2 = 0.99$). Antocyanins and *p*-coumaryl-hexoside were semi-quantified based on calibration curves built with cyanidin-3-*O*-glucoside and *p*-coumaric acid with absorbance at 520 and 320 nm, respectively. Data were processed using the ChemStation software (version A.10) purchased with the chromatograph. Analyses were carried out in triplicate and peak area values were averaged.

2.6.2. Nanoflow HPLC-ESI MS/MS Analysis

Identification of the phenolic compounds were confirmed by nanoflow-HPLC ESI MS/MS analysis, which was performed using an Ultimate 3000 ultra-high performance liquid chromatography instrument (Dionex/Thermo Scientific, San Jose, CA, USA), online coupled with a Q Exactive Orbitrap (Thermo Scientific) mass spectrometer, using previously detailed conditions [12]. The mass spectrometer switched between positive and negative ionization polarity in 1 s and scanned the 120–1200 m/z range, and operated in data-dependent acquisition for MS/MS, fragmenting up to 5 most intense signals in 1 s with 10 s of dynamic exclusion. Spectra were elaborated using the Xcalibur Software 3.1 version (Thermo Scientific).

2.7. Volatile Organic Compounds (VOCs) Analysis

2.7.1. Sample Preparation and HS SPME Procedure

The HS SPME conditions of analysis were optimized assaying strawberry samples obtained from a local supermarket. Profiling of VOCs was performed by HS SPME/GC-MS according to Zorrilla-Fontanesi et al. [13], but utilizing a DVB/CAR/PDMS (50/30 mm) fiber, with 50 °C as the extraction temperature and 20 min as the extraction time. Concerning the sample preparation, 1 g of "Candonga" strawberry sample was put into a 20 mL screw-on cap HS vial and mixed into 0.3 g of NaCl. To ensure the analytical reproducibility, 1.5 µL each sample was spiked with 20 ppm of 2-octanol taken from a stock solution, used as the internal standard (IS). Vials were then sealed with a Teflon septum and an aluminum cap (Chromacol, Hertfordshire, UK) and stirred. The equilibration time and temperature were 10 min and 40 °C, respectively. The extraction and injection phases were automatically performed using an autosampler MPS 2 (Gerstel, Mülheim, Germany). Afterwards, the HS-SPME fiber was automatically introduced into the vial's septum for 20 min to allow VOCs to be adsorbed onto the fiber surface.

2.7.2. Gas Chromatography-Quadrupole Mass Spectrometry Analysis (GC-qMS)

VOC analysis was performed using a gas chromatograph model GC 7890A coupled to a mass spectrometer 5975 C (system from Agilent Technologies, CA, USA). The HS SPME fiber was inserted for 10 min into the injector port of the GC instrument. VOCs were thermally desorbed and directly transferred to a capillary column HP-Innowax. Oven temperature conditions were initially set at 50 °C for 3 min, then increased to 160 °C at 5 °C min^{-1}, held at 160 °C for 1 min, ramped to 250 °C at 10 °C min^{-1} and stable at 250 °C for 2 min. VOCs were analyzed at an ionization energy of 70 eV and detected by mass selective detector. The detector operated in a mass range between 30 and 300 u with a scanning speed of 2.7 scans/s. VOCs were identified by mass spectra through matching with the standard NIST05/Wiley07 libraries, by comparing the retention indices (RI) (as Kovats indices) with literature data and pure standards when available. Each sample was analyzed in triplicate with a randomized sequence in which blanks were also recorded. For each volatile component, the peak area was calculated from the total ion chromatogram (TIC) and semi-quantified by relative comparison with the peak area of the IS (Relative Peak Area, RPA%).

2.8. Statistical Data Analysis

For each harvest, the effect of the ripening stage (half-red or red) on respiration rate, TA, TSS, pH, AA, TPC and VOCs was evaluated by performing a one-way Analysis of variance (ANOVA)for $p \leq 0.05$. The mean values (n = 3) were separated using the least significant difference (LSD) test ($p \leq 0.05$), and Statgraphics Centurion (version 18.1.12, Warrenton, VA, USA) was used for statistical analyses.

To highlight the VOCs correlated to half-red and red strawberries, a principal component analysis (PCA) was carried out using the software Statistica version 6.0. (Statsoft Inc., Tulsa, OK, USA).

3. Results and Discussion

3.1. *Quality Traits in Half-Red and Red Strawberries*

Respiration rate did not significantly differ between the two ripening stages, presenting values of about 15.10 ± 1.3 mL CO_2 kg^{-1} h^{-1}, in line with those observed in the classification reported by Kader [14] (Table 1). As concerns the physical parameters, the one-way ANOVA showed statistically significant different values ($p \leq 0.0001$) between the ripening stages at H1, H2 and H3. In detail, TA in red strawberries was lower than that detected in half-red samples in all harvest times. These results were consistent with the lower pH values detected in half-red strawberries compared to those of the red samples at all the harvest times (Table 1). The decrease in TA with the consequent increase in pH during the ripening has been previously explained by the conversion of organic acids into sugars in the course of the respiration process [15]. The ripening stage significantly influenced TSS at each harvest time, showing higher values (13.8, 14.5 and 4.4% more in H1, H2 and H3, respectively) in red samples than in half-red ones (Table 1). Similar results of TA, pH and TSS have been already reported by Correia et al. [16] and Aguero et al. [17] in "Candonga" strawberries harvested in spring at full ripening.

Table 1. Physical and chemical parameters measured in strawberries cv "Sabrosa" at two different ripening stages (half-red or red) and at three harvest times (H1, H2, H3).

Parameters	Harvest Time														
	H1					H2					H3				
	Ripening Stage														
	Red		Half-Red		*p*	Red		Half-Red		*p*	Red		Half-Red		*p*
Respiration rate (mL CO_2 kg^{-1} h^{-1})	16.74		14.70		ns	15.33		13.61		ns	16.50		14.00		ns
Titratable acidity (% citric acid)	0.78	b	0.99	a	****	0.86	b	1.00	a	****	0.76	b	0.97	a	****
pH	3.56	a	3.31	b	****	3.44	a	3.27	b	***	3.54	a	3.46	b	*
Total soluble solids (°Brix)	9.61	a	8.37	b	***	9.83	a	8.50	b	****	10.17	a	9.73	b	**
Antioxidant activity (mg Trolox 100 g^{-1} fw)	314.72	a	345.06	b	***	308.12	b	339.11	a	*	287.5	b	347.25	a	***
Total phenols (mg GAE 100 g^{-1} fw)	175.91	a	207.29	b	*	191.46	b	218.05	a	*	183.42	b	207.84	a	*

For each parameter the mean values followed by different letters (a, b) are significantly different ($p \leq 0.05$) according to least significant difference (LSD) test. Significance: ns = not significant; **** significant for $p \leq 0.0001$; *** significant for $p \leq 0.001$; ** significant for $p \leq 0.01$; * significant for $p \leq 0.05$. GAE is Gallic Acid Equivalent.

The ripening stage of the strawberries significantly influenced AA, as half-red samples presented higher values (345.06 ± 4.2 mg Trolox 100 g^{-1} fw) than red berries (287.46 ± 14.2 mg Trolox 100 g^{-1} fw), considering the mean of the three harvest times (Table 1). These findings are in line with previous determination of DPPH radical scavenging capacity [18,19].

Analysis of TPC conducted on the strawberry samples at all the three harvest times mirrored the trend of AA, confirming that the AA values are directly related to the TPC, as already recorded in several fruit and vegetable crops [20–22]. Specifically, the half-red fruits displayed a TPC of 13.9% higher than the one detected on the red samples (Table 1), according to previous reports [18,19].

3.2. *Phenolic Compounds in Half-Red and Red Strawberries*

The comprehensive pattern of polyphenols in the strawberry appears very complex, as compounds belonging to several classes, including hydroxycinnamic acid derivatives, flavonoids, and anthocyanins are variously represented. Despite numerous studies carried out to characterize strawberry metabolites, the current inventory of phenolic compounds emerging from the literature is controversial because of the large number of variety of cultivars available and the diversity of analytical methods employed. Nevertheless, the fraction is substantially dominated by a few compounds, whereas a multitude of other metabolites occur at a minor abundance [23,24]. The RP-HPLC separation of phenolic compounds in strawberry methanol extracts has been monitored at multiple wavelengths. Typical RP-HPLC chromatograms of extracts from red (left A) and half-red (right B) "Candonga" strawberries are shown in Figure 2. In particular, in Figure 2, the HPLC chromatograms recorded at 280 nm for the general detection of phenolic compounds, at 360 nm for the selective detection of flavonoids and EA and at 520 nm for the diagnostic detection of anthocyanins are compared. The main HPLC peaks were assigned in Table 2, based on the converging indications coming from previous identification of strawberry phenolics [23–25] UV-Vis spectra acquired with the DAD, high-resolution MS and MS/MS spectra. In agreement with previous data, *p*-coumaryl-hexoside (peak 1, P1) was the most abundant hydroxycinnamic acid derivative of strawberry, better detected at 320 nm (not shown) because of the characteristic absorbance band centered at 315 nm. In half-red strawberries, P1 was 3-/4-fold lower than in the red samples. Similar to other strawberry cultivars, the anthocyanin profile of the strawberry is substantially conserved among the cultivars [25]. Pelargonidin-3-*O*-rutinoside (P4) was the prevalent anthocyanin both in red and half-red strawberry samples, followed by less abundant pelargonidin-3-*O*-glucoside (P3) and cyanidin-3-*O*-glucoside (P2). However, the concentration of anthocyanins in unripe fruits was nearly half than the red counterpart. The chromatogram at 360 nm was dominated by quercetin-3-*O*-glucuronide (P7), which was more abundant in Half-red-H1 and Half-red-H2 than in the ripened counterpart. Strawberries have been generally described as fruits rich in EA; however, the content of free EA can vary in a wide range, depending on the cultivar as well as on a series of abiotic factors [26]. In the current strawberry samples, free EA was detected at concentrations lower than the limit of quantification, co-eluting with kaempferol-3-*O*-glucoside (peak 6, P6), as also confirmed with a separate injection of pure EA. MS and MS/MS analysis allowed to establish the presence of free EA and assess that it was slightly more intense in half-red than in ripe strawberry [25] (data not shown). No glycosylated derivatives of EA were detected among the main compounds by RP-HPLC, while EA conjugates (i.e., pentoside and hexoside) were detected both in half-red and red strawberry by ion extraction in the MS runs (data not shown). Several minor signals of relatively high molecular weight compounds detected by LC-MS were likely ellagitannins that could be converted into EA by processing or chemical hydrolysis [25]. A detailed characterization of strawberry ellagitannins is challenging and requires dedicated investigations [24,27].

Table 2. Assignment and semi-quantitative determination of phenolic compounds measured in strawberries cv "Sabrosa" at two different ripening stages (half-red or red) and at three harvest times (H1, H2, H3).

Polyphenols	Code	H1					H2					H3				
		Half-Red		Red		p	Half-Red		Red		p	Half-Red		Red		p
p-coumaryl hexoside	P1	9.6	b	28.5	a	****	11.4	b	25.9	a	****	12.7	b	29.0	a	****
cyanidin-3-*O*-glucoside	P2	1.8	b	3.3	a	****	2.0	b	3.8	a	***	1.5	b	3.1	a	***
pelargonidin 3-*O*-glucoside	P3	25.1	b	48.7	a	****	19.8	b	42.6	a	****	22.4	b	43.6	a	****
pelargonidin 3-*O*-rutinoside	P4	3.1	b	5.6	a	****	2.4	b	4.5	a	***	3.5	b	5.1	a	***
quercetin-3-*O*-glucoside	P5	5.1	a	4.2	b	****	4.1	a	3.6	b	*	3.3	a	3.2	b	*
kaempferol-3-*O*-glucoside	P6	2.1	b	2.3	a	*	2.5	a	2.1	b	***	2.1	b	2.3	a	*
quercetin-3-*O*-glucuronide	P7	20.2	a	17.0	b	***	18.6	a	15.8	b	****	12.5	b	14.1	a	***
kaempferol-3-*O*-glucuronide	P8	2.9	b	3.1	a	**	3.0	b	3.4	a	**	2.5	b	2.8	a	**

For each parameter the mean values followed by different letters (a, b) are significantly different ($p \leq 0.05$) according to least significant difference (LSD) test. Significance: ns = not significant; **** significant for $p \leq 0.0001$; *** significant for $p \leq 0.001$; ** significant for $p \leq 0.01$; * significant for $p \leq 0.05$.

The semi-quantitative figures determined for the individual phenolic compounds were in line with most of the previous determinations reported by other authors [8].

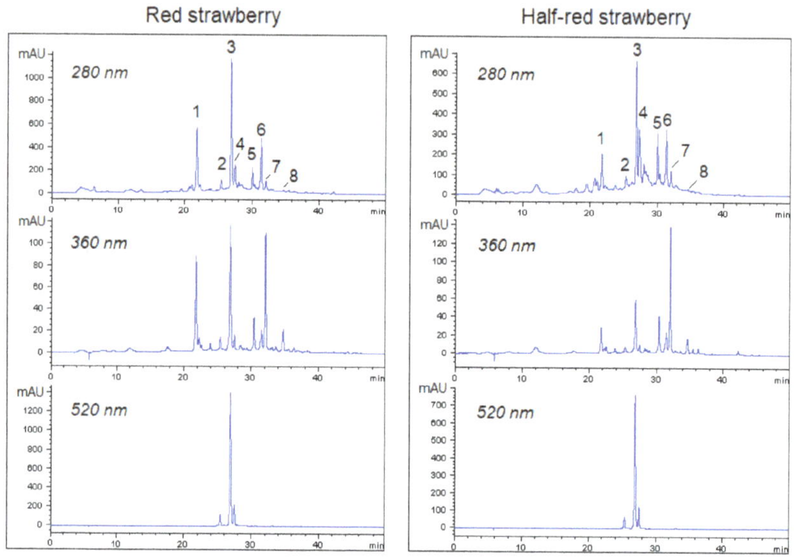

Figure 2. RP-HPLC chromatograms of extracts from red (**left**) and half-red (**right**) "Candonga" strawberries. Numbers (1–8) above the peaks correspond, respectively, to the codes P1–P8 reported on Table 2.

3.3. *VOCs Compounds in Half-Red and Red Strawberries*

3.3.1. Comparative Determination of VOCs in the "Candonga" Strawberry Samples at Two Different Ripening Stages

Overall, fifty-seven volatile compounds were identified by HS-SPME GC-MS analysis in the "Candonga" strawberry samples at two different ripening stages (half-red and red), and at three different harvest times which consisted of esters (26), aldehydes (5), alcohols (5), acids (9), terpenes (5), furanones (3), lactones (3) and others (1).

One-way ANOVA performed on the semi-quantitative data (RPA%) evidenced significant qualitative and quantitative changes in the profile of VOCs over the course of the two maturation stages, as reported in Table 3. On the other hand, Table S1 includes the abbreviation code, the experimental and literature Kovats indexes and the identification methods for the assigned VOCs.

According to previous studies, ester compounds, responsible for the strawberry fruity and floral aroma, were the most abundant chemical class, with 18 and 26 different components accounting for about 37% and 49% of all the volatile compounds detected in the half-red and red fruit, respectively [3]. The majority of these esters have been previously reported in "Candonga" strawberries [28,29]. The most representative esters in half-red strawberries were methyl butyrate (E2) (37%), methyl hexanoate (E9) (27%) and *trans*-2-hexen-1-ol acetate (E16) (20%), which increased up to about 40% of the total amount of esters in the red fruit. At ripening, together with γ-decalactone (L2), E16 became the component with the highest concentrations, individually constituting nearly 20% of the total content of the VOCs. These findings suggest that E16 and L2 could be considered as key indicators of full maturity in this variety (Table 3).

Table 3. Volatile compounds obtained in strawberries cv "Sabrosa" at two different ripening stages (half-red or red) and at three harvest times (H1, H2, H3).

Volatile Componds		Ripening Stage														
		H1					H2					H3				
	Code	Red		Half-Red		*p*	Red		Half-Red		*p*	Red		Half-Red		*p*
Methyl propionate	E1	1.92	a	0.91	b	*	2.65	a	0.94	b	****	2.57	a	0.90	b	****
Methyl butyrate	E2	682.95	a	246.38	b	***	806.88	a	256.15	b	****	812.99	a	266.61	b	****
Methyl isovalerate	E3	6.65	a	4.45	b	**	7.70	a	3.97	b	****	7.89	a	4.04	b	****
Ethyl butyrate	E4	46.52	a	4.77	b	*	63.08	a	4.24	b	****	61.06	a	4.29	b	****
Isopropyl butyrate	E5	58.07	a	3.88	b	****	73.39	a	3.62	b	****	75.59	a	4.08	b	****
Butyl acetate	E6	4.18	a	0.00	b	**	4.77	a	0.00	b	***	5.99	a	0.00	b	***
Methyl pentanoate	E7	18.08	a	0.00	b	****	28.53	a	0.00	b	****	28.44	a	0.00	b	****
Ethyl pentanoate	E8	1.75	a	0.00	b	**	4.51	a	0.00	b	****	4.63	a	0.00	b	****
Methyl hexanoate	E9	472.43	a	188.74	b	**	444.88	a	181.44	b	****	465.00	a	182.94	b	****
Butyl butyrate	E10	48.35	a	0.00	b	****	48.95	a	0.00	b	****	49.57	a	0.00	b	****
Ethyl hexanoate	E11	219.48	a	0.00	b	****	227.68	a	0.00	b	****	278.43	a	0.00	b	****
Isoamyl butyrate	E12	3.84	a	0.00	b	**	5.76	a	0.00	b	****	5.07	a	0.00	b	****
Hexyl acetate	E13	473.52	a	35.15	b	****	446.02	a	35.86	b	****	441.31	a	35.77	b	****
Methyl 2-hexenoate	E14	21.51	a	3.86	b	****	21.07	a	3.48	b	****	24.98	a	11.61	b	****
cis-3-Hexen-1-ol acetate	E15	39.92	a	4.63	b	****	39.40	a	3.97	b	****	39.57	a	3.82	b	****
trans-2-Hexen-1-ol acetate	E16	2724.77	a	142.74	b	****	2664.89	a	141.33	b	****	2688.80	a	141.61	b	****
Methyl octanoate	E17	42.43	a	4.64	b	****	41.15	a	4.73	b	****	41.56	a	4.72	b	****
trans-2-Hexen-1-ol propionate	E18	93.04	a	4.49	b	****	91.79	a	4.51	b	****	91.15	a	4.72	b	****
n-Hexyl isobutyrate	E19	520.24	a	6.75	b	****	515.75	a	6.89	b	****	511.47	a	6.76	b	****
trans-2-Hexenyl butyrate	E20	1114.73	a	21.94	b	****	1107.06	a	21.32	b	****	1080.35	a	21.21	b	****
Methyl 3-(methylthio) propionate	E21	93.01	a	2.19	b	****	91.28	a	2.33	b	****	92.29	a	2.18	b	****
Hexyl hexanoate	E22	4.29	a	1.18	b	****	4.10	a	1.29	b	****	4.48	a	1.40	b	****
n-Octyl isobutyrate	E23	11.63	a	0.00	b	****	11.31	a	0.00	b	****	11.42	a	0.00	b	****
Octyl 2-methylbutyrate	E24	4.03	a	0.00	b	****	4.00	a	0.00	b	****	4.00	a	0.00	b	****
Methyl 3-hydroxyhexanoate	E25	4.62	a	1.67	b	****	4.64	a	1.74	b	****	4.07	a	1.73	b	****
Benzyl acetate	E26	15.53	a	7.49	b	****	15.66	a	7.66	b	****	15.93	a	7.65	b	****
Hexanal	Ald1	19.41	a	11.47	b	*	33.11	a	11.96	b	****	33.88	a	12.69	b	****
2-Hexenal	Ald2	532.42	a	234.98	b	***	503.36	a	221.54	b	****	523.03	a	219.89	b	****
Nonanal	Ald3	29.57	a	2.22	b	****	26.25	a	2.46	b	****	26.75	a	2.64	b	****
Benzaldehyde	Ald4	76.65	a	5.12	b	****	77.97	a	5.22	b	****	77.14	a	5.37	b	****
Dodecanal	Ald5	8.24	a	3.42	b	****	8.08	a	3.66	b	****	8.14	a	3.60	b	****
1-Hexanol	Al1	209.84	a	27.21	b	****	209.38	a	28.10	b	****	215.20	a	27.69	b	****
trans-3-Hexen-1-ol	Al2	11.12	a	2.02	b	****	11.08	a	2.11	b	****	11.89	a	2.18	b	****

Table 3. *Cont.*

Volatile Componds		Ripening Stage														
		H1					H2					H3				
	Code	Red		Half-Red		p	Red		Half-Red		p	Red		Half-Red		p
cis-3-Hexen-1-ol	Al3	11.69	a	4.92	b	****	12.38	a	4.96	b	****	12.79	a	4.95	b	****
trans-2-Hexen-1-ol	Al4	419.31	a	110.29	b	****	404.16	a	111.24	b	****	402.86	a	110.18	b	****
1-Octanol	Al5	3.47	a	1.31	b	****	3.03	a	1.31	b	****	3.32	a	1.69	b	****
Propanoic acid	Ac1	37.29	a	1.56	b	****	37.87	a	1.69	b	****	37.72	a	1.78	b	****
2-Methylpropionic acid	Ac2	70.11	a	1.30	b	****	70.47	a	1.30	b	****	71.67	a	1.33	b	****
Butyric acid	Ac3	24.85	a	8.13	b	****	24.15	a	8.20	b	****	24.27	a	8.31	b	****
2-Methylbutanoic acid	Ac4	80.09	a	12.99	b	****	88.70	a	12.82	b	****	733.19	a	12.06	b	****
Hexanoic acid	Ac5	1423.08	a	332.86	b	****	1424.30	a	331.78	b	****	1426.72	a	332.20	b	****
Heptanoic acid	Ac6	33.53	a	4.26	b	****	33.57	a	4.22	b	****	33.89	a	4.32	b	****
Octanoic acid	Ac7	35.88	a	5.99	b	****	35.50	a	6.05	b	****	35.56	a	6.19	b	****
Nonanoic acid	Ac8	56.54	a	19.18	b	****	56.05	a	19.35	b	****	56.06	a	19.15	b	****
Decanoic acid	Ac9	28.59	a	2.52	b	****	29.25	a	2.50	b	****	29.71	a	2.54	b	****
Linalool	T1	111.83	a	62.76	b	****	112.66	a	62.20	b	****	112.18	a	62.92	b	****
β-Farnesene	T2	11.12	a	0.00	b	****	8.88	a	0.00	b	****	11.75	a	0.00	b	****
α-Terpineol	T3	98.36	a	10.26	b	****	98.21	a	10.86	b	****	98.37	a	10.57	b	****
β-Damascenone	T4	1.47	a	0.92	b	****	1.47	a	0.92	b	****	1.46	a	0.91	b	****
Nerolidol	T5	130.56	a	6.56	b	****	130.15	a	6.21	b	****	130.22	a	6.29	b	****
Mesifurane	F1	140.90	a	12.89	b	****	133.47	a	12.81	b	****	1837.09	a	12.65	b	****
Furaneol	F2	42.61	a	3.28	b	****	41.99	a	3.26	b	****	42.37	a	3.19	b	****
trans-γ-Jasmolactone	F3	46.65	a	0.00	b	****	46.69	a	0.00	b	****	46.48	a	0.00	b	****
γ-Octalactone	L1	4.31	a	0.00	b	****	4.44	a	0.00	b	****	4.57	a	0.00	b	****
γ-Decalactone	L2	2689.87	a	262.49	b	****	2631.41	a	260.20	b	****	2660.37	a	260.14	b	****
γ-Dodecalactone	L3	60.16	a	11.62	b	****	61.88	a	11.00	b	****	59.96	a	11.64	b	****
Acetophenone	O1	4.42	a	0.00	b	****	4.47	a	0.00	b	****	4.52	a	0.00	b	****

For each parameter the mean values followed by different letters (a, b) are significantly different ($p \leq 0.05$) according to least significant difference (LSD) test. Significance: ns = not significant; **** significant for $p \leq 0.0001$; *** significant for $p \leq 0.001$; ** significant for $p \leq 0.01$; * significant for $p \leq 0.05$.

Butyl acetate (E6), methyl pentanoate (E7), ethyl pentanoate (E8), butyl butyrate (E10), ethyl hexanoate (E11), isoamyl butyrate (E12), octyl isobutyrate (E23) and octyl 2-methylbutyrate (E24) were detected only in red berries (Table 3). In fruit, enzymatic biosynthesis of volatile esters through the esterification of alcohols and acyl-CoA, derived from both fatty acid and amino acid metabolism, occurs during the late ripening steps [30]. The enzyme responsible for the final step of ester formation is alcohol acyltransferase (AAT), which in strawberries can exhibit a 16-fold increase from the half-red to the full red ripeness degree [31]. A consistent correlation occurs among the expression of an AAT gene, AAT activity and concentrations of some volatile esters [31]. The availability of substrates for the biosynthesis of volatile esters is believed to be a limiting step in the production of esters, leading to different flavor profiles at each ripeness stage [30–32].

Five aldehyde compounds were common to the half-red and red stages, although occurring at a significantly higher concentration in full red berries (Table 3). The formation of C6 aldehydes originates from the lipid oxidation pathway, which involves lipoxygenase (LOX) and hydroperoxide lyase (HPL) enzymes. During ripening, the content of the two C6 aldehydes, namely hexanal (Ald1) and 2-hexenal (Ald 2), increases, paralleling the 25–200% increment in the activity of LOX and HPL enzymes and the availability of total linolenic acid as the precursor, which rises up to 200% [32].

Nonanal (Ald3) and decanal (Ald5), arising from the autoxidation of oleic acid, and benzaldehydes (Ald4), deriving from Strecker degradation of aromatic amino acids, increase during oxidative processes [33].

Gene expression studies reported the existence of a specific enzyme, O-methyltransferase, that is responsible for the synthesis of mesifurane (F1). Additionally, a *Fragaria × ananassa* quinine oxidoreductase (FaQR) was documented to be involved in the formation of furaneol (F2). F1 and F2 increased considerably during ripening of strawberries, as the enzymes belonging to their biosynthetic pathway showed the maximum activity at the full red stage, allowing potential targets to engineer the strawberry flavor from breeding populations [31].

3.3.2. Selection of the VOCs Correlated to Ripening Stage of Strawberry by Principal Component Analysis

Principal component analysis (PCA) took into account AA, TPC, the semi-quantitative data of both individual phenolic compounds (Table 2) and VOCs (Table 3), to infer possible significant associations of strawberries at the two different ripening stages (half-red or red) and at the three harvesting times (Figure 3).

The two components accounted for 94.8% of the variation in the dataset, since PC1 and PC2 explained 90.3% and 4.5% of the total variance, respectively. In the PCA plot, all red samples (Red-H1, Red-H2 and Red-H3) appeared in the left part of the score plot, while all the half-red strawberries (Half-red-H1, Half-red-H2 and Half-red-H3) were positioned on the right part of the graph, revealing a different distribution of the samples in the PCA quadrants (Figure 3A). Specifically, Half-red-H1 and Half-red-H2 samples clustered closely in the bottom right part of the plot, having positive PC1 and negative PC2 values (Figure 3A). These samples showed a significant correlation with three phenolic compounds (P5, P6 and P7) and with AC (Figure 3B). The direct association of Half-red-H1 and Half-red-H2 samples with the P5, P6 and P7 metabolites likely reflects the decline in the synthesis of flavonoids during the last stage of ripening, as demonstrated for other fruits. The decrease in flavonoids might correspond to their utilization for the downstream biosynthesis of other metabolites, or to the covalent association with other cellular components [34]. Furthermore, a higher contribution of flavonoids compared to other classes of phenolic compounds could explain the correlation with AA.

On the contrary, Half-red-H3 fruit presented positive PC1 and PC2 scores (Figure 3A) and appeared statistically associated only with TPC (Figure 3B).

The strawberry samples classified as Red-H1 and Red-H2, with negative PC1 and PC2 scores, were grouped closely in the bottom left part of the PCA plot (Figure 3A) and covaried with the same volatile metabolites: 39 VOCs comprising 14 esters (E9, E10, E13,

E15, E16, E17, E18, E19 E20, E21, E23, E24, E25, E26), 4 aldehydes (Ald2, Ald3, Ald4, Ald5), 3 alcohols (Al1, Al2, Al4), 8 acids (Ac1, Ac2, Ac3, Ac5, Ac6, Ac7, Ac8, Ac9), 4 terpenes (T1, T3, T4, T5), 2 furanones (F2, F3), 3 lactones (L1, L2, L3) 1 other (O1) and 3 phenolic compounds (P2, P3 and P8) (Figure 3B).

In the PCA score plot, Red-H3 strawberries presented negative PC1 and positive PC2, and were placed in the high left section of the PCA plot (Figure 3A). This sample showed a significant correlation with 2 phenolic compounds (P1 and P4) and 18 VOCs, including 12 esters (E1, E2, E3, E4, E5, E6, E7, E8, E11, E12, E14, E22), 1 aldehydes (Ald1), 2 alcohols (Al3 and Al5), 2-methylbutyric acid (Ac4), β-farnesene (T2) and mesifurane (F1) (Figure 3B).

The correlation of anthocyanins (P2-P4) with red samples reflects the evident accumulation of these metabolites during ripening, which has been determined for the strawberry on a biosynthetic basis [35]. Interestingly, the increased production of P1 (i.e., *p*-coumaryl-glucoside, a conjugated monolignol) parallels the decline in flavonoids, since the *p*-coumaroyl-CoA precursor is common to the metabolic routes leading to either flavonoid or monolignol biosynthetic pathways. Thus, the downregulation of the flavonoid pathway corresponds to increased levels of monolignols deriving from the conversion of *p*-coumaroyl-CoA [36].

PCA clearly highlighted that only the red samples (Red-H1, Red-H2 and Red-H3) were directly correlated to volatile components (Figure 3B), in agreement with previous reports demonstrating a considerable increase both in number and content of VOCs during the ripening of strawberries [3,30–32]. In particular, Red-H1 and Red-H2 samples were statistically associated with a higher number of volatile metabolites (39) compared to Red-H3 strawberries (18) (Figure 3B).

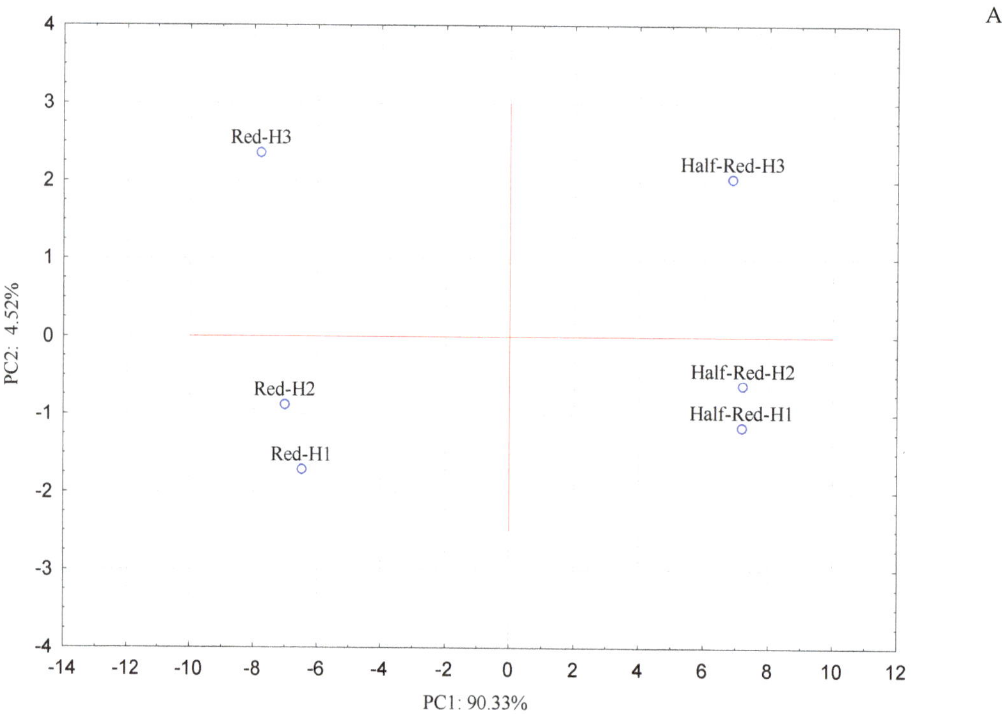

Figure 3. *Cont.*

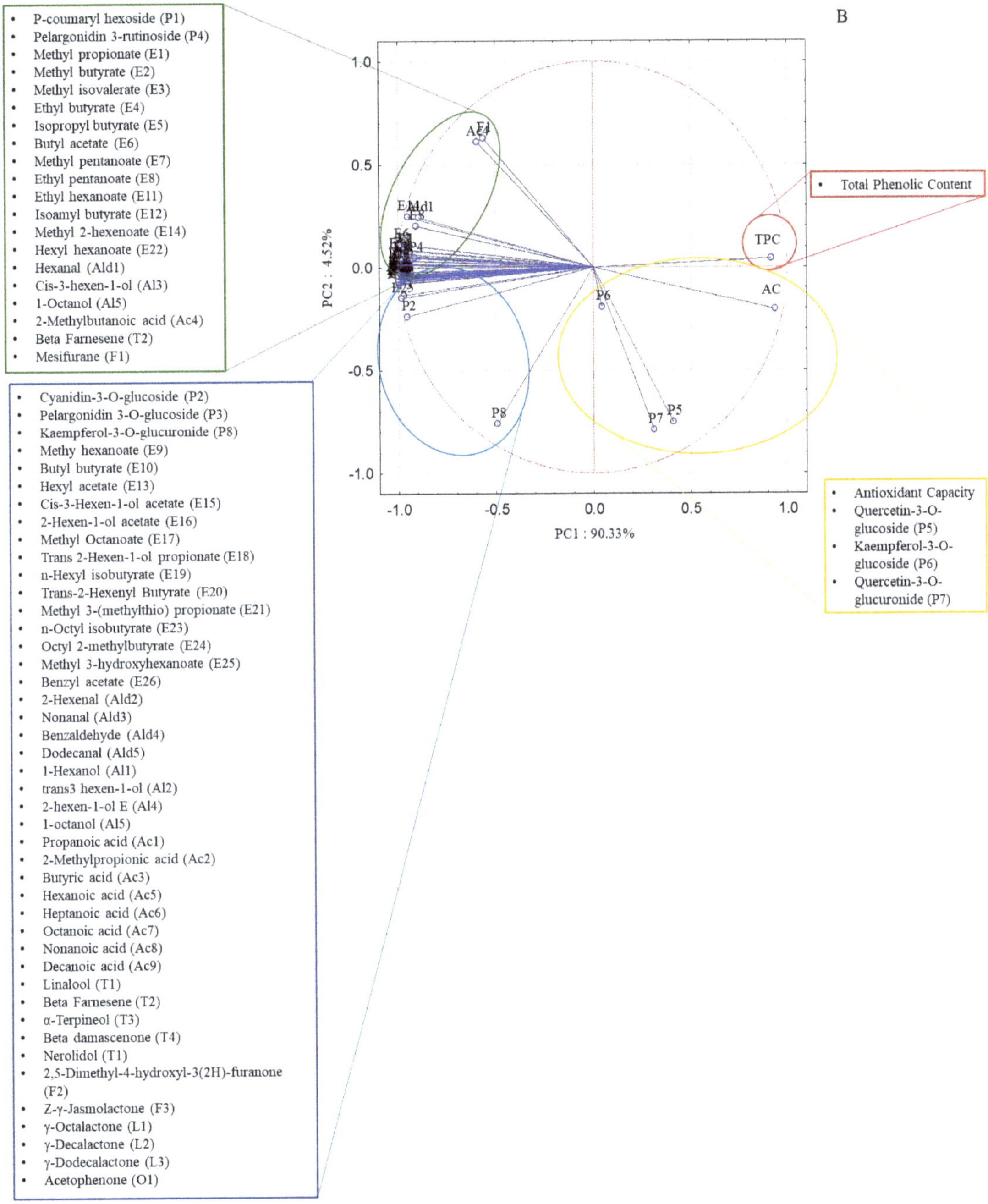

Figure 3. Principal component analysis (PCA) score (**A**) and loading (**B**) plots.

Ester compounds were the most affected by fruit ripening, and they actually characterize mature berries as they are responsible for fruity aromas. Nevertheless, only some of them may be useful as indicators of full ripeness [3]. Among the 14 esters positively associated with the Red-H1 and Red-H2 samples, *trans*-2-hexen-1-ol acetate (E16) and *trans*-2-hexenyl butyrate (E20) were the most abundant esters at the red stage (Table 3), while methyl hexanoate (E9) has been considered as a volatile marker related to the degree of maturity in different strawberry cultivars. Moreover, since it has been reported to be show its highest content in the full red strawberry, E9 has been closely associated with the flavor of mature strawberries [3]. On the other hand, methyl octanoate (E17) has been tentatively identified in the "Candonga" strawberries as responsible for the tropical/pineapple/citrus/green odor [29]. Finally, hexyl acetate (E13) has been previously described among the main contributors to the overall appreciation of strawberry quality [37].

Red-H1 and Red-H2 samples were statistically correlated to the aldehydes Ald2, Ald3, Ald4, Ald5 (Table 3; Figure 3B). Specifically, nonanal (Ald3) and benzaldehyde (Ald4) have been recently identified as volatiles, which enhance sweetness independently of the sugar content together with γ-decalactone (L2) and γ-dodecalactone (L3), which are among the three lactones directly correlated to the Red-H1 and Red-H2 samples (Figure 3B) [38,39]. Specifically, γ-decalactone (L2) gives the major contribution to the desirable "fruity", "sweet" or "peach-like" aroma in strawberries [40]. In general, several lactone compounds are commonly described as key odorants, determining the pleasant and fruity notes to strawberries [28].

Considering the nine carboxylic acids (Ac1, Ac2, Ac3, Ac5, Ac6, Ac7, Ac8 and Ac9) positively associated to the Red-H1 and Red-H2 samples, hexanoic acid (Ac5) was among the highest abundant odorants in strawberries at the red stage (Table 3). Ac5 has been identified as a key aroma compound of strawberry fruit [37].

Concerning the terpenes group, linalool (T1) also increased with ripeness (Table 3). To this purpose, previous studies pointed at this terpene as an indicator to measure the ripeness stage of fresh strawberry fruits, offering a valuable indication of ideal harvest timing. Nevertheless, it is important to underline that these findings might not be transferable to other cultivars [5].

Finally, among the volatile metabolites positively related to the Red-H1 and Red-H2 strawberries, there was furaneol (F2) already reported in several strawberry varieties, including "Candonga" [29]. F2 is considered a key aroma volatile in strawberry; it increases with ripening, conferring the characteristic caramel-like, sweet, floral, and fruity odor [29,41].

Regarding the 12 esters directly correlated to the Red-H3 strawberries (Figure 3B), methyl butyrate (E2), ethyl butyrate (E4), ethyl pentanoate (E8) and ethyl hexanoate (E11) have been reported as prominent components in several strawberry cultivars and historically recognized as crucial to the characteristic fruity notes of mature strawberries [41,42].

Among the volatile metabolites positively associated with the Red-H3 samples, there were also octanol (Al5), previously reported in fruits at late stages of maturation, and mesifurane (F1), which is known to increase during the fruit ripening imparting a particular caramel or cotton candy-like odor in red strawberries [5,37].

Recent studies have established that specific VOCs can enhance both the flavor and the sweetness perception in fresh strawberries, evaluating the overall acceptance of strawberry fruit by the statistical correlations among sensory attributes, such as 'sweet' and 'aromatic', assessed by a consumers' panel, and analytical data, including the VOCs profile [38,42]. Regardless of the sugar-acid balance, VOCs with a positive impact on the consumers' acceptance generally are some esters, including E10, E11 and E13, some aldehydes, such as Ald3, terpenes and lactones, which in our study were positively associated with the Red-H1 and Red-H2 strawberries, while negative coefficients related to liking occur with branched esters, including E5 and E12, and mesifurane (F1), which are directly correlated to the Red-H3 samples (Figure 3B) [38,42]. According to these findings, the Red-H1 and Red-H2 harvesting times could be preferred by the consumers compared to the Red-H3 fruit.

4. Conclusions

In this study, the changes in fruit quality traits, content of VOCs and phenolic compounds in strawberries (cv Sabrosa), commercially referred to as "Candonga", were evaluated at three consecutive harvesting times. Fruits collected at different times did not present significant differences among the determined qualitative traits, which, on the other hand, significantly differentiated the two ripening stages. Multivariate analysis carried out considering all the chemical data highlighted that Red-H1 and Red-H2 samples were similar to each other, while Red-H3 significantly differed, probably indicating overripe or an early stage of fruit involution. Some volatiles with a positive impact on the consumers' preference, including E10, E11, E13, Ald3, terpenes and lactones, were positively associated with the Red-H1 and Red-H2 samples, while volatiles with a detrimental contribution to the aroma traits, such as E5, E12 and F1, were directly correlated to the Red-H3 samples. However, although no univocal optimal patterns of volatile metabolites can be associated with a high level of acceptance, and taking into account that the contribution of individual VOCs to a typical aroma relies on both the aroma threshold value (ATV) and its concentration, a correlation between sensory parameters and the VOC patterns could help discover associations within the multiple factors of complex biosystems, such as fruit. For these reasons, the characterization of the VOCs related to the sensory perception of red or half-red samples of "Candonga" strawberries requires the integration with forthcoming experiments based on sensorial perception by a panel test.

In conclusion, the explorative results reported here might be of interest for producers, to identify putative markers useful for the objective assessment of the ripening stage, to optimize the strawberry harvest with the final aim to meet consumers' liking.

Supplementary Materials: The following are available online at https://www.mdpi.com/article/10.3390/foods10123102/s1, Table S1: Volatile metabolites detected in "Sabrosa" strawberries and their identification codes.

Author Contributions: Conceptualization, R.C., G.P., M.C. and B.P.; methodology, R.C., G.P. and F.S.; formal analysis, C.L., F.S., B.D.G., S.P. and M.P.; investigation, R.C. and G.P.; data curation, R.C., G.P., F.S., M.C. and M.P.; writing—original draft preparation, R.C., G.P. and M.C.; writing—review and editing, R.C., G.P., M.P., M.C. and B.P.; visualization, R.C. and G.P.; supervision, R.C., G.P. and M.C.; funding acquisition, B.P. and M.C. All authors have read and agreed to the published version of the manuscript.

Funding: This research was funded by the Project PON «R&I» 2014–2020—Azione II—"E-crops—Technologies for Digital and Sustainable Agriculture" funded by the Italian Ministry of University and Research (MUR) under the PON Agrifood Program (Contract ARS01_01136).

Data Availability Statement: The datasets generated for this study are available on request to the corresponding author.

Acknowledgments: Giuseppe Sicuro of APOFRUIT ITALIA Massimo Franchi and Mariella Quarto of CNR-ISPA for the technical and administrative support.

Conflicts of Interest: The authors declare no conflict of interest.

References

1. Zhao, J.; Liu, J.; Wang, F.; Wang, S.; Feng, H.; Xie, X.; Hao, F.; Zhang, L.; Fang, C. Volatile constituents and ellagic acid formation in strawberry fruits of selected cultivars. *Food Res. Int.* **2020**, *138*, 109767. [CrossRef] [PubMed]
2. Yan, J.W.; Ban, Z.J.; Lu, H.Y.; Li, D.; Poverenov, E.; Luo, Z.S.; Li, L. The aroma volatile repertoire in strawberry fruit: A review. *J. Sci. Food Agric.* **2018**, *98*, 4395–4402. [CrossRef] [PubMed]
3. Padilla-Jiménez, S.M.; Angoa-Pérez, M.V.; Mena-Violante, H.G.; Oyoque-Salcedo, G.; Montañez-Soto, J.L.; Oregel-Zamudio, E. Identification of Organic Volatile Markers Associated with Aroma during Maturation of Strawberry Fruits. *Molecules* **2021**, *26*, 504. [CrossRef] [PubMed]
4. Sheng, L.; Ni, Y.; Wang, J.; Chen, Y.; Gao, H. Characteristic- aroma-component-based evaluation and classification of strawberry varieties by aroma type. *Molecules* **2021**, *26*, 6219. [CrossRef] [PubMed]
5. Padilla-Jiménez, S.M.; Angoa-Pérez, M.V.; Mena-Violante, H.G.; Oyoque-Salcedo, G.; Renteria-Ortega, M.; Oregel-Zamudio, E. Changes in the Aroma of Organic Blackberries (*Rubus fruticosus*) During Ripeness. *Anal. Chem. Lett.* **2019**, *9*, 64–73. [CrossRef]

6. Kader, A.A. Flavor quality of fruits and vegetables. *J. Sci. Food Agric.* **2008**, *88*, 1863–1868. [CrossRef]
7. Li, H.; Brouwer, B.; Oud, N.; Verdonk, J.C.; Tikunov, Y.; Woltering, E.; Schouten, R.; Pereira da Silva, F. Sensory, GC-MS and PTR-ToF-MS profiling of strawberries varying in maturity at harvest with subsequent cold storage. *Postharvest Biol. Technol.* **2021**, *182*, 111719. [CrossRef]
8. Aaby, K.; Mazur, S.; Nes, A.; Skrede, G. Phenolic compounds in strawberry (*Fragaria* x *ananassa* Duch.) fruits: Composition in 27 cultivars and changes during ripening. *Food Chem.* **2012**, *132*, 86–97. [CrossRef] [PubMed]
9. Kader, A.A. Methods of gas mixing, sampling and analysis. In *Postharvest Technology of Horticultural Crops*; Kader, A.A., Ed.; University of California Agriculture and Natural Resources: Oakland, CA, USA, 2002; pp. 145–148.
10. Cefola, M.; Pace, B.; Sergio, L.; Baruzzi, F.; Gatto, M.A.; Carito, A.; Linsalata, V.; Cascarano, N.A.; Di Venere, D. Postharvest performance of fresh-cut 'Big Top' nectarine as affected by dipping in chemical preservatives and packaging in modified atmosphere. *Int. J. Food Sci. Technol.* **2014**, *49*, 1184–1195. [CrossRef]
11. Fadda, A.; Pace, B.; Angioni, A.; Barberis, A.; Cefola, M. Suitability for ready-to-eat processing and preservation of six green and red baby leaves cultivars and evaluation of their antioxidant value during storage and after the expiration date. *J. Food Process. Preserv.* **2016**, *40*, 550–558. [CrossRef]
12. Picariello, G.; Sciammaro, L.; Siano, F.; Volpe, M.G.; Puppo, M.C.; Mamone, G. Comparative analysis of C-glycosidic flavonoids from *Prosopis* spp. and *Ceratonia siliqua* seed germ flour. *Food Res. Int.* **2017**, *99*, 730–738. [CrossRef]
13. Zorrilla-Fontanesi, Y.; Rambla, J.L.; Cabeza, A.; Medina, J.J.; Sánchez-Sevilla, J.F.; Valpuesta, V.; Botella, M.A.; Granell, A.; Amaya, I. Genetic analysis of strawberry fruit aroma and identification of O-methyltransferase FaOMT as the locus controlling natural variation in mesifurane content. *Plant Physiol.* **2012**, *159*, 851–870. [CrossRef]
14. Kader, A.A. Postharvest biology and technology: An overview. In *Postharvest Technology of Horticultural Crops*; Kader, A.A., Ed.; University of California Agriculture and Natural Resources: Oakland, CA, USA, 2002; pp. 39–47.
15. Janurianti, N.M.D.; Utama, I.M.S.; Gunam, I.B.W. Colour and quality of strawberry fruit (*Fragaria* x *ananassa* Duch.) at different levels of maturity. *SEAS* **2021**, *5*, 22–28. [CrossRef]
16. Correia, P.J.; Pestana, M.; Martinez, F.; Ribeiro, E.; Gama, F.; Saavedra, T.; Palencia, P. Relationships between strawberry fruit quality attributes and crop load. *Sci. Hortic.* **2011**, *130*, 398–403. [CrossRef]
17. Agüero, J.J.; Salazar, S.M.; Kirschbaum, D.S.; Jerez, E.F. Factors affecting fruit quality in strawberries grown in a subtropical environment. *Int. J. Fruit Sci.* **2015**, *15*, 223–234. [CrossRef]
18. Hwang, H.; Kim, Y.J.; Shin, Y. Influence of ripening stage and cultivar on physicochemical properties, sugar and organic acid profiles, and antioxidant compositions of strawberries. *Food Sci. Biotechnol.* **2019**, *28*, 1659–1667. [CrossRef] [PubMed]
19. Kim, Y.J.; Shin, Y. Antioxidant profile, antioxidant activity, and physicochemical characteristics of strawberries from different cultivars and harvest locations. *J. Appl. Biol. Chem.* **2015**, *58*, 587–595. [CrossRef]
20. Kandoliya, U.K.; Bajaniya, V.K.; Bhadja, N.K.; Bodar, N.P.; Golakiya, B.A. Antioxidant and nutritional components of eggplant (*Solanum melongena* L.) fruit grown in Saurastra region. *Int. J. Curr. Microbiol. Appl. Sci.* **2015**, *4*, 806–813.
21. Cefola, M.; Carbone, V.; Minasi, P.; Pace, B. Phenolic profiles and postharvest quality changes of fresh-cut radicchio (*Cichorium intybus* L.): Nutrient value in fresh vs. stored leaves. *J. Food Comp. Anal.* **2016**, *51*, 76–84. [CrossRef]
22. Capotorto, I.; Innamorato, V.; Cefola, M.; Cervellieri, S.; Lippolis, V.; Longobardi, F.; Logrieco, A.F.; Pace, B. High CO_2 short-term treatment to preserve quality and volatiles profile of fresh-cut artichokes during cold storage. *Postharvest Biol. Technol.* **2020**, *160*, 111056. [CrossRef]
23. Aaby, K.; Ekeberg, D.; Skrede, G. Characterization of phenolic compounds in strawberry (*Fragaria* × *ananassa*) fruits by different HPLC detectors and contribution of individual compounds to total antioxidant capacity. *J. Agric. Food Chem.* **2007**, *30*, 4395. [CrossRef]
24. La Barbera, G.; Capriotti, A.L.; Cavaliere, C.; Piovesana, S.; Samperi, R.; Zenezini Chiozzi, R.; Laganà, A. Comprehensive polyphenol profiling of a strawberry extract (*Fragaria* × *ananassa*) by ultra-high-performance liquid chromatography coupled with high-resolution mass spectrometry. *Anal. Bioanal. Chem.* **2017**, *409*, 2127–2142. [CrossRef]
25. Truchado, P.; Larrosa, M.; García-Conesa, M.T.; Cerdá, B.; Vidal-Guevara, M.L.; Tomás-Barberán, F.A.; Espín, J.C. Strawberry processing does not affect the production and urinary excretion of urolithins, ellagic acid metabolites, in humans. *J. Agric. Food Chem.* **2012**, *60*, 5749–5754. [CrossRef] [PubMed]
26. Williner, M.R.; Pirovani, M.E.; Güemes, D.R. Ellagic acid content in strawberries of different cultivars and ripening stages. *J. Sci. Food Agric.* **2003**, *83*, 842–845. [CrossRef]
27. Seeram, N.P.; Lee, R.; Scheuller, H.S.; Heber, D. Identification of phenolic compounds in strawberries by liquid chromatography electrospray ionization mass spectroscopy. *Food Chem.* **2006**, *97*, 1–11. [CrossRef]
28. González-Domínguez, R.; Sayago, A.; Akhatou, I.; Fernández-Recamales, Á. Volatile profiling of strawberry fruits cultivated in a soilless system to investigate cultivar-dependent chemical descriptors. *Foods* **2020**, *9*, 768. [CrossRef] [PubMed]
29. Ubeda, C.; San-Juan, F.; Concejero, B.; Callejón, R.M.; Troncoso, A.M.; Morales, M.L.; Ferreira, V.; Hernández-Orte, P. Glycosidically bound aroma compounds and impact odorants of four strawberry varieties. *J. Agric. Food Chem.* **2012**, *60*, 6095–6102. [CrossRef] [PubMed]
30. Zhu, X.; Li, Q.; Li, J.; Luo, J.; Chen, W.; Li, X. Comparative study of volatile compounds in the fruit of two banana cultivars at different ripening stages. *Molecules* **2018**, *23*, 2456. [CrossRef] [PubMed]
31. Song, J.; Forney, C.F. Flavour volatile production and regulation in fruit. *Can. J. Plant Sci.* **2008**, *88*, 537550. [CrossRef]

32. Ozcan, G.; Barringer, S. Effect of enzymes on strawberry volatiles during storage, at different ripeness level, in different cultivars and during eating. *J. Food Sci.* **2011**, *71*, C324–C333. [CrossRef]
33. Xu, L.; Yu, X.; Li, M.; Chen, J.; Wang, X. Monitoring oxidative stability and changes in key volatile compounds in edible oils during ambient storage through HS-SPME/GC–MS. *Int. J. Food Prop.* **2017**, *20*, S2926–S2938. [CrossRef]
34. Arena, M.E.; Postemsky, P.; Curvetto, N.R. Accumulation patterns of phenolic compounds during fruit growth and ripening of Berberis buxifolia, a native Patagonian species. *N. Z. J. Bot.* **2012**, *50*, 15–28. [CrossRef]
35. Song, J.; Du, L.; Li, L.; Kalt, W.; Palmer, L.C.; Fillmore, S.; Zhang, Y.; Zhang, Z.; Li, X. Quantitative changes in proteins responsible for flavonoid and anthocyanin biosynthesis in strawberry fruit at different ripening stages: A targeted quantitative proteomic investigation employing multiple reaction monitoring. *J. Proteom.* **2015**, *122*, 1–10. [CrossRef]
36. Yeh, S.Y.; Huang, F.C.; Hoffmann, T.; Mayershofer, M.; Schwab, W. FaPOD27 functions in the metabolism of polyphenols in strawberry fruit (*Fragaria* sp.). *Front. Plant Sci.* **2014**, *5*, 518. [CrossRef]
37. Parra-Palma, C.; Úbeda, C.; Gil, M.; Ramos, P.; Castro, R.I.; Morales-Quintana, L. Comparative study of the volatile organic compounds of four strawberry cultivars and it relation to alcohol acyltransferase enzymatic activity. *Sci. Hortic.* **2019**, *251*, 65–72. [CrossRef]
38. Ulrich, D.; Olbricht, K. A search for the ideal flavor of strawberry—Comparison of consumer acceptance and metabolite patterns in *Fragaria* × *ananassa* Duch. *J. Appl. Bot. Food Qual.* **2016**, *89*, 223–234.
39. Fan, Z.; Plotto, A.; Bai, J.; Whitaker, V.M. Volatiles Influencing sensory attributes and bayesian modeling of the soluble solids–sweetness relationship in strawberry. *Front. Plant Sci.* **2021**, *12*, 640704. [CrossRef]
40. Oh, Y.; Barbey, C.R.; Chandra, S.; Bai, J.; Fan, Z.; Plotto, A.; Pillet, J.; Folta, K.M.; Whitaker, V.M.; Lee, S. Genomic characterization of the fruity aroma gene, FaFAD1, reveals a gene dosage effect on γ-decalactone production in strawberry (*Fragaria* × *ananassa*). *Front. Plant. Sci.* **2021**, *12*, 639345. [CrossRef] [PubMed]
41. Jetti, R.R.; Yang, E.; Kurnianta, A.; Finn, C.; Qian, M.C. Quantification of selected aroma-active compounds in strawberries by headspace solid-phase microextraction gas chromatography and correlation with sensory descriptive analysis. *J. Food Sci.* **2007**, *72*, S487–S496. [CrossRef] [PubMed]
42. Fan, Z.; Hasing, T.; Johnson, T.S.; Garner, D.M.; Schwieterman, M.L.; Barbey, C.R.; Colquhoun, T.A.; Sims, C.A.; Resende, M.F.R.; Whitaker, V.M. Strawberry sweetness and consumer preference are enhanced by specific volatile compounds. *Hortic. Res.* **2021**, *8*, 66. [CrossRef]

foods

MDPI

Article

Rapid and Non-Destructive Techniques for the Discrimination of Ripening Stages in Candonga Strawberries

Michela Palumbo [1,2], Rosaria Cozzolino [3,*], Carmine Laurino [3], Livia Malorni [3], Gianluca Picariello [3], Francesco Siano [3], Matteo Stocchero [4], Maria Cefola [1,*], Antonia Corvino [1], Roberto Romaniello [2] and Bernardo Pace [1]

1 Institute of Sciences of Food Production, National Research Council (CNR), c/o CS-DAT, Via Michele Protano, 71121 Foggia, Italy; michela.palumbo@ispa.cnr.it (M.P.); antonia.corvino@ispa.cnr.it (A.C.); bernardo.pace@ispa.cnr.it (B.P.)
2 Department of Agriculture, Food, Natural Resources, and Engineering (DAFNE), University of Foggia, Via Napoli 25, 71121 Foggia, Italy; roberto.romaniello@unifg.it
3 Institute of Food Science, National Research Council (CNR), Via Roma 64, 83100 Avellino, Italy; carmine.laurino@isa.cnr.it (C.L.); livia.malorni@isa.cnr.it (L.M.); gianluca.picariello@isa.cnr.it (G.P.); francesco.siano@isa.cnr.it (F.S.)
4 Department of Women's and Children's Health, University of Padova, 35131 Padova, Italy; matteo.stocchero@unipd.it
* Correspondence: rosaria.cozzolino@isa.cnr.it (R.C.); maria.cefola@ispa.cnr.it (M.C.); Tel.: +39-0825-299381 (R.C.); +39-0881-630-201 (M.C.)

Abstract: Electronic nose (e-nose), attenuated total reflection-Fourier transform infrared (ATR-FTIR) spectroscopy and image analysis (IA) were used to discriminate the ripening stage (half-red or red) of strawberries (cv Sabrosa, commercially named Candonga), harvested at three different times (H1, H2 and H3). Principal component analysis (PCA) performed on the e-nose, ATR-FTIR and IA data allowed us to clearly discriminate samples based on the ripening stage, as in the score space they clustered in distinct regions of the plot. Moreover, a correlation analysis between the e-nose sensor and 57 volatile organic compounds (VOCs), which were overall detected in all the investigated fruit samples by headspace solid-phase microextraction coupled to gas chromatography-mass spectrometry (HS-SPME/GC-MS), allowed us to distinguish half-red and red strawberries, as the e-nose sensors gave distinct responses to samples with different flavours. Three suitable broad bands were individuated by PCA in the ATR-FTIR spectra to discriminate half-red and red samples: the band centred at 3295 cm^{-1} is generated by compounds that decline, whereas those at 1717 cm^{-1} and at 1026 cm^{-1} stem from compounds that accumulate during ripening. Among the chemical parameters (titratable acidity, total phenols, antioxidant activity and total soluble solid) assayed in this study, only titratable acidity was somehow correlated to ATR-FTIR and IA patterns. Thus, ATR-FTIR spectroscopy and IA might be exploited to rapidly assess titratable acidity, which is an objective indicator of the ripening stage.

Keywords: *Fragaria* × *ananassa* Duch.; e-nose; ATR-FTIR; image analysis; multivariate analysis

Citation: Palumbo, M.; Cozzolino, R.; Laurino, C.; Malorni, L.; Picariello, G.; Siano, F.; Stocchero, M.; Cefola, M.; Corvino, A.; Romaniello, R.; et al. Rapid and Non-Destructive Techniques for the Discrimination of Ripening Stages in Candonga Strawberries. *Foods* **2022**, *11*, 1534. https://doi.org/10.3390/foods11111534

Academic Editor: Liguang Xu

Received: 22 April 2022
Accepted: 20 May 2022
Published: 24 May 2022

1. Introduction

Strawberries (*Fragaria* × *ananassa* Duch.) are worldwide popular fruits, highly appreciated for their colour, delicious taste, pleasant flavour and for their richness in nutrients, including flavonoids, anthocyanins, vitamin C and ellagic acid [1–5].

Usually, strawberries can be harvested at several ripening stages defined by colour changes, as red colour break, half-red and full red, depending on the final uses of the fruit [6]. However, since strawberries are non-climacteric fruits, it is necessary to harvest them at the optimum ripening stage to achieve the maximum quality in terms of taste, colour, consistency, size and shape.

Strawberry odour, which is strictly related to the consumer's preference and to the sensory quality of the fruit, is the result of a complex mixture of multiple volatile organic compounds (VOCs) [4,5,7]. The profile of VOCs rapidly changes during ripening, but it is also affected by several other factors, including cultivar and pre- and postharvest handlings [7,8].

Currently, the harvest time is principally evaluated by counting the days after flowering, as well as by subjective tasting or visual evaluation of fruit colour, texture or plant canopy structure. All these methods on their own or in combination lack accuracy and objective evaluation [9]. Chromatographic techniques (i.e., gas chromatography) coupled with different detectors (e.g., mass spectrometer) are suitable analytical techniques to estimate fruit ripeness and quality on a molecular basis. Although these tools present several advantages in terms of reliability, selectivity and ability to detect the classes of compounds contained in the sample, in general they are destructive, time consuming, expensive, require skilled personnel, polluting and unable to analyse the entire batch of products, since only a small more or less representative sample of fruits can be inspected [6].

In recent years, time-/cost-effective, non-destructive, contactless, user-friendly and green non-targeted methods have been increasingly introduced to perform fingerprinting studies aimed at monitoring the degree of ripening and the quality traits of fruit on an objective basis. The practical advantages of these methods consist of minimizing potential losses for growers and packers, as well as fast decay for the end consumer. The quality parameters or the index of ripening assessed by non-destructive methods can also be designed to predict the optimal harvest time [9].

Fourier transform infrared (FTIR) spectroscopy [10–12], image analysis (IA) [13–15] and electronic nose (e-nose) [16,17] are among the main approaches that have been developed to assess real time the state of ripeness and the internal and external quality appearance of fruit.

Specifically, IA technique has proved to be a successful contactless tool in the quality assessment of fruits and vegetables. This technology acquires images of the whole visible surface of the products, extracts the most discriminative external features (shape, colour and defects) and processes the data by regression or classification models and algorithms to predict chemical and physical properties of the samples [18].

For example, IA and machine learning techniques have been used to recognize mature strawberries from shape and colour characteristics extracted from the images acquired [14]. Moreover, ref. [19] proposed an innovative low-cost computer vision based on convolutional neural networks for strawberry detection to implement and develop a contactless and non-destructive robot for fruit harvesting.

Fruity aroma is another key indicator of strawberry quality, which is strictly related to the fruit ripeness. VOCs and phenolic compounds have been recently profiled in "Candonga" strawberries, comparing fruits harvested at half-red and red ripening stages [20]. Some VOCs, namely butyl butyrate, ethyl hexanoate, hexyl acetate, nonanal, terpenes and lactones, were individuated as putative markers of the maturity phase at harvest.

E-nose represents one of the most promising, fast, easy-to-handle, low cost and non-destructive methods, as an alternative to conventional ones (i.e., headspace solid-phase microextraction (HS SPME) sampling followed by gas chromatography-mass spectrometry (GC-MS)) for the detection of food odour, aimed at discriminating and classifying food matrices with different aroma fingerprints. E-nose devices comprise an array of chemical sensors, which offer selectivity toward varying classes of VOCs, collecting aroma information of a sample as a whole. This technique can simulate the human olfactory sense and generate fingerprints of VOCs in real time [21]. E-nose sensory data combined with suitable chemometric tools have been successfully applied to detect changes in VOC content in several food matrices for several purposes, including food quality, safety and fraud detection as well as to discriminate the fruit based on its degree of ripening [7,21–23].

E-nose equipped with metal oxide gas sensors was successfully employed to characterize the volatile patterns of Festival and Florida Radiance strawberry cultivars at five ripening stages: white, half red, three-quarter red, full ripe and overripe [16]. Results of this study could provide data reproducibility of 90% (±10%), demonstrating the potentiality of this technology to monitor strawberry maturity and fruit quality. Ref. [17] reported the use of a new self-developed e-nose system to detect strawberry freshness during postharvest. This system consisted of different metal oxide semiconductor sensors linked to a data acquisition system and to a computer with pattern recognition software. The e-nose response values detected during the storage were used to build both a partial least squares (PLS) discriminant analysis and a support vector machine (SVM) model, demonstrating that SVM was a better model than the PLS one, providing an accuracy of 96.3% and 94.9% for the training and testing sets, respectively.

FTIR coupled with chemometrics has been extensively explored as a powerful, rapid and time-/cost-effective analytical technique for studying the chemical composition of fruit, also offering specific information about their dynamic changes. More recently, attenuated total reflection (ATR)-FTIR techniques have been exploited to simultaneously determine attributes of quality, bioactive compounds and antioxidant capacity of ten strawberries cultivars harvested at seven different stages of ripening [12]. FTIR has also been successfully exploited to monitor strawberry spoilage [10].

In this research paper, the applicability of rapid and non-destructive techniques, including e-nose, ATR-FTIR and IA, combined with chemometric methods, for the fast and reliable discrimination of two ripening stages (half-red and red) in "Candonga" strawberries, the most cultivated in Europe, was investigated. The final aim was to establish the most suitable rapid methods for assessing strawberry ripening, through correlation with one or more objective quality indicators of ripening (VOCs, titratable acidity (TA), total phenols (TP), antioxidant activity (AA) and total soluble solid (TSS)).

2. Materials and Methods

2.1. Plant Material

Candonga strawberries (*Fragaria* × *ananassa* Duch.) var. Sabrosa (Planitalia s.r.l., Policoro, Italy), packed into PET trays (Carton Pack SpA, Rutigliano, Italy), were provided by a cooperative company of fresh fruit (Apofruit Italia Soc. Coop., Scanzano Jonico, Italy) at three different consecutive harvest times, on 21 May, 27 May and 1 June, indicated as H1, H2 and H3, respectively, and at two maturity stages, thereinafter indicated as half-red (in ripening phase, fully expanded and 50% red) and red (in ripening phase, fully expanded and 100% red), consistent with the visual criteria reported by [20] (Figure S1). Total soluble solids and titratable acidity at harvest were about 8.8 ± 0.7 °Brix and 0.9% for half-red strawberries and 9.8 ± 0.2 °Brix and 0.7% for red ones. The same samples have been previously subjected to the analysis of VOCs and individual phenolic compounds and to the determination of several chemical parameters, including titratable acidity (TA), total phenols (TP), antioxidant activity (AA) and total soluble solid (TSS) as reported in [20]. In this work, these samples have been analysed by e-nose, ATR-FTIR and IA.

2.2. Electronic Nose (E-Nose)

Comprehensive profiles of VOCs from the headspace of "Candonga" strawberry samples were obtained using a commercial portable electronic nose (e-nose, PEN 3, Airsense Analytics Inc., Schwerin, Germany, including the Win Muster software). The sensor array was fitted out with 10 metal oxide semiconductor (MOS) sensors characterized by different thicknesses and chemical compositions, to offer selectivity toward different classes of volatile, as described by [24]. Owing to the high operative temperatures (200–500 °C), VOCs delivered to the surface of the sensors were completely converted to carbon dioxide and water, causing a variation in the resistance. The response of the MOS sensors, revealed as resistivity (O), was established on the changes in conductivity, due to the adsorption of gas molecules, and on the subsequent surface reactions. For sample preparation, 4 g of each sample were placed in 45 mL airtight glass vials and closed with a screw cap with poly (1,1,2,2-tetrafluoroethylene) (PTFE)/silicone septum. To get the headspace equilibrium, each vial was held for 30 min at 30 °C and detections were performed at a temperature of 22 ± 2 °C and a relative humidity (RH) of 50 ± 5%. To minimize the drift in MOS sensors, each sampling involved a measurement process of 80 s to get the stable signals followed by a 70 s cleaning process to normalize the sensor responses, in line with preliminary experiments. The sample gas was driven into the sensor chamber at a flow rate of 400 mL/min. For each sample, analyses were carried out in 11 technical replicates and the signals per second were collected. The signals of each sensor were expressed by the ratio G/G0, where G and G0 indicate the conductance of the sensors exposed to sample gas and to the clean gas, respectively. Successively, to clean the system between two sequential analyses, a second pump transports the filtered air to the sensor array for 400 s with a flow rate of 600 mL/min. E-nose data were registered by the pattern recognition software (WinMuster, v.1.6., Airsense Analytics GmbH, Schwerin, Germany) and the average response of each sensor in the range 70–75 s (area under the curve) was submitted to statistical data analysis.

2.3. Attenuated Total Reflection-Fourier Transform Infrared (ATR-FTIR) Spectroscopy

Sets of at least 10 strawberries of each maturity stage were randomly sampled, homogenized with a blender, frozen in dry ice and finally lyophilized. The resulting solids were finely powdered using a ceramic mortar. ATR-FTIR spectra of powders were acquired using a Spectrum 400 spectrophotometer (PerkinElmer, Waltham, MA, USA), with a Deuterated Triglycine Sulfate (DTGS) detector. The sampling station was equipped with an overhead ATR accessory. The spectra were recorded in the 650–4000 cm^{-1} region at 32 scans/spectrum range and with resolution of 4 cm^{-1}. Analyses were performed in triplicate and averaged spectra were processed using the PE Spectrum software (PerkinElmer, Milan, Italy, version 10.5.1).

2.4. Image Analysis (IA)

At each harvest time and for each maturity stage (half-red or red), a lot containing 12 strawberries for each replicate was imaged using a Digital Camera AP-3200T-PGE (JAI Ltd., Yokohama, Japan) positioned inside a Photo studio box HPB-60D (HAVOX®, Vendôme, France). In total, for each maturity stage, four replicates were considered, for a total of 48 berries. The camera sensor was composed of three CMOS RGB-sensors, delivering a spatial resolution of 3.2 MPixels @12 frame/second and a colour depth of 24 bit/pixel. The objective used was a 12 mm focal and F1.8 (KOWA Lens mod. LM12NC3 1/2) allowing a field of view (FOV) of (35 × 30 cm). Illumination was provided by two led ramps with 120 lamps (HAVOX HPB-60, 5500K, 13,000 ± 100 lumen CRI 93+). A ColourChecker Passport Photo 2 (X-rite Italy srl, Prato Italy) with 24 patches of known colour was placed in the camera FOV as the chromatic reference.

Images were pre-processed and analysed using the Image Processing Toolbox of Matlab software R2021b (MathWorks Inc., Natick, MA, USA). The algorithm enabled us to acquire the raw images and pre-process them by cropping the unnecessary image border and separating the three colour-components, red, green and blue (RGB), from the original raw images. The background was thresholded using the R image, since having the highest contrast with the black background. The coarse segmentation of the strawberries was carried out by a threshold method [25]. On the obtained binary images, a morphological filter was applied to erode the strawberry edge and a flood filling operation was performed to overcome the thresholding defects. Using this primary mask, a secondary segmentation was performed to individuate the green and red areas. To this topic an enhanced image was obtained by subtracting the G image to R image. The R image was thresholded and a secondary mask was gained, obtaining the red area. The green area was simply obtained by subtracting the secondary to the primary mask. Finally, the pixel count of each area was performed to calculate the red and green percentage area.

2.5. Total Soluble Solids, Titratable Acidity, Antioxidant Activity and Total Phenols

Total soluble solids (TSS) and titratable acidity (TA) were measured on about 100 g of homogenized strawberries (for each replicate and maturity stage) as reported by [20].

For each replicate and at each ripening stage, about 100 g of strawberries were homogenized to obtain the fruit juice on which the measurements of pH have been performed. The TSS content was determined using a digital refractometer (DBR35-XS Instruments, Carpi, Italy) and results were expressed in °Brix.

The analysis of antioxidant activity (AA) and the total phenols (TP) was performed with the spectrophotometric method previously described by [20] using a spectrophotometer UV-1800, (Shimadzu, Kyoto, Japan).

In detail, antioxidants were extracted in 20 mL methanol/water solution (80:20 v/v) for 2 min, using a homogenizer (T-25 digital ULTRA-TURRAX®—IKA, Staufen, Germany). Then, AA was measured on the methanol extract using the DPPH (1,1-diphenyl-2-picrylhydrazyl) assay, while TP were determined by Folin–Ciocalteu's method.

2.6. Analysis of Volatile Compounds (VOCs)

Identification and semi-quantification of VOCs was carried out by HS-SPME/GC-MS as previously reported by [20] using a DVB/CAR/PDMS (50/30 µm) fibre and 50 °C and 20 min as extraction temperature and time, respectively. VOC analysis was performed by using the GC system, model GC 7890A, Agilent (Agilent Technologies, CA, USA), coupled to the mass spectrometer 5975C (Agilent). Semi-quantitative data of each volatile (relative peak area, RPA%) were measured in respect to the peak area of 2-octanol, used as the internal standard [20]. Peak areas of the identified VOCs were calculated from the total ion chromatogram (TIC).

2.7. Statistical Data Analysis

Data were investigated by univariate and multivariate data analysis. In detail, in univariate data analysis a linear mixed effect model (LME), in which the maturity stage was assumed as the fixed factor and the technical replicate as the random factor, was considered, in order to evaluate the effect of the maturity stage. False discovery rate (FDR) was controlled by the Benjamini–Hochberg procedure assuming a level $\delta = 0.05$. Moreover, the receiver operating characteristic (ROC) curve was used as a tool to estimate the capability to distinguish the maturity stage based on the data. Multivariate data analysis was based on principal component analysis (PCA) [26].

The correlations among e-nose responses and VOCs and between chemical data and IA and ATR-FTIR data were explored by correlation analysis. In particular, the correlation matrices based on the Pearson correlation coefficient were explored by heatmap, using a hierarchical clustering procedure that used Euclidean distance and Ward's method. A significance level $\alpha = 0.05$ was assumed for the correlation coefficients. The optimal number of clusters was determined by silhouette analysis. Data analysis was carried out by in house R-function executed using the R 4.0.4 platform (R Foundation for Statistical Computing, Vienna, Austria).

3. Results and Discussion

3.1. E-Nose Discrimination of the Ripening Stage of "Candonga" Strawberries and Correlation Analysis with VOC Pattern

Four lots of 12 "Candonga" strawberries were obtained pooling the fruit collected for each harvest time (H1, H2 and H3) at the two different ripening stages (half-red and red), obtaining a total of 12 samples for the half-red and 12 samples for the red fruit. Since 11 technical replicates were performed for each sample, a data set composed of 264 observations and 10 variables (sensor responses) was investigated by univariate data analysis and the results are reported in Table 1. All sensors, except for S4 and S10, resulted in us being able to distinguish the two maturity stages.

Table 1. Univariate data analysis: ID indicates the sensor name, in columns half-red and red the median (5th–95th) percentile is reported for the two maturity stages, p is the p-value of the factor maturity stage obtained by LME, FC is the fold change calculated as the ratio between the median of red and the median of half-red, AUC is the area under the ROC curve and CI 95% is the confidence interval of AUC at the level of 95%.

ID	Half-Red	Red	p	FC	AUC	CI 95%
S1	0.439 [0.415–0.456]	0.408 [0.364–0.429]	<0.001	0.929	0.946	0.922–0.969
S2	3.812 [3.416–4.185]	4.710 [3.924–5.541]	<0.001	0.809	1.236	0.950–0.989
S3	0.430 [0.409–0.443]	0.389 [0.359–0.414]	<0.001	1.105	0.905	0.942–0.98
S4	1.075 [1.061–1.084]	1.077 [1.067–1.085]	0.40	0.998	1.002	0.506–0.646
S5	0.421 [0.398–0.443]	0.380 [0.356–0.406]	<0.001	1.108	0.903	0.940–0.98
S6	3.907 [3.691–4.225]	4.198 [3.845–4.669]	0.003	0.931	1.074	0.810–0.903
S7	1.322 [1.234–1.445]	1.930 [1.506–2.198]	<0.001	0.685	1.460	1.000–1.000
S8	5.713 [5.349–6.303]	6.370 [5.671–6.974]	<0.001	0.897	1.115	0.847–0.924
S9	1.800 [1.642–1.95]	2.283 [2.042–2.555]	<0.001	0.788	1.269	1.000–1.000
S10	1.196 [1.160–1.224]	1.204 [1.172–1.226]	0.40	0.994	1.006	0.526–0.669

The same data set (264 observations and 10 variables) was mean centred and investigated by PCA. A model with two principal components with $R^2 = 0.95$ and $Q^2 = 0.81$ was obtained. In particular, in the score scatter plot of the model shown in Figure 1, where half-red samples are indicated in green and red samples are reported in red, the data variation due to technical replicates (indicated by the circles) was smaller than the biological variability (the medians of the technical replicates of the same sample are indicated by the triangles). Moreover, the samples at different ripening stages were easily classified into two groups based on their degree of ripeness (red and half-red). Interestingly, half-red samples were spread over a smaller area than the region of the plot relevant to the red strawberries. This result indicates a larger variability among the red than the half-red fruit.

The PCA results indicated that there was a potential relationship among the e-nose signals and the ripeness of fruit, as the e-nose device responded sensitively and selectively to the modification in the patterns of aroma VOCs from strawberries at different ripening stages.

Overall, 57 volatile compounds were previously identified by HS-SPME/GC-MS analysis of the "Candonga" strawberry samples at two different ripening stages (half-red and red) and at three different harvest times (H1, H2 and H3) (Table S1) [20]. As reported by [20], PCA performed on volatiles, quality traits and phenolic com-pounds highlighted that only the red samples were directly correlated to volatile components, as VOCs clearly increased both in number and amount during ripening. In particular, volatiles with positive impact on the consumers' acceptance, including butyl butyrate, ethyl hexanoate, hexyl acetate, nonanal, terpenes and lactones, were positively associated to the red-H1 and red-H2 strawberries, while volatiles with negative coefficients related to consumer liking, including isopropyl butyrate, isoamyl butyrate and mesifurane, were directly correlated to the red-H3 samples

All these VOCs were correlated with the data obtained by e-nose on the same fruit samples. The heatmap representing the correlation between VOCs and e-nose is reported in Figure 2.

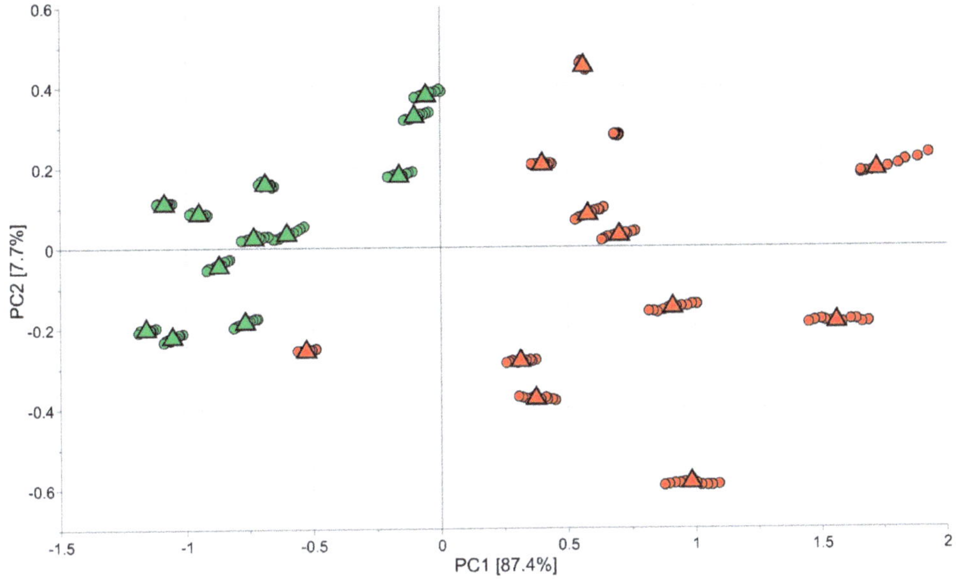

Figure 1. PCA model of the e-nose data: circles represent the observations (technical replicates), while triangles indicate the medians of the technical replicates of the same sample; the green colour is used for observations belonging to the half-red group and the red colour for those of the red group.

Figure 2. Heatmap: a different colour code was used to represent the clusters discovered by cluster analysis; ρ is the Pearson correlation coefficient.

Cluster analysis allowed us to gather the VOCs in two groups and the e-nose sensors in three clusters. Specifically, the first group consisted of all the VOCs with the exception of 2-methylbutanoic acid (Ac4) and mesifurane (F1), which formed the second cluster (Figure 2). On the other hand, the first group of the e-nose sensors included S2, S6, S7, S8 and S9, the second involved S1, S3 and S5 and the third cluster contained S4 and S10. Specifically, the first group of VOCs exhibited a highly positive association with the sensors

S2 (broad range), S6 (broad-methane), S7 (sensitive to many terpenes), S8 (broad-alcohol) and S9 (responsive to sulphur and aromatic metabolites) and a highly inverse correlation with the sensors S1 (sensitive to aromatic compounds), S3 (sensitive to ammonia and aromatic compounds) and S5 (sensitive to hydrocarbons and aromatic compounds). The correlations with the sensors S4 (sensitive to hydrogen) and S10 (sensitive to methane) were negligible, in agreement with the univariate data analysis (Table 1). Moreover, the sensors of the first cluster (S2, S6, S7, S8 and S9) presented a higher response to the red fruit, which may be caused by the increase in methane and alcohol, in line with previous studies [16]. On the other hand, the sensors of the second cluster (S1, S3 and S5) showed a higher signal to the half-red samples, allowing us to discriminate between the two ripening stages (Table 1, Figure 2).

The compounds Ac4 and F1 were positively correlated to the sensors S7 and S9 (Figure 2), and their correlation coefficient showed smaller absolute values with respect to the other VOCs, maintaining the same sign. Interestingly, in our previous study these two compounds showed a significant correlation only with red-H3 strawberries [20]. Specifically, mesifurane (F1), which is known to increase during fruit ripening and to confer caramel or cotton candy-like notes to ripe strawberries, has been reported to have a negative impact on the consumers' acceptance [27]. Overall, our results highlighted that e-nose responses were in line with HS-SPME/GC-MS analysis. In addition, e-nose sensors were able to discriminate "Candonga" strawberries at the two different ripening stages by differently reacting to specific VOCs from the different fruit samples.

3.2. ATR-FTIR Discrimination of the Ripening Stage of "Candonga" Strawberries and Correlation Analysis with Chemical Data

Spectral sub-ranges identifying the main strawberry constituents could be detected within the exemplary ATR-FTIR full range spectra (4000–600 cm^{-1}) of half-red and red samples (Figure 3A). The broad band centred at 3295 cm^{-1} and those at 2928 and 2891 cm^{-1} corresponded to the νOH and νCH vibrational modes, respectively. The band at 1717 cm^{-1}, which is diagnostic of the νC=O vibrational mode of esters or organic acids, revealed the methyl galacturonates of strawberry pectins. The bands at around 1400 cm^{-1} arose from the δCH_2 vibrational modes [11].

A magnified view of the spectral region in the fingerprint section (1200 and 800 cm^{-1}) containing the specific bands of sugars is shown in Figure 3B. The main vibrational modes in this region include the δC–O–C of the glycosidic linkage, δCOH, and νC–C. Clearly, this region is characterized by overlapping bands resulting from the contribution of all the different carbohydrates [11]. The various intensity of the bands, which emerges with stronger evidence from the second derivative of spectra in the 1200–800 cm^{-1} range (Figure 3C), reflected the dynamic concentration changes of the various sugars during strawberry ripening.

Six different samples of "Candonga" strawberries were obtained pooling the collected fruits for each harvest time and the two different ripening stages, gaining a total of three samples for the half-red and three samples for the red fruit. Five technical replicates were performed for each sample obtaining a data set composed of 30 observations and 3401 variables (wavenumber). The PCA model obtained after standard normal variate (SNV) transformation and by mean centring the data showed two principal components with $R^2 = 0.947$ and $Q^2 = 0.941$. The score scatter plot of the model is reported in Figure 4A. PCA1, explaining the 79.8% of the total variance, discriminates the samples belonging to the half-red group from those of the red group. As a result, the loading of the first component can be investigated to discover the IR spectral regions capable to distinguish the two groups. Specifically, Figure 4B allows us to observe that the average spectrum is coloured according to the loading value. Moreover, the broad band centred at 3295 cm^{-1}, the band at 1717 cm^{-1} and the band at 1026 cm^{-1} were the most related to ripening. The compounds responsible for the signals of the first band decrease their concentration during fruit ripening, while the others increase their concentration along with strawberry maturation. The broad band centred at 3295 cm^{-1} corresponds to the νOH vibrational mode of organic acids, while the band 1717 cm^{-1} can be ascribed to the νC=O vibrational mode of esters or organic acids. Moreover, the 1200 to 900 cm^{-1} spectral range, which includes the band at 1026 cm^{-1}, is known as the "fingerprint region" of sugars [11]. The main vibrational modes absorbing in this region correspond to the δC–O–C of the glycosidic linkage, the δCOH, and the νC–C of sugars and organic acids, both affected by the ripening stage of strawberries. Indeed, during fruit ripening the concentration of organic acids tends to drop, while that of sugars (sucrose, glucose and fructose) is likely to increase [28].

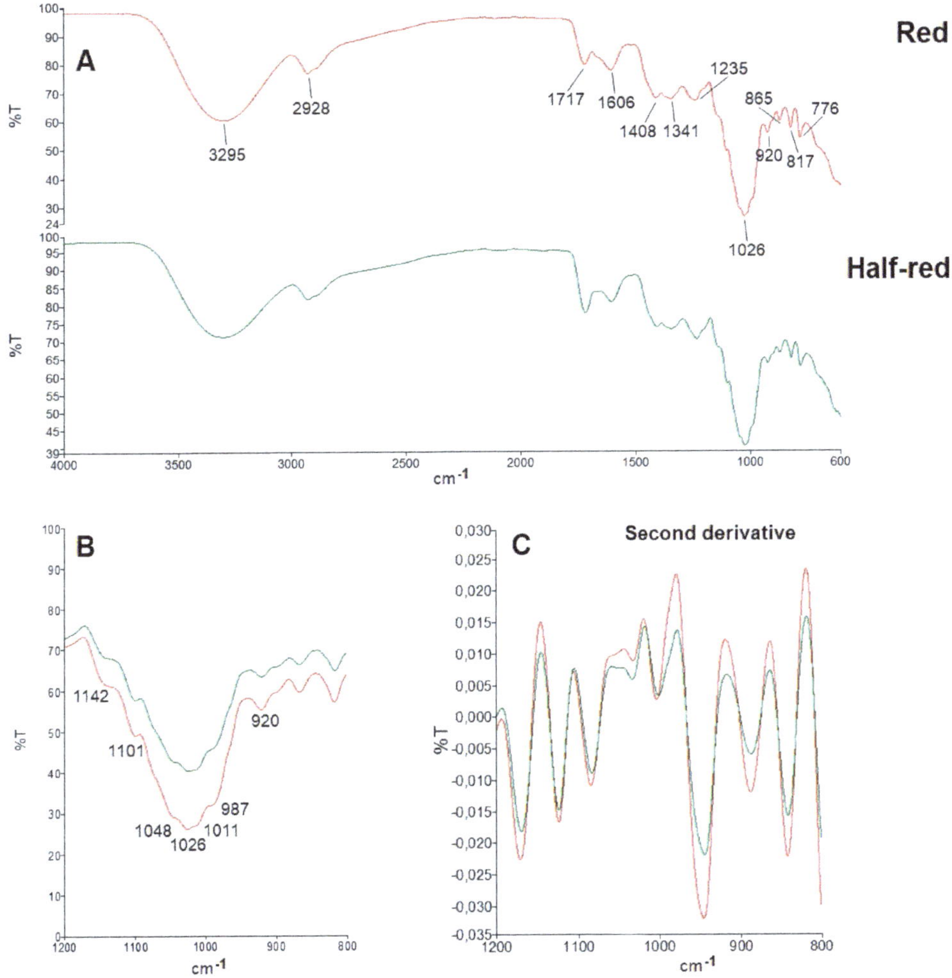

Figure 3. (Panel (**A**)): Typical ATR-FTIR spectrum in the 4000–600 cm^{-1} range of freeze-dried red (red line) and half-red (green line) strawberries. (Panel (**B**)): Magnified view of the spectral fingerprint region (1200–800 cm^{-1}), dominated by vibrational modes of sugars. (Panel (**C**)): Second derivative of spectra in the 1200–800 cm^{-1}.

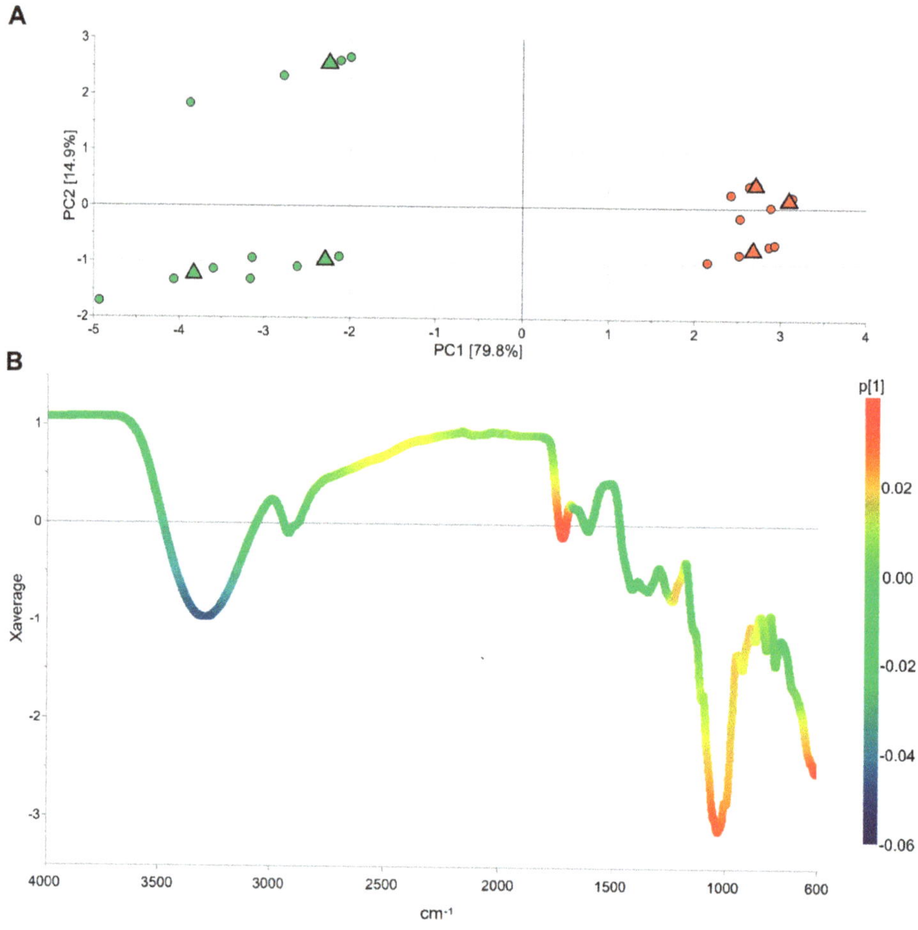

Figure 4. PCA model of the ATR-FTIR data: score scatter plot (panel (**A**)) and average profile of the SNV data coloured according to the first loading vector (panel (**B**)); in panel A, circles represent the technical replicates, triangles indicate the medians of the technical replicates of the same sample and green is used for observations belonging to the half-red group while red is used for those of the red group.

The relationships between ATR-FTIR data and the chemical parameters including TA, TP, AA and TSS, detected as reported in [20] (Table S2), were assessed by the heatmap of Figure 5. Two clusters were discovered for the chemical data. Specifically, TA, TP and AA behave similarly being positively correlated to the IR signals in the regions 3650–2700 cm^{-1} and 1500–600 cm^{-1}, while TSS showed an opposite trend. Only TA resulted to be significantly correlated to the ATR-FTIR data and, in particular, to the two bands in the region 3650–2700 cm^{-1} and with the peaks in the regions 1500–1280 cm^{-1} and 1160–700 cm^{-1}. In other words, TA resulted to be the chemical attribute more related to IR data and the IR spectra mirrored the behaviour of TA. Generally, TA decreases during ripening due to the conversion of organic acids into sugars [29], and it is an objective indicator of maturity stage. Consequently, its prediction by ATR-FTIR might be a valid solution for a fast and non-destructive assessment of the ripening stage in strawberries.

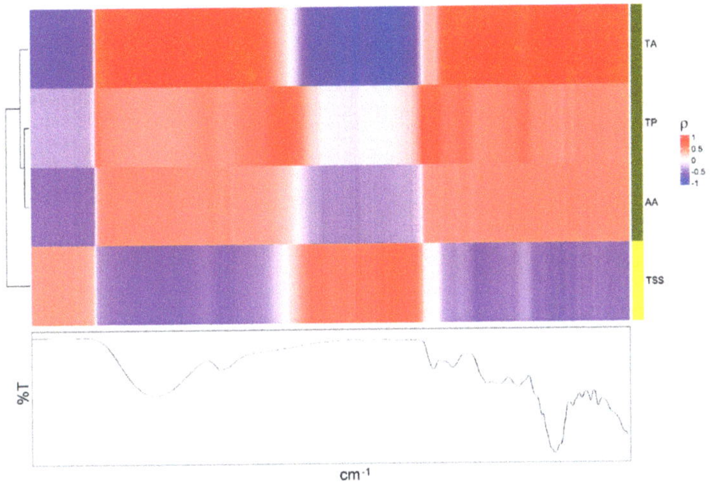

Figure 5. Correlations between ATR-FTIR and chemical data: heatmap; a different colour code was used to represent the clusters discovered by cluster analysis; ρ is the Pearson correlation coefficient; in the bottom, the average IR spectrum is reported.

3.3. Image Analysis Discrimination of the Ripening Stage of "Candonga" Strawberries and Correlation Analysis with Chemical Data

Four samples of 12 "Candonga" strawberries were obtained pooling the collected fruit at each harvest time and at two different ripening stages, achieving a total of 12 samples for the half-red and 12 samples for the red fruit. Eight technical replicates were performed for each sample obtaining a data set composed of 192 observations and five variables (measuring colour attributes). All colour parameters obtained by strawberry image analysis allowed us to significantly discriminate the half-red and the red samples, as reported in Table 2.

Table 2. Univariate data analysis of image data: ID indicates the feature name, in columns half-red and red the median (5th–95th) percentile is reported for the two maturity stages, *p* is the *p*-value of the factor maturity stage obtained by LME, FC is the fold change calculated as the ratio between the median of red and the median of half-red, AUC is the area under the ROC curve and CI 95% is the confidence interval of AUC at the level of 95%.

ID	Half-Red	Red	*p*	FC	AUC	CI
*L**	15.71 [13.61–16.81]	10.58 [9.78–11.40]	<0.001	0.676	1.00	1.00–1.00
*a**	25.78 [23.28–27.46]	20.23 [19.02–21.58]	<0.001	0.787	1.00	1.00–1.00
*b**	16.40 [14.03–17.72]	10.16 [9.14–11.31]	<0.001	0.621	1.00	1.00–1.00
Chroma	30.51 [27.23–32.56]	22.62 [21.20–24.36]	<0.001	0.741	1.00	1.00–1.00
Hue-angle	0.56 [0.53–0.59]	0.46 [0.44–0.48]	<0.001	0.826	1.00	1.00–1.00

Indeed, for all the colour parameters considered the half-red samples showed median values significantly higher than red ones at each harvest time (fold changes greater than 1). This result is clearly visible also looking at the score scatter plot obtained by PCA analysis (autoscaled data, two principal components, R^2 = 0.999), using as variable the colour data (Figure 6). Red and half-red strawberries clustered in two different regions, at the right-hand side and at the left-hand side along the first component, respectively.

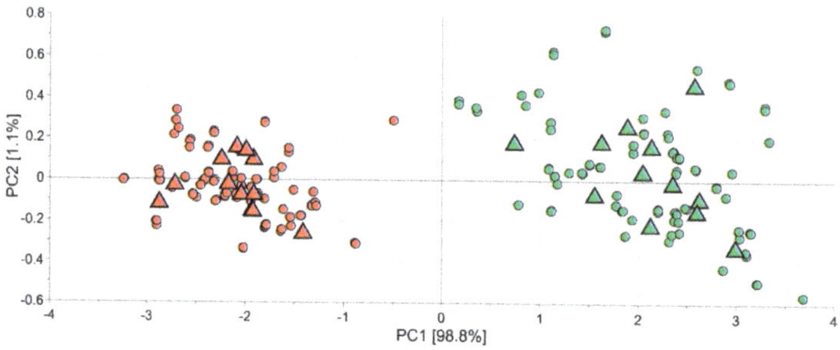

Figure 6. PCA model of the image data: circles represent the observations (technical replicates), while triangles indicate the medians of the technical replicates of the same sample; green is used for observations belonging to the half-red group and red for those of the red group.

The relationships between image data and the chemical attributes (TA, TP, AA and TSS) were explored by heatmap (Figure 7). As in the case of ATR-FTIR data, two clusters were discovered for the chemical data. Specifically, TA, TP and AA behave similarly being positively correlated to the colour parameters, while TSS showed an opposite trend, being negatively correlated to the colour parameters. On the other hand, the colour parameters were grouped into two clusters, one formed by Chroma and a* and the other including b*, L* and Hue-angle. Only TA resulted to be significantly correlated to the image data that mirrored its trend. The development of fruit colour during ripening is considered a maturity index and when its intensity increases, the ripening attributes also improve. As the fruit colour development enhances, the TSS content increases proportionately. On the contrary, the trend of the fruit acidity decreases with the enhancement of the colour of fruit during ripening [30]. This result suggests that the prediction of TA by the colour parameters obtained by IA of the strawberry represent a valid method to rapidly assess the ripening stage.

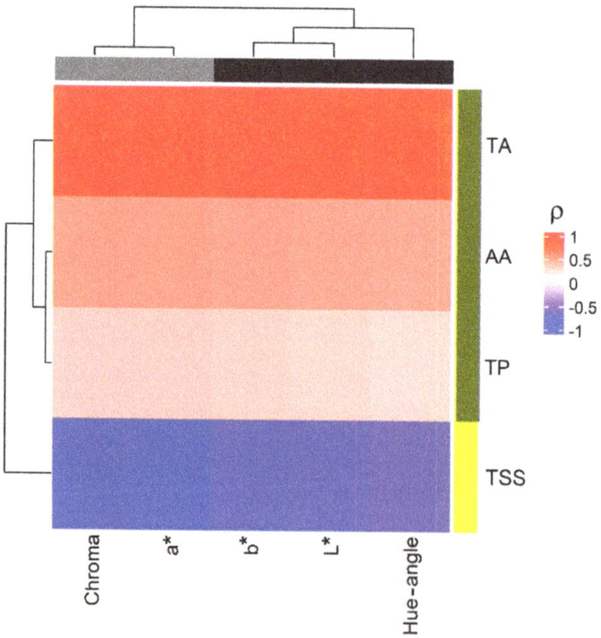

Figure 7. Correlations between image data and chemical data: heatmap; a different colour code was used to represent the clusters discovered by cluster analysis; ρ is the Pearson correlation coefficient.

4. Conclusions

E-nose, ATR-FTIR and image analysis were used as rapid and non-destructive techniques to discriminate the ripening stage (half-red or red) of strawberries cv Sabrosa, commercially named Candonga, harvested at three different times. The correlation analysis between e-nose sensor responses and VOC data demonstrated that HS-SPME/MS-e-nose experimental data contain enough information to allow the discrimination of strawberry samples based on their degree of ripening. Moreover, among the chemical indicators of ripening, TA was correlated to both ATR-FTIR and IA data. Since TA usually decreases during ripening, its assessment by ATR-FTIR or IA might provide a suitable indicator for a fast and non-destructive evaluation of the ripening stage in strawberries.

Further investigations to improve the results obtained herein, performed using a larger sample size and cultivars as well as samples of different strawberry crop seasons, will be carried out with the aim to build predictive models of ripening, which can offer several advantages in terms of rapidity, low cost and easy-to-handle analysis for the industries of the sector. Models built considering multiple techniques (i.e., ATR-FTIR and IA) will also be considered in the perspective of improving the reliability of the models. The final goal is to obtain significant models usable in portable devices for a real-time and on-site prediction of the most suitable harvest time.

Supplementary Materials: The following supporting information can be downloaded at: https://www.mdpi.com/article/10.3390/foods11111534/s1, Table S1: Volatiles compounds obtained in strawberries cv Sabrosa, commercially named Candonga at two different ripening stages (half-red and red) and at three harvest times (H1, H2 and H3); Table S2: Chemical parameters measured in strawberries cv Sabrosa, commercially named Candonga, at two different ripening stages (half-red and red) and at three harvest times (H1, H2 and H3); Figure S1. Half-red (in ripening phase, fully expanded and 50% red), and Red (in ripening phase, fully expanded and 100% red) Candonga strawberries used in the experiment.

Author Contributions: Conceptualization, writing, revision, R.C., M.C. and B.P.; formal analysis, data processing, C.L., F.S., L.M., G.P., M.P. and A.C.; data curation, methodology, validation, writing—original draft, M.S. and R.R.; funding acquisition, project administration, M.C. and B.P. All authors have read and agreed to the published version of the manuscript.

Funding: This research was funded by the Project PON «R&I» 2014–2020—Azione II—"E-crops—Technologies for Digital and Sustainable Agriculture" (grant number ARS01_01136) from European Union, and the Italian Ministry of Education University.

Institutional Review Board Statement: Not applicable.

Informed Consent Statement: Not applicable.

Data Availability Statement: Not applicable.

Acknowledgments: The authors thank Massimo Franchi of CNR-ISPA for the technical support in the laboratory and Giuseppe Sicuro and Gianluca Faliero of Apofruit Italia Soc. Coop. Agricola (Scanzano Jonico, Italy) for agronomy support during fruit harvest.

Conflicts of Interest: The authors declare no conflict of interest.

References

1. Yan, J.W.; Ban, Z.J.; Lu, H.Y.; Li, D.; Poverenov, E.; Luo, Z.S.; Li, L. The aroma volatile repertoire in strawberry fruit: A review. *J. Sci. Food Agric.* **2018**, *98*, 4395–4402. [CrossRef]
2. Weng, S.; Yu, S.; Guo, B.; Tang, P.; Liang, D. Non-destructive detection of strawberry quality using multi-features of hyperspectral imaging and multivariate methods. *Sensors* **2020**, *20*, 3074. [CrossRef]
3. Zhao, J.; Liu, J.; Wang, F.; Wang, S.; Feng, H.; Xie, X.; Hao, F.; Zhang, L.; Fang, C. Volatile constituents and ellagic acid formation in strawberry fruits of selected cultivars. *Food Res. Int.* **2020**, *138*, 109767. [CrossRef]
4. Padilla-Jiménez, S.M.; Angoa-Pérez, M.V.; Mena-Violante, H.G.; Oyoque-Salcedo, G.; Montañez-Soto, J.L.; Oregel-Zamudio, E. Identification of Organic Volatile Markers Associated with Aroma during Maturation of Strawberry Fruits. *Molecules* **2021**, *26*, 504. [CrossRef]

5. Sheng, L.; Ni, Y.; Wang, J.; Chen, Y.; Gao, H. Characteristic-aroma-component-based evaluation and classification of straw-berry varieties by aroma type. *Molecules* **2021**, *26*, 6219. [CrossRef]
6. Saad, A.; Azam, M.M.; Amer, B. Quality Analysis Prediction and Discriminating Strawberry Maturity with a Hand-held Vis-NIR Spectrometer. *Food Anal. Methods* **2022**, *15*, 689–699. [CrossRef]
7. Aghilinategh, N.; Dalvand, M.J.; Anvar, A. Detection of ripeness grades of berries using an electronic nose. *Food Sci. Nutr.* **2020**, *19*, 4919–4928. [CrossRef]
8. Li, H.; Brouwer, B.; Oud, N.; Verdonk, J.C.; Tikunov, Y.; Woltering, E.; Schouten, R.; Pereira da Silva, F. Sensory, GC-MS and PTR-ToF-MS profiling of strawberries varying in maturity at harvest with subsequent cold storage. *Postharvest Biol. Technol.* **2021**, *182*, 111719. [CrossRef]
9. Li, B.; Lecourt, J.; Bishop, G. Advances in non-destructive early assessment of fruit ripeness towards defining optimal time of harvest and yield prediction—A review. *Plants* **2018**, *7*, 3. [CrossRef]
10. Dong, D.; Zhao, C.; Zheng, W.; Wang, W.; Zhao, X.; Jiao, L. Analyzing Strawberry Spoilage via its Volatile Compounds Using Longpath Fourier Transform Infrared Spectroscopy. *Sci. Rep.* **2013**, *3*, 2585. [CrossRef]
11. Cassani, L.; Santos, M.; Gerbino, E.; del Rosario Moreira, M.; Gomez-Zavaglia, A. A Combined Approach of Infrared Spectroscopy and Multivariate Analysis for the Simultaneous Determination of Sugars and Fructans in Strawberry Juices During Storage. *J. Food Sci.* **2018**, *83*, 631–638. [CrossRef]
12. Minutti-López Sierra, P.; Gallardo-Velázquez, T.; Osorio-Revilla, G.; Meza-Márquez, O.G. Chemical composition and antioxidant capacity in strawberry cultivars (Fragaria x ananassa Duch.) by FT-MIR spectroscopy and chemometrics. *CyTA J. Food* **2019**, *17*, 724–732. [CrossRef]
13. Oo, L.M.; Aung, N.Z. A simple and efficient method for automatic strawberry shape and size estimation and classification. *Biosyst. Eng.* **2018**, *170*, 96–107. [CrossRef]
14. Xin, L.; Li, J.; Tang, J. A deep learning method for recognizing elevated mature strawberries. In Proceedings of the 33rd Youth Academic Annual Conference of Chinese Association of Automation (YAC), Nanjing, China, 18–20 May 2018; IEEE: Piscataway, NJ, USA, 2018; pp. 1072–1077. [CrossRef]
15. Zhang, Q.; Xiangjun, Z.; Guichao, L.; Yanhui, S. Image Feature Extraction and Online Grading Method for Weight and Shape of Strawberry. *J. Syst. Simul.* **2019**, *31*, 7. [CrossRef]
16. Du, X.; Bai, J.; Plotto, A.; Baldwin, E.; Whitaker, V.; Rouseff, R. Electronic nose for detecting strawberry fruit maturity. In Proceedings of the Florida State Horticultural Society, Crystal River, FA, USA, 6–8 June 2010; Florida State Horticultural Society: Alexandria, VA, USA, 2010; Volume 123, pp. 259–263.
17. Xing, M.; Sun, K.; Liu, Q.; Pan, L.; Tu, K. Development of novel electronic nose applied for strawberry freshness detection during storage. *Int. J. Food Eng.* **2018**, *14*, 7–8. [CrossRef]
18. Palumbo, M.; Pace, B.; Cefola, M.; Montesano, F.F.; Serio, F.; Colelli, G.; Attolico, G. Self-configuring CVS to discriminate rocket leaves according to cultivation practices and to correctly attribute visual quality level. *Agronomy* **2021**, *11*, 1353. [CrossRef]
19. Lamb, N.; Chuah, M.C. A strawberry detection system using convolutional neural networks. In Proceedings of the IEEE International Conference on Big Data (Big Data), Seattle, WA, USA, 10–13 December 2018; IEEE: Piscataway, NJ, USA, 2018; pp. 2515–2520. [CrossRef]
20. Cozzolino, R.; Pace, B.; Palumbo, M.; Laurino, C.; Picariello, G.; Siano, F.; De Giulio, B.; Pelosi, S.; Cefola, M. Profiles of Volatile and Phenolic Compounds as Markers of Ripening Stage in Candonga Strawberries. *Foods* **2021**, *10*, 3102. [CrossRef]
21. Galvan, D.; Aquino, A.; Effting, L.; Mantovani, A.C.G.; Bona, E.; Conte-Junior, C.A. E-sensing and nanoscale-sensing devices associated with data processing algorithms applied to food quality control: A systematic review. *Crit. Rev. Food Sci. Nutr.* **2021**, *29*, 1–41. [CrossRef]
22. Nategh, N.A.; Dalvand, M.J.; Anvar, A. Detection of toxic and non-toxic sweet cherries at different degrees of maturity using an electronic nose. *J. Food Meas. Charact.* **2021**, *15*, 1213–1224. [CrossRef]
23. Wang, D.; Zhang, M.; Mujumdar, A.S.; Yu, D. Advanced Detection Techniques Using Artificial Intelligence in Processing of Berries. *Food Eng. Rev.* **2022**, *14*, 176–199. [CrossRef]
24. Shi, J.; Nian, Y.; Da, D.; Xu, X.; Zhou, G.; Zhao, D.; Li, C. Characterization of flavor volatile compounds in sauce spareribs by gas chromatography-mass spectrometry and electronic nose. *LWT Food Sci. Technol.* **2020**, *124*, 109182–109190. [CrossRef]
25. Gonzalez, R.C.; Woods, R.E.; Eddins, S.L. *Digital Image Processing Using MATLAB*; Pearson Prentice Hall: Upper Saddle River, NJ, USA, 2004.
26. Jolliffe, I.T. Principal Component Analysis. In *Springer Series in Statistics*, 2nd ed.; Springer: New York, NY, USA, 2002.
27. Ulrich, D.; Olbricht, K. A search for the ideal flavor of strawberry—Comparison of consumer acceptance and metabolite patterns in Fragaria × ananassa Duch. *J. Appl. Bot. Food Qual.* **2016**, *89*, 223–234. [CrossRef]
28. Bae, H.; Yun, S.K.; Yoon, I.K.; Nam, E.Y.; Kwon, J.H.; Jun, J.H. Assessment of organic acid and sugar composition in apricot, plumcot, plum, and peach during fruit development. Journal of applied botany and food quality. *J. Appl. Bot.* **2014**, *87*, 24–29. [CrossRef]
29. Janurianti, N.M.D.; Utama, I.M.S.; Gunam, I.B.W. Colour and quality of strawberry fruit (Fragaria x ananassa Duch.) at different levels of maturity. *SEAS* **2021**, *5*, 22–28. [CrossRef]
30. Kaur, H.; Sawhney, B.K.; Jawandha, S.K. Evaluation of plum fruit maturity by image processing techniques. *J. Food Sci. Technol.* **2018**, *55*, 3008–3015. [CrossRef]

foods

MDPI

Article

Assessment of "Sugranineteen" Table Grape Maturation Using Destructive and Auto-Fluorescence Methods

Najwane Hamie [1], Luigi Tarricone [2], Vincenzo Verrastro [3], Giuseppe Natrella [1], Michele Faccia [1] and Giuseppe Gambacorta [1,*]

[1] Department of Soil, Plant and Food Science, University of Bari Aldo Moro, 70126 Bari, Italy; najwane.hamie@uniba.it (N.H.); giuseppe.natrella@uniba.it (G.N.); michele.faccia@uniba.it (M.F.)
[2] Research Centre for Viticulture and Enology, Council for Agricultural Research and Economics, CREA, 70010 Turi, Italy; luigi.tarricone@crea.gov.it
[3] Department of Mediterranean Organic Agriculture, Mediterranean Agronomic Institute of Bari, 70010 Valenzano, Italy; verrastro@iamb.it
* Correspondence: giuseppe.gambacorta@uniba.it; Tel.: +39-080-5442-942

Abstract: The optimal harvesting of table grapes is commonly determined based on technological and phenolic indices analyzed over the course of its maturity. The classical techniques used for these analyses are destructive, time-consuming, and work for a limited number of samples that may not represent the heterogeneity of the vineyard. This study aimed to follow the ripening season of table grapes using non-destructive tools as a rapid and accurate alternative for destructive techniques. Grape samples were collected from a Sugranineteen vineyard during the ripening season to measure the basic maturity indices via wet chemistry, and total polyphenols, anthocyanins, and flavonoids were evaluated by spectrophotometry. Fluorescent readings were collected from intact clusters with a portable optical sensor (Multiplex® 3, Force-A, France) that generates indices correlated to different maturity parameters. Results revealed strong relationships between the Multiplex® indices ANTH_RG and FERARI and the skin anthocyanin content, with R^2 values equal to 0.9613 and 0.8713, respectively. The NBI_R index was also related to total anthocyanins (R^2 = 0.8032), while the SFR_R index was linked to the titratable acidity (R^2 = 0.6186), the sugar content (R^2 = 0.7954), and to the color index of red grapes (CIRG) (R^2 = 0.7835). Results demonstrated that Multiplex® 3 can be applied on intact clusters as an effective non-destructive tool for a rapid estimation of table grapes' anthocyanin content.

Keywords: table grapes; optimal harvesting; technological maturity; phenolic maturity; non-destructive tools; Multiplex® 3

Citation: Hamie, N.; Tarricone, L.; Verrastro, V.; Natrella, G.; Faccia, M.; Gambacorta, G. Assessment of "Sugranineteen" Table Grape Maturation Using Destructive and Auto-Fluorescence Methods. *Foods* **2022**, *11*, 663. https://doi.org/10.3390/foods11050663

Academic Editors: Maria Cefola, Bernardo Pace and Liguang Xu

Received: 13 January 2022
Accepted: 23 February 2022
Published: 24 February 2022

1. Introduction

The maturity assessment of table grapes (*Vitis vinifera* L.) is necessary to determine the optimal harvesting time, which is a critical point for the postharvest handling period. Both destructive and non-destructive methods are applied in viticulture for crop monitoring and evaluation [1]. A common practice is to harvest the grapes based on technological and phenolic maturity indices that are measured using destructive laboratory analyses [2]. The technological maturity mainly includes the measurement of sugar content, titratable acidity, and the pH value of grape juice. The phenolic maturity reflects the ripeness of berry skin, pulp, and seeds, considering their phenolic compositions, and is expressed either as total polyphenols or skin anthocyanin content [3,4]. The total phenolic compounds are usually extracted from grape skins and estimated via spectrophotometry methods, such as the Folin–Ciocalteu assay [5]. The detailed profile of anthocyanins that are responsible for the red color in mature berries [6] is commonly identified by high pressure liquid chromatography (HPLC) [7]. Headspace solid phase microextraction (HS-SPME) followed by gas chromatography mass spectrometry (GC-MS) is applied for the analysis and quantification of some polyphenols in wine and grapes [8].

Although effective and precise, all these techniques are expensive, destructive, time-consuming, and consider a limited number of samples [9,10]. These problems demonstrate the limitations of laboratory analyses to properly estimate the grape status in the vineyard and to reflect the spatial and temporal heterogeneity of grapes throughout the maturation period. Consequently, researchers are increasingly oriented toward non-destructive techniques which are fast, accurate, and enable real-time analyses of fruit quality and maturity for a large number of samples [11–13]. Particularly, the optical methods were advantageously used for precision viticulture to solve the problem of grape heterogeneity in vineyards [12,14,15]. Among the most recent optical techniques, a method based on fruit auto-fluorescence has been successfully used for the monitoring of grape maturity [16,17]. The fluorescence indices were proved to reflect the epidermal phenolic content in wine grape leaves and skin anthocyanin content in grape berries [16,18,19]. Multiplex® (FORCE-A, Orsay, France) is a commercial hand-held optical sensor that applies the chlorophyll fluorescence screening technique [15]. Several successful applications of Multiplex® sensor were reported for the non-destructive determination of grape anthocyanins, the assessment of the spatial variability of grape color in the vineyard, and the ripening evaluation of different wine grape cultivars [12,17,20,21].

While most studies were applied to wine grapes, the present investigation aimed to assess the ripening of red table grapes using both destructive and non-destructive fluorescence-based methods. In practice, the objective was to evaluate the feasibility of using the fluorescence-based method to measure skin anthocyanin content along with maturity and to study the possible relationship between laboratory results and in-site fluorescence readings collected by Multiplex® 3 from intact clusters.

2. Materials and Methods

2.1. Plant Material

The experiment was carried out during the ripening season of 2019 on *Vitis vinifera* L. cv. Sugranineteen (Scarlotta Seedless® brand) grown in a commercial organic vineyard (Azienda Agricola Romanazzi S.r.l., Castellaneta Marina, Taranto, Italy) in Castellaneta Marina, south Italy. Scarlotta Seedless is a late-season seedless cultivar characterized by sweet, crisp, red- to dark-red-colored and oval-shaped berries. Grapevines were grafted onto a 1103 Paulsen rootstock, spaced 2.5 × 2.5 m, and trained to the Y-shaped trellis system and covered by plastic film to protect canopy and clusters from hail, wind, and rainfall. The plastic film was characterized by a high solar total transmissivity coefficient (83.7%), while the photosynthetically active radiation (PAR) total transmissivity coefficients was 81.8% and the long wave infrared (LWIR) transmissivity coefficient was 53.6%. The growing techniques were implemented according to the viticulture practices of organic table grapes in Italy, and the harvest time was determined according to the commercial maturity specifications set by the parent company Sun World International, LCC, Bakersfield, CA, USA.

2.2. Grape Sampling and In-Field Measurements

The monitoring of grape ripening was performed weekly from the onset of veraison on 1 August 2019 (day of the year, DOY = 213) until the harvest time on 7 October 2019 (DOY = 280). The experimental site was divided into two blocks of different irrigation systems: Scarlotta Block 1, SB1 (farmer irrigation) and Scarlotta Block 2, SB2 (controlled irrigation by an Internet of Things sensor-based system). Each block constituted 98 vines planted into 7 adjacent rows. At each sampling time, 3 berries per vine (294 berries per block) were removed from different parts of the clusters free of visible damages, with their pedicel still attached. In parallel, 2 fluorescence readings (196 measurements per block) were collected by Multiplex® 3 from clusters attached to both sides of the grapevine (Section 2.3). Collected berries were stored at −20 °C until total polyphenols and skin anthocyanins extraction and quantification. Additionally, another 10 clusters were randomly collected 5 times during maturation to conduct the basic wet chemistry analysis and decide the optimal harvesting day (Section 2.4).

2.3. Optical Sensor and Indices

The fluorescence measurements of grapes were collected by the hand-held optical sensor Multiplex® (FORCE-A, Orsay, France), which is described in detail elsewhere [17]. In this study, we used the version Multiplex® 3 (MP3) that generates 12 signals produced by the combination of different excitation light emitting diodes (LED), i.e., UV and red–blue–green (RGB), and three photodiode detectors for fluorescence recordings, i.e., yellow (YF), red (RF), and far red (FRF). Real-time measurements were acquired directly in the vineyard on intact clusters. Among the parameters that MP3 can measure, this study focuses on the measurement of skin anthocyanins using 2 indices: ANTH_RG, which is based on the chlorophyll fluorescence excited with red (R) and green (G) lights, and FERARI (fluorescence excitation ratio anthocyanin relative index), which is based on the far-red chlorophyll fluorescence under red excitation [12,17]. The other studied indices include the FLAV index, which is proportional to skin flavonols, the simple fluorescent ratio (SFR_R), which is related to the chlorophyll content in leaves and berries, and the nitrogen balance index (NBI_R), which accounts for both epidermal phenolics and chlorophyll contents [22–25].

The mathematical expressions of the used fluorescence indices are defined as:

$$\mathrm{ANTH\_RG} = \log\left(\mathrm{FRF\_R}/\mathrm{FRF\_G}\right), \tag{1}$$

$$\mathrm{FERARI} = \log\left(5000/\mathrm{FRF\_R}\right), \tag{2}$$

$$\mathrm{FLAV} = \log\left(\mathrm{FRF\_R}/\mathrm{FRF\_UV}\right), \tag{3}$$

$$\mathrm{SFR\_R} = \mathrm{FRF\_R}/\mathrm{RF\_R}, \tag{4}$$

$$\mathrm{NBI\_R} = \mathrm{FRF\_UV}/\mathrm{FRF\_R}, \tag{5}$$

where FRF_R, FRF_G, and FRF_UV are far-red fluorescence under red, green, and UV excitation, respectively. RF_R is red fluorescence under red excitation. Mx fluorescence signals were corrected for residual electronic offsets and normalized to a fluorescence standard (blue plastic foil, FORCE-A, Orsay, France).

2.4. Chemical Analysis

The technological maturity of Sugranineteen was monitored starting from DOY = 246 to decide the optimal harvesting date. Three replicates of grape juice were extracted from 20 berries per replicate and then filtered after centrifugation (10 °C, 3000× g, 15 min). Total soluble solids (TSS) were determined using a DBR 95 digital refractometer (XS Instruments, Carpi, Italy) and the pH was measured with a pH meter (Eutech Instruments, Breda, The Netherlands, XS pH 2700). The titratable acidity (TA) was determined by titrating the juice with 0.1 N NaOH to an endpoint of 7.0 pH, and results are expressed as g/L tartaric acid equivalent. Berry color was collected from 294 berries per block using a portable spectrophotometer (CM–700d, Konica Minolta, Inc., Tokio, Japan). Results are expressed as a color index of red grapes (CIRG), calculated from the CIELAB color coordinates, i.e., hue angle (h), lightness (L*), and chroma (C*), and defined as $((180 - h)/(L^* + C^*))$ [26]. This index enables an objective color evaluation of red grape cultivars at different ripening stages [27].

2.5. Analysis of Total Polyphenols, Anthocyanins, and Flavonoids

The assessment of total polyphenols, anthocyanins, and flavonoids during ripening was performed on skin extracts. Nine berries of different color tones (fully, medium, and slightly colored) were weighed using an electronic balance (Gibertini, Milan, Italy) before the skins were peeled and infused in 25 mL of ethanol–chloride solution (70:30:1;

$C_2H_5OH:H_2O:HCl$) for 24 h in darkness. Extracts were then filtered through 0.45 μm filter papers and stored at −20 °C until spectrophotometric analyses. Total polyphenols were quantified using the Folin–Ciocalteu method on 1:10 diluted skin extracts with distilled water. In an Eppendorf tube, 100 μL of distilled water, 100 μL of diluted skin extract, and 100 μL of Folin–Ciocalteu reagent were homogenized and incubated for 5 min at room temperature. Another incubation for 90 min was performed after adding 500 μL of 10% of sodium carbonate to the mixture, and the absorbance was then recorded at 750 nm using the DU® 800 UV/Vis spectrophotometer (Beckman Coulter, Brea, CA, USA). The quantification of total anthocyanins and flavonoids was performed on 1:25 diluted skin extracts with the ethanol–chloride solution. In this case, the absorbance was determined within the UV-Vis spectrum range of 230–700 nm, and peaks were identified using the graphical method [28]. The obtained results of different parameters were expressed as mg/kg of grape berries (skin and flesh).

2.6. Antioxidant Activity

The antioxidant activity of Sugranineteen was assessed with ABTS and DPPH assays on grape juice obtained from nine weighed berries per sample and subjected to centrifugation (8000 RPM for 5 min at 10 °C) and filtration using 0.45 μm filters. The ABTS (2,2′-azino-bis(3-ethylbenzothiazoline-6-sulfonic acid)) assay was carried out according to a method reported elsewhere, with minor modifications [29]. This method is based on the ability of antioxidants to scavenge the $ABTS^+$. The stock solution was obtained by reacting 7 mM of ABTS solution (0.0960 g/25 mL H_2O) with 400 μL of potassium persulphate ($K_2S_2O_8$) for 16 h at room temperature and in darkness. An aqueous solution of 100 mM $ABTS^+$ was then prepared and diluted with water to an absorbance of 0.80 ± 0.005 at 734 nm. The absorbance of 950 μL of diluted $ABTS^+$ solution added to 50 μL of filtered juice was then recorded at 734 nm after a sharp 8 min of incubation at room temperature and in darkness.

The DPPH (2,2-diphenyl-1-picrylhydrazyl) assay was performed based on a pre-explained method, with small adjustments [30]. A mother solution of DPPH 0.08 mM was produced by solubilizing 0.0031 g of DPPH powder in 100 mL of ethanol. The solution of DPPH was diluted with ethanol to an absorbance of 0.80 ± 0.003 at 517 nm to perform the spectrophotometric analysis. A mixture of 50 μL of filtered juice and 950 μL of diluted DPPH solution was incubated for 30 min at room temperature and in darkness to record the absorbance of the mixture at 517 nm. The obtained results of both assays are expressed as μM Trolox equivalent antioxidant capacity (TEAC)/g of grapes.

2.7. HPLC-DAD Anthocyanin Analysis

The anthocyanin composition of grape skins extracts was determined using an HPLC Waters 600 E device (Waters Inc., Milford, MA, USA) equipped with a quaternary pump, a photodiode array detector, and an injection valve with a 20 μL loop [31]. Skin extracts were filtered through 0.45 μm nylon membrane and diluted 1:2 with 10% formic acid. Samples were then injected into a NovaPack C18 (150 × 3.9 mm, 4 μm particle size, Waters Inc.) column maintained at 30 °C and eluted at a flow rate of 1 mL/min with 10% formic acid (solvent A) and acetonitrile (solvent B). The gradient program of solvent A was the following: 0–10 min 95%, 10–20 min 87%, 20–30 min 85%, 30–45 min 75%, and 45–50 min 95%. Anthocyanins were detected at 520 nm and quantitative analysis was performed according to an external standard method based on a calibration curve obtained by injecting different concentrations of malvidin-3-O-glucoside solutions (R^2 = 0.9991). Anthocyanin compounds were identified by comparing the elution pattern and data reported in previous studies [32–34]. The results were expressed as mg/kg of malvidin-3-O-glucoside equivalents in grape berries.

2.8. Statistical Analysis

All destructive determinations were made in triplicate. Data processing and statistical analysis were carried out on Microsoft® Excel and Minitab 20.3 (Minitab, LLC, State College, PA, USA) software. A one-way analysis of variance (ANOVA) with Tukey's honest significant difference (HSD) test were performed for each studied parameter at a p-value ≤ 0.05.

3. Results

3.1. Analysis of Ripeness

The technological maturity of grapes was monitored starting from 3 September (DOY = 246) until harvest (Table 1). The pH of grape juice fluctuated with slight differences among blocks and maturation days and averaged 3.47 for SB1 and 3.61 for SB2 at harvest. The sugar content had an increasing trend, while the acidity was decreasing during maturation and reached at harvest an average TSS of 17.9 °Brix and an acidity value of 4.86 g/L with no significant differences among the blocks.

Table 1. Changes in quality attributes of Sugranineteen grapes during ripening (mean values ± SD).

Quality Parameters	Sample	Maturation Time (DOY)				
		246	254	263	278	280 (Harvest)
pH	SB1	$^{A}3.35^{d} \pm 0.01$ *	$^{A}3.51^{b} \pm 0.01$	$^{A}3.56^{a} \pm 0.01$	$^{A}3.51^{b} \pm 0.01$	$^{B}3.47^{c} \pm 0.01$
	SB2	$^{A}3.34^{c} \pm 0.01$	$^{A}3.49^{b} \pm 0.02$	$^{B}3.44^{b} \pm 0.02$	$^{B}3.45^{b} \pm 0.01$	$^{A}3.61^{a} \pm 0.01$
TSS (°Brix)	SB1	$^{A}16.6^{b} \pm 0.1$	$^{A}16.9^{b} \pm 0.2$	$^{A}18.1^{a} \pm 0.1$	$^{A}18.4^{a} \pm 0.2$	$^{A}17.8^{a} \pm 0.1$
	SB2	$^{B}15.6^{cd} \pm 0.3$	$^{A}16.5^{bc} \pm 0.5$	$^{B}14.8^{d} \pm 0.3$	$^{B}17.5^{ab} \pm 0.1$	$^{A}18.1^{a} \pm 0.2$
Titratable acidity (g/L)	SB1	$^{B}7.23^{a} \pm 0.58$	$^{A}4.83^{b} \pm 0.03$	$^{B}4.26^{c} \pm 0.15$	$^{A}4.89^{b} \pm 0.19$	$^{A}4.78^{b} \pm 0.08$
	SB2	$^{A}8.18^{a} \pm 0.09$	$^{A}5.03^{b} \pm 0.30$	$^{A}4.54^{b} \pm 0.04$	$^{A}4.83^{b} \pm 0.05$	$^{A}4.94^{b} \pm 0.10$
CIRG	SB1	$^{B}5.30^{a} \pm 0.07$	$^{A}5.47^{a} \pm 0.32$	$^{A}4.80^{b} \pm 0.14$	$^{A}5.35^{a} \pm 0.14$	$^{A}5.09^{ab} \pm 0.16$
	SB2	$^{A}5.89^{a} \pm 0.21$	$^{A}5.59^{ab} \pm 0.19$	$^{B}4.47^{c} \pm 0.08$	$^{A}5.29^{b} \pm 0.08$	$^{A}5.32^{b} \pm 0.20$

* On the same row, means with different right superscripts differ significantly ($p \leq 0.05$). Between the SB1 (farmer irrigation) and SB2 (sensor-based controlled irrigation) pair, means with different left superscripts differ significantly ($p \leq 0.05$).

The CIRG averaged at harvest 5.09 and 5.32 for SB1 and SB2, respectively, reflecting the red color of the berries. The harvest was conducted on 7 October (DOY = 280) according to the measured parameters and the commercial maturity standards of Sugranineteen.

3.2. Polyphenols and Antioxidant Acitivy

The total amount of polyphenols, flavonoids, and anthocyanins measured using spectrophotometry was expressed as mg/kg of berry FW (Table 2). Total polyphenols fluctuated in a constant range during the ripening season between 586 and 996 mg/kg. Both blocks had almost the same trend, with slightly higher amounts of polyphenols for SB2, which showed an average concentration at harvest of 851 mg/kg compared to 738 mg/kg in SB1. Total flavonoids followed the same trend, as total polyphenols and fluctuated within a constant range until harvest. Flavonoids had significantly higher concentrations in SB1 during maturation, but at harvest, the content was slightly higher in SB2 (508.8 versus 468.7 mg/kg). In contrast, anthocyanins had an evident increasing trend over the entire maturation period, except for DOY 263, and had almost a similar trend among blocks, with mostly higher values for SB2. At harvest, SB2 reached an average value of 145 mg/kg, compared to 123 mg/kg for SB1.

Table 2. Analyses of polyphenols and antioxidant activity of Sugranineteen grapes during ripening (mean values ± SD).

Parameters	Sample	Maturation Time (DOY)							
		213	225	234	246	254	263	278	280 (Harvest)
Total polyphenols (mg/kg)	SB1	${}^{B}715.3^{b} \pm 9.2$ *	${}^{B}911.7^{a} \pm 56.6$	${}^{A}669.2^{b} \pm 44.9$	${}^{A}846.9^{ab} \pm 95.5$	${}^{A}746.5^{b} \pm 33.9$	${}^{A}710.1^{b} \pm 30.0$	${}^{A}813.8^{ab} \pm 59.1$	${}^{B}737.5^{b} \pm 54.9$
	SB2	${}^{A}819.9^{b} \pm 40.7$	${}^{A}995.9^{a} \pm 16.2$	${}^{A}684.0^{c} \pm 16.2$	${}^{A}913.6^{ab} \pm 89.0$	${}^{A}731.4^{bc} \pm 54.6$	${}^{B}586.3^{d} \pm 6.8$	${}^{A}785.6^{b} \pm 36.1$	${}^{A}851.4^{ab} \pm 43.4$
Flavonoids (mg/kg)	SB1	${}^{A}405.7^{c} \pm 7.4$	${}^{A}542.6^{a} \pm 35.1$	${}^{A}386.4^{c} \pm 38.4$	${}^{A}480.8^{ab} \pm 55.2$	${}^{A}436.8^{b} \pm 9.3$	${}^{A}376.4^{c} \pm 22.4$	${}^{A}469.4^{b} \pm 36.0$	${}^{A}468.7^{b} \pm 28.0$
	SB2	${}^{B}379.3^{b} \pm 11.2$	${}^{B}485.0^{a} \pm 13.7$	${}^{A}365.7^{b} \pm 17.9$	${}^{A}473.7^{a} \pm 20.8$	${}^{B}376.5^{b} \pm 34.5$	${}^{B}289.8^{c} \pm 1.7$	${}^{A}454.4^{a} \pm 24.8$	${}^{A}508.8^{a} \pm 33.0$
Anthocyanins (mg/kg)	SB1	${}^{B}44.8^{e} \pm 7.4$	${}^{A}86.2^{bc} \pm 7.5$	${}^{B}63.2^{d} \pm 14.6$	${}^{A}84.9^{bc} \pm 14.3$	${}^{A}82.4^{c} \pm 1.6$	${}^{A}63.9^{d} \pm 0.1$	${}^{B}93.4^{b} \pm 6.4$	${}^{B}123.0^{a} \pm 4.1$
	SB2	${}^{A}65.7^{d} \pm 11.4$	${}^{B}65.6^{d} \pm 9.7$	${}^{A}91.5^{c} \pm 10.8$	${}^{A}99.9^{c} \pm 1.6$	${}^{A}84.5^{c} \pm 19.2$	${}^{B}58.6^{d} \pm 2.1$	${}^{A}133.7^{b} \pm 4.3$	${}^{A}144.6^{a} \pm 1.7$
Antioxidant activity									
ABTS (μM/g)	SB1	${}^{A}3.6^{a} \pm 0.1$	${}^{A}3.5^{a} \pm 0.1$	${}^{A}2.7^{b} \pm 0.2$	${}^{A}2.0^{c} \pm 0.1$	${}^{A}2.9^{b} \pm 0.2$	${}^{A}3.7^{a} \pm 0.1$	${}^{A}4.1^{a} \pm 0.6$	${}^{A}3.4^{a} \pm 0.2$
	SB2	${}^{B}2.9^{c} \pm 0.1$	${}^{A}3.3^{b} \pm 0.1$	${}^{B}2.1^{d} \pm 0.1$	${}^{A}2.1^{d} \pm 0.1$	${}^{B}1.9^{d} \pm 0.1$	${}^{B}3.1^{bc} \pm 0.1$	${}^{A}3.7^{a} \pm 0.2$	${}^{A}3.8^{a} \pm 0.2$
DPPH (μM/g)	SB1	${}^{A}0.9^{c} \pm 0.1$	${}^{A}1.2^{c} \pm 0.2$	${}^{A}1.7^{b} \pm 0.1$	${}^{A}1.6^{b} \pm 0.1$	${}^{A}1.8^{b} \pm 0.1$	${}^{A}2.0^{ab} \pm 0.1$	${}^{A}2.4^{a} \pm 0.4$	${}^{A}2.2^{ab} \pm 0.3$
	SB2	${}^{A}1.1^{d} \pm 0.1$	${}^{A}1.2^{d} \pm 0.1$	${}^{B}1.4^{cd} \pm 0.1$	${}^{A}1.6^{c} \pm 0.1$	${}^{B}1.2^{d} \pm 0.1$	${}^{A}2.0^{b} \pm 0.1$	${}^{A}2.1^{b} \pm 0.1$	${}^{A}2.5^{a} \pm 0.1$

* On the same row, means with different right superscripts differ significantly ($p \leq 0.05$). Between the SB1 (farmer irrigation) and SB2 (sensor-based controlled irrigation) pair, means with different left superscripts differ significantly ($p \leq 0.05$).

Regarding the antioxidant activity, results were expressed as the μM Trolox equivalent antioxidant capacity (TEAC)/kg of grapes (Table 2). This was analyzed using two different assays, ABTS and DPPH. Despite the fluctuation of values, both methods revealed a similar increasing trend throughout maturation. ABTS expressed the antioxidant activity with higher concentration than DPPH, varying between 1.9 and 4.1 μM/kg. At harvest, no significant difference was detected between the two blocks, and the values averaged 3.4 and 3.8 μM/kg for SB1 and SB2, respectively. Results from DPPH followed a range of values between 0.9 and 2.5 μM/kg. At harvest, concentrations were also similar between the two blocks, averaging 2.2 μM/kg for SB1 and 2.5 μM/kg for SB2.

3.3. Anthocyanin Profile

A total of 13 anthocyanin compounds were identified and quantified using HPLC-DAD, as shown in Figure 1, corresponding to the SB2 sample at harvest time. The anthocyanin compositions of grape skin extracts are reported in Table 3. The most abundant anthocyanins at harvest were peonidin-3-O-glucoside (37.5 and 42.4 mg/kg) and malvidin-3-O-glucoside (17.4 and 24.7 mg/kg), followed by cyanidin-3-O-glucoside (14.9 and 8.8 mg/kg), peonidin-3-O-coumaroyl-glucoside (6.9 and 9.2 mg/kg), and *trans*-malvidin-3-O-coumaroyl-glucoside (3.9 and 7.3 mg/kg) in SB1 and SB2, respectively. All these compounds were significantly higher in SB2, except for cyanidin-3-O-glucoside.

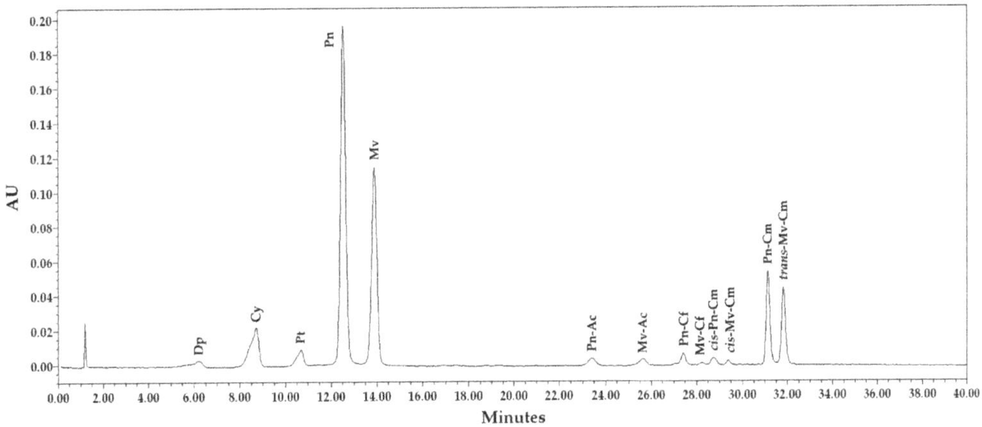

Figure 1. HPLC-DAD anthocyanin profile of SB2 Sugranineteen at DOY 280 (harvest). Dp, delphinidin-3-O-glucoside; Cy, cyanidin-3-O-glucoside; Pt, petunidin-3-O-glucoside; Pn, peonidin-3-O-glucoside; Mv, malvidin-3-O-glucoside; Pn-Ac, peonidin-3-O-acetyl-glucoside; Mv-Ac, malvidin-3-O-acetyl-glucoside; Pn-Cf, peonidin-3-O-caffeoyl-glucoside; Mv-Cf, malvidin-3-O-caffeoyl-glucoside; *cis*-Pn-Cm, *cis*-peonidin-3-O-coumaroyl-glucoside; *cis*-Mv-Cm, *cis*-malvidin-3-O-coumaroyl-glucoside; Pn-Cm, peonidin-3-O-coumaroyl-glucoside; *trans*-Mv-Cm, *trans*-malvidin-3-O-coumaroyl-glucoside.

Another observation is that the amounts of almost all the identified compounds were increasing with maturation, despite the fluctuation of values among the maturation days. At harvest, the total amount of anthocyanins reached 87.7 mg/kg for SB1, which is significantly lower than the content of SB2, which averaged 101.6 mg/kg of grapes.

Table 3. Anthocyanin composition of Sugranineteen grapes during ripening (mg/kg, mean values ± SD).

Anthocyanins	Sample	Maturation Time (DOY)							
		213	225	234	246	254	263	278	280 (Harvest)
Dp	SB1	${}^{B}0.5^{d} \pm 0.2$ *	${}^{A}2.0^{a} \pm 0.2$	${}^{B}1.0^{c} \pm 0.1$	${}^{B}1.4^{b} \pm 0.2$	${}^{A}1.7^{ab} \pm 0.1$	${}^{A}0.7^{cd} \pm 0.2$	${}^{B}0.3^{d} \pm 0.1$	${}^{A}1.2^{bc} \pm 0.1$
	SB2	${}^{A}1.4^{c} \pm 0.1$	${}^{A}1.9^{b} \pm 0.3$	${}^{A}1.7^{bc} \pm 0.3$	${}^{A}2.7^{a} \pm 0.4$	${}^{B}0.7^{d} \pm 0.2$	${}^{A}0.5^{d} \pm 0.2$	${}^{A}1.5^{bc} \pm 0.5$	${}^{A}1.4^{c} \pm 0.1$
Cy	SB1	${}^{B}2.2^{d} \pm 0.4$	${}^{A}2.9^{cd} \pm 0.4$	${}^{A}1.9^{d} \pm 0.3$	${}^{A}2.6^{cd} \pm 0.7$	${}^{A}3.1^{c} \pm 0.4$	${}^{A}1.2^{e} \pm 0.1$	${}^{A}10.6^{b} \pm 0.1$	${}^{A}14.9^{a} \pm 0.3$
	SB2	${}^{A}5.9^{b} \pm 1.5$	${}^{A}2.5^{c} \pm 0.2$	${}^{A}2.4^{c} \pm 0.5$	${}^{A}2.2^{c} \pm 0.1$	${}^{B}1.0^{d} \pm 0.3$	${}^{B}0.7^{d} \pm 0.1$	${}^{B}5.3^{b} \pm 1.7$	${}^{B}8.8^{a} \pm 0.6$
Pt	SB1	${}^{B}1.0^{d} \pm 0.2$	${}^{A}2.5^{a} \pm 0.2$	${}^{B}1.3^{cd} \pm 0.2$	${}^{B}2.1^{ab} \pm 0.2$	${}^{A}2.2^{a} \pm 0.1$	${}^{A}1.6^{c} \pm 0.1$	${}^{A}1.3^{cd} \pm 0.2$	${}^{B}1.9^{b} \pm 0.1$
	SB2	${}^{A}2.4^{b} \pm 0.1$	${}^{A}2.3^{b} \pm 0.3$	${}^{A}2.4^{b} \pm 0.3$	${}^{A}3.5^{a} \pm 0.5$	${}^{B}0.9^{c} \pm 0.3$	${}^{B}0.3^{d} \pm 0.2$	${}^{A}2.0^{b} \pm 0.6$	${}^{A}2.5^{b} \pm 0.1$
Pn	SB1	${}^{B}16.4^{cd} \pm 3.9$	${}^{A}21.5^{c} \pm 2.3$	${}^{A}21.5^{c} \pm 4.5$	${}^{A}19.5^{cd} \pm 5.0$	${}^{A}16.3^{d} \pm 0.1$	${}^{A}11.4^{e} \pm 0.1$	${}^{A}31.7^{b} \pm 1.3$	${}^{B}37.5^{a} \pm 0.5$
	SB2	${}^{A}34.2^{b} \pm 3.3$	${}^{B}15.4^{d} \pm 1.6$	${}^{A}26.9^{bc} \pm 4.8$	${}^{A}16.7^{d} \pm 0.9$	${}^{B}7.1^{e} \pm 2.1$	${}^{A}10.0^{e} \pm 2.1$	${}^{B}22.6^{c} \pm 3.2$	${}^{A}42.4^{a} \pm 0.5$
Mv	SB1	${}^{B}10.5^{c} \pm 1.5$	${}^{A}23.0^{a} \pm 1.3$	${}^{B}15.8^{bc} \pm 3.7$	${}^{B}23.1^{a} \pm 2.4$	${}^{A}21.0^{ab} \pm 2.1$	${}^{A}15.6^{bc} \pm 0.8$	${}^{A}11.5^{c} \pm 1.8$	${}^{B}17.4^{b} \pm 2.1$
	SB2	${}^{A}23.0^{b} \pm 2.1$	${}^{A}19.7^{bc} \pm 3.0$	${}^{A}24.1^{b} \pm 2.9$	${}^{A}32.5^{a} \pm 1.6$	${}^{B}8.4^{c} \pm 2.5$	${}^{A}14.1^{c} \pm 2.4$	${}^{A}15.8^{c} \pm 5.0$	${}^{A}24.7^{b} \pm 0.9$
Pn-Ac	SB1	${}^{A}0.2^{c} \pm 0.1$	${}^{A}0.3^{d} \pm 0.1$	${}^{A}0.4^{c} \pm 0.1$	${}^{A}0.7^{b} \pm 0.1$	${}^{A}0.8^{b} \pm 0.1$	${}^{A}0.6^{bc} \pm 0.1$	${}^{A}0.8^{b} \pm 0.1$	${}^{A}1.1^{a} \pm 0.1$
	SB2	${}^{A}0.7^{b} \pm 0.3$	${}^{A}0.6^{b} \pm 0.2$	${}^{A}0.5^{c} \pm 0.1$	${}^{A}0.9^{b} \pm 0.1$	${}^{B}0.4^{c} \pm 0.1$	${}^{B}0.3^{c} \pm 0.1$	${}^{A}0.9^{ab} \pm 0.3$	${}^{A}1.2^{a} \pm 0.1$
Mv-Ac	SB1	${}^{B}0.3^{c} \pm 0.1$	${}^{A}0.8^{ab} \pm 0.1$	${}^{B}0.5^{bc} \pm 0.1$	${}^{B}0.8^{ab} \pm 0.2$	${}^{A}0.9^{a} \pm 0.1$	${}^{A}0.9^{a} \pm 0.1$	${}^{B}0.4^{ab} \pm 0.1$	${}^{B}0.6^{b} \pm 0.1$
	SB2	${}^{A}0.8^{bc} \pm 0.3$	${}^{A}0.6^{c} \pm 0.1$	${}^{A}1.0^{b} \pm 0.1$	${}^{A}1.5^{a} \pm 0.1$	${}^{B}0.3^{d} \pm 0.1$	${}^{A}0.7^{c} \pm 0.1$	${}^{A}0.8^{bc} \pm 0.2$	${}^{A}1.1^{b} \pm 0.1$
Pn-Cf	SB1	${}^{B}0.1^{e} \pm 0.1$	${}^{A}0.7^{cd} \pm 0.1$	${}^{A}0.3^{de} \pm 0.1$	${}^{A}0.5^{d} \pm 0.1$	${}^{A}0.9^{c} \pm 0.1$	${}^{A}0.5^{cd} \pm 0.1$	${}^{A}1.3^{b} \pm 0.1$	${}^{A}1.6^{a} \pm 0.1$
	SB2	${}^{A}0.6^{bc} \pm 0.3$	${}^{A}0.7^{b} \pm 0.1$	${}^{A}0.4^{c} \pm 0.1$	${}^{A}0.7^{b} \pm 0.1$	${}^{B}0.3^{b} \pm 0.1$	${}^{A}0.7^{b} \pm 0.1$	${}^{A}1.0^{ab} \pm 0.3$	${}^{B}1.2^{a} \pm 0.1$
Mv-Cf	SB1	${}^{A}0.3^{ab} \pm 0.1$	nd	${}^{A}0.1^{b} \pm 0.1$	${}^{A}0.2^{ab} \pm 0.1$	nd	${}^{A}0.4^{a} \pm 0.1$	nd	nd
	SB2	${}^{A}0.3^{b} \pm 0.1$	nd	${}^{A}0.2^{b} \pm 0.1$	${}^{A}0.2^{b} \pm 0.1$	${}^{A}0.1^{b} \pm 0.1$	${}^{A}0.8^{a} \pm 0.2$	nd	${}^{A}0.3^{b} \pm 0.1$
Cis-Pn-Cm	SB1	${}^{B}0.2^{b} \pm 0.1$	${}^{A}0.6^{a} \pm 0.1$	${}^{A}0.3^{b} \pm 0.1$	${}^{B}0.6^{a} \pm 0.1$	${}^{A}0.6^{a} \pm 0.1$	${}^{A}0.6^{a} \pm 0.1$	${}^{A}0.6^{a} \pm 0.1$	${}^{A}0.7^{a} \pm 0.1$
	SB2	${}^{A}0.6^{bc} \pm 0.2$	${}^{A}0.6^{b} \pm 0.1$	${}^{A}0.5^{bc} \pm 0.1$	${}^{A}1.0^{a} \pm 0.1$	${}^{A}0.4^{b} \pm 0.1$	${}^{B}0.3^{c} \pm 0.1$	${}^{A}0.8^{ab} \pm 0.2$	${}^{A}0.9^{ab} \pm 0.1$
cis-Mv-Cm	SB1	${}^{A}0.1^{a} \pm 0.1$	${}^{B}0.2^{a} \pm 0.1$	${}^{A}0.2^{a} \pm 0.1$	${}^{B}0.3^{a} \pm 0.1$	${}^{A}0.3^{a} \pm 0.1$	${}^{A}0.3^{a} \pm 0.1$	${}^{A}0.1^{a} \pm 0.1$	${}^{B}0.2^{a} \pm 0.1$
	SB2	${}^{A}0.3^{b} \pm 0.1$	${}^{A}0.3^{b} \pm 0.1$	${}^{A}0.2^{b} \pm 0.1$	${}^{A}0.6^{a} \pm 0.1$	${}^{B}0.1^{b} \pm 0.1$	${}^{A}0.2^{bc} \pm 0.1$	${}^{A}0.3^{b} \pm 0.1$	${}^{A}0.4^{ab} \pm 0.1$
Pn-Cm	SB1	${}^{B}1.5^{f} \pm 0.3$	${}^{A}2.9^{e} \pm 0.2$	${}^{B}2.7^{e} \pm 0.8$	${}^{B}3.7^{cd} \pm 0.9$	${}^{A}4.5^{c} \pm 0.1$	${}^{A}3.7^{d} \pm 0.1$	${}^{A}6.0^{b} \pm 0.1$	${}^{B}6.9^{a} \pm 0.1$
	SB2	${}^{A}5.6^{b} \pm 1.3$	${}^{A}2.9^{c} \pm 0.4$	${}^{A}3.9^{bc} \pm 0.6$	${}^{A}5.0^{b} \pm 0.3$	${}^{B}2.5^{c} \pm 0.7$	${}^{B}2.8^{c} \pm 0.2$	${}^{A}5.7^{b} \pm 1.8$	${}^{A}9.2^{a} \pm 0.2$
trans-Mv-Cm	SB1	${}^{B}1.2^{e} \pm 0.1$	${}^{A}3.9^{c} \pm 0.2$	${}^{B}3.5^{cd} \pm 0.9$	${}^{B}6.2^{ab} \pm 0.9$	${}^{A}6.8^{a} \pm 0.9$	${}^{A}5.5^{b} \pm 0.1$	${}^{B}2.5^{d} \pm 0.4$	${}^{B}3.9^{c} \pm 0.7$
	SB2	${}^{A}4.7^{cd} \pm 2.1$	${}^{A}3.8^{cd} \pm 0.7$	${}^{A}5.4^{c} \pm 0.5$	${}^{A}9.9^{a} \pm 0.3$	${}^{B}2.9^{d} \pm 0.9$	${}^{A}5.1^{c} \pm 0.7$	${}^{A}5.3^{c} \pm 1.5$	${}^{A}7.3^{b} \pm 0.2$
Total	SB1	${}^{B}34.5^{d} \pm 5.8$	${}^{A}61.5^{bc} \pm 4.8$	${}^{A}49.4^{cd} \pm 10.9$	${}^{B}61.6^{bc} \pm 10.6$	${}^{A}59.0^{c} \pm 2.7$	${}^{A}42.8^{d} \pm 1.3$	${}^{A}67.0^{b} \pm 4.0$	${}^{B}87.7^{a} \pm 2.4$
	SB2	${}^{A}80.5^{b} \pm 15.3$	${}^{A}51.3^{c} \pm 7.0$	${}^{A}69.5^{b} \pm 9.9$	${}^{A}77.3^{b} \pm 4.5$	${}^{B}25.1^{e} \pm 7.5$	${}^{B}36.6^{d} \pm 0.9$	${}^{A}61.8^{bc} \pm 19.7$	${}^{A}101.6^{a} \pm 0.4$

* On the same row, means with different right superscripts differ significantly ($p \leq 0.05$). Between the SB1 (farmer irrigation) and SB2 (sensor-based controlled irrigation) pair, means with different left superscripts differ significantly ($p \leq 0.05$). nd, not detected. For anthocyanins code, see the note to Figure 1.

3.4. *Changes in Cluster Fluorescence during Maturation*

The variation of Multiplex® indices during the maturation period is shown in Figure 2. The collection of fluorescent readings was performed in parallel with berry sampling from DOY 213 to 280. The nitrogen balance index (NBI_R) that is correlated to the epidermal phenolics and chlorophyll decreased until DOY 254, then increased until DOY 278 and dropped again at harvest. The simple fluorescent ratio SFR_R that is correlated to the chlorophyll content in berries had a decreasing trend without fluctuation until DOY 280 (harvest time). In contrast, the indices correlated to the anthocyanin contents AHTH_RG and the FERARI index had a similar trend over the entire maturation period, with higher ranges for the former compared to the FERARI indices, and were increasing toward harvest time.

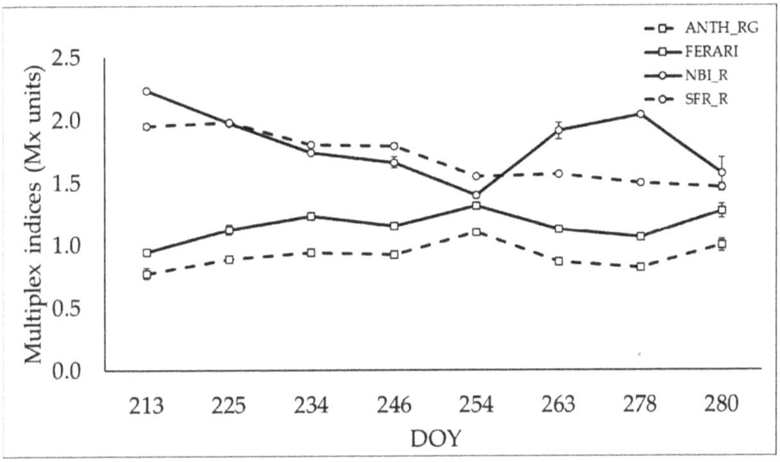

Figure 2. Changes in fluorescent indices (Mx units) of Sugranineteen clusters during ripening (day of the year, DOY).

3.5. *Relationship between Destructive and Fluorescent Measurements*

Regression model analyses were conducted to study the relationship between the different grape maturity parameters analyzed destructively and the fluorescent measurements collected by Multiplex® 3 along the maturation season. Regression equations with the corresponding coefficient of determination (R^2) are shown in Figure 3 and Table S1. The indices ANTH_RG and FERARI showed a significant relationship with skin anthocyanins ($p < 0.001$) and fitted a positive linear model, with R^2 values equal to 0.9613 and 0.8743, respectively. The NBI_R index that is correlated to the epidermal phenolics and chlorophyll revealed significant negative linear relationships with anthocyanins ($R^2 = 0.8032$; $p < 0.001$) and flavonoids ($R^2 = 0.4773$; $p = 0.039$). Among the technological maturity parameters, TA, CIRG, and TSS were significantly related to the simple fluorescent index SFR_R that is associated with the chlorophyll content of berries, with R^2 values equal to 0.6186 ($p = 0.012$), 0.7835 ($p = 0.005$), and 0.7954 ($p = 0.001$), respectively. The FLAV index that is correlated to flavonols did not show a significant relationship when compared with flavonoid content (Table S1).

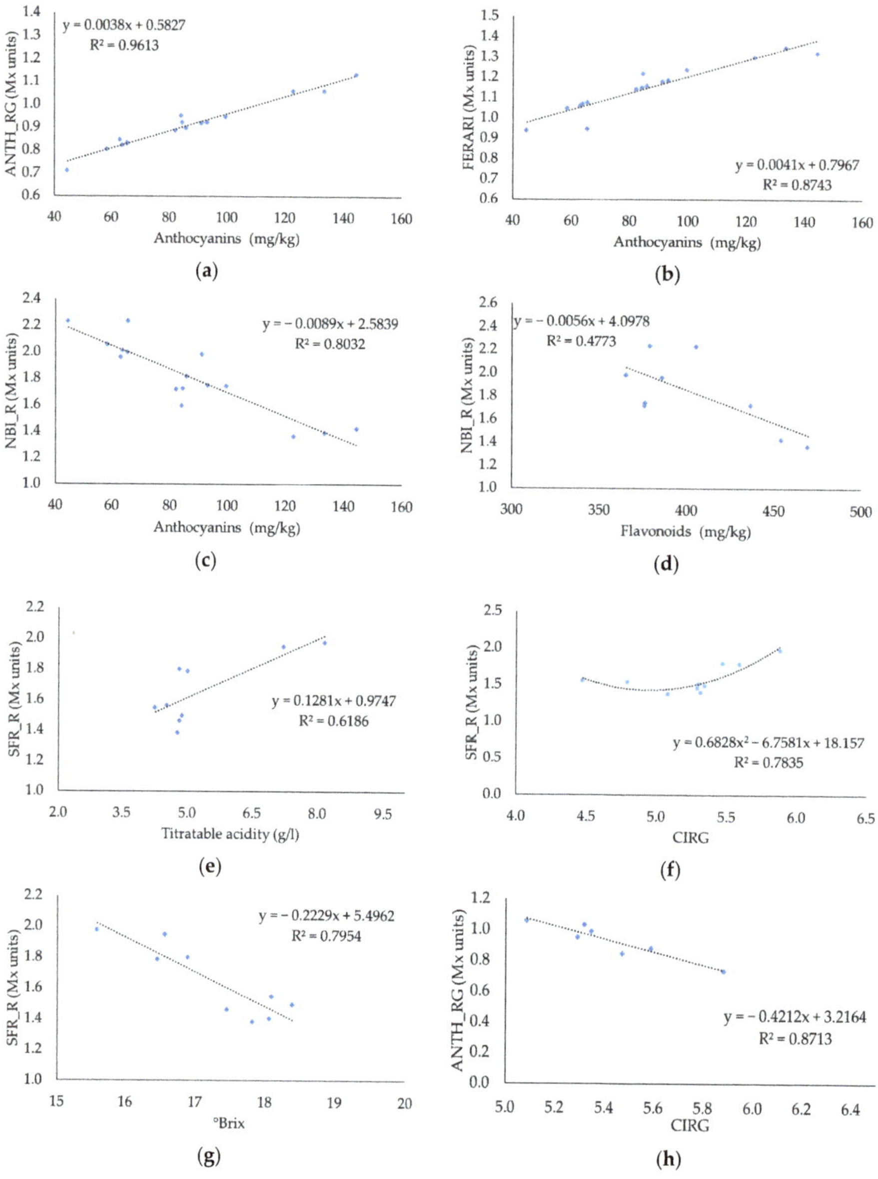

Figure 3. Regression models reflecting the relationships between different grape maturity parameters and the fluorescent measurements collected by Multiplex® 3 along the maturation season: (**a**) relationship between ANTH_RG (Mx units) and anthocyanins (mg/kg) (n = 16); (**b**) relationship between FERARI (Mx units) and anthocyanins (mg/kg) (n = 16); (**c**) relationship between NBI_R (Mx units) and anthocyanins (mg/kg) (n = 16); (**d**) relationship between NBI_R (Mx units) and flavonoids (mg/kg) (n = 9); (**e**) relationship between SFR_R (Mx units) and titratable acidity (g/L) (n = 9); (**f**) relationship between SFR_R (Mx units) and CIRG (n = 10); (**g**) relationship between SFR_R (Mx units) and TSS (°Brix) (n = 9)); (**h**) relationship between ANTH_RG (Mx units) and CIRG (n = 7). Each dot represents the mean value of triplicates of laboratory analyses and the relative 196 in-vineyard-collected fluorescent measurements.

4. Discussion

The optimal harvesting of Sugranineteen table grapes was assessed on DOY 280 based on the monitored technological maturity indices. The main indicator of the technological maturity was the TSS of grape juice that averaged at harvest 17.9 °Brix, with a corresponding low titratable acidity level (4.86 g/L). This sugar content was in line with the quality standards of Sun World International, LCC, that indicate a minimum TSS of 15.5 °Brix and the EU regulation (Reg. CE 1221/08) that claims a minimum of 14 °Brix for seedless table grapes. The skin color had a CIRG of 5.20 that reflects the red to dark-red color of the berries [35], which is the characteristic color of fully ripe Sugranineteen grapes. At harvest, all the measured indices were higher in SB2 (sensor-based controlled irrigation) than in SB1 (farmer irrigation), but no significant differences were detected except for pH (Table 1).

Regarding the phenolic maturity, total polyphenols and flavonoids fluctuated during the ripening period, while the anthocyanins had an increasing trend until harvest. The main reason behind this fluctuation, despite collecting the berries from the same grapevines at each sampling time, is the spatial and temporal heterogeneity of grape ripening in the vineyard [9,15]. This is a challenging characteristic of grapes in general because it affects the accurate monitoring of the maturity status, especially in large vineyards. The total anthocyanins reached at harvest was 144.6 mg/kg in SB2, and this was expected due to the high amounts of flavonoids (508.8 mg/kg) and total polyphenols (851.4 mg/kg). A study conducted on seedless table grapes from the Apulia region to quantify their total flavonoids revealed comparable results in 2007 for the black mid- and late-season cultivars Summer Royal and Autumn Royal, with 540 and 450 mg/kg, respectively. All the measured phenolic compounds were significantly higher in SB2 than in SB1 at harvest (Table 2), indicating a major maturation degree of grapes in controlled irrigation conditions and demonstrating the chemical composition (Table 1). The phenolic compounds are responsible for the organoleptic and qualitative characteristics of the fruit. Grapes with high levels of polyphenolic compounds appear redder or darker due to the presence of anthocyanins, which are colored flavonoid-type polyphenols mainly concentrated in the fruit epidermal tissues [36,37]. Except for the irrigation system, both blocks were grown using the same agricultural practices and environmental conditions. The sensor-based controlled irrigation system installed in SB2 provided an adequate amount of water to the grapevines, while the block SB1 was irrigated according to the farmer management. Following other studies, these results show the soil water content as a limiting factor affecting the overall quality of grape production [38]. The effect of different environmental factors on the accumulation of the phenolic compounds and their biosynthesis were previously demonstrated [39–41]. In general, the concentrations of polyphenols, anthocyanins, and flavonoids are influenced by the geographical site, climate conditions, soil fertility, cultivation practices, and pest management [42]. In addition to the environmental factors, the content of polyphenols varied widely between grape cultivars and growing seasons [42–45]. For instance, a study performed on Sugranineteen in 2014 showed a lower content of total polyphenols (451.4 mg/kg) and total anthocyanins (110.89 mg/kg) at harvest time [46] when compared to the reported results of season 2019 in Table 2.

In addition to being recognized as food sources of polyphenols [47], grapes are also known to be beneficial for human health, thanks to their antioxidant activity [48,49]. The present study has applied the two assays ABTS and DPPH to evaluate the antioxidant activity of Sugranineteen grapes. In parallel with total polyphenols, results fluctuated during the ripening season and were higher in SB2 than in SB1, although no statistical differences were detected. The antioxidant activity evaluated using the ABTS assay in SB2 reached 3.8 μM Trolox/kg at DOY 280, compared to 2.5 μM Trolox/kg when evaluated using the DPPH assay. Results showed that the antioxidant activity was better expressed by the ABTS method compared to DPPH, since the former revealed higher concentrations of antioxidant activity when testing the same samples along the maturation period. In fact, the ABTS assay is based on the ability of the sample to inhibit $ABTS^{+}$, while the Trolox is used as a reference antioxidant standard, and the DPPH assay is a method that evaluates the ability of the sample to scavenge against the stable chromogenic radical DPPH.

The anthocyanin-3-O-glucosides and their acetyl, caffeoyl, and coumaroyl derivatives, eluted in the order Dp < Cy < Pt < Pn < Mv, were consistent with previous reports [43,50–52]. The most abundant anthocyanins throughout the maturation period were peonidin-3-O-glucoside and malvidin-3-O-glucoside, followed by cyanidin-3-O-glucoside, peonidin-3-O-coumaroyl-glucoside, and *trans*-malvidin-3-O-coumaroyl-glucoside. A similar profile of anthocyanin peaks was detected in the red Autumn Royal cultivar [43]. These results follow other studies, where it was found that the most abundant anthocyanins present in pink- and red-colored cultivars were peonidin-3-O-glucoside, whereas malvidin-3-O-glucoside, cyanidin-3-O-glucoside, and petunidin-3-O-glucoside forms were abundant in red-black cultivars [51,53]. This means that different proportions of individual anthocyanin compounds can affect the skin color of grapes [54].

Regarding the optical fluorescent measurements, the ANTH_RG and FERARI indices that are correlated to skin anthocyanins [17] were increasing toward harvest time. In parallel, the simple fluorescence index SFR_R and the nitrogen balance index NBI_R that are correlated to the chlorophyll content of the berries [17] were decreasing with the maturation time (Figure 2). These results reflect the normal variations that happened during the growing season, where the chlorophyll content that characterizes the green berries decreases and the anthocyanin content responsible for the color of berries rises with the ripening season. The trends of ANTH_RG, FERARI, SFR_R, and NBI_R during maturation were similar to the changes in the fluorescence ratios of Thompson Seedless grapes [55] and other wine grape varieties [21].

The comparison between destructive and non-destructive methods applied to assess the quality of table grapes was fruitful. Results revealed strong relationships among the technological and phenolic maturity indices measured in the laboratory and the fluorescent readings collected with the optical sensor Multiplex® 3. This means that it has the ability to quantify the skin anthocyanin contents and other maturity indices of grape berries immediately and rapidly in-vineyard without collecting samples or using time-consuming technologies. Previous studies have described many calibrations of the Multiplex® for anthocyanins estimation in wine grapes [12,17,20,24,56–58]. The most important relationships were found between the total anthocyanins estimated by spectrophotometry and the ANTH_RG and FEARI indices with R^2 values equal to 0.9613 and 0.8743, respectively. These indices were also highly correlated to other wine grapes anthocyanins, such as Aleatico [12], Pinot Noir, Pinot Meunier, and Chardonnay [15]. In addition to these two indices, the NBI_R index that reflects the epidermal chlorophyll and anthocyanins was also correlated to total anthocyanins (R^2 = 0.8032) and flavonoids, but with a low coefficient of determination (R^2 = 0.4773). Regarding the technological maturity parameters, the simple fluorescent index SFR_R was correlated to the titratable acidity (R^2 = 0.6186), CIRG (R^2 = 0.7835), and total soluble solids (R^2 = 0.7954); this was also detected in a previous study focused on wine grape cultivars [17]. Therefore, even though Multiplex® 3 provides indices correlated to phenolic parameters, these results also revealed their good relationship with the technological indices of maturity. Only the FLAV index did not have a relationship with flavonoids, and the same results were found when it was compared to flavonols [12].

While almost all the studies were applied to wine grapes, these results revealed the new regression equations characteristic of Sugranineteen table grapes. The obtained models were generated from data collected from the veraison season (August 2019) until full maturity (October 2019); thus, they can be used for different stages of grape maturation. The relationship between the total anthocyanins evaluated through wet chemistry and the ANTH_RG index fitted a positive linear relationship. The same trend was also detected between the total anthocyanins and ANTH_RG index in Malvasia Rosa cultivar [21], while it was not the case of Tempranillo wine grape berries evaluated in the laboratory, which had a negative exponential model [20]. The prediction models depend on the cultivar and its morphological differences among grape varieties (size, weight, berry-skin thickness, and color) that may influence the fluorescent signals acquired by the Multiplex® [59]. Another study suggests that the ANTH_RG index depends on the general anthocyanin profile and every single compound of the cultivar [58]. Moreover, the cultural practices and meteorological conditions that change among seasons can induce water stress status in the vines, thus affecting berry size and anthocyanin synthesis [12,60,61]. Therefore, more studies are required to characterize the ripening development of different table grape cultivars under different seasons and conditions when analyzed using Multiplex®.

5. Conclusions

The non-destructive assessment of grape maturity has recently become a promising technique in viticulture. The results demonstrated relationships between the different quality parameters analyzed destructively and the optical non-destructive fluorescent readings of Sugranineteen table grapes. The main finding was the regression equation developed between the ANTH_RG index from Multiplex® 3 and the skin anthocyanin content, which helps to estimate this latter rapidly in-vineyard without damaging the plant material. While most of the previous studies were conducted on wine grape cultivars, this study assessed the whole maturity season of table grapes. This is a first step toward further promising studies to adjust the application of fluorescent techniques for the better estimation of the maturity status of different cultivars of table grape.

Supplementary Materials: The following supporting information can be downloaded at: https://www.mdpi.com/article/10.3390/foods11050663/s1, Table S1: Relationships between the Multiplex® 3 ratios and the different quality parameters of Sugranineteen table grapes performed in this study. The ratios are described according to the manufacturer (Force A, Orsay, France).

Author Contributions: Conceptualization, L.T. and G.G.; methodology, N.H., and G.N.; software, N.H. and G.N.; validation, L.T., V.V. and G.G.; formal analysis, N.H. and L.T.; investigation, N.H., L.T., V.V., G.N., M.F. and G.G.; data curation, N.H., V.V. and G.N.; writing—original draft preparation, N.H.; writing—review and editing, L.T, V.V., M.F. and G.G.; supervision, G.G. All authors have read and agreed to the published version of the manuscript.

Funding: This research received no external funding.

Data Availability Statement: Not applicable.

Conflicts of Interest: The authors declare no conflict of interest.

References

1. Krstic, M. *Growing Quality Grapes to Winery Specification: Quality Measurement and Management Options for Grapegrowers*; Winetitles: Adelaide, Australia, 2003.
2. Nogales-Bueno, J.; Hernández-Hierro, J.M.; Rodríguez-Pulido, F.J.; Heredia, F.J. Determination of technological maturity of grapes and total phenolic compounds of grape skins in red and white cultivars during ripening by near infrared hyperspectral image: A preliminary approach. *Food Chem.* **2014**, *152*, 586–591. [CrossRef]
3. Ferrer-Gallego, R.; Hernández-Hierro, J.M.; Rivas-Gonzalo, J.C.; Escribano-Bailón, M.T. Influence of climatic conditions on the phenolic composition of *Vitis vinifera* L. cv. Graciano. *Anal. Chim. Acta* **2012**, *732*, 73–77. [CrossRef] [PubMed]
4. Meléndez, E.; Ortiz, M.; Sarabia, L.; Íñiguez, M.; Puras, P. Modelling phenolic and technological maturities of grapes by means of the multivariate relation between organoleptic and physicochemical properties. *Anal. Chim. Acta* **2013**, *761*, 53–61. [CrossRef] [PubMed]
5. Singleton, V.L.; Rossi, J.A. Colorimetry of total phenolics with phosphomolybdic-phosphotungstic acid reagents. *Am. J. Enol. Vitic.* **1965**, *16*, 144–158.
6. Boulton, R. The copigmentation of anthocyanins and its role in the color of red wine: A critical review. *Am. J. Enol. Vitic.* **2001**, *52*, 67–87.
7. Xu, Y.; Simon, J.E.; Welch, C.; Wightman, J.D.; Ferruzzi, M.G.; Ho, L.; Passinetti, G.M.; Wu, Q. Survey of polyphenol constituents in grapes and grape-derived products. *J. Agric. Food Chem.* **2011**, *59*, 10586–10593. [CrossRef]
8. Vinas, P.; Campillo, N.; Martínez-Castillo, N.; Hernández-Córdoba, M. Solid-phase microextraction on-fiber derivatization for the analysis of some polyphenols in wine and grapes using gas chromatography–mass spectrometry. *J. Chromatogr. A* **2009**, *1216*, 1279–1284. [CrossRef]
9. Kontoudakis, N.; Esteruelas, M.; Fort, F.; Canals, J.M.; De Freitas, V.; Zamora, F. Influence of the heterogeneity of grape phenolic maturity on wine composition and quality. *Food Chem.* **2011**, *124*, 767–774. [CrossRef]

10. Lorrain, B.; Ky, I.; Pechamat, L.; Teissedre, P.L. Evolution of analysis of polyhenols from grapes, wines, and extracts. *Molecules* **2013**, *18*, 1076–1100. [CrossRef]
11. Costa, G.; Noferini, M.; Fiori, G.; Torrigiani, P. Use of Vis/NIR spectroscopy to assess fruit ripening stage and improve management in post-harvest chain. *Fresh Prod.* **2009**, *1*, 35–41.
12. Tuccio, L.; Remorini, D.; Pinelli, P.; Fierini, E.; Tonutti, P.; Scalabrelli, G.; Agati, G. Rapid and non-destructive method to assess in the vineyard grape berry anthocyanins under different seasonal and water conditions. *Aust. J. Grape Wine Res.* **2011**, *17*, 181–189. [CrossRef]
13. Giovenzana, V.; Beghi, R.; Malegori, C.; Civelli, R.; Guidetti, R. Wavelength selection with a view to a simplified handheld optical system to estimate grape ripeness. *Am. J. Enol. Vitic.* **2014**, *65*, 117–123. [CrossRef]
14. Bramley, R.G.V. Understanding variability in winegrape production systems 2. Within vineyard variation in quality over several vintages. *Aust. J. Grape Wine Res.* **2005**, *11*, 33–42. [CrossRef]
15. Cerovic, Z.G.; Moise, N.; Agati, G.; Latouche, G.; Ghozlen, N.B.; Meyer, S. New portable optical sensors for the assessment of winegrape phenolic maturity based on berry fluorescence. *J. Food Compos. Anal.* **2008**, *21*, 650–654. [CrossRef]
16. Kolb, C.A.; Kopecký, J.; Riederer, M.; Pfündel, E.E. UV screening by phenolics in berries of grapevine (*Vitis vinifera*). *Funct. Plant Biol.* **2003**, *30*, 1177–1186. [CrossRef] [PubMed]
17. Ben Ghozlen, N.; Cerovic, Z.G.; Germain, C.; Toutain, S.; Latouche, G. Non-destructive optical monitoring of grape maturation by proximal sensing. *Sensors* **2010**, *10*, 10040–10068. [CrossRef] [PubMed]
18. Kolb, C.A.; Pfündel, E.E. Origins of non-linear and dissimilar relationships between epidermal UV absorbance and UV absorbance of extracted phenolics in leaves of grapevine and barley. *Plant Cell Environ.* **2005**, *28*, 580–590. [CrossRef]
19. Agati, G.; Meyer, S.; Matteini, P.; Cerovic, Z.G. Assessment of anthocyanins in grape (*Vitis vinifera* L.) berries using a noninvasive chlorophyll fluorescence method. *J. Agric. Food Chem.* **2007**, *55*, 1053–1061. [CrossRef]
20. Baluja, J.; Diago, M.; Goovaerts, P.; Tardaguila, J. Assessment of the spatial variability of anthocyanins in grapes using a fluorescence sensor: Relationships with vine vigour and yield. *Precis. Agric.* **2012**, *13*, 457–472. [CrossRef]
21. Savi, S.; Poni, S.; Moncalvo, A.; Frioni, T.; Rodschinka, I.; Arata, L.; Gatti, M. Destructive and optical non-destructive grape ripening assessment: Agronomic comparison and cost-benefit analysis. *PLoS ONE* **2019**, *14*, e0216421. [CrossRef]
22. Agati, G.; Cerovic, Z.G.; Pinelli, P.; Tattini, M. Light-induced accumulation of ortho-dihydroxylated flavonoids as non-destructively monitored by chlorophyll fluorescence excitation techniques. *Environ. Exp. Bot.* **2011**, *73*, 3–9. [CrossRef]
23. Pedrós, R.; Goulas, Y.; Jacquemoud, S.; Louis, J.; Moya, I. FluorMODleaf: A new leaf fluorescence emission model based on the PROSPECT model. *Remote Sens. Environ.* **2010**, *114*, 155–167. [CrossRef]
24. Ben Ghozlen, N.; Moise, N.; Latouche, G.; Martinon, V.; Mercier, L.; Besançon, E.; Cerovic, Z. Assessment of grapevine maturity using a new portable sensor: Non-destructive quantification of anthocyanins. *J. Int. Sci. Vigne Vin* **2010**, *44*, 1–8.
25. Cerovic, Z.G.; Goutouly, J.P.; Hilbert, G.; Destrac-Irvine, A.; Martinon, V.; Moise, N. Mapping winegrape quality attributes using portable fluorescence-based sensors. *Frutic* **2009**, *9*, 301–310.
26. Carreño, J.; Martínez, A.; Almela, L.; Fernández-López, J. Proposal of an index for the objective evaluation of the colour of red table grapes. *Food Res. Int.* **1995**, *28*, 373–377. [CrossRef]
27. Fernández-López, J.A.; Almela, L.; Muñoz, J.A.; Hidalgo, V.; Carreño, J. Dependence between colour and individual anthocyanin content in ripening grapes. *Food Res. Int.* **1998**, *31*, 667–672. [CrossRef]
28. Gambacorta, G.; Antonacci, D.; La Gatta, M.; Faccia, M.; La Gatta, B.; Pati, S.; Coletta, A.; La Notte, E. Phenolic composition of Aglianico and Nero di Troia grapes and wines as affected by cover cropping and irrigation. *Ital. J. Food Sci.* **2011**, *23*, 381.
29. Re, R.; Pellegrini, N.; Proteggente, A.; Pannala, A.; Yang, M.; Rice-Evans, C. Antioxidant activity applying an improved ABTS radical cation decolorization assay. *Free Radic. Biol. Med.* **1999**, *26*, 1231–1237. [CrossRef]
30. Brand-Williams, W.; Cuvelier, M.E.; Berset, C. Use of a free radical method to evaluate antioxidant activity. *LWT Food Sci. Technol.* **1995**, *28*, 25–30. [CrossRef]
31. Coletta, A.; Trani, A.; Faccia, M.; Punzi, R.; Dipalmo, T.; Crupi, P.; Antonacci, D.; Gambacorta, G. Influence of viticultural practices and winemaking technologies on phenolic composition and sensory characteristics of Negroamaro red wines. *Int. J. Food Sci.* **2013**, *48*, 2215–2227. [CrossRef]
32. Revilla, E.; Ryan, J.M. Analysis of several phenolic compounds with potential antioxidant properties in grape extracts and wines by high-performance liquid chromatography-photodiode array detection without sample preparation. *J. Chromatogr. A* **2000**, *881*, 461–469. [CrossRef]
33. Brar, H.S.; Singh, Z.; Swinny, E. Dynamics of anthocyanin and flavonol profiles in the 'Crimson Seedless' grape berry skin during development and ripening. *Sci. Hortic.* **2008**, *117*, 349–356. [CrossRef]
34. De la Cruz, A.A.; Hilbert, G.; Rivière, C.; Mengin, V.; Ollat, N.; Bordenave, L.; Decroocq, S.; Delaunay, J.C.; Delrot, S.; Mérillon, J.M. Anthocyanin identification and composition of wild Vitis spp. accessions by using LC–MS and LC–NMR. *Anal. Chim. Acta* **2012**, *732*, 145–152. [CrossRef] [PubMed]
35. Carreño, J.; Martínez, A.; Almela, L.; Fernández-López, J.A. Measuring the color of table grapes. *Color Res. Appl.* **1996**, *21*, 50–54. [CrossRef]
36. Crupi, P.; Palattella, D.; Corbo, F.; Clodoveo, M.L.; Masi, G.; Caputo, A.R.; Battista, F.; Tarricone, L. Effect of pre-harvest inactivated yeast treatment on the anthocyanin content and quality of table grapes. *Food Chem.* **2021**, *337*, 128006. [CrossRef]

37. Brouillard, R.; Figueiredo, P.; Elhabiri, M.; Dangles, O. Molecular interactions of phenolic compounds in relation to the colour of fruit and vegetables. In *Phytochemistry of Fruit and Vegetables*; Tomás-Barberán, F.A., Robins, R.J., Eds.; Oxford Science Publications: Oxford, UK, 1997; pp. 29–49.
38. Pérez-Álvarez, E.P.; Molina, D.I.; Vivaldi, G.A.; García-Esparza, M.J.; Lizama, V.; Álvarez, I. Effects of the irrigation regimes on grapevine cv. Bobal in a Mediterranean climate: I. Water relations, vine performance and grape composition. *Agric Water Manag.* **2021**, *248*, 106772. [CrossRef]
39. Bergqvist, J.; Dokoozlian, N.; Ebisuda, N. Sunlight exposure and temperature effects on berry growth and composition of Cabernet Sauvignon and Grenache in the Central San Joaquin Valley of California. *Am. J. Enol. Vitic.* **2001**, *52*, 1–7.
40. Mori, K.; Sugaya, S.; Gemma, H. Decreased anthocyanin biosynthesis in grape berries grown under elevated night temperature condition. *Sci. Hortic.* **2005**, *105*, 319–330. [CrossRef]
41. Fujita, A.; Soma, N.; Goto-Yamamoto, N.; Mizuno, A.; Kiso, K.; Hashizume, K. Effect of shading on proanthocyanidin biosynthesis in grape berry. *J. Jpn. Soc. Hort. Sci.* **2007**, *76*, 112–119. [CrossRef]
42. De la Cerda-Carrasco, A.; López-Solís, R.; Nuñez-Kalasic, H.; Peña-Neira, Á.; Obreque-Slier, E. Phenolic composition and antioxidant capacity of pomaces from four grape varieties (*Vitis vinifera* L.). *J. Sci. Food Agric.* **2015**, *95*, 1521–1527. [CrossRef]
43. Crupi, P.; Coletta, A.; Anna Milella, R.; Perniola, R.; Gasparro, M.; Genghi, R.; Antonacci, D. HPLC-DAD-ESI-MS Analysis of Flavonoid Compounds in 5 Seedless Table Grapes Grown in Apulian Region. *J. Food Sci.* **2012**, *77*, C174–C181. [CrossRef] [PubMed]
44. Vilanova, M.; Santalla, M.; Masa, A. Environmental and genetic variation of phenolic compounds in grapes (*Vitis vinifera*) from northwest Spain. *J. Agric. Sci.* **2009**, *147*, 683–697. [CrossRef]
45. Hornedo-Ortega, R.; González-Centeno, M.R.; Chira, K.; Jourdes, M.; Teissedre, P.L. Phenolic Compounds of Grapes and Wines: Key Compounds and Implications in Sensory Perception. In *Winemaking-Stabilization, Aging Chemistry and Biochemistry*; IntechOpen: Rijeka, Croatia, 2020.
46. Admane, N.; Genovese, F.; Altieri, G.; Tauriello, A.; Trani, A.; Gambacorta, G.; Verrastro, V.; Di Renzo, G.C. Effect of ozone or carbon dioxide pre-treatment during long-term storage of organic table grapes with modified atmosphere packaging. *LWT* **2018**, *98*, 170–178. [CrossRef]
47. Macheix, J.; Fleuriet, A.; Billot, J. The main phenolics of fruits. In *Fruit Phenolics*, 1st ed.; CRC Press: Boca Raton, FL, USA, 1990; pp. 1–103.
48. Milella, R.A.; Antonacci, D.; Crupi, P.; Incampo, F.; Carrieri, C.; Semeraro, N.; Colucci, M. Skin extracts from 2 Italian table grapes (Italia and Palieri) inhibit tissue factor expression by human blood mononuclear cells. *J. Food Sci.* **2012**, *77*, H154–H159. [CrossRef]
49. Coletta, A.; Berto, S.; Crupi, P.; Cravero, M.C.; Tamborra, P.; Antonacci, D.; Daniele, P.G.; Prenesti, E. Effect of viticulture practices on concentration of polyphenolic compounds and total antioxidant capacity of Southern Italy red wines. *Food Chem.* **2014**, *152*, 467–474. [CrossRef]
50. Heier, A.; Blaas, W.; Droß, A.; Wittkowski, R. Anthocyanin analysis by HPLC/Esi-MS. *Am. J. Enol. Vitic.* **2002**, *53*, 78–86.
51. Ryan, J.M.; Revilla, E. Anthocyanin composition of Cabernet Sauvignon and Tempranillo grapes at different stages of ripening. *J. Agric. Food Chem.* **2003**, *51*, 3372–3378. [CrossRef]
52. Wang, H.; Race, E.J.; Shrikhande, A.J. Anthocyanin transformation in Cabernet Sauvignon wine during aging. *J. Agric. Food Chem.* **2003**, *51*, 7989–7994. [CrossRef]
53. Liang, Z.; Owens, C.L.; Zhong, G.Y.; Cheng, L. Polyphenolic profiles detected in the ripe berries of *Vitis vinifera* germplasm. *Food Chem.* **2011**, *129*, 940–950. [CrossRef]
54. Ferrara, G.; Mazzeo, A.; Matarrese, A.M.S.; Pacucci, C.; Punzi, R.; Faccia, M.; Trani, A.; Gambacorta, G. Application of abscisic acid (S-ABA) and sucrose to improve colour, anthocyanin content and antioxidant activity of cv. Crimson Seedless grape berries. *Aust. J. Grape Wine Res.* **2015**, *21*, 18–29. [CrossRef]
55. Bahar, A.; Kaplunov, T.; Zutahy, Y.; Daus, A.; Lurie, S.; Lichter, A. Auto-fluorescence for analysis of ripening in Thompson Seedless and colour in Crimson Seedless table grapes. *Aust. J. Grape Wine Res.* **2012**, *18*, 353–359. [CrossRef]
56. Bramley, R.; Le Moigne, M.; Evain, S.; Ouzman, J.; Florin, L.; Fadaili, E.; Hinze, C.; Cerovic, Z. On-the-go sensing of grape berry anthocyanins during commercial harvest: Development and prospects. *Aust. J. Grape Wine Res.* **2011**, *17*, 316–326. [CrossRef]
57. Agati, G.; D'Onofrio, C.; Ducci, E.; Cuzzola, A.; Remorini, D.; Tuccio, L.; Lazzini, F.; Mattii, G. Potential of a multiparametric optical sensor for determining in situ the maturity components of red and white *Vitis vinifera* wine grapes. *J. Agric. Food Chem.* **2013**, *61*, 12211–12218. [CrossRef]
58. Ferrandino, A.; Pagliarani, C.; Carlomagno, A.; Novello, V.; Schubert, A.; Agati, G. Improved fluorescence-based evaluation of flavonoid in red and white winegrape cultivars. *Aust. J. Grape Wine Res.* **2017**, *23*, 207–214. [CrossRef]
59. Pinelli, P.; Romani, A.; Fierini, E.; Agati, G. Prediction models for assessing anthocyanins in grape berries by fluorescence sensors: Dependence on cultivar, site and growing season. *Food Chem.* **2018**, *244*, 213–223. [CrossRef] [PubMed]
60. Ojeda, H.; Andary, C.; Kraeva, E.; Carbonneau, A.; Deloire, A. Influence of pre-and postveraison water deficit on synthesis and concentration of skin phenolic compounds during berry growth of *Vitis vinifera* cv. *Shiraz. Am. J. Enol. Vitic.* **2002**, *53*, 261–267.
61. Deluc, L.G.; Quilici, D.R.; Decendit, A.; Grimplet, J.; Wheatley, M.D.; Schlauch, K.A.; Mérillon, J.M.; Cushman, J.C.; Cramer, G.R. Water deficit alters differentially metabolic pathways affecting important flavor and quality traits in grape berries of Cabernet Sauvignon and Chardonnay. *BMC Genom.* **2009**, *10*, 212. [CrossRef]

foods

MDPI

Article

Large-Scale, High-Throughput Phenotyping of the Postharvest Storage Performance of 'Rustenburg' Navel Oranges and the Development of Shelf-Life Prediction Models

Abiola Owoyemi [1,2], Ron Porat [1,*], Amnon Lichter [1], Adi Doron-Faigenboim [3], Omri Jovani [4,5], Noam Koenigstein [4] and Yael Salzer [5]

1 Department of Postharvest Science of Fresh Produce, ARO, The Volcani Institute, Rishon LeZion 7528809, Israel; abiola.owoyemi@mail.huji.ac.il (A.O.); vtlicht@volcani.agri.gov.il (A.L.)
2 Robert H. Smith Faculty of Agricultural, Food and Environmental Sciences, Hebrew University of Jerusalem, Rehovot 76100, Israel
3 Genomics and Bioinformatics Unit, ARO, The Volcani Institute, Rishon LeZion 7528809, Israel; adif@volcani.agri.gov.il
4 Department of Industrial Engineering, Tel Aviv University, P.O. Box 39040, Tel Aviv 6997801, Israel; omrijovani@gmail.com (O.J.); noamk@tauex.tau.ac.il (N.K.)
5 Department of Growing, Production and Environmental Engineering, ARO, The Volcani Institute, Rishon LeZion 7528809, Israel; salzer@volcani.agri.gov.il
* Correspondence: rporat@volcani.agri.gov.il; Tel.: 972-50-6220615

Citation: Owoyemi, A.; Porat, R.; Lichter, A.; Doron-Faigenboim, A.; Jovani, O.; Koenigstein, N.; Salzer, Y. Large-Scale, High-Throughput Phenotyping of the Postharvest Storage Performance of 'Rustenburg' Navel Oranges and the Development of Shelf-Life Prediction Models. *Foods* **2022**, *11*, 1840. https://doi.org/10.3390/foods11131840

Academic Editors: Bernardo Pace and Maria Cefola

Received: 23 May 2022
Accepted: 16 June 2022
Published: 22 June 2022

Abstract: We conducted a large-scale, high-throughput phenotyping analysis of the effects of various pre-harvest and postharvest features on the quality of 'Rustenburg' navel oranges, in order to develop shelf-life prediction models to enable the use of the First Expired, First Out logistics strategy. The examined pre-harvest features included harvest time and yield, and the examined postharvest features included storage temperature, relative humidity during storage and duration of storage. All together, we evaluated 12,000 oranges (~4 tons) from six different orchards and conducted 170,576 measurements of 14 quality parameters. Storage time was found to be the most important feature affecting fruit quality, followed by storage temperature, harvest time, yield and humidity. The examined features significantly affected ($p < 0.001$) fruit weight loss, firmness, decay, color, peel damage, chilling injury, internal dryness, acidity, vitamin C and ethanol levels, and flavor and acceptance scores. Four regression models were evaluated for their ability to predict fruit quality based on pre-harvest and postharvest features. Extreme gradient boosting (XGBoost) combined with a duplication approach was found to be the most effective approach. It allowed for the prediction of fruit-acceptance scores among the full data set, with a root mean square error (RMSE) of 0.217 and an R^2 of 0.891.

Keywords: citrus; intelligent logistics; modeling; orange; postharvest

1. Introduction

Orange (*Citrus sinensis*) is the world's fifth largest fruit crop (after banana, watermelons, apples and grapes), with an annual global production of 75.4 million tons [1]. Oranges, similar to other citrus fruits, are very popular and consumers appreciate their delicate, fruity and refreshing flavor, as well as their high nutritional value [2].

'Rustenburg' navel orange, which originated in South Africa, is a high-quality late-season orange that is seedless, relatively easy to peel, has a rich fruity flavor and is excellent for fresh consumption [3]. For these reasons, it is a widely cultivated late-season cultivar in several citrus-growing countries, including Israel, and 'Rustenburg' fruits are commercially stored for relatively long periods of up to 3–4 months, in order to extend the marketing season [4].

The postharvest storage performance of oranges may be affected by various pre-harvest and postharvest features [5]. Pre-harvest features such as climate conditions, cultivation practices, harvest time, choice of rootstock, tree age and yield may affect postharvest quality [6,7]. For example, during ripening, the fruit undergoes continuous metabolic changes, such as the accumulation of sugars and a decrease in acidity levels, which cause changes in fruit flavor and nutritional quality over the course of the ripening period [8]. Furthermore, early-harvested oranges are relatively sensitive to postharvest chilling injury and more tolerant to postharvest decay, whereas late-harvested fruits are more tolerant to chilling, but more sensitive to microbial spoilage [9]. It has also been reported that the maturity stage at harvest may affect peel pitting during storage [10]. Other pre-harvest features, such as tree age and yield, may affect the sugar and acidity levels, firmness and physiological behaviors, manifested as respiration and the rate of ethylene production [11].

The postharvest storage performance of oranges is greatly influenced by the environmental conditions under which the fruits are stored, especially temperature and relative humidity (RH) [12]. Temperature is the most important environmental factor affecting fruit quality after harvest, as low storage temperatures reduce respiration, water loss and pathogen growth, but can also cause chilling injury (CI) [12]. The optimal RH for postharvest storage of most fruit species is between 90–95%. Lower RH levels lead to greater water loss and RH levels that are too high may lead to condensation and the enhanced growth of pathogens [13]. Overall, it is recommended that oranges be stored at 3–8 °C and 90–95% RH [12].

Currently, the postharvest storage management of fruits and vegetables is principally governed by the First In, First Out (FIFO) logistics strategy, meaning that marketing decisions are based solely on storage time irrespective of the initial quality of the produce and its remaining potential shelf life [14,15]. Although the FIFO approach is straightforward and easy to implement, it is based on the assumption that all products arriving at the cold-storage facility on a particular date have the same shelf-life potential, which all too often is not the actual case [14]. In contrast, the adoption of the more advanced First Expired, First Out (FEFO) logistics strategy would enable more efficient inventory management based on shelf-life predictions for each particular batch of produce, to ensure that only high-quality produce will reach the distinct marketing destinations [15]. However, the adoption of an intelligent FEFO logistics-management system requires the development of accurate shelf-life prediction models that will provide reliable information regarding the remaining shelf-life of each batch of produce.

In recent years, various models have been developed to predict numerous postharvest storage traits, including respiration rate, microbial growth, physical, chemical, and sensorial characteristics, and storage life [16–19]. Advanced machine learning and artificial intelligence technologies now allow the development of even more accurate forecasting and prediction models for important agriculture outputs [20,21]. Nonetheless, the development of accurate and advanced shelf-life prediction models requires the acquisition of large amounts of postharvest storage data and the consideration of all of the factors that may affect produce quality, including various pre-harvest and postharvest features [22,23].

Overall, the main objective of the current research was to conduct a large-scale, high-throughput phenotyping analysis of the postharvest storage performance of 'Rustenburg' navel oranges, in order to develop shelf-life prediction models to enable the implementation of the FEFO method. To that end, we evaluated 12,000 oranges (~4 tons) harvested from six different orchards and conducted 170,576 measurements of 14 quality parameters.

2. Materials and Methods

2.1. Plant Material

'Rustenburg' navel oranges were harvested from six commercial orchards on different dates between 21 February 2021 and 6 April 2022, as described in Table 1. The rational for choosing different harvest times was to examine the postharvest performances of fruit with different maturity indices, i.e., different total soluble solids (TSS), acidity, peel color, etc. The day after harvest, the fruits were treated in a commercial citrus packinghouse. This treatment included washing, waxing, the application of fungicides, sorting and packaging according to common commercial practices. Then, the fruits were transferred to the Department of Postharvest Science, ARO, The Volcani Institute, where they were placed in cold storage rooms as described below.

Table 1. Harvest times and yields for the six different orchards used for the current experiment.

	Harvest Time (Weeks from Blooming)	Yield (Ton/Hectare)
Harvest 1 (21 February 2021)	48	48
Harvest 2 (24 February 2021)	48	44
Harvest 3 (28 February 2021)	49	18
Harvest 4 (17 March 2021)	51	35
Harvest 5 (25 March 2021)	52	18
Harvest 6 (6 April 2021)	54	33

2.2. Postharvest Storage Conditions

Fruits from each orchard were stored at 90% RH and at 3 different storage temperatures: 2, 5 and 8 °C. Fruits from Harvest 5 (Table 1) were also stored at 5 °C under high humidity (RH = 95%) and low humidity conditions (RH = 70%). The high RH was achieved by using a PARKOO PH06LB ultrasonic humidifier (ANIA, Kfar Aviv, Israel) and the low RH was achieved by using an IVLTD08 dehumidifier (Vogel Refrigeration Services, Ltd., Rishon LeZion, Israel). The normal RH at 5 °C without modifications was ~90%. Each harvest included 60 10 kg cartons of oranges (20 cartons were stored at each temperature), while the Harvest 5 included 100 10 kg cartons of oranges (20 cartons stored at each temperature and/or RH condition). Overall, the experiment included 400 10 kg cartons of oranges (i.e., a total of 4 tons of oranges).

2.3. Evaluations of Fruit Quality

Evaluations of fruit quality were conducted at Time 0 and at weekly intervals over a period of 20 weeks (~4.5 months). The quality evaluations were conducted after one additional week of storage under shelf-life conditions at 20 °C. The different quality evaluations are described below.

2.3.1. Firmness

Firmness was tested with a texture analyzer (TA-XT plus, Stable Micro Systems, Surrey, UK) with a 50-kg load cell, using a 75 mm (diam.) cylindrical probe. The machine compressed the samples (15 replicates for each treatment and storage time) in the equatorial zone until 5% deformation at a speed of 1 $mm \cdot s^{-1}$. Results were expressed as the force required to induce that level of deformation.

2.3.2. Weight Loss

Weight loss was evaluated by weighing the produce at Time 0 and then again after the different storage periods. Weight-loss data are expressed as percentages of weight lost relative to the initial weight.

2.3.3. Peel Color

Peel color was determined by measuring lightness (L*), chroma (C*) and hue angle (H°) values, using a Minolta Chromo Meter, Model CR-400 (Minolta, Tokyo, Japan). The presented data are means of 15 measurements.

2.3.4. Peel Damage, Decay and Internal Dryness

Peel damage, the incidence of decay and internal dryness were evaluated for the different storage periods following the manual sorting of the produce. Results are expressed as the percentage of infected fruits among the total amount of produce.

2.3.5. Total Soluble Solids (TSS) and Titratable Acidity (TA)

The total soluble solids (TSS) contents of the extracted juice were determined with a Model PAL-1 digital refractometer (Atago, Tokyo, Japan), and acidity levels (%) were measured by titration to pH 8.3 with 0.1 M NaOH using a Model CH-9101 automatic titrator (Metrohm, Herisau, Switzerland). Each measurement was replicated three times, with juice collected from three fruits used each time.

2.3.6. Vitamin C

The vitamin C (ascorbic acid) content of the orange juice was determined by titration with 2,6-dichlorophenolindophenol, as described previously [24]. Levels of ascorbic acid were determined by comparing the titration volumes of the fruit juices with those of a standard solution containing 0.1% ascorbic acid.

2.3.7. Ethanol Levels

Juice ethanol concentrations were determined as described by Davis and Chace [25]. Generally, 10 mL aliquots of juice were incubated at 37 °C for 30 min in 25 mL Erlenmeyer flasks. In parallel, Erlenmeyer flasks containing 10 mL of 100 $\mu L\ L^{-1}$ ethanol were used as internal standards for quantity evaluations. After the incubation, 2 mL gas samples were withdrawn from the Erlenmeyers' headspaces into syringes and ethanol levels were determined with a Varian 3300 gas chromatograph. The results are means of three replications; each replicate included juice collected from three different fruits.

2.3.8. Flavor

Flavor evaluations were conducted by three trained judges, who used a 1–9 hedonic scale, in which 1 = very bad and 9 = excellent.

2.3.9. Acceptance Scores

Visual and final acceptance scores were assigned by three trained judges using a 5-grade scale, in which 1 = very bad, 2 = poor, 3 = fair, 4 = good and 5 = excellent.

2.4. Statistical Analysis

Feature-importance values were analyzed using mean Shapley additive explanation (SHAP) values [26] and the mean decrease accuracy (MDA) method [27]. The ClusViz tool was used for heatmap graphical representation [28]. Pearson correlation values were calculated using the open-source R software (available from: http://www.r-project.org; accessed on 10 April 2022).

2.5. Quality-Prediction Models

2.5.1. Data-Set Preparation

The experiment produced 400-labeled data points; each data point represented one carton in the experiment. The input features were two pre-harvest variables (harvest date and yield) and three postharvest variables (storage time, temperature, humidity). The main model's output (i.e., label) was the 5-grade scale final acceptance score assigned to each carton.

2.5.2. Prediction Models

We tested four different linear and non-linear regression models for their ability to predict fruit-acceptance scores. These models are described below.

Multiple Linear Regression (MLR)—This basic model attempts to establish a linear relationship between the features and the label [29]. The MLR uses two or more independent features to predict the outcome of a dependent output label by fitting a hyperplane. The model finds the optimal parameters that minimize the mean squared error for the predicted quality scores.

Support Vector Regression (SVR)—SVR is a generalization of support vector machine (SVM) for regression tasks [30,31]. Unlike other models, SVR attempts to predict the label within a small range of allowed error. In other words, while MLR punishes every prediction error, SVR tolerates small errors as long as they fall within a predefined range. SVR models often employ kernels, which enable non-linearity in the input space. Non-linearity is achieved by transforming the data to a higher dimensional space, in which the relation between the inputs and label will be a linear one. In this work, a radial basis function (RBF) kernel was found to produce the best results, hence the results below relate to the use of SVR with an RBF kernel.

Random Forests (RF)—RF is a supervised ensemble method that is widely used for regression problems [32]. The RF model employs multiple regression trees (i.e., forests) to reduce the variance error [33]. For each tree, the model introduces different subsets of samples and features with replacements, also known as the bagging approach [34]. At prediction time (inference), each individual tree predicts a different value and the average of all of the predictions is used. The tree structure enables non-linearity and by averaging multiple predictions of different trees, RF often produces more accurate results than SVR or MLR.

Extreme Gradient Boosting (XGBoost)—This is a state-of-the-art ensemble method that has become popular in recent years for tabular data predictions [35]. Similar to RF, this model is based on regression trees. However, unlike RF, it uses a boosting approach instead of bagging. In boosting, the trees are built sequentially, with each tree trying to minimize the remaining error of all previous trees [36].

2.5.3. Evaluation of the Models

A great deal of laboratory work was done to evaluate the 12,000 oranges and produce a data set of 400-labeled points. However, machine-learning models usually require many more data to learn and make predictions [37]. The common 80–20% train-set test-set split is risky when the data set is very small, since it is more likely to introduce bias. Hence, a K fold cross-validation method was used [38], with five folds and 20 repetitions producing 100 samples. If needed, pre-processing parameter scaling was applied to each iteration separately. Two metrics were used for the model evaluations: root mean squared error (RMSE) to compare the competing alternatives and the coefficient of determination (R^2) to measure the amount of variance explained by each model.

2.5.4. Duplication as a Way to Deal with Unbalanced Data Sets

In this study, the target label (i.e., acceptance score) is a 5-grade scale variable, with 5 denoting that the produce is very suitable for marketing and 1 denoting produce unlikely to be purchased. The advanced FEFO logistics strategy is expected to enable an efficient inventory management based on shelf-life predictions, and thus has great interest in predicting low-scoring fruit. However, low scores were relatively rare in the current data set, as only 14.75% of the samples in the total data set had scores of 3 ("fair") or less. To cope with this challenge, training set samples with scores that were equal or lower than 3.25 were duplicated. Overall, six modes of duplication were compared: no duplication (i.e., 0) and 1 to 5 duplications.

3. Results

3.1. *Effects of Pre-Harvest and Postharvest Features on the Quality of 'Rustenburg' Navel Oranges*

To collect the high-throughput phenotyping data required to develop shelf-life prediction models, we examined the effects of various pre-harvest and postharvest features on the quality of 'Rustenburg' navel oranges. The examined pre-harvest features included harvest time (with six different harvest times ranging from 21 February 2021 to 6 April 2021) and orchard yield, which ranged between 18 and 48 Ton/Hectare (Table 1). The examined postharvest features included storage temperature (2, 5 or 8 °C), RH level during storage (70, 90 or 95%) and duration of storage, with the fruits being evaluated weekly over a period of 20 weeks.

The effects of the different features on the postharvest storage performance of 'Rustenberg' navel oranges, including weight loss, firmness, color, decay, peel damage, CI and internal dryness, are presented in Figure 1.

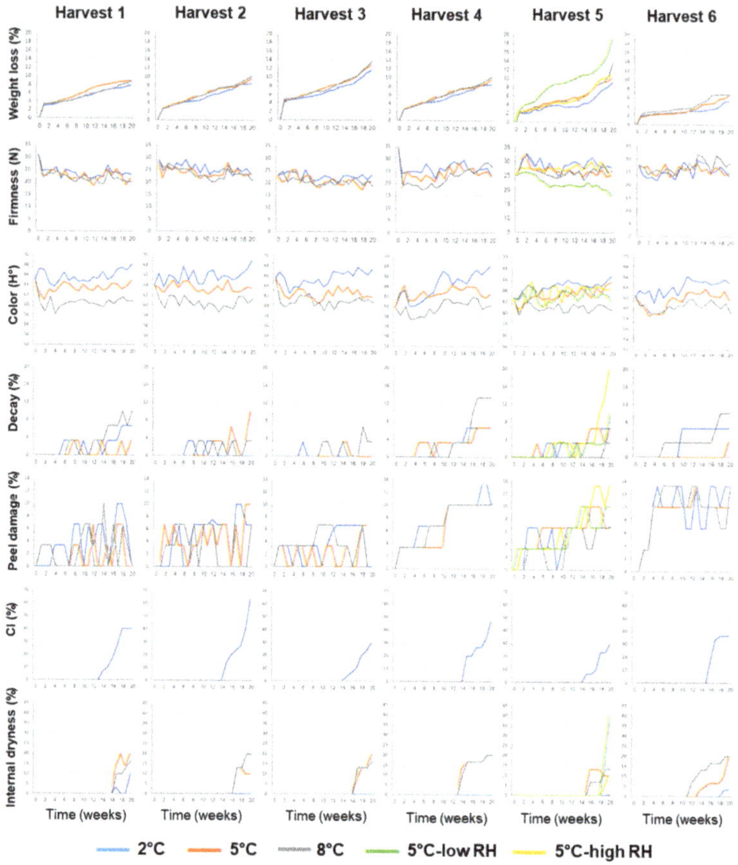

Figure 1. Effects of various pre-harvest and postharvest features on the postharvest storage quality of 'Rustenburg' navel oranges harvested from six different orchards. The harvest times and yields of the different orchards are presented in Table 1. The postharvest features included storage temperature (2, 5 or 8 °C) and storage RH level (70, 90 or 95%). Fruit quality was evaluated every week after one additional week of storage under shelf conditions (at 20 °C) for a period of 20 weeks.

It can be seen that weight loss gradually increased during storage under all of the examined conditions, but reached its highest levels following storage under low-RH conditions. Firmness levels remained more or less stable over the course of the postharvest storage period, but gradually decreased under low-RH conditions.

It is worth noting that the fruits from all of the orchards had a yellowish pale color (hue angle between 59–61°) following storage at the low temperature of 2 °C, a normal orange color (hue angle between 62–64°) following storage at 5 °C and a deep orange/reddish color (hue angle between 66–69°) following storage at 8 °C (Figures 1 and 2a). It is also worth noting that CI symptoms developed only upon storage at the low temperature of 2 °C, whereas internal dryness symptoms developed mainly during storage at the moderate and higher temperatures of 5 °C and 8 °C (Figures 1 and 2b,c).

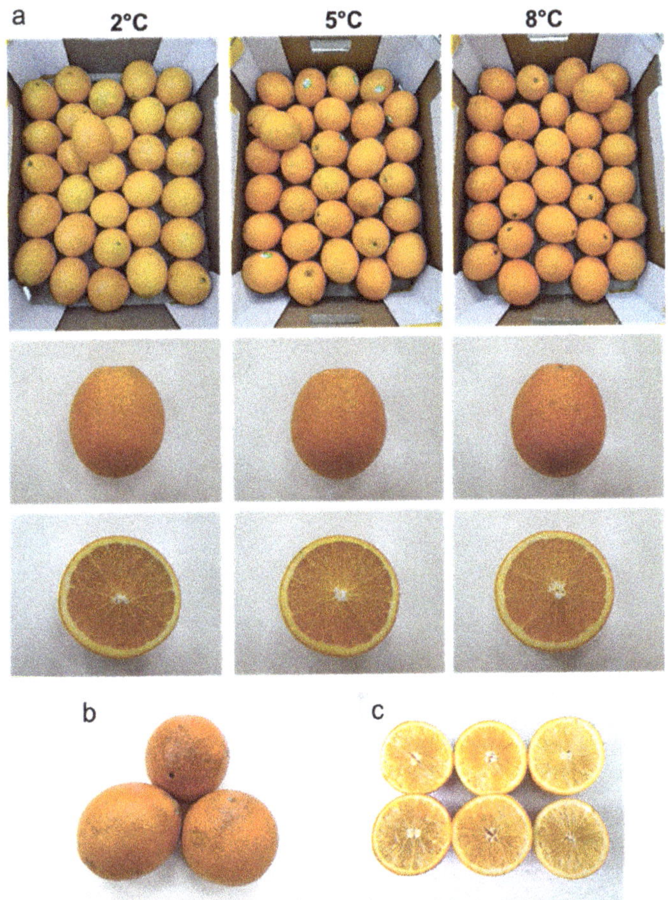

Figure 2. Photographs of 'Rustenburg' navel oranges. (**a**) Fruit stored at 2, 5 and 8 °C, (**b**) chilling injury, and (**c**) internal dryness symptoms.

The effects of the various pre-harvest and postharvest features on the biochemical composition of the juice of 'Rustenburg' navel oranges, including TSS, acidity and vitamin C and ethanol levels, as well as flavor acceptance, are presented in Figure 3. The TSS levels remained relatively stable during storage, whereas acidity and vitamin C levels decreased and ethanol levels increased during storage. Fruit flavor acceptability remained relatively high (flavor acceptance score $\geq$ 7.0, on a scale of 1 to 9) for 14–18 weeks for the different orchards and treatments, but then tended to decrease, doing so somewhat more rapidly at the higher storage temperature of 8 °C.

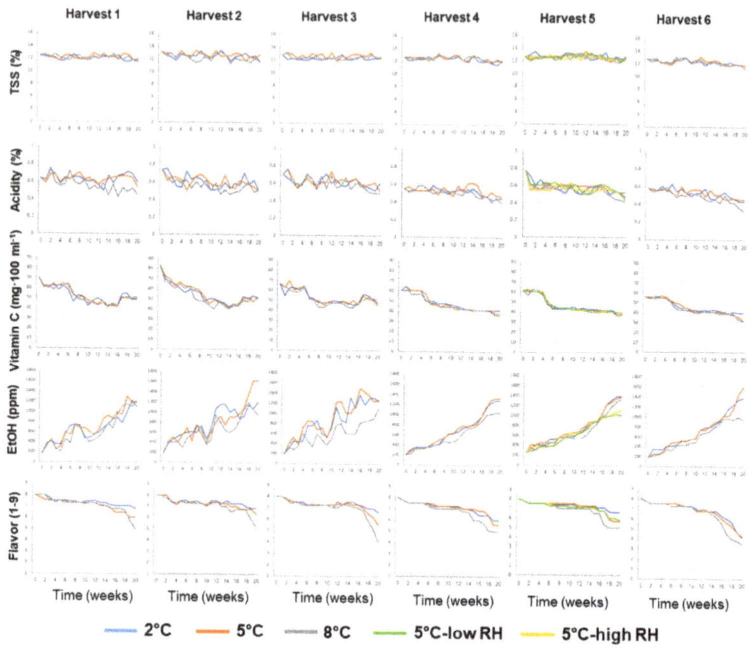

Figure 3. Effects of various pre-harvest and postharvest features on the biochemical composition and flavor of 'Rustenburg' navel oranges harvested from six different orchards. The harvest times and yields of the different orchards are presented in Table 1. The postharvest features included storage temperature (2, 5 or 8 °C) and storage RH (70, 90 or 95%). Fruit quality was evaluated every week after one additional week of storage under shelf conditions (20 °C) for a period of 20 weeks.

Overall, fruit-acceptance scores remained relatively high (acceptance score ≥ 4, on a scale of 1 to 5) for 14–19 weeks, depending on the various pre-harvest and postharvest features (Figure 4). It is worth noting that the fruits stored at 5 °C maintained high acceptance scores for longer periods when compared to the fruits stored at a lower temperature of 2 °C or a higher temperature of 8 °C.

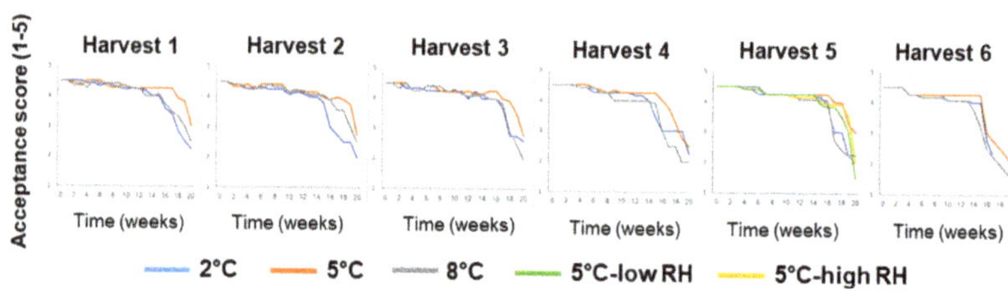

Figure 4. Effects of various pre-harvest and postharvest features on the acceptance scores of 'Rustenburg' navel oranges harvested from six different orchards. The harvest times and yields of the different orchards are presented in Table 1. The postharvest features included storage temperature (2, 5 or 8 °C) and storage RH (70, 90 or 95%). Fruit quality was evaluated every week after one additional week of storage under shelf conditions (at 20 °C) for a period of 20 weeks.

To evaluate the importance of the various pre-harvest and postharvest features on the quality of 'Rustenburg' navel oranges, we used various statistical analysis tools, including SHAP and RF MDA methods (Figure 5a,b). A feature importance analysis using mean SHAP values revealed that storage time was the most important feature affecting fruit quality, followed by storage temperature, harvest time, yield and humidity conditions

(Figure 5a). Furthermore, RF MDA analysis revealed that harvest time had major effects on peel damage and vitamin C levels, and that orchard yield had major effects on decay, firmness and TSS. Regarding the postharvest features, storage time had major effects on ethanol accumulation, fruit flavor and acceptance scores; storage temperature had major effects on CI and peel color; humidity conditions during storage had a major effect on fruit weight loss (Figure 5b). It is worth noting that the different pre-harvest and postharvest features affected different postharvest quality parameters. For example, storage time mainly affected flavor, ethanol levels and acceptance scores, whereas storage temperature mainly affected color and CI (Figure 5b).

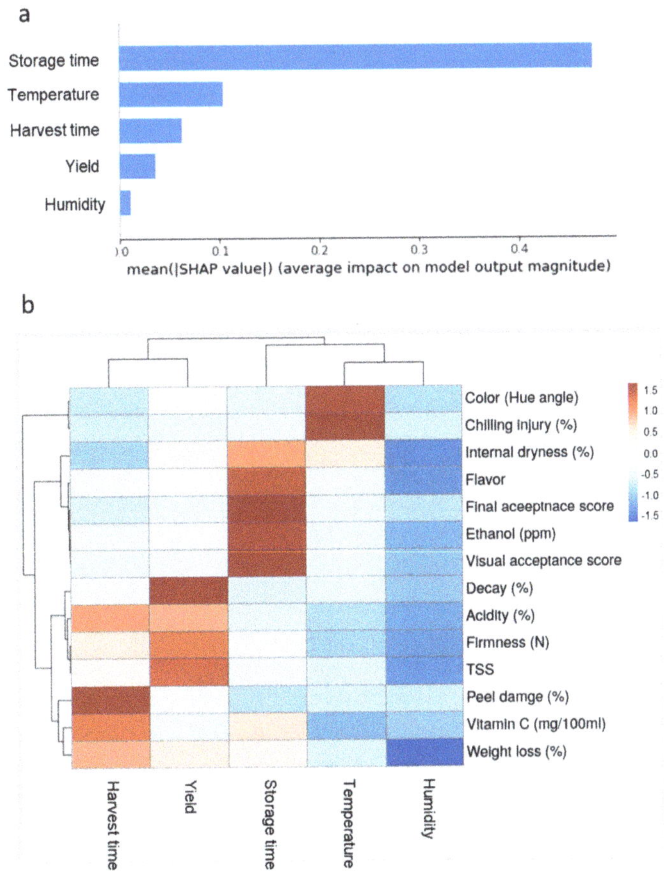

Figure 5. Importance of the various pre-harvest and postharvest features on the quality of 'Rustenburg' navel oranges. (**a**) Feature importance analysis using the SHAP method. (**b**) Feature importance and heat-map analyses using the RF MDA method.

ANOVA revealed that the harvest date significantly affected acidity, vitamin C levels, decay, flavor, firmness, weight loss, peel damage, color and final acceptance scores, and orchard yields significantly affected acidity, vitamin C, flavor, weight loss and peel damage ($p < 0.001$; Table 2). Among the postharvest features, storage time significantly affected all of the examined parameters, with the exception of TSS. Storage temperature significantly affected all the parameters except for decay and TSS, and storage humidity significantly affected firmness and weight loss (Table 2).

Table 2. Effects of different pre-harvest and postharvest features on the quality attributes of 'Rustenburg' navel oranges. The presented data are ANOVA p values. Gray shading indicates statistical significance ($p \leq 0.001$).

	Harvest Time	Storage Time	Yield	Storage Temperature	Humidity
Acidity	6.55×10^{42}	3.06×10^{33}	2.01×10^{10}	9.42×10^{15}	-
Vitamin C	3.15×10^{44}	4.20×10^{210}	1.54×10^{13}	7.17×10^{17}	0.001
Internal dryness	-	6.98×10^{106}	0.002	1.08×10^{13}	-
Decay	4.86×10^{6}	2.97×10^{35}	0.001	0.034	0.005
Flavor	7.92×10^{17}	4.37×10^{268}	3.55×10^{5}	3.08×10^{16}	-
Final acceptance score	2.82×10^{7}	0	0.003	3.03×10^{7}	-
Firmness	7.56×10^{69}	5.58×10^{9}	-	5.11×10^{16}	2.07×10^{21}
Ethanol	0.004	0	0.002	1.76×10^{17}	-
Visual acceptance score	-	1.47×10^{279}	0.035	1.09×10^{10}	-
Weight loss	2.33×10^{27}	4.82×10^{270}	1.58×10^{16}	3.77×10^{22}	6.08×10^{12}
Peel damage	3.24×10^{31}	3.95×10^{16}	8.33×10^{6}	3.28×10^{6}	0.019
Hue angle	3.47×10^{8}	4.15×10^{12}	0.001	2.33×10^{238}	0.006
TSS	-	-	-	-	-
Chilling injury	-	1.46×10^{44}	0.030	5.12×10^{45}	0.012

Pearson correlations between the various quality parameters revealed positive correlations among flavor, visual and final acceptance scores and acidity and vitamin C levels, as well as negative correlations between these attributes and changes in ethanol levels, weight loss, decay, internal dryness, peel damage, color and CI (Figure 6; red and blue boxes, respectively). We also observed positive correlations between ethanol content, decay, weight loss and internal dryness, as well as negative correlations between weight loss and firmness (Figure 6; pink and black boxes, respectively).

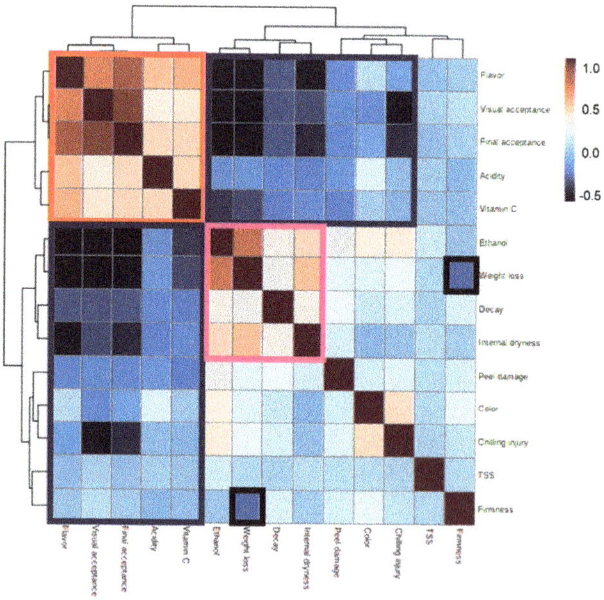

Figure 6. Pearson correlations and heat-map analysis of the various fruit-quality parameters. Red and pink boxes indicate parameters that are positively correlated with one another, and blue and black boxes indicate parameters that are negatively correlated with one another.

Finally, Pearson correlations revealed positive correlations between final acceptance scores and acidity, vitamin C, flavor and visual acceptance scores, and negative correlations between final acceptance scores and changes in TSS, firmness, color, peel damage, decay, CI, internal dryness, weight loss and ethanol accumulation (Figure 7).

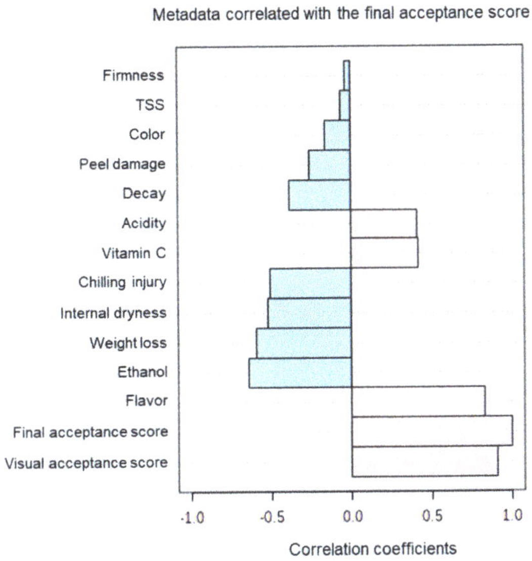

Figure 7. Pearson correlations between the various fruit-quality parameters and the final acceptance scores.

3.2. Quality-Prediction Models

Four regression models, including one linear model (MLR) and three non-linear models (SVR, RF and XGBoost), were evaluated for their abilities to predict acceptance scores based on two pre-harvest and three postharvest features (Table 3). Since the data set was not balanced, the models were evaluated twice. First, we evaluated the entire test sets (referred to as the "full set") and then we evaluated a subset of the full set that consisted only of samples with acceptance scores equal or lower than 3.25 (i.e., low-quality samples, referred to here as the "low-score subset"). RMSE and R^2 were calculated using a cross-validation method, which produced 100 samples, and the Wilcoxon non-parametric signed-rank test was applied to the results. As shown in Table 3, the XGBoost regression model outperformed the other models ($p < 0.01$). The observed RMSE values of the XGBoost model for the full set and the low-score subset were 0.195 and 0.380, respectively; and the R^2 values were 0.914 and 0.305, respectively. As expected, the RMSE and R^2 model prediction values were higher for the full set than for the low-score subset.

Table 3. RMSE and R^2 values for the MLR, SVR, RF and XGBoost regression models for the full data set and the low-score subset.

	Full Set		Low-Score Subset	
Algorithm	**RMSE**	**R^2**	**RMSE**	**R^2**
MLR	0.444	0.566	0.831	−2.314
SVR	0.312	0.784	0.662	−1.107
RF	0.200	0.910	0.404	0.217
XGBoost	0.195	0.914	0.380	0.305

To further improve the XGBoost prediction model, especially for the low-score subset, we used a duplication approach in which low-quality samples in the training set were duplicated for each repetition and fold. Overall, six duplication modes were compared:

no duplication (i.e., zero) and one to five duplications. The RMSE and R^2 of the full set and low-score subset were measured again for each duplication mode (Table 4). The duplications had a relatively small diminishing effect on the RMSE and R^2 of the full set (RMSE increased from 0.195 to 0.217; R^2 decreased from 0.914 to 0.891), but substantially reduced the RMSE of the low-score subset from 0.380 to 0.329 and increased the R^2 of the low-score subset from 0.305 to 0.479.

Table 4. RMSE and R^2 values for the XGBoost regression model for the full data set and the low-score subset with 0 to 5 duplications.

Number of Duplications	Full Set		Low-Score Subset	
	RMSE	R^2	RMSE	R^2
0	0.195	0.914	0.380	0.305
1	0.201	0.907	0.358	0.385
2	0.206	0.903	0.344	0.433
3	0.210	0.898	0.337	0.456
4	0.214	0.894	0.333	0.469
5	0.217	0.891	0.329	0.479

The FIFO logistic management method is based on the notion that storage time is the most crucial feature for predicting fruit quality. This was supported by the SHAP analysis, which revealed that storage time was the most important feature affecting the postharvest quality of oranges (Figure 5a). However, the results suggest that the other pre-harvest and postharvest features had additional effects on fruit quality. To evaluate the relative importance of the other examined pre-harvest and postharvest features in the prediction model, we conducted post hoc analyses to explore the contributions of various feature subsets on fruit-quality predictions. The data subgroups included: (1) storage time only, (2) storage time and pre-harvest features (harvest time and yield), (3) storage time and postharvest features (temperature and humidity) and (4) storage time and all pre-harvest and postharvest features (harvest time, yield, storage time, storage temperature and storage humidity).

XGBoost models with five duplications of the training sets were used to compare these subgroups. RMSE and R^2 values for the full set and the low-score subset are presented in Table 5. For the full data set, the addition of either pre-harvest (Subgroup 2) or postharvest data (Subgroup 3) lowered the RMSE from 0.371 (Subgroup 1; i.e., storage time) to 0.304 and 0.340, respectively, and increased the R^2 from 0.687 (Subgroup 1), based on storage time alone, to 0.790 (Subgroup 2) and 0.739 (Subgroup 3). The inclusion of all features (Subgroup 4) further reduced RMSE to 0.217 and increased R^2 to 0.891. For the low-score subset, the neither did the addition of either pre-harvest (Subgroup 2) or postharvest data (Subgroup 3) improve the RMSE and R^2 values. However, we observed an improved RMSE of 0.329 and R^2 of 0.479 when all of the pre-harvest and postharvest features were included in the model (Subgroup 4), relative to storage time alone (Subgroup 1), with an RMSE of 0.473 and an R^2 of −0.074.

Table 5. The RMSE and R^2 values for the XGBoost regression model with five duplications for the full data set and low-score subset, using four different feature subgroups.

Subgroup	Full Set		Low-Score Subset	
	RMSE	R^2	RMSE	R^2
1. Storage time	0.371	0.687	0.473	−0.074
2. Storage time + pre-harvest features (harvest time, yield)	0.304	0.790	0.464	−0.036
3. Storage time + postharvest features (temperature, humidity)	0.340	0.739	0.535	−0.376
4. Storage time + pre-harvest (harvest time, yield) + postharvest features (temperature, humidity)	0.217	0.891	0.329	0.479

4. Discussion

The main goal of the current study was to conduct a large-scale, high-throughput phenotyping analysis of the effects of various pre-harvest and postharvest features on the quality of 'Rustenburg' navel oranges, in order to develop quality-prediction models to enable the implementation of the more efficient FEFO logistic management system [14,15].

Performing the current postharvest storage evaluations of 'Rustenburg' navel oranges yielded two main outcomes: (1) they contribute to our current understanding of the importance of various pre-harvest and postharvest features on the quality of navel oranges, and (2) they allowed for the generation of quality-prediction models for oranges.

The SHAP analysis revealed that storage time is the most important feature affecting the postharvest quality of oranges, but also showed that other pre-harvest and postharvest features also have meaningful effects on fruit quality (Figure 5a). Furthermore, feature-importance analysis conducted using RF MDA revealed that different pre-harvest and postharvest features affect different aspects of fruit quality. For example, storage temperature mainly affects color and CI, whereas RH mainly affects weight loss (Figure 5b). More specifically, ANOVA revealed that the harvest time, storage time and storage temperature features significantly affected most examined fruit quality parameters, whereas tree yield significantly affected acidity, vitamin C, flavor, weight loss and peel damage, and humidity significantly affected fruit firmness and weight loss (Table 2). Thus, we conclude that all of the examined pre-harvest and postharvest features eventually influence the postharvest quality of navel oranges and that the most comprehensive quality-prediction models will include as many factors as possible.

Similar findings regarding the importance of both pre-harvest and postharvest features for shelf-life predictions were recently reported for nectarines. That work found that pre-harvest features, such as irrigation methods and fruit load, and postharvest features, such as storage temperature and storage RH, all have major effects on the postharvest quality of nectarines [39]. Other studies involving the development of shelf-life prediction models for strawberries, rocket leaves and mushrooms have reported that storage temperature is the most important feature affecting shelf life and quality [40–42]. However, another study pointed to the importance of RH for preserving the postharvest quality of strawberries [43].

The calculation of Pearson correlations between various orange fruit-quality parameters and final acceptance scores revealed that the fruit quality parameters that were positively correlated with high acceptance scores were acidity, vitamin C level and flavor, and that these parameters were also positively correlated with one another (Figures 6 and 7). In contrast, changes in firmness, weight loss, decay, internal dryness, peel damage, CI, ethanol accumulation and fruit color were all negatively correlated with the fruit acceptance scores (Figures 6 and 7).

The second main goal of the current study was to develop fruit quality prediction models. To that end, we examined four regression models and found that RF and XGBoost provided very effective quality predictions when the full experimental data set was used. The XGBoost regression model had a low RMSE value of just 0.195 for an acceptance score scale of one to five, as well as a high R^2 value of 0.914 (Table 3). A more or less similar RMSE value of 0.184 and an R^2 value of 0.911 were recently reported for the development of a shelf life prediction model of table grapes using an optimized radial basis function (RBF) and neural network [16]. However, unfortunately, the collected data were not fairly balanced, as low acceptance scores (equal or below three on a scale of one to five) accounted for only 14.75% of the total data set and, accordingly, the observed RMSE and R^2 values for the low-scoring data set were only 0.380 and 0.305, respectively. Thus, the tested regression models provided much better quality predictions for fruits with higher acceptance scores than for fruits with lower acceptance scores.

To address this obstacle, we added a further duplication of the low-score subset and found that doing so helped to reduce the observed RMSE from 0.380 to 0.329 and increased the R^2 from 0.305 to 0.479 (Table 4). Nonetheless, a further improvement of the quality predictions of low-acceptance-score fruits will require the performance of additional experiments and the generation of larger data sets of fruit with low acceptance scores.

Finally, post hoc analyses revealed that quality prediction models based on storage-time data alone were much less accurate than quality prediction models based on storage time together with the other examined pre-harvest and postharvest features (Table 5). Thus, the development of accurate quality-prediction models required for the implementation of the FEFO intelligent logistic management system necessitates the collection and use of as much pre-harvest and postharvest data as possible.

5. Conclusions

The main objective of the current research was to conduct a large-scale, high-throughput phenotyping analysis of the postharvest storage performance of 'Rustenburg' navel oranges, in order to develop shelf-life prediction models to enable the implementation of the FEFO logistic management method. The key findings were that the storage time appeared to be the most important feature affecting fruit quality, but also other features, including harvest time, storage temperature, yield and humidity, majorly influenced fruit postharvest quality. Based on the achieved data, we examined the efficacy of different regression models for their ability to predict fruit quality, and found that using XGBoost combined with a duplication approach provided the most effective approach, and allowed the prediction of fruit-acceptance scores among the full data set, with an RMSE of 0.217 and an R^2 of 0.891. The development of accurate shelf-life prediction models should now allow the implementation of the FEFO logistic management system for more efficient inventory management and reduction of losses.

Author Contributions: All authors contributed substantially to this work. Conceptualization, R.P., A.L., N.K. and Y.S.; Data curation, A.O. and O.J.; Formal analysis, A.D.-F., O.J. and N.K.; Funding acquisition, R.P., A.L. and Y.S.; Investigation, A.O., O.J. and R.P.; Methodology, R.P., N.K. and Y.S.; Project administration, R.P. and Y.S.; Software, A.D.-F. and O.J.; Supervision, R.P., A.L., N.K. and Y.S.; Visualization, A.O. and R.P.; Writing–original draft, R.P. and O.J; Writing–review and editing, A.O., R.P., A.L., O.J., N.K. and Y.S. All authors have read and agreed to the published version of the manuscript.

Funding: This research was funded by The Israel Innovation Authority, grant number 70076.

Data Availability Statement: The data presented in this study are available on request from the corresponding author.

Conflicts of Interest: The authors declare no conflict of interest.

References

1. Fruit: World Production by Type 2020 | Statista, (n.d.). Available online: https://www.statista.com/statistics/264001/worldwide-production-of-fruit-by-variety/ (accessed on 19 May 2022).
2. Ranganna, S.; Govindarajan, V.S.; Ramana, K.V.R.; Kefford, J.F. Citrus fruits–Varieties, chemistry, technology, and quality evaluation. Part II. Chemistry, technology, and quality evaluation. A. Chemistry. *Crit. Rev. Food Sci. Nutr.* **1983**, *18*, 313–386. [CrossRef] [PubMed]
3. Marloth, R.H.; Basson, W.J. Relative Performance of Washington Navel Orange Selections and other Navel Varieties. *J. Hortic. Sci.* **1959**, *34*, 133–141. [CrossRef]
4. Gao, Y.; Liu, Y.; Kan, C.; Chen, M.; Chen, J. Changes of peel color and fruit quality in navel orange fruits under different storage methods. *Sci. Hortic.* **2019**, *256*, 108522. [CrossRef]
5. Kader, A.A. A Perspective on Postharvest Horticulture (1978–2003). *HortScience* **2003**, *38*, 1004–1008. [CrossRef]
6. Arpaia, M. Preharvest factors influencing postharvest quality of tropical and subtropical fruit. *HortScience* **1994**, *29*, 982–985. [CrossRef]
7. Tyagi, S.; Sahay, S.; Imran, M.; Rashmi, K.; Mahesh, S.S. Pre-harvest Factors Influencing the Postharvest Quality of Fruits: A Review. *Curr. J. Appl. Sci. Technol.* **2017**, *23*, 1–12. [CrossRef]
8. Tadeo, J.L.; Ortiz, J.M.; Estellés, A. Sugar changes in Clementine and orange fruit during ripening. *J. Hortic. Sci.* **1987**, *62*, 531–537. [CrossRef]
9. Schirra, M.; D'Hallewin, G.; Cabras, P.; Angioni, A.; Garau, V.L. Seasonal Susceptibility of Tarocco Oranges to Chilling Injury As Affected by Hot Water and Thiabendazole Postharvest Dip Treatments. *J. Agric. Food Chem.* **1998**, *46*, 1177–1180. [CrossRef]

10. Alferez, F.; Zacarías, L. Influence of fruit maturity in the susceptibility of Navelina oranges to develop postharvest non-chilling peel pitting. *Food Sci. Technol. Int.* **2014**, *20*, 183–191. [CrossRef]
11. Khalid, S.; Malik, A.U.; Khan, A.S.; Khan, M.N.; Ullah, M.I.; Abbas, T.; Khalid, M.S. Tree age and fruit size in relation to postharvest respiration and quality changes in 'Kinnow' mandarin fruit under ambient storage. *Sci. Hortic.* **2017**, *220*, 183–192. [CrossRef]
12. Kader, A.A.; Arpaia, M.L. Postharvest handling systems: Subtropical fruit. In *Postharvest Technology of Horticultural Crops*, 3rd ed.; University of California, Agriculture and Natural Resources: Oakland, CA, USA, 2002; pp. 375–384.
13. Paull, R. Effect of temperature and relative humidity on fresh commodity quality. *Postharvest Biol. Technol.* **1999**, *15*, 263–277. [CrossRef]
14. Hertog, L.A.T.M.; Uysal, I.; McCarthy, U.; Verlinden, B.M.; Nicolaï, B.M. Shelf life modelling for first-expired-first-out warehouse management. *Philos. Trans. R. Soc. A Math. Phys. Eng. Sci.* **2014**, *372*, 20130306. [CrossRef] [PubMed]
15. Jedermann, R.; Nicometo, M.; Uysal, I.; Lang, W. Reducing food losses by intelligent food logistics. *Philos. Trans. R. Soc. London. Ser. A: Math. Phys. Eng. Sci.* **2014**, *372*, 20130302. [CrossRef] [PubMed]
16. Li, Y.; Chu, X.; Fu, Z.; Feng, J.; Mu, W. Shelf life prediction model of postharvest table grape using optimized radial basis function (RBF) neural network. *Br. Food J.* **2019**, *121*, 2919–2936. [CrossRef]
17. Song, Y.; Hu, Q.; Wu, Y.; Pei, F.; Kimatu, B.M.; Su, A.; Yang, W. Storage time assessment and shelf-life prediction models for postharvest Agaricus bisporus. *LWT* **2019**, *101*, 360–365. [CrossRef]
18. Jalali, A.; Linke, M.; Geyer, M.; Mahajan, P.V. Shelf life prediction model for strawberry based on respiration and transpiration processes. *Food Packag. Shelf Life* **2020**, *25*, 100525. [CrossRef]
19. Salehi, F. Recent Advances in the Modeling and Predicting Quality Parameters of Fruits and Vegetables during Postharvest Storage: A Review. *Int. J. Fruit Sci.* **2020**, *20*, 506–520. [CrossRef]
20. Cammarota, G.; Ianiro, G.; Ahern, A.; Carbone, C.; Temko, A.; Claesson, M.J.; Gasbarrini, A.; Tortora, G. Gut microbiome, big data and machine learning to promote precision medicine for cancer. *Nat. Rev. Gastroenterol. Hepatol.* **2020**, *17*, 635–648. [CrossRef]
21. Neethirajan, S. The role of sensors, big data and machine learning in modern animal farming. *Sens. Bio-Sens. Res.* **2020**, *29*, 100367. [CrossRef]
22. La Scalia, G.; Nasca, A.; Corona, O.; Settanni, L.; Micale, R. An Innovative Shelf Life Model Based on Smart Logistic Unit for an Efficient Management of the Perishable Food Supply Chain. *J. Food Process Eng.* **2017**, *40*, e12311. [CrossRef]
23. Chaudhuri, A.; Dukovska-Popovska, I.; Subramanian, N.; Chan, H.K.; Bai, R. Decision-making in cold chain logistics using data analytics: A literature review. *Int. J. Logist. Manag.* **2018**, *29*, 839–861. [CrossRef]
24. Hiromi, K.; Kuwamoto, C.; Ohnishi, M. A rapid sensitive method for the determination of ascorbic acid in the excess of 2,6-dichlorophenolindophenol using a stopped-flow apparatus. *Anal. Biochem.* **1980**, *101*, 421–426. [CrossRef]
25. Davis, P.L.; Chace, W.G. Determination of alcohol in citrus juice by gas chromatographic analysis of headspace. *Hortscience* **1969**, *2*, 168–169.
26. Lundberg, S.M.; Lee, S.U. A unified approach to interpreting model predictions. *Adv. Neural. Info Process. Syst.* **2017**, *30*, 1–10. [CrossRef]
27. Louppe, G.; Wehenkel, L.; Sutera, A.; Geurts, P. Understanding variable importances in Forests of randomized trees. *Adv. Neural Inf. Process. Syst.* **2013**, *26*, 431–439.
28. Metsalu, T.; Vilo, J. ClustVis: A web tool for visualizing clustering of multivariate data using Principal Component Analysis and heatmap. *Nucleic Acids Res.* **2015**, *43*, W566–W570. [CrossRef]
29. Lewis-Beck, C.; Lewis-Beck, M. *Applied Regression: An Introduction*; Sage Publications: Thousand Oaks, CA, USA, 2015.
30. Drucker, H.; Surges, C.J.C.; Kaufman, L.; Smola, A.; Vapnik, V. Support vector regression machines. *Adv. Neural Info Process Syst.* **1997**, *1*, 155–161.
31. Awad, M.; Khanna, R. Support vector regression. In *Efficient Learning Machines*; Apress; Springer: New York, NY, USA, 2015; pp. 67–80. [CrossRef]
32. Breiman, L. Random Forests. *Mach. Learn.* **2001**, *45*, 5–32. [CrossRef]
33. Breiman, L.; Friedman, J.H.; Olshen, R.A.; Stone, C.J. *Classification and Regression Trees*; Chapman & Hall/CRC Press: Boca Raton, FL, USA, 2017. [CrossRef]
34. Breiman, L. Bagging predictors. *Mach. Learn.* **1996**, *24*, 123–140. [CrossRef]
35. Chen, T.; Guestrin, C. XGBoost: A Scalable Tree Boosting System. In Proceedings of the 22nd ACM SIGKDD International Conference on Knowledge Discovery and Data Mining, San Francisco, CA, USA, 13–17 August 2016; pp. 785–794.
36. Polikar, R. Ensemble based systems in decision making. *IEEE Circuits Syst. Mag.* **2006**, *6*, 21–44. [CrossRef]
37. Lin, Y.; Guan, Y.; Asudeh, A.; Jagadish, H.V.; Michigan, Y.L.U.; Michigan, Y.G.U.; Illinois, A.A.U.; Michigan, H.V.J.U. Identifying insufficient data coverage in databases with multiple relations. *Proc. VLDB Endow.* **2020**, *13*, 2229–2242. [CrossRef]
38. Efron, B. Estimating the error rate of a prediction rule: Improvement on cross-validation. *J. Amer. Statist. Assoc.* **1983**, *78*, 316–331. [CrossRef]
39. Casagrande, E.; Génard, M.; Lurol, S.; Charles, F.; Plénet, D.; Lescourret, F. A process-based model of nectarine quality development during pre- and post-harvest. *Postharvest Biol. Technol.* **2021**, *175*, 111458. [CrossRef]
40. Amodio, M.L.; Derossi, A.; Mastrandrea, L.; Colelli, G. A study of the estimated shelf life of fresh rocket using a non-linear model. *J. Food Eng.* **2015**, *150*, 19–28. [CrossRef]

41. Wang, W.; Hu, W.; Ding, T.; Ye, X.; Liu, D. Shelf-life prediction of strawberry at different temperatures during storage using kinetic analysis and model development. *J. Food Process. Preserv.* **2018**, *42*, e13693. [CrossRef]
42. Niu, Y.; Yun, J.; Bi, Y.; Wang, T.; Zhang, Y.; Liu, H.; Zhao, F. Predicting the shelf life of postharvest Flammulina velutipes at various temperatures based on mushroom quality and specific spoilage organisms. *Postharvest Biol. Technol.* **2020**, *167*, 111235. [CrossRef]
43. Ktenioudaki, A.; O'Donnell, C.P.; do Nascimento Nunes, M.C. Modelling the biochemical and sensory changes of strawberries during storage under diverse relative humidity conditions. *Postharvest Biol. Technol.* **2019**, *154*, 148–158. [CrossRef]

foods

MDPI

Review

Advances in Postharvest Storage and Preservation Strategies for *Pleurotus eryngii*

Yuxi Guo, Xuefeng Chen, Pin Gong *, Ruotong Wang, Zhuoya Qi, Zhenfang Deng, Aoyang Han, Hui Long, Jiating Wang, Wenbo Yao, Wenjuan Yang, Jing Wang and Nan Li

School of Food and Biological Engineering, Shaanxi University of Science and Technology, Xi'an 710021, China
* Correspondence: gongpin@sust.edu.cn; Tel.: +86-13772196479

Abstract: The king oyster mushroom (*Pleurotus eryngii*) is a delicious edible mushroom that is highly prized for its unique flavor and excellent medicinal properties. Its enzymes, phenolic compounds and reactive oxygen species are the keys to its browning and aging and result in its loss of nutrition and flavor. However, there is a lack of reviews on the preservation of *Pl. eryngii* to summarize and compare different storage and preservation methods. This paper reviews postharvest preservation techniques, including physical and chemical methods, to better understand the mechanisms of browning and the storage effects of different preservation methods, extend the storage life of mushrooms and present future perspectives on technical aspects in the storage and preservation of *Pl. eryngii*. This will provide important research directions for the processing and product development of this mushroom.

Keywords: *Pleurotus eryngii*; king oyster mushroom; quality influential factors; preservation

Citation: Guo, Y.; Chen, X.; Gong, P.; Wang, R.; Qi, Z.; Deng, Z.; Han, A.; Long, H.; Wang, J.; Yao, W.; et al. Advances in Postharvest Storage and Preservation Strategies for *Pleurotus eryngii*. *Foods* **2023**, *12*, 1046. https://doi.org/10.3390/foods12051046

Academic Editors: Maria Cefola and Bernardo Pace

Received: 23 December 2022
Revised: 22 February 2023
Accepted: 22 February 2023
Published: 1 March 2023

1. Introduction

The king oyster mushroom (*Pleurotus eryngii*) is a high-quality, large, fleshy umbrella mushroom that is widely grown in many parts of the world [1]. It is grown in Europe, the Middle East and China [2]. *Pl. eryngii* has been intensively studied as a medicinal mushroom, a part of traditional diet and medicine, for its unique flavor, nutrition and biological functions [1]. In addition, *Pl. eryngii* has a wide market for its easily cultivated, high-yielding, and delicious product that can be cooked directly [3]. Simultaneously, *Pl. eryngii* is rich in protein, carbohydrates, unsaturated fatty acids, vitamins and other nutrients. It is also low in fat with high nutritional and medicinal value, which results in its high economic value. Its dried product contains 14.85% protein, 4.46% fat, 15.51% crude fiber, 43.15% carbohydrates and 18 amino acids. It is also rich in polysaccharides [1,4] and has good therapeutic effects, such as its activities against viruses and hypoglycemia and its ability to lower cholesterol, promote intestinal digestion, prevent cardiovascular disease and improve immunity [1,5].

Postharvest quality is a major concern for mushroom growers. *Pl. eryngii* is a highly perishable commodity and is not suitable for prolonged storage or transport over long distances [6,7]. In recent years, *Pl. eryngii* has become popular with consumers owing to its crunchy texture and nutritious nature; it meets the demand for a healthy lifestyle [8]. However, unexpected softening in texture and browning caused by polyphenol oxidase always occurs during storage, which significantly increases the challenges of postharvest storage and preservation and significantly increases the cost of transporting the king oyster mushroom [8]. From the relevant postharvest preservation studies that have been conducted on *Pl. eryngii*, drying not only has a positive impact on physical properties, such as shrinkage, dehydration capacity and color, but also on the components that exert antioxidant and health-promoting properties. These include preservation technologies, such as modified atmosphere packaging (MAP) [9,10], γ-radiation [3], 1-methylcyclopropene (1-MCP) nanopackaging and polysaccharide nanoparticle preservation [11], that can maintain their texture and nutrient content and extend the storage period. In addition, physical methods, such as microwave hot-air drying, vacuum freeze drying, solar drying and steam bleaching, can effectively reduce the loss of nutrients and reduce the intensity of respiration during storage and preservation. However, the current preservation techniques only have a small effect on the primary nutrients, and it is not known whether other nutrients are affected. Chemical methods, such as essential oils (EOs) and coating preservation [12], can delay the water loss and softening of *Pl. eryngii* to some extent and inhibit their respiration rate, thus resulting in successful storage and preservation (Table 1).

Table 1. Methods of *Pl. eryngii* storage and preservation.

Treatments	Process Parameters	Storage Days	Preservation Effect	Ref.
Modified atmosphere packaging	Storage temperature: 4 °C Storage relative humidity: 95% Grouping processing: -high carbon dioxide packaging (HCP: 20% CO_2 + 15% O_2) -low carbon dioxide packaging (LCP: 30% O_2 + 2% CO_2) -high nitrogen packaging (HNP: 85% N_2, 15% O_2)	10 d	Optimal processing: HCP: 20% CO_2 + 15% O_2 -High total phenolic content -Darkening delaying effect	[13]
High carbon dioxide and low oxygen storage	Storage temperature: 4 °C Storage relative humidity: 95% Grouping processing: -2% O_2 + 30% CO_2 -Air	5 d	Optimal processing: 2% O_2 + 30% CO_2 -Inhibition of serine protease activity	[9]
	Storage temperature:4 °C Storage relative humidity:95% Grouping processing: -2% O_2 -2% O_2 + 10% CO_2 -2% O_2 + 30% CO_2 -1% O_2 + 50% CO_2 -Air	5 d	Optimal processing: 2% O_2 + 30% CO_2 -O_2^- production rate: 50.7% -Improve enzyme activity (SOD)	[10]
1-MCP treatment combined with nano-packaging	Storage temperature:4 ± 1 °C Storage relative humidity: 90–95% Grouping processing: -Untreated -1-MCP (0.3 μL L^{-1}, 24 h) -Nano-packaging -1-MCP (0.3 μL L^{-1}, 24 h) + nano-packaging.	12 d	Optimal processing: 1-MCP + nanopackaging -Texture enhancement -Delay respiration rate -Soluble protein improved -Avoid the accumulation of activated oxygen and enhance antioxidant activity (PPO, SOD and CAT)	[11]
A novel phase change material	Storage temperature: 22 °C ± 2 °C Preparation of PCM: 0.01% nano-TiO_2, 2.09% K_2SO_4, 1.72% maltitol, and 0.50% superabsorbent polymer Grouping processing: -Novel PCM (−2 °C) -Ice (−2 °C) -Equal mass of water	5 d	Optimal processing: the novel PCM (−2 °C) -Total flavonoid contents: 37.31% higher than control -Free amino acids: the contents of Glu, Phe and Pro were 1.95-fold, 1.34-fold and 2.07-fold higher than those in control, respectively; electrolyte leakage: 17.94% lower than that in control -Antioxidant activity enhancement (GDH, POD, SOD and CCO)	[14]

Table 1. *Cont.*

Treatments	Process Parameters	Storage Days	Preservation Effect	Ref.
Gamma irradiation	Storage temperature: 5 ± 1 °C Group: 0, 1, 2, 3 kGy	28 d	Optimal processing: 1 kGy -Uniform color with no fungus spoilage and blemishes -Scanning electron microscopy: comparable micro-structure to that of the control	[3]
MARDB (microwave hot-air flow rolling dry-blanching)	Storage temperature: 4 °C MARDB pretreatment: constant microwave power: 3 W/g, the speed of the rolling bed: 5 rpm Hot-air drying treatment: speed of rolling bed: 5 rpm, drying temperature of the material: 60 °C Group processing: -After pretreatment, cooled to 60 °C in the air and dried. -After pretreatment, packed in plastic bags, sealed and placed in the refrigerator of 4 °C	12 d	Optimal processing: microwave hot-air flow rolling dry-blanching for 9 min -Maintaining quality parameters -Maintain moisture ratio -Reducing water holding capacity and water binding capacity	[15]
Temperature-controlled cold rooms	Relative humidity: 87 ± 5% Packing material: PE Group: 2 °C low temperature 4 °C low temperature 8 °C low temperature	18 d	Optimal processing: 2 °C low temperature -High total phenolic content -Darkening delaying effect -Membrane lipid peroxidation is low	[8]
Distilled water coating, CS coating, PA-g-CS I (low grafting 125degree) coating, PA-g-CS II (medium grafting degree) coating, PA-g-CS III (high grafting degree) coating	Treatment Time: 30 s Storage temperature: 4 ± 1 °C Relative humidity: 95% Group: -Control (distilled water coating) group -CS coating group -PA-g-CS I (low grafting degree) coating group -PA-g-CS II (medium grafting degree) coating group -PA-g-CS III (high grafting degree) coating group	15 d	Optimal processing: PA-g-CS III (high grafting rate) coating group -Maintain high quality -Lower membrane lipid peroxidation -Antioxidant activity enhancement (SOD, APX, GR, CAT) -Microstructure: PA-g-CS coating group has a less entangled fiber structure and smaller pores.	[12]
Lactic acid fermentation	Group: -Storage temperature: 20 °C Sauerkraut process: 2% salt, 1% crystal sugar, and 0.1% Lactic Acid Bacteria Powder Starter -Storage temperature: 4 °C Kimchi process: 4% solar salt, 2% sugar and 0.1% Lactic Acid Bacteria Powder Starter -Storage temperature: 30 °C Pickle process: 50 mM acetic acid, 2.06 M NaCl and 2% sugar and 0.1% Lactic Acid Bacteria Powder Starter -Storage temperature: 20–25 °C Control heavy salting process: Saturated brine (450 mL, 25%, approximately)	30 d	Optimal processing: Control heavy salting process -Microbial counts changes: no count of lactic acid bacteria and Enterobacterial was detected; yeasts and molds were able to survive at 30 days -Inhibit the action of microorganisms: pH and titratable acidity: nearly unchanged -Nitrite concentration: relatively low and stable	[16]

Table 1. *Cont.*

Treatments	Process Parameters	Storage Days	Preservation Effect	Ref.
Natural freezing (NF, −20 °C) or individually quick-frozen (IQF) (−62.5 °C and speed 8.23 m/s) methods	Storage temperature: −20 °C Group: -NF, thawed by NT at room temperature -NF, thawed by FT at 4 °C -NF, thawed by MT at 620 W. -IQF, thawed by NT at room temperature -IQF, thawed by FT at 4 °C -IQF, thawed by MT at 620 W	—	Optimal processing: IQF, thawed by NT at room temperature -Thawing curve: takes less time to reach 4 °C -Water holding capacity: significantly higher than that of NF; thawing loss: significantly lower than that of NF -Cutting force analysis: high hardness -Sensory evaluation of thawed mushroom: superior to NF samples in all aspects; IQF least affected the quality after thawing	[17]
freezing or canning	Group: Storage temperature: −25 °C Freezing and Canning	—	Optimal processing: Boletus edulis, Freezing Preservation effect: -The coefficients for converting total nitrogen to protein: 4.18	[18]
PPP@chitosan nanoparticles	Storage temperature: 37 °C Group: -PPP 1.5 mg/mL -PPP 3 mg/mL -PPP 4.5 mg/mL -PPP 6 mg/mL	5 d	Optimal processing: PPP 3 mg/mL -Inhibit the activity of *E. coli* O157:H7 on food surfaces. Antimicrobial activity: pork: The number of *E. coli* O157:H7 decreased by 99.02% and 99.11% cucumber: the number of *E. coli* O157:H7 decreased by 99.48% and 99.77%	[19]

To fully preserve the nutrient contents of *Pl. eryngii*, increase its shelf life and better promote the interests of the whole king oyster mushroom industry, this paper reviews the primary manifestations of quality deterioration of these mushrooms, the quality changes of *Pl. eryngii* in postharvest storage, the mechanism of browning and the storage effects of different preservation methods, and provides a reference for the development of green preservation processes for *Pl. eryngii*.

2. Deterioration of the Quality of *Pl. eryngii*

The deterioration of the quality of *Pl. eryngii* after harvesting severely limits its commercial value and hinders the development of the mushroom industry. The deterioration in mushroom quality is characterized by the reduction in sensory and nutritional quality, which is owing to a combination of internal and external factors. Currently, research on the deterioration of the quality of the mushroom has focused on water loss, weight loss, postharvest morphological changes, changes in textural characteristics, color-specific changes, loss of nutrition and flavor and microbial infection.

2.1. *Loss of Water and Weight*

Fresh *Pl. eryngii* has a moisture content of up to 90% (wet basis), but its loss of moisture during storage can easily lead to weight loss, which is an important factor in the quality of fresh mushrooms [20]. A study showed that the weight loss of *Pl. eryngii* stored at 4 °C and 25 °C increased to 0.69% and 3.41%, respectively, ($p < 0.01$) compared with that of *Pl. eryngii* on day 0 throughout the storage period [21]. This is primarily owing to the exudation of cell contents, the sudden increase in the content of malondialdehyde (MDA) and the effect of related enzymes, such as superoxide dismutase (SOD), peroxidase (POD), and catalase (CAT). When the weight loss reaches 3.41% of its fresh mass, *Pl. eryngii* is considered decayed and unusable as food [21].

2.2. Altered Textural Properties

King oyster mushrooms are subject to aging during storage, which results in a rapid loss of hardness and contamination by microorganisms, thus explaining their short shelf life [22]. Studies have demonstrated that after 12 days of storage at 4 °C, the hardness of the mushroom decreases from 9.024 N to approximately 3.132 N. After 6 days of storage at 25 °C, the hardness of the mushroom decreases sharply to 3.11 N. Studies have shown that when the hardness of stored mushrooms decreases to less than 3.11 N, microorganisms appear on their surface, which causes them to deteriorate [21]. During storage at low temperatures, the fresh appearance of the *Pl. eryngii* is always accompanied by deterioration owing to lignification [8]. Lignification not only leads to a toughening of the *Pl. eryngii* texture and a significant reduction in nutrients but also promotes lipid peroxidation and deterioration of the king oyster mushroom substrate [8].

2.3. Change in Color Characteristics

Of all the quality properties that drive consumer purchasing behavior, color is the most evident dimension of quality. Among the parameters of mushroom browning, the L* value is often used to reflect the color change of the mushroom; a higher L* value indicates less browning and higher quality [21]. A study showed that the L* of *Pl. eryngii* stored at 4 °C during the first 6 days did not change significantly. The L* value of *Pl. eryngii* stored at 25 °C decreased from 94.15 to 75.33 from day 0 to day 6, respectively, indicating severe deterioration of *Pl. eryngii* [21]. When $L^* \leq 82$, the mushroom is of poor quality and not acceptable to the consumer. In addition to the L* value, the browning index (BI) can also be used to measure the degree of browning on the surface of king oyster mushrooms. Studies have shown that the degree of browning of the mushroom continues to increase with storage time. After 9 days of storage at 4 °C, there was slight browning on the surface of the mushroom. On day 12 of storage at 4 °C, the BI value on the surface of the mushroom reached 5.33-flod, which was no longer acceptable to the consumer compared to day 0. Compared with storage at 4 °C, the surface of the mushrooms stored at 25 °C reached a severe degree of browning after 6 days [21].

2.4. Loss of Nutrition

Sugars and soluble proteins in the king oyster mushroom are the primary nutrients that support ongoing metabolic activity during the postharvest phase. A reduction in protein or sugar is an important indicator of deterioration [23]. Li et al. demonstrated that the cellular oxidation of *Pl. eryngii* increased with storage time, resulting in more reactive oxygen species (ROS), which caused a decrease in reducing sugars owing to oxidation. In addition, the total free amino acid content is consumed during the pre-storage period to maintain the metabolic functions of the mushrooms. The contents of amino acids generally continue to decrease during the first 3 days of storage and do not start to increase until after 3 days. In addition to this, the fat content decreases as the storage time increases because the fat stored in the fat cells is gradually hydrolyzed by lipase into fatty acids and glycerol, which are then oxidized in other tissues [21].

3. Factors That Affect the Storage Quality of *Pl. eryngii*

3.1. Moisture

The tissue concentration of active polyphenol oxidase (PPO) and phenolic compounds, pH, temperature, water activity and oxygen accessibility are the most important factors that influence the rate of enzymatic browning in freshly harvested *Pl. eryngii* [4,24,25], which are highly susceptible to mechanical damage and microbial infection owing to their high contents of water (approximately 89%), lack of cuticle and presence of microorganisms on them [26]. Secondly, water loss or transpiration is an important physiological process that affects the primary quality characteristics of fresh mushrooms, such as marketable weight, appearance and texture, depending on the ambient and relative temperature and humidity [27]. Fresh *Pl. eryngii* has a very limited shelf life of 1–3 days at ambient temperature and 4–7 days at 4 °C [28]. With the increase in storage time, the apparent degradation of *Pl. eryngii* after harvesting gradually decreases in moisture, changes in internal enzymatic activity and bacterial enzymatic activity, which manifests as browning, texture softening and loss of flavor, which seriously affects its nutritional and commercial value [22] (Figure 1).

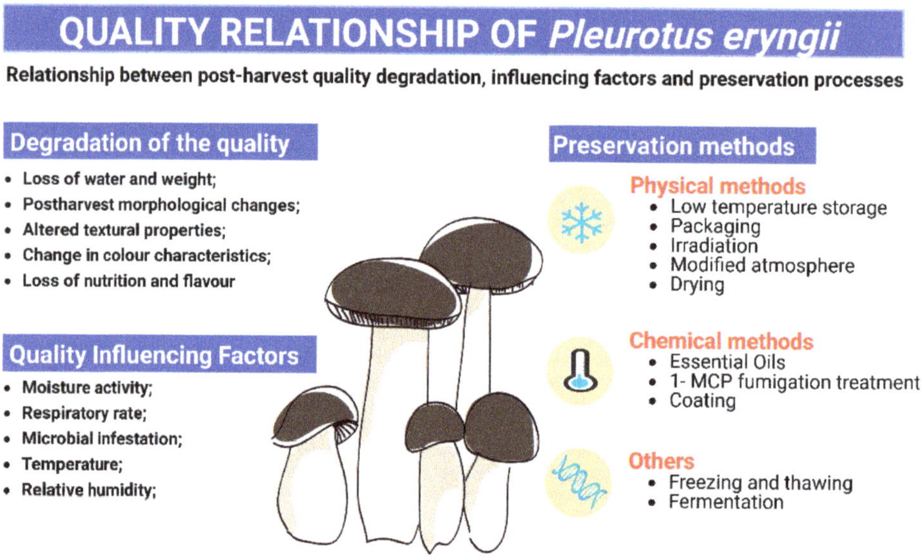

Figure 1. Relationship between post-harvest quality degradation, influencing factors and preservation processes of *Pl. eryngii*. Created with BioRender.com.

3.2. Respiratory Rate

The respiration rate and energy status of *Pl. eryngii* are key factors that influence postharvest senescence [29–31]. During storage at 25 °C, the respiratory intensity of freshly cut mushrooms increased rapidly with time, reaching 1382 CO_2 mg/(kg·h) at 12 h and 3526 CO_2 mg/(kg·h) at 72 h when more than 50% of the surface of the mushroom became brown and basically lost its edible value [29,30]. First, in terms of respiration rate, postharvest storage is an abiotic stress on *Pl. eryngii* since storage conditions are very different from those of growth. This storage leads to an inhibition of electron transfer in the mitochondria and an increase in the production of ROS [10]. As the levels of ROS surpass the cell's antioxidant capacity, oxidative stress develops and mediates structural damage to lipids, membranes, proteins and DNA [32]. These results demonstrate that mitochondrial membrane enzymes implicated in mitochondrial respiratory metabolism, such as cytochrome C oxidase (CCO), will be destroyed and their function severely diminished [33,34]. To better preserve the mushrooms, the relationship between ROS and respiratory metabolism in *Pl. eryngii* is currently a hot topic in postharvest preservation research [10]. Secondly, in terms of energy metabolism, it has been shown that an inadequate supply of ATP is closely associated with a variety of postharvest symptoms, such as chilling injury, browning, yellowing and decay [35]. An adequate supply of ATP inhibits the accumulation of ROS and maintains membrane integrity, thereby delaying the aging and deterioration of *Pl. eryngii* [36]. Under postharvest abiotic stress, the energy status plays an important role in mitigating oxidative damage and maintaining organoleptic properties [36,37]. Therefore, the energy maintenance of *Pl. eryngii* during postharvest storage needs to be given high priority (Figure 1).

3.3. Microbial Infection

Freshly harvested *Pl. eryngii* is highly susceptible to mechanical damage and microbial infection owing to its high content of water (approximately 89%), the absence of cuticle protection and the presence of many microorganisms on its surface [4,24,25]. Decay in *Pl. eryngii* is usually induced by the tolaasin toxin in *Pseudomonas tolaasii*, which results in brown spots and yellow to dark brown lesions on the cap of the fungus [38–40]. In addition, other bacteria, such as *Pseudomonas azotoformans*, *Pseudomonas brenneri* and *Ewingella americana*, have been reported to be able to cause decay in *Pl. eryngii* [41]. *Listeria monocytogenes* has also been isolated from *Pl. eryngii* farm environments, which highlights the importance of monitoring the production chain from substrate production to harvesting, processing and packaging [42–44] (Figure 1).

3.4. Temperature and Relative Humidity

The various nutrients, such as polysaccharides, aldehydes and phenolic compounds, quality characteristics and microbial reproduction in the king oyster mushroom are influenced by temperature and relative humidity. Temperature fluctuations during storage can activate a variety of oxidative enzymes, enhance physiological activity, affect respiration and transpiration and increase the post-ripening period of stored *Pl. eryngii* [27], while temperature is also an important factor in determining the rate of enzymatic browning [26]. Therefore, in general, when storing *Pl. eryngii*, the shelf life is usually extended by reducing the storage temperature and increasing the ambient humidity. A storage temperature of 4–6 °C and relative humidity (RH) of approximately 95% is generally used (Figure 1).

4. Methods for Storing and Preserving *Pl. eryngii*

4.1. Physical Methods and Mechanism

4.1.1. Modified Atmosphere Packaging (MAP)

MAP is used to control the proportion of nitrogen, oxygen, carbon dioxide and ethylene in the gas, humidity, temperature (above the freezing threshold) and air pressure of the gas in the gas conditioning warehouse, thereby inhibiting the amount of cellular respiration and reducing the metabolic rate, so that the mushrooms are nearly dormant, thus preserving them over a long-term period [45]. As a complement to storage temperature control, MAP has been found to be a simple, economical and effective postharvest preservation technique for commodities [46]. Four core parameters need to be considered when designing MAP, including the product characteristics, the permeability of the packaging material, gas concentration (carbon dioxide and oxygen) and temperature dependence [47,48]. The quality of the MAP of *Pl. eryngii* is related to texture, microbial count, whiteness, color variation and organoleptic characteristics, which are essential for the analysis of spoilage rates and thus influence consumer acceptance [49]. Figure 2 summarizes the mechanism of action of MAP preservation and the changes in *Pl. eryngii* morphology from existing MAP studies.

To investigate the total quality index of king oyster mushrooms treated with different gas mixtures of MAP after harvesting, Wan-Mohtar et al. [13] investigated this and showed that high CO_2 packaging (HCP) (20% CO_2 and 15% O_2) retained the best qualities of king oyster mushrooms. HCP recorded the highest total phenolic content (TPC) and showed the highest effectiveness in maintaining the color and odor of *Pl. eryngii* compared with the control and low CO_2 packaging (LCP: 2% CO_2 and 30% O_2). Briones et al. [50] suggested that the use of 2.5–5% CO_2 and 5–10% O_2 would result in optimal storage conditions for mushrooms. For safety reasons, it is recommended that O_2 should not be less than 2% under MAP conditions [51]. Research by Jafri et al. [52] that utilized 10% O_2 + 5% CO_2 for the MAP treatment of king oyster mushrooms showed that this model was more effective at retaining quality characteristics and higher organoleptic ratings compared with other samples, which could be maintained for a storage period of 25 days. The treated mushrooms showed minimal changes in weight loss, pH and total soluble solids. Free radical scavenging activity and the total polyphenol contents were maintained at 85% and 91%, respectively [52]. The effect of MAP on the enzymatic activity and shelf life of king oyster mushrooms stored at 20–25 °C and 90–95% RH for 5 days was investigated by Li et al. The results indicated that 2% O_2 + 30% CO_2 significantly prolonged the shelf life of the mushrooms compared with the control. A total of 2% O_2 + 30% CO_2 mixture was more suitable for maintaining the organoleptic properties of the mushrooms and delaying the increase in MDA and O_2 production during storage. In addition, the activities of SOD, POD and CAT were significantly higher than those of the control. Treatment with 2% O_2 + 30% CO_2 reduced lipid peroxidation and enhanced the activity of antioxidant enzymes but had little effect on the CCO activity of the mushrooms [10].

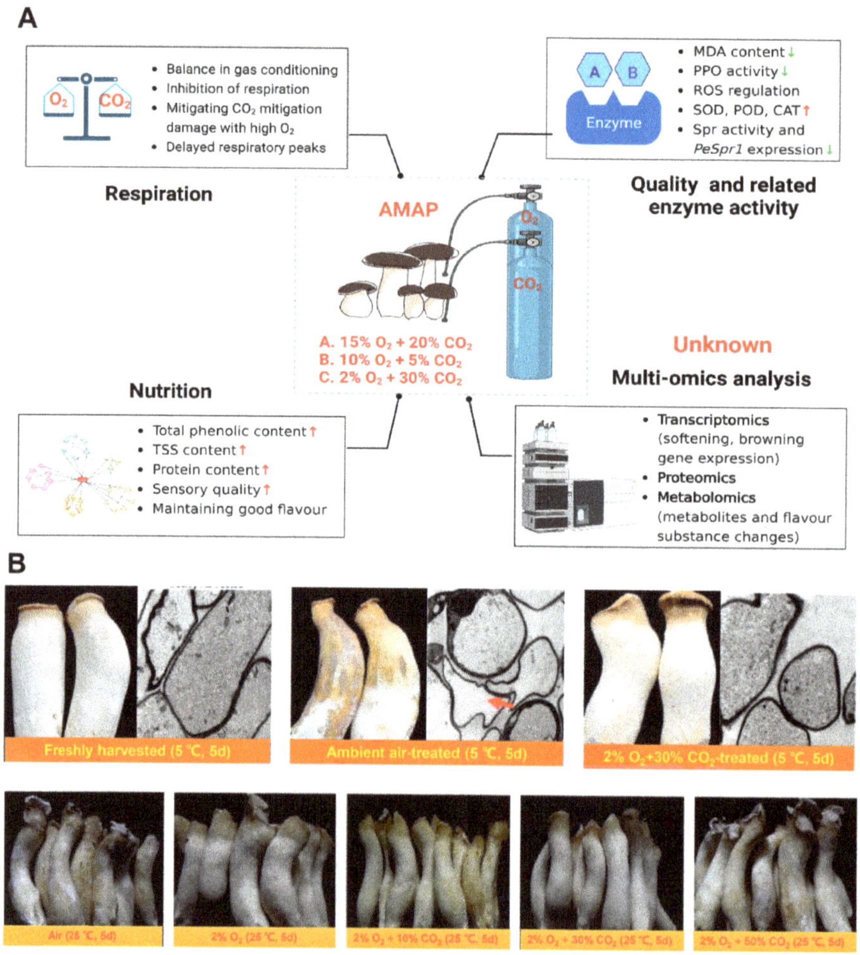

Figure 2. Mechanism of action of MAP preservation of *Pl. eryngii* (**A**) and changes in *Pl. eryngii* morphology from existing MAP research (**B**). (**A**) Created with BioRender.com. (**B**) cited from [9,10] The arrow in the figure points to the destruction of the cell structure.

The molecular mechanisms of postharvest senescence also merit attention. Zhang et al. [9] showed that the shelf-life of the mushrooms was prolonged after 2% O_2 + 30% CO_2 treatment and that the cell morphology was normal with no obvious aberrations, and the cytoplasmic distribution was as uniform as that of freshly harvested mushrooms, which significantly inhibited cell abnormalities, serine protease activity and *PeSpr1* expression. However, there is a lack of research on the flavor and nutrient changes caused by metabolic substances during the gas conditioning process of king oyster mushrooms, which is crucial for acceptability by consumers. Further studies on the transcriptome, proteome, metabolome and multi-omics of this mushroom after gas conditioning treatment should be strengthened to provide a theoretical basis for the gas conditioning preservation mechanism (Figure 2).

4.1.2. Special Packaging

Phase change materials (PCMs) are substances that absorb latent heat through phase changes and play an important role in short-duration cold chain transport [53]. Li et al. [14] developed a new water-based PCM and showed that king oyster mushrooms treated with the new PCM accumulated the most phenolics and flavonoids in all three groups, which mitigated the deterioration of its appearance during storage (Figure 3). The measurements of free amino acids demonstrated that the new PCM treatment increased the levels of phenylalanine, glutamic acid (Glu) and proline (Pro) by creating low-temperature conditions, thus improving the nutritional quality and flavor attributes and delaying the postharvest aging of king oyster mushrooms. In addition, the new PCM treatment maintained an adequate energy supply to the mushroom by activating the activities of succinate dehydrogenase, CCO and ATPases, thus reducing the catabolism of Pro and Glu.

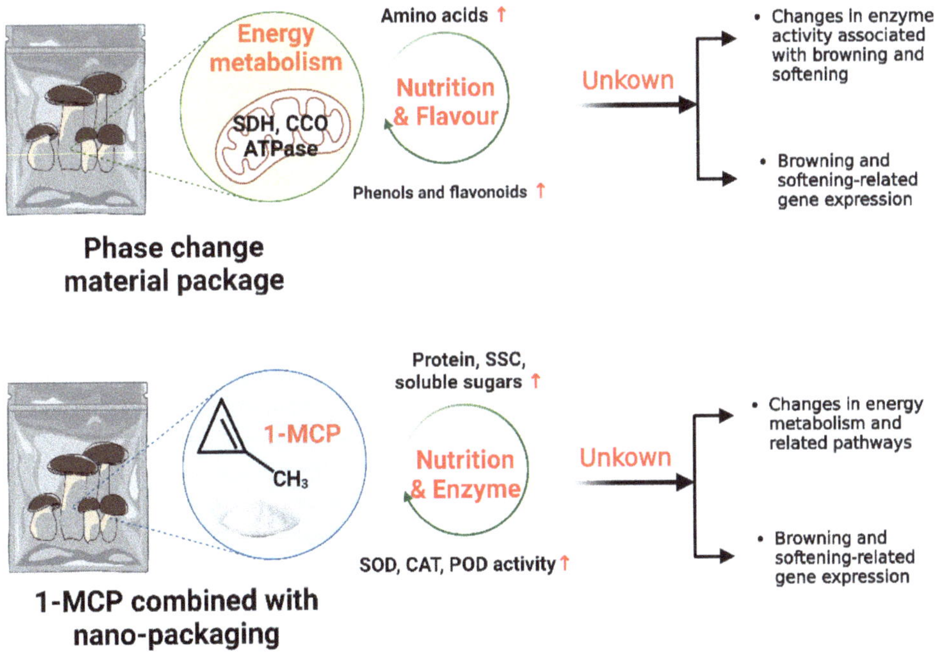

Figure 3. Mechanism of action of special packaging preservation of *Pl. eryngii*. Created with BioRender.com.

The application of nano-packaging can extend the life of postharvest edible mushrooms and maintain their original color and taste [54,55]. 1-MCP, a type of cyclopropene, has been widely used and shown to inhibit the action of ethylene in respiratory senescent fruit by competitively binding to ethylene receptors [56]. Xu et al. indicated that 1-MCP combined with nanopackaging treatment was effective at suppressing the increase in respiratory intensity, weight loss, MDA content and PPO activity of *Pl. eryngii* at 4 °C, delaying the decrease in soluble protein content, maintaining soluble sugar and soluble solid content and increasing the activities of SOD and POD, thereby maintaining the postharvest quality of king oyster mushrooms and extending the storage time [11]. The efficiency of the combined treatment was superior to that of the sole packaging with 1-MCP or nano compared with the untreated samples [11].

Currently, nanopackaging studies on Enoki mushrooms (*Flammulina velutipe*) are relatively thorough [57–60] and complete in terms of basic physicochemical indicators, reactive oxygen metabolism, energy metabolism, proteomics and metabolomics to elaborate the storage quality, primarily browning and softening, of Enoki mushrooms extreme mechanisms of action. The study on king oyster mushrooms can also be studied in-depth in this respect in terms of a single packaging technique, which expands its intrinsic preservation mechanisms in terms of energy and multi-omics expression.

4.1.3. Low-Temperature Storage

Low-temperature storage is a common way to store and preserve edible mushrooms. Low temperatures can inhibit enzyme activity, reduce physiological metabolic activity, reduce the respiratory intensity and inhibit the growth and reproduction of microorganisms (Figure 4). Li et al. [8] conducted a related study on this in 2015 and showed that the optimal treatment was 2 °C and that toughening occurred twice throughout the storage process. This treatment maintained high textural properties for 18 days, with higher contents of chitin and higher activities of phenylalanine ammonia lyase (PAL), CAT and POD, and maintained a high content of total phenolics and lower membrane lipid peroxidation. This also suggests that toughening may be primarily caused by oxidation and can affect the quality of the mushrooms after harvesting [8]. A further complementary study on the same preservation method by Li et al. in 2021 compared quality parameters, chemical composition, MDA concentration and metabolic enzyme activity during storage at 4 °C for 12 days and at 25 °C for 6 days. The best treatment measure was found to be the treatment group stored at 4 °C for 12 days, which maintained high quality, high nutritional characteristics, a high content of total phenolics, progressively higher enzyme activity and low membrane lipid peroxidation. Simultaneously, increased activities of laccase, lipoxygenase and PAL and the accumulation of MDA, as well as polysaccharide degradation, were the primary factors that contributed to the deterioration of the king oyster mushrooms during storage [21].

Freezing prevents the growth of microorganisms and preserves the texture of tissues and the nutritional value of food [61]. Long-term freezing (fast or slow) is the appropriate way to preserve mushrooms for the long term [61]. It involves the extensive exposure of cells to low temperatures and dehydration. Jiang et al. [62] evaluated the metabolite content of substrates to improve the understanding of changes in the nutritional composition of king oyster mushrooms during short-term slow frozen storage. The study showed that the optimal treatment was a storage temperature of −30 °C for the caps, which maintained a high nutritional value. The content of polysaccharides, proteins and amino acids in the cap increased and then decreased, while the content of all measured substances in the stalk slowly decreased. The activity of α-amylase decreased; that of POD increased, and the contents of reducing sugars and vitamin C continuously decreased with the extension of the freezing time [62].

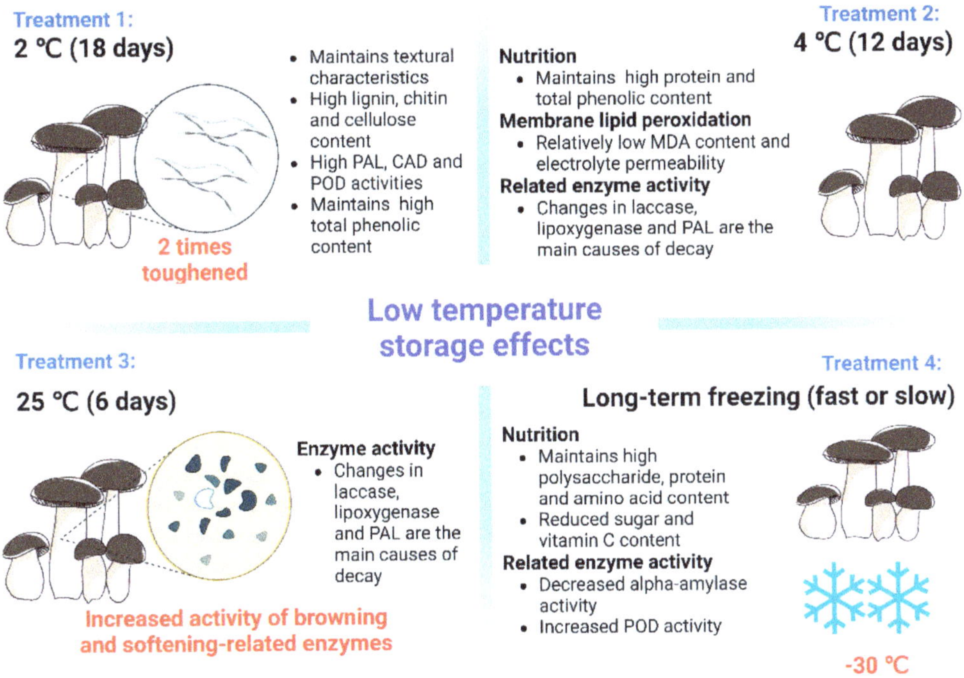

Figure 4. Mechanism of action of *Pl. eryngii* for low-temperature storage and preservation. Created with BioRender.com.

4.1.4. Irradiation

The application of improved postharvest techniques, such as food irradiation, can improve marketability and extend storage life, and the technology is now widely commercialized [63,64]. The technical suitability and nutritional safety of irradiated foods have been well studied [31,64]. Low doses of irradiation of fresh produce can provide hygienic safety and affect different physiological processes, such as enzyme activity and respiration, thus significantly improving postharvest storage [22,65,66]. Akram et al. investigated the quality attributes of irradiated king oyster mushrooms. The study showed that the best treatment measure was irradiation at 1 kGy and that the L-value (brightness) of this group increased after irradiation and remained high throughout storage, maintaining a good appearance as indicated by homogeneous color and the absence of fungal decay and blemishes, good hardness and microstructure and low weight loss [3]. Irradiation at 1 kGy was the most effective for extended postharvest storage and had additional advantages [3] (Figure 5). However, irradiation treatment requires a high level of skill on the part of the operator and still requires significant consideration of its cost.

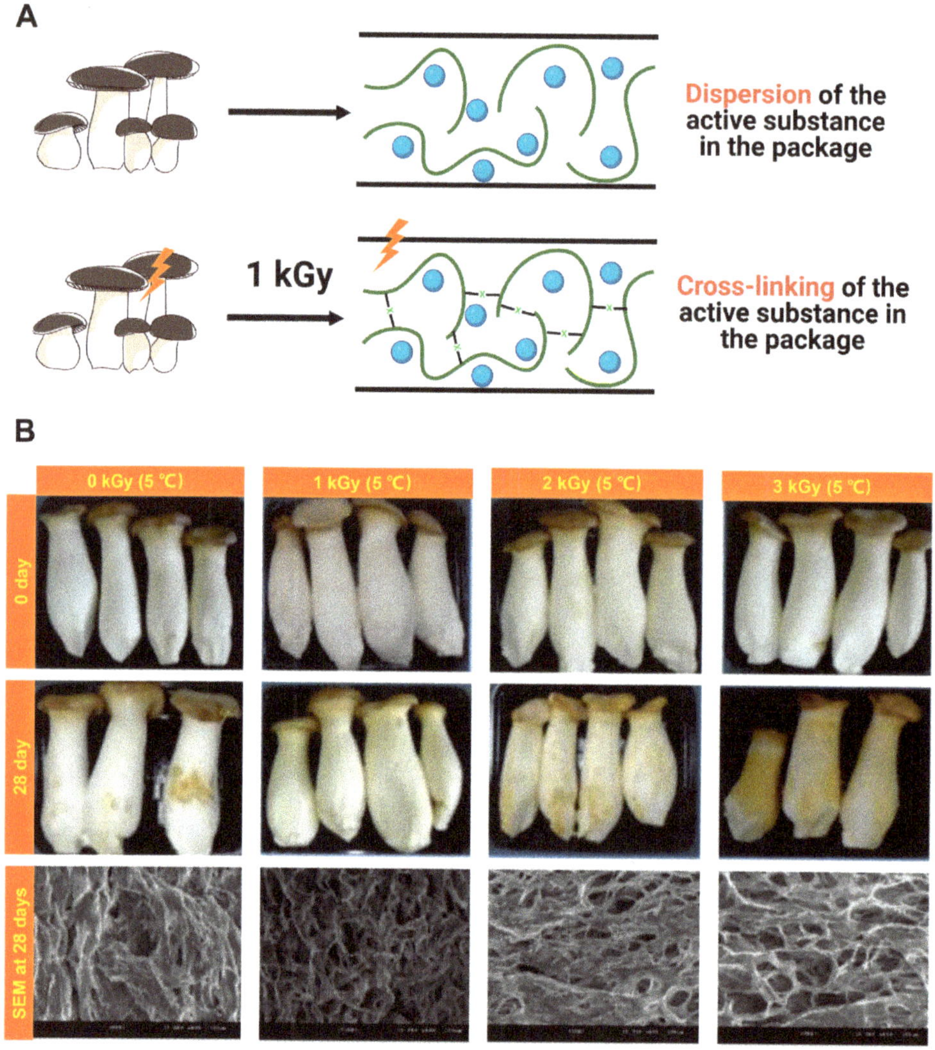

Figure 5. Mechanism of action of irradiation preservation of *Pl. eryngii* (**A**) and changes in *Pl. eryngii* morphology and SEM images from existing irradiation research (**B**). (**A**) Created with BioRender.com. (**B**) Cited from [3] ©Copyright 2012, Elsevier.

4.1.5. Drying

Drying is a typical approach to food preservation based on the principle that the water activity of the product should be minimized to a defined level to ensure microbiological and physicochemical stabilization; it has been used for many years to improve the shelf life of food commodities [67,68]. Hot blanching is receiving increasing attention as a pretreatment method to improve drying quality [69]. The current scalding process, which is conducted by direct interaction between the sample and a medium, such as hot water and steam [70], can significantly ($p < 0.05$) reduce the total number of bacteria, improve drying efficiency and reduce the level of browning of the sample during drying [71]. However, shortcomings of water and steam blanching have been reported, including the loss of nutrients, such as vitamins, proteins and polysaccharides, and uneven blanching [70]. The results of Tolera et al. showed that the optimal treatment was a solar drying method with an infiltration concentration of 5%, which reduced the moisture by 7.74% and maintained the following proximal component contents: crude protein content 25.13% db, crude fat 2.27% db, total ash 10.17% db, crude fiber 10.26% db and carbohydrates 44.42% db. The purpose of microwave hot-air flow rolling dry-blanching (MARDB) pretreatment is to improve the drying efficiency and quality of the king oyster mushroom [72]. Microwaving can alter the microstructure during the drying-hot blanching process, which could affect the drying characteristics, water state and migration [72]. Su et al. [15] revealed that optimal pretreatment (9 min) with MARDB significantly improved the quality indicators, such as color, water content and polysaccharide content of *Pl. eryngii*, shortened the drying time and completely deactivated PPO and POD. T2 relaxation spectra and microstructural analysis indicated that the primary reason for the improved drying efficiency at the optimal MARDB time was the resistance to free water migration and reduction in the pore structure. Excessive hot blanching (12 min) prolongs the drying time and leads to a reduction in whiteness and the contents of polysaccharides and phenolics [15]. Ucar et al. [73] freeze-dried *Pl. eryngii* at −20 °C, which maintained a better color and preserved the textural properties to prevent softening. However, the cost is relatively high when it comes to industrial production [73].

4.2. Chemical Methods and Mechanism

4.2.1. Essential Oil Treatment

EOs are natural volatiles obtained by distillation and have the characteristic aroma of the plants from which they are extracted [74]. An EO acts on the biochemical processes of the mushroom and inhibits or increases the concentration of enzymes and secondary metabolites associated with the preservation of quality [75,76]. Manjari et al. [77] conducted an experiment to study the effect of different essential oils on the enzymatic activity of stored *Pl. eryngii*. The results showed that the best treatment was peppermint oil (10 µL), which maintained high contents of total phenolics, TPC (0.286 mg/g), PAL (0.038 µM/g), PPO (0.042 U/mg) and POD (0.38 U/mg). The higher levels of TPC and PAL in the *Pl. eryngii* treated with EO and the lower levels of PPO and POD in the treated samples compared with those of the control indicated that the EO treatment had a positive effect on the quality of the harvested mushrooms [77]. This preservative technique will help to extend the shelf life of the harvested substrates. Studies have reported that EOs have a significant antibacterial effect [78], but there is a lack of available research on the antibacterial effect and mechanism of action of EOs on *Pl. eryngii* during storage.

4.2.2. Coating

In recent years, many different types of edible coatings have been successfully explored and further developed for the postharvest storage of mushrooms [79]. Chitosan is a biodegradable polymer that occurs naturally and can be applied as an edible coating to suppress changes in the quality of mushrooms during storage [80]. Liu et al. [12] investigated a solution of protocatechuic acid grafted chitosan (PA-g-CS) with an antioxidant potential as a possible new edible coating material for the postharvest storage of king oyster mushrooms [12]. The results showed that the best treatment was the PA-g-CS III (high grafting rate) coating group, which was able to maintain good textural properties, low membrane lipid peroxidation, high activities of SOD, ascorbate peroxidase (APX), glutathione reductase (GR) and CAT and low activity of PPO [12].

There is good current acceptance of edible coating films in mushroom preservation, but there is still a need to expand the use of edible coating solutions of natural plant origin in king oyster mushrooms. Moreover, the mechanism of action of coated film preservation in *Pl. eryngii* merits further study, such as the use of a multi-omics approach to elucidate the expression of relevant browning and softening genes, protein up-/downregulation and flavor changes during storage caused by differential metabolites. In addition to this, changes in energy owing to respiration after film coating for preservation need to be considered to elucidate the mechanisms of preservation (Figure 6).

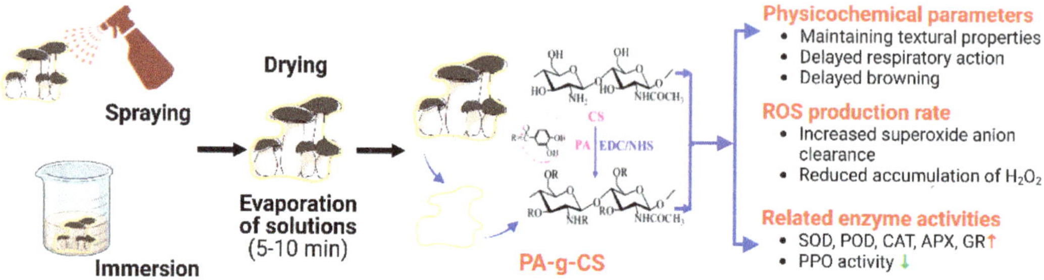

Figure 6. Mechanism of action of *Pl. eryngii* for coating. Created with BioRender.com.

4.3. Others

4.3.1. Different Freeze–Thaw Treatments

In contrast to slow block freezing, the single-piece quick freezing method uses cryogenic gases to rapidly reduce the temperature of mushrooms to the freezing point, which maintains cellular integrity with little change in nutritional quality and organoleptic properties [81]. In general, frozen foods need to be thawed before processing and consumption, and thawing has a direct or indirect effect on the quality of the product. Therefore, freezing and defrosting are equally important to consumers. There are several methods of defrosting that are frequently used by consumers, such as natural air convection defrosting (NT), flow-through defrosting (FT) or microwave defrosting (MT) [81]. Li et al. used the natural freezing (NF, −20 °C) or single freezing (−62.5 °C, speed 8.23 m/s) methods to freeze cut king oyster mushrooms, and three thawing methods, including flowing water (FT, 4 °C), microwaving (MT, 620 W) and natural air convection (NT, 20 ± 5 °C), to thaw the mushrooms [17]. The results of the study showed that the best treatment measure was individual quick freezing and thawing with NT at room temperature, which was able to maintain cell integrity, preserve the texture of king oyster mushrooms and maintain high water holding capacity, low thawing losses, good color and good flavor. As a result, the method minimizes changes in the quality of frozen king oyster mushrooms [17].

4.3.2. Fermentation

As one of the oldest processing techniques, lactic acid fermentation is recognized as a highly valuable processing approach to retain and improve the safety, nutritional and sensory characteristics of vegetables [82]. In addition, varieties of lacto-fermented vegetables are often classified by their composition and method of preparation. For example, sauerkraut, kimchi, such as that made from cucumber and olive, and kimchi are the most investigated lacto-fermented vegetables, predominantly for their commercial importance [83,84]. Today, pure fermentations of lactic acid bacteria (LAB) are widely used on a commercial scale for these commodities. As a result, this technique offers advantages over traditional methods, including shortened fermentation cycles, the elimination of non-lactic acid contaminants, and rapid fermentation at higher temperatures. It also ensures hygienic conditions and maintains consistency for better quality and flavor. *Lactobacillus plantarum* is an important member of the LAB family and is commonly used to ferment vegetables [85]. Zheng et al. studied the preservation of king oyster mushrooms using three typical lactic acid fermentation processes, including sauerkraut, pickling and kimchi, with *L. plantarum* as the fermentation agent. This study showed that controlling the heavy salt pickling process inhibited microbial growth and reproduction and rendered most microorganisms inactive. These LAB rapidly colonize the mushroom substrate and quickly control spoilage and pathogenic microorganisms [16]. The final fermentation product contained high levels of LAB (>7 Log CFU/g). In addition, the nitrite concentration in the final fermentation product was below the current maximum level permitted in China (<20 mg/kg). The results indicate that the lactic acid fermentation method is effective and safe for the preservation of king oyster mushrooms [16].

4.3.3. Polysaccharide Nanoparticle Preservation

Chitosan nanoparticles are used to encapsulate bioactive substances owing to their good biocompatibility, high efficiency of encapsulation, safety and non-toxic properties [86]. Therefore, if chitosan-based nanoparticles are used in combination with antimicrobial agents, such nanoparticles may induce synergistic effects between chitosan and antimicrobial agents [87]. Microbial contamination usually occurs on the surface of food products. When nanoparticles are sprayed directly onto the food surface, vesicles and uneven distribution can occur, thus weakening the antimicrobial effect. The morphological transformation from nanoparticles to nanofibers is considered a feasible approach because nanofibers have a larger specific surface area and disperse more effectively. Pomegranate peel polyphenol (PPP), a natural, safe and green antimicrobial agent, was introduced and embedded in chitosan to form stable nanoparticles. PPP chitosan nanoparticles (PPP-CNPs) were further electrospun into king oyster mushroom polysaccharide (PEP)-based nanofibers. The optimal treatment measure of PPP 3 mg/mL was obtained by Cai et al. [19]. This group was able to maintain small nanoparticle size and uniform nanoparticle dispersion, maintain optimum stability, produce tighter nanofibers, improve the thermal stability of PEP nanofibers, inhibit the activity of *E. coli* O157: H7 on the food surface, maintain good color quality and obtain the highest encapsulation rate of 23.71 ± 0.51% [19]. However, the safety of this type of preservation technology in industrial applications that produce king oyster mushrooms still requires additional evaluation.

5. Challenges and Future Trends

There are currently relatively few effective means to commercially preserve *Pl. eryngii*. They primarily include low-temperature storage, gas preservation and vacuum drying, and their ability to preserve the mushrooms is highly inadequate for the needs of industrialization. Therefore, it is important to study the mechanisms that cause the quality of *Pl. eryngii* to deteriorate and implement new preservation techniques to extend the shelf life of these mushrooms. Based on the mechanism of the deterioration in the quality of king oyster mushrooms, future research should focus on the following aspects.

(1) To further investigate the mechanisms of quality fission during storage and preservation, such as browning, softening and lignification, and to use multi-omics techniques to study the potential molecular mechanisms of gene regulation in different preservation methods. This approach should help to address the problem of postharvest quality deterioration of king oyster mushroom strains at the molecular level.

(2) Research on the mechanisms of nutrient retention and flavor transfer during storage and the effects of different preservation methods on the biological activity and quality characteristics of king oyster mushrooms should be strengthened to improve the quality characteristics of king oyster mushrooms after preservation while extending its shelf life and greatly enhancing its commercial value.

(3) Among the methods of preserving *Pl. eryngii*, relatively little research has been conducted on the use of radiation, ozone and film coatings to preserve these mushrooms. There is still a need to explore the effects of these traditional methods of preserving edible mushrooms on *Pl. eryngii* and the mechanism of preservation, as well as the development of new green preservatives, based on natural types of bioactive substances.

(4) In the future, a combination of new and traditional technologies can be used to improve the postharvest quality of *Pl. eryngii*, such as combining radiation treatment with 1-MCP in concert with nanopackaging treatment, developing cold sterilization equipment, creating safe and efficient sterilization processes, such as irradiation,

microwave, low-pressure electrostatic field and low-temperature plasma sterilization equipment and processes, and decreasing the deterioration of the quality of *Pl. eryngii* during storage and distribution.

Author Contributions: Y.G.: Writing—original draft, Conceptualization. X.C.: Visualization, Investigation, Writing—review and editing. P.G.: Supervision. Z.D., Z.Q., R.W. and A.H.: Investigation and Formal analysis. H.L. and J.W. (Jiating Wang): Data curation, Formal analysis, Investigation. W.Y. (Wenbo Yao), W.Y. (Wenjuan Yang), J.W. (Jing Wang) and N.L.: Formal analysis, Investigation. All authors have read and agreed to the published version of the manuscript.

Funding: This work was supported in part by grants from the National Key Research and Development Program [No. 2021YFD1600400], the General Plan of Shaanxi Province [No. 2020GY-236, 2022NY-035], the Key Industrial Chain Projects of the Shaanxi Province-Agricultural Field [2021ZDLNY04-01, 2022ZDLNY04-05], Industrialization projects of the Education Department of Shaanxi Province [22JC021], the Project from Weiyang Technology Bureau (202131), the Project from the Xi'an City Innovation Plan-Agricultural Field (21NYYF0022), the project from Qinchuang Yuan "Scientist& Engineers" Team (S2022-ZC-QCYK-0011) and the Project from the Ningxia Zhong Ning Goji Industry Innovation Research Institute (ZNGQCX-A-2020003).

Institutional Review Board Statement: Not applicable.

Informed Consent Statement: Not applicable.

Data Availability Statement: Not applicable.

Acknowledgments: Thanks to the College of Food and Biological Engineering, School of Biology and Medicine, Shaanxi University of Science and Technology for their support.

Conflicts of Interest: The authors declare no conflict of interest.

References

1. Zhang, B.; Li, Y.; Zhang, F.; Linhardt, R.J.; Zeng, G.; Zhang, A. Extraction, structure and bioactivities of the polysaccharides from Pleurotus eryngii: A review. *Int. J. Biol. Macromol.* **2020**, *150*, 1342–1347. [CrossRef]
2. Choi, U.-K.; Bajpai, V.K.; Lee, N.-H. Influence of calcinated starfish powder on growth, yield, spawn run and primordial germination of king oyster mushroom (*Pleurotus eryngii*). *Food Chem. Toxicol.* **2009**, *47*, 2830–2833. [CrossRef]
3. Akram, K.; Ahn, J.-J.; Yoon, S.-R.; Kim, G.-R.; Kwon, J.-H. Quality attributes of Pleurotus eryngii following gamma irradiation. *Postharvest Biol. Technol.* **2012**, *66*, 42–47. [CrossRef]
4. Guo, Y.; Chen, X.; Gong, P.; Li, Z.; Wu, Y.; Zhang, J.; Wang, J.; Yao, W.; Yang, W.; Chen, F. Advances in the mechanisms of polysaccharides in alleviating depression and its complications. *Phytomedicine* **2023**, *109*, 154566–154574. [CrossRef]
5. Guo, Y.; Chen, X.; Gong, P. Classification, structure and mechanism of antiviral polysaccharides derived from edible and medicinal fungus. *Int. J. Biol. Macromol.* **2021**, *183*, 1753–1773. [CrossRef]
6. Mesgari, M.; Aalami, A.H.; Sathyapalan, T.; Sahebkar, A. A comprehensive review of the development of carbohydrate macromolecules and copper oxide nanocomposite films in food nanopackaging. *Bioinorg. Chem. Appl.* **2022**, *2022*, 2565320–2565328. [CrossRef]
7. Al-Dairi, M.; Pathare, P.B.; Al-Yahyai, R.; Opara, U.L. Mechanical damage of fresh produce in postharvest transportation: Current status and future prospects. *Trends Food Sci. Technol.* **2022**, *124*, 195–207. [CrossRef]
8. Li, D.; Qin, X.; Tian, P.; Wang, J. Toughening and its association with the postharvest quality of king oyster mushroom (*Pleurotus eryngii*) stored at low temperature. *Food Chem.* **2016**, *196*, 1092–1100. [CrossRef]
9. Zhang, L.; Gao, J.; Hu, H.; Li, P. The activity and molecular characterization of a serine proteinase in Pleurotus eryngii during high carbon dioxide and low oxygen storage. *Postharvest Biol. Technol.* **2015**, *105*, 1–7. [CrossRef]
10. Li, P.; Zhang, X.; Hu, H.; Sun, Y.; Wang, Y.; Zhao, Y. High carbon dioxide and low oxygen storage effects on reactive oxygen species metabolism in Pleurotus eryngii. *Postharvest Biol. Technol.* **2013**, *85*, 141–146. [CrossRef]
11. Xu, F.; Liu, Y.; Shan, X.; Wang, S. Evaluation of 1-methylcyclopropene (1-MCP) treatment combined with nano-packaging on quality of pleurotus eryngii. *J. Food Sci. Technol.* **2018**, *55*, 4424–4431. [CrossRef]
12. Liu, J.; Meng, C.-G.; Wang, X.-C.; Chen, Y.; Kan, J.; Jin, C.-H. Effect of Protocatechuic Acid-Grafted-Chitosan Coating on the Postharvest Quality of *Pleurotus eryngii*. *J. Agric. Food Chem.* **2016**, *64*, 7225–7233. [CrossRef]
13. Wan-Mohtar, W.A.A.Q.I.; Klaus, A.; Cheng, A.; Salis, S.A.; Halim-Lim, S.A. Total quality index of commercial oyster mushroom *Pleurotus sapidus* in modified atmosphere packaging. *Br. Food J.* **2019**, *121*, 1871–1883. [CrossRef]
14. Li, D.; Wang, D.; Fang, Y.; Belwal, T.; Li, L.; Lin, X.; Xu, Y.; Chen, H.; Zhu, M.; Luo, Z. Involvement of energy metabolism and amino acid metabolism in quality attributes of postharvest Pleurotus eryngii treated with a novel phase change material. *Postharvest Biol. Technol.* **2021**, *173*, 111427. [CrossRef]

15. Su, D.; Lv, W.; Wang, Y.; Wang, L.; Li, D. Influence of microwave hot-air flow rolling dry-blanching on microstructure, water migration and quality of pleurotus eryngii during hot-air drying. *Food Control.* **2020**, *114*, 107228–207236. [CrossRef]
16. Zheng, H.-G.; Chen, J.-C.; Ahmad, I. Preservation of King Oyster Mushroom by the use of different fermentation processes. *J. Food Process. Preserv.* **2018**, *42*, e13396. [CrossRef]
17. Li, T.; Lee, J.-W.; Luo, L.; Kim, J.; Moon, B. Evaluation of the effects of different freezing and thawing methods on the quality preservation of Pleurotus eryngii. *Appl. Biol. Chem.* **2018**, *61*, 257–265. [CrossRef]
18. Bernaś, E.; Jaworska, G. Effect of preservation method on amino acid content in selected species of edible mushroom. *LWT* **2012**, *48*, 242–247. [CrossRef]
19. Cai, M.; Zhang, G.; Li, C.; Chen, X.; Cui, H.; Lin, L. Pleurotus eryngii polysaccharide nanofiber containing pomegranate peel polyphenol/chitosan nanoparticles for control of *E. coli* O157:H7. *Int. J. Biol. Macromol.* **2021**, *192*, 939–949. [CrossRef]
20. Lagnika, C.; Zhang, M.; Mothibe, K.J. Effects of ultrasound and high pressure argon on physico-chemical properties of white mushrooms (*Agaricus bisporus*) during postharvest storage. *Postharvest Biol. Technol.* **2013**, *82*, 87–94. [CrossRef]
21. Li, R.; Zheng, Q.; Lu, J.; Zou, Y.; Lin, J.; Guo, L.; Ye, S.; Xing, Z. Chemical composition and deterioration mechanism of Pleurotus tuoliensis during postharvest storage. *Food Chem.* **2021**, *338*, 127731–127740. [CrossRef]
22. Zhang, K.; Pu, Y.-Y.; Sun, D.-W. Recent advances in quality preservation of postharvest mushrooms (*Agaricus bisporus*): A review. *Trends Food Sci. Technol.* **2018**, *78*, 72–82. [CrossRef]
23. Meng, D.-M.; Zhang, Y.-X.; Yang, R.; Wang, J.; Zhang, X.-H.; Sheng, J.-P.; Wang, J.-P.; Fan, Z.-C. Arginase participates in the methyl jasmonate-regulated quality maintenance of postharvest *Agaricus bisporus* fruit bodies. *Postharvest Biol. Technol.* **2017**, *132*, 7–14. [CrossRef]
24. Villaescusa, R.; Gil, M. Quality improvement of Pleurotus mushrooms by modified atmosphere packaging and moisture absorbers. *Postharvest Biol. Technol.* **2003**, *28*, 169–179. [CrossRef]
25. Brennan, M.; Le Port, G.; Gormley, R. Post-harvest Treatment with Citric Acid or Hydrogen Peroxide to Extend the Shelf Life of Fresh Sliced Mushrooms. *LWT* **2000**, *33*, 285–289. [CrossRef]
26. Fernandes, Â; Antonio, A.L.; Oliveira, M.B.P.; Martins, A.; Ferreira, I.C. Effect of gamma and electron beam irradiation on the physico-chemical and nutritional properties of mushrooms: A review. *Food Chem.* **2012**, *135*, 641–650. [CrossRef]
27. Singh, P.; Langowski, H.-C.; Wani, A.A.; Sängerlaub, S. Recent advances in extending the shelf life of fresh Agaricus mushrooms: A review. *J. Sci. Food Agric.* **2010**, *90*, 1393–1402. [CrossRef]
28. Ares, G.; Parentelli, C.; Gámbaro, A.; Lareo, C.; Lema, P. Sensory shelf life of shiitake mushrooms stored under passive modified atmosphere. *Postharvest Biol. Technol.* **2006**, *41*, 191–197. [CrossRef]
29. Wang, H.; Qian, Z.; Ma, S.; Zhou, Y.; Patrick, J.W.; Duan, X.; Jiang, Y.; Qu, H. Energy status of ripening and postharvest senescent fruit of litchi (*Litchi chinensis* Sonn.). *BMC Plant Biol.* **2013**, *13*, 55. [CrossRef]
30. Azevedo, S.; Cunha, L.M.; Caldas-Fonseca, S. Modelling the influence of time and temperature on the respiration rate of fresh oyster mushrooms. *Food Sci. Technol. Int.* **2015**, *21*, 593–603. [CrossRef]
31. Marçal, S.; Sousa, A.S.; Taofiq, O.; Antunes, F.; Morais, A.M.; Freitas, A.C.; Barros, L.; Ferreira, I.C.; Pintado, M. Impact of postharvest preservation methods on nutritional value and bioactive properties of mushrooms. *Trends Food Sci. Technol.* **2021**, *110*, 418–431. [CrossRef]
32. Prasad, S.B.; Rosangkima, G.; Kharbangar, A. Structural and biochemical changes in mitochondria after cisplatin treatment of Dalton's lymphoma-bearing mice. *Mitochondrion* **2010**, *10*, 38–45. [CrossRef]
33. Choksi, K.B.; Nuss, J.E.; Boylston, W.H.; Rabek, J.P.; Papaconstantinou, J. Age-related increases in oxidatively damaged proteins of mouse kidney mitochondrial electron transport chain complexes. *Free. Radic. Biol. Med.* **2007**, *43*, 1423–1438. [CrossRef]
34. Sedlák, E.; Fabian, M.; Robinson, N.C.; Musatov, A. Ferricytochrome c protects mitochondrial cytochrome c oxidase against hydrogen peroxide-induced oxidative damage. *Free. Radic. Biol. Med.* **2010**, *49*, 1574–1581. [CrossRef]
35. Aghdam, M.S.; Jannatizadeh, A.; Luo, Z.; Paliyath, G. Ensuring sufficient intracellular ATP supplying and friendly extracellular ATP signaling attenuates stresses, delays senescence and maintains quality in horticultural crops during postharvest life. *Trends Food Sci. Technol.* **2018**, *76*, 67–81. [CrossRef]
36. Li, L.; Sun, H.; Kitazawa, H.; Wang, X. Effects of a high O_2 dynamic-controlled atmosphere technology on the browning of postharvest white mushroom (*Agaricus bisporus*) in relation to energy metabolism. *Food Sci. Technol. Int.* **2017**, *23*, 385–395. [CrossRef]
37. Wu, Y.; Hu, Q.; Li, Z.; Pei, F.; Mariga, A.M.; Yang, W. Effect of nanocomposite-based packaging on microstructure and energy metabolism of *Agaricus bisporus*. *Food Chem.* **2019**, *276*, 790–796. [CrossRef]
38. González, A.J.; González-Varela, G.; Gea, F.J. Brown Blotch Caused by *Pseudomonas tolaasii* on Cultivated *Pleurotus eryngii* in Spain. *Plant Dis.* **2009**, *93*, 667. [CrossRef]
39. Okorley, B.A.; Sossah, F.L.; Dai, D.; Xu, S.; Liu, Z.; Song, B.; Sheng, H.; Fu, Y.; Li, Y. Resistance Sources to Brown Blotch Disease (*Pseudomonas tolaasii*) in a Diverse Collection of Pleurotus Mushroom Strains. *Pathogens* **2019**, *8*, 227. [CrossRef]
40. Sajben, E.; Manczinger, L.; Nagy, A.; Kredics, L.; Vágvölgyi, C. Characterization of pseudomonads isolated from decaying sporocarps of oyster mushroom. *Microbiol. Res.* **2011**, *166*, 255–267. [CrossRef]
41. Yun, Y.-B.; Park, S.-W.; Cha, J.-S.; Kim, Y.-K. Biological characterization of various strains of Pseudomonas tolaasii that causes brown blotch disease. *J. Korean Soc. Appl. Biol. Chem.* **2013**, *56*, 41–45. [CrossRef]

42. Chen, M.; Wu, Q.; Zhang, J.; Guo, W.; Wu, S.; Yang, X. Prevalence and Contamination Patterns of *Listeria monocytogenes* in *Flammulina velutipes* Plants. *Foodborne Pathog. Dis.* **2014**, *11*, 620–627. [CrossRef] [PubMed]
43. Pennone, V.; Lehardy, A.; Coffey, A.; Mcauliffe, O.; Jordan, K. Diversity of *Listeria monocytogenes* strains isolated from *Agaricus bisporus* mushroom production. *J. Appl. Microbiol.* **2018**, *125*, 586–595. [CrossRef]
44. Murugesan, L.; Kucerova, Z.; Knabel, S.J.; Laborde, L.F. Predominance and Distribution of a Persistent Listeria monocytogenes Clone in a Commercial Fresh Mushroom Processing Environment. *J. Food Prot.* **2015**, *78*, 1988–1998. [CrossRef] [PubMed]
45. Loredana, L.; Francesca, M.; Florinda, F.; Filomena, N.; Paola, O.; Donatella, A. Effect of argon-enriched modified atmosphere on the over quality and bioactive compounds of ready-to-use broccoli rabe (*Brassica rapa sylvestris* L. var. *esculenta)* during the storage. *Food Sci. Technol. Int.* 2021; 29, 84–94. [CrossRef]
46. Olveira-Bouzas, V.; Pita-Calvo, C.; Vázquez-Odériz, M.L.; Romero-Rodríguez, M. Evaluation of a modified atmosphere packaging system in pallets to extend the shelf-life of the stored tomato at cooling temperature. *Food Chem.* **2021**, *364*, 130309–130316. [CrossRef]
47. Mortazavi, S.M.H.; Kaur, M.; Farahnaky, A.; Torley, P.J.; Osborn, A.M. The pathogenic and spoilage bacteria associated with red meat and application of different approaches of high CO_2 packaging to extend product shelf-life. *Crit. Rev. Food Sci. Nutr.* **2021**, *1*, 1–22. [CrossRef]
48. Fonseca, S.C.; Oliveira, F.A.; Brecht, J.K. Modelling respiration rate of fresh fruits and vegetables for modified atmosphere packages: A review. *J. Food Eng.* **2002**, *52*, 99–119. [CrossRef]
49. Mohapatra, D.; Bira, Z.M.; Frias, J.M.; Kerry, J.P.; Rodrigues, F.A. Probabilistic shelf life assessment of white button mushrooms through sensorial properties analysis. *LWT* **2011**, *44*, 1443–1448. [CrossRef]
50. Lopezbriones, G.; Varoquaux, P.; Bureau, G.; Pascat, B. Modified atmosphere packaging of common mushroom. *Int. J. Food Sci. Technol.* **1993**, *28*, 57–68. [CrossRef]
51. Varoquaux, P.; Gouble, B.; Barron, C.; Yildiz, F. Respiratory parameters and sugar catabolism of mushroom (*Agaricus bisporus* Lange). *Postharvest Biol. Technol.* **1999**, *16*, 51–61. [CrossRef]
52. Jafri, M.; Jha, A.; Bunkar, D.S.; Ram, R.C. Quality retention of oyster mushrooms (*Pleurotus florida*) by a combination of chemical treatments and modified atmosphere packaging. *Postharvest Biol. Technol.* **2013**, *76*, 112–118. [CrossRef]
53. Kasaai, M.R. Bio-nano-composites containing at least two components, chitosan and zein, for food packaging applications: A review of the nano-composites in comparison with the conventional counterparts. *Carbohydr. Polym.* **2022**, *280*, 119027–119035. [CrossRef]
54. Fernandes, B.C.N.; Paulo, B.B.; Guimarães, M.C.; Sarantopoulos, C.I.G.D.L.; Melo, N.R.; Prata, A.S. Prospection of the use of encapsulation in food packaging. *Compr. Rev. Food Sci. Food Saf.* **2022**, *21*, 2309–2334. [CrossRef] [PubMed]
55. Almasi, H.; Oskouie, M.J.; Saleh, A. A review on techniques utilized for design of controlled release food active packaging. *Crit. Rev. Food Sci. Nutr.* **2021**, *61*, 2601–2621. [CrossRef] [PubMed]
56. Xu, F.; Wang, S.; Xu, J.; Liu, S.; Li, G. Effects of combined aqueous chlorine dioxide and UV-C on shelf-life quality of blueberries. *Postharvest Biol. Technol.* **2016**, *117*, 125–131. [CrossRef]
57. Donglu, F.; Wenjian, Y.; Kimatu, B.M.; Mariga, A.M.; Liyan, Z.; Xinxin, A.; Qiuhui, H. Effect of nanocomposite-based packaging on storage stability of mushrooms (*Flammulina velutipes*). *Innov. Food Sci. Emerg. Technol.* **2016**, *33*, 489–497. [CrossRef]
58. Shi, C.; Wu, Y.; Fang, D.; Pei, F.; Mariga, A.M.; Yang, W.; Hu, Q. Effect of nanocomposite packaging on postharvest senescence of Flammulina velutipes. *Food Chem.* **2018**, *246*, 414–421. [CrossRef] [PubMed]
59. Donglu, F.; Wenjian, Y.; Kimatu, B.M.; Xinxin, A.; Qiuhui, H.; Liyan, Z. Effect of nanocomposite packaging on postharvest quality and reactive oxygen species metabolism of mushrooms (*Flammulina velutipes*). *Postharvest Biol. Technol.* **2016**, *119*, 49–57. [CrossRef]
60. Fang, D.; Yang, W.; Deng, Z.; An, X.; Zhao, L.; Hu, Q. Proteomic Investigation of Metabolic Changes of Mushroom (*Flammulina velutipes*) Packaged with Nanocomposite Material during Cold Storage. *J. Agric. Food Chem.* **2017**, *65*, 10368–10381. [CrossRef]
61. Veberic, R.; Stampar, F.; Schmitzer, V.; Cunja, V.; Zupan, A.; Koron, D.; Mikulic-Petkovsek, M. Changes in the Contents of Anthocyanins and Other Compounds in Blackberry Fruits Due to Freezing and Long-Term Frozen Storage. *J. Agric. Food Chem.* **2014**, *62*, 6926–6935. [CrossRef]
62. Jiang, S.; Wang, S.; Sun, Y.; Ma, Y. Nutrients responses of Pleurotus ostreatus to slow frozen storage in the short term. *RSC Adv.* **2014**, *4*, 47200–47205. [CrossRef]
63. Ji, J.; Allahdad, Z.; Sarmast, E.; Salmieri, S.; Lacroix, M. Combined effects of microencapsulated essential oils and irradiation from gamma and X-ray sources on microbiological and physicochemical properties of dry fermented sausages during storage. *LWT* **2022**, *159*, 113180–113189. [CrossRef]
64. Cardoso, R.V.; Fernandes, Â.; Barreira, J.C.; Verde, S.C.; Antonio, A.; González-Paramás, A.M.; Barros, L.; Ferreira, I.C. Effectiveness of gamma and electron beam irradiation as preserving technologies of fresh *Agaricus bisporus* Portobello: A comparative study. *Food Chem.* **2019**, *278*, 760–766. [CrossRef] [PubMed]
65. Shi, D.; Yin, C.; Fan, X.; Yao, F.; Qiao, Y.; Xue, S.; Lu, Q.; Feng, C.; Meng, J.; Gao, H. Effects of ultrasound and gamma irradiation on quality maintenance of fresh Lentinula edodes during cold storage. *Food Chem.* **2022**, *373*, 131478–131488. [CrossRef] [PubMed]
66. Subramaniam, S.; Jiao, S.; Zhang, Z.; Jing, P. Impact of post-harvest processing or thermal dehydration on physiochemical, nutritional and sensory quality of shiitake mushrooms. *Compr. Rev. Food Sci. Food Saf.* **2021**, *20*, 2560–2595. [CrossRef] [PubMed]

67. Politowicz, J.; Lech, K.; Lipan, L.; Figiel, A.; Carbonell-Barrachina, Á.A. Volatile composition and sensory profile of shiitake mushrooms as affected by drying method. *J. Sci. Food Agric.* **2018**, *98*, 1511–1521. [CrossRef] [PubMed]
68. Özünlü, O.; Ergezer, H. Possibilities of using dried oyster mushroom (*Pleurotus ostreatus*) in the production of beef salami. *J. Food Process. Preserv.* **2021**, *45*, e15117–e15125. [CrossRef]
69. Nowacka, M.; Laghi, L.; Rybak, K.; Rosa, M.D.; Witrowa-Rajchert, D.; Tylewicz, U. Water state and sugars in cranberry fruits subjected to combined treatments: Cutting, blanching and sonication. *Food Chem.* **2019**, *299*, 125122–125131. [CrossRef]
70. Xiao, H.-W.; Bai, J.-W.; Sun, D.-W.; Gao, Z.-J. The application of superheated steam impingement blanching (SSIB) in agricultural products processing—A review. *J. Food Eng.* **2014**, *132*, 39–47. [CrossRef]
71. Muhammad, A.I.; Chen, W.; Liao, X.; Xiang, Q.; Liu, D.; Ye, X.; Ding, T. Effects of Plasma-Activated Water and Blanching on Microbial and Physicochemical Properties of Tiger Nuts. *Food Bioprocess Technol.* **2019**, *12*, 1721–1732. [CrossRef]
72. Xu, Y.; Xiao, Y.; Lagnika, C.; Li, D.; Liu, C.; Jiang, N.; Song, J.; Zhang, M. A comparative evaluation of nutritional properties, antioxidant capacity and physical characteristics of cabbage (*Brassica oleracea* var. *Capitate* var L.) subjected to different drying methods. *Food Chem.* **2020**, *309*, 124935–124943. [CrossRef]
73. Ucar, T.M.; Karadag, A. The effects of vacuum and freeze-drying on the physicochemical properties and in vitro digestibility of phenolics in oyster mushroom (*Pleurotus ostreatus*). *J. Food Meas. Charact.* **2019**, *13*, 2298–2309. [CrossRef]
74. Souza, E.L.; Lundgren, G.A.; Oliveira, K.R.; Berger, L.R.R.; Magnani, M. An Analysis of the Published Literature on the Effects of Edible Coatings Formed by Polysaccharides and Essential Oils on Postharvest Microbial Control and Overall Quality of Fruit. *Compr. Rev. Food Sci. Food Saf.* **2019**, *18*, 1947–1967. [CrossRef]
75. Zhang, W.; Jiang, H.; Rhim, J.-W.; Cao, J.; Jiang, W. Effective strategies of sustained release and retention enhancement of essential oils in active food packaging films/coatings. *Food Chem.* **2022**, *367*, 130671–130679. [CrossRef] [PubMed]
76. Sharifi-Rad, J.; Sureda, A.; Tenore, G.C.; Daglia, M.; Sharifi-Rad, M.; Valussi, M.; Tundis, R.; Sharifi-Rad, M.; Loizzo, M.R.; Ademiluyi, A.O.; et al. Biological Activities of Essential Oils: From Plant Chemoecology to Traditional Healing Systems. *Molecules* **2017**, *22*, 70. [CrossRef] [PubMed]
77. Anon, M.; Chandra, R. Effect of Different Essential Oils on Enzymatic Activity of Oyster Mushroom (*Pleurotus florida*). *Curr. Sci.* **2021**, *121*, 1357–1360. [CrossRef]
78. Guo, Y.; Chen, X.; Gong, P.; Wang, R.; Han, A.; Deng, Z.; Qi, Z.; Long, H.; Wang, J.; Yao, W.; et al. Advances in the role and mechanisms of essential oils and plant extracts as natural preservatives to extend the postharvest shelf life of edible mushrooms. *Foods* **2023**, *12*, 801. [CrossRef]
79. Kerch, G. Chitosan films and coatings prevent losses of fresh fruit nutritional quality: A review. *Trends Food Sci. Technol.* **2015**, *46*, 159–166. [CrossRef]
80. Guo, Y.; Chen, X.; Gong, P.; Guo, J.; Deng, D.; He, G.; Ji, C.; Wang, R.; Long, H.; Wang, J.; et al. Effect of shiitake mushrooms polysaccharide and chitosan coating on softening and browning of shiitake mushrooms (*Lentinus edodes*) during postharvest storage. *Int. J. Biol. Macromol.* **2022**, *218*, 816–827. [CrossRef]
81. Hassoun, A.; Shumilina, E.; Di Donato, F.; Foschi, M.; Simal-Gandara, J.; Biancolillo, A. Emerging Techniques for Differentiation of Fresh and Frozen–Thawed Seafoods: Highlighting the Potential of Spectroscopic Techniques. *Molecules* **2020**, *25*, 4472. [CrossRef]
82. Tabaszewska, M.; Gabor, A.; Jaworska, G.; Drożdż, I. Effect of fermentation and storage on the nutritional value and contents of biologically-active compounds in lacto-fermented white asparagus (*Asparagus officinalis* L.). *LWT* **2018**, *92*, 67–72. [CrossRef]
83. Muhialdin, B.J.; Zawawi, N.; Razis, A.F.A.; Bakar, J.; Zarei, M. Antiviral activity of fermented foods and their probiotics bacteria towards respiratory and alimentary tracts viruses. *Food Control.* **2021**, *127*, 108140. [CrossRef] [PubMed]
84. Bagheripoor-Fallah, N.; Mortazavian, A.; Hosseini, H.; Khoshgozaran-Abras, S.; Rad, A.H.; Hosseyni, H. Comparison of Molecular Techniques with other Methods for Identification and Enumeration of Probiotics in Fermented Milk Products. *Crit. Rev. Food Sci. Nutr.* **2015**, *55*, 396–413. [CrossRef] [PubMed]
85. Beganović, J.; Kos, B.; Pavunc, A.L.; Uroić, K.; Jokić, M.; Šušković, J. Traditionally produced sauerkraut as source of autochthonous functional starter cultures. *Microbiol. Res.* **2014**, *169*, 623–632. [CrossRef]
86. Hashad, R.A.; Ishak, R.A.; Fahmy, S.; Mansour, S.; Geneidi, A.S. Chitosan-tripolyphosphate nanoparticles: Optimization of formulation parameters for improving process yield at a novel pH using artificial neural networks. *Int. J. Biol. Macromol.* **2016**, *86*, 50–58. [CrossRef] [PubMed]
87. Cui, H.; Bai, M.; Rashed, M.M.; Lin, L. The antibacterial activity of clove oil/chitosan nanoparticles embedded gelatin nanofibers against *Escherichia coli* O157:H7 biofilms on cucumber. *Int. J. Food Microbiol.* **2018**, *266*, 69–78. [CrossRef] [PubMed]

foods

MDPI

Review

Emerging Postharvest Technologies to Enhance the Shelf-Life of Fruit and Vegetables: An Overview

Michela Palumbo [1,2], Giovanni Attolico [3], Vittorio Capozzi [2], Rosaria Cozzolino [4,*], Antonia Corvino [2], Maria Lucia Valeria de Chiara [1,2], Bernardo Pace [2,*], Sergio Pelosi [2], Ilde Ricci [2], Roberto Romaniello [1] and Maria Cefola [2]

1 Department of Science of Agriculture, Food and Environment, University of Foggia, Via Napoli, 25, 71122 Foggia, Italy
2 Institute of Sciences of Food Production, National Research Council of Italy (CNR), c/o CS-DAT, Via Michele Protano, 71121 Foggia, Italy
3 Institute on Intelligent Industrial Systems and Technologies for Advanced Manufacturing, National Research Council of Italy (CNR), Via G. Amendola, 122/O, 70126 Bari, Italy
4 Institute of Food Science, National Research Council (CNR), Via Roma 64, 83100 Avellino, Italy
* Correspondence: rosaria.cozzolino@isa.cnr.it (R.C.); bernardo.pace@ispa.cnr.it (B.P.); Tel.: +39-0825-299111 (R.C.); +39-0881-630210 (B.P.)

Abstract: Quality losses in fresh produce throughout the postharvest phase are often due to the inappropriate use of preservation technologies. In the last few decades, besides the traditional approaches, advanced postharvest physical and chemical treatments (active packaging, dipping, vacuum impregnation, conventional heating, pulsed electric field, high hydrostatic pressure, and cold plasma) and biocontrol techniques have been implemented to preserve the nutritional value and safety of fresh produce. The application of these methodologies after harvesting is useful when addressing quality loss due to the long duration when transporting products to distant markets. Among the emerging technologies and contactless and non-destructive techniques for quality monitoring (image analysis, electronic noses, and near-infrared spectroscopy) present numerous advantages over the traditional, destructive methods. The present review paper has grouped original studies within the topic of advanced postharvest technologies, to preserve quality and reduce losses and waste in fresh produce. Moreover, the effectiveness and advantages of some contactless and non-destructive methodologies for monitoring the quality of fruit and vegetables will also be discussed and compared to the traditional methods.

Keywords: active packaging; cold plasma; dipping; E-nose; high hydrostatic pressure; image analysis; innovative postharvest technologies; pulsed electric field; vacuum impregnation; near-infrared spectroscopy

Citation: Palumbo, M.; Attolico, G.; Capozzi, V.; Cozzolino, R.; Corvino, A.; de Chiara, M.L.V.; Pace, B.; Pelosi, S.; Ricci, I.; Romaniello, R.; et al. Emerging Postharvest Technologies to Enhance the Shelf-Life of Fruit and Vegetables: An Overview. *Foods* **2022**, *11*, 3925. https://doi.org/10.3390/foods11233925

Academic Editor: Arun K. Bhunia

Received: 2 November 2022
Accepted: 28 November 2022
Published: 5 December 2022

1. Introduction

Fresh fruit and vegetables are important sources of essential vitamins, minerals, and antioxidants. In recent decades, consumers have become more likely to consume these products since they are aware of their potential preventive effects against some non-communicable diseases. In addition, increasing safety (chemical, toxicological and microbial) and traceability are important aspects for all the players in the supply chain (from the farm to consumers) [1]. Food quality may be defined as the combination of several physical and chemical attributes (appearance, texture, flavor, and nutritional value) which have a crucial impact on determining the degree of consumer acceptability [2]. The quality of fruit and vegetables is mostly based on the evaluation of different external attributes (size, shape, color, gloss, firmness, texture, and taste) and internal factors (chemical, physical and microbial) dealing with nutritional traits, safety, and sustainability [3].

Harvested commodities are metabolically active and highly perishable. They undergo quality losses due to the ripening and senescence processes, which are often associated with the development of spoilage microorganisms and other undesired phenomena, which must be controlled to preserve the quality and increase the shelf-life of the product during storage [4,5]. Furthermore, high water activity and the presence of nutritional factors associated with these matrices can also favor the growth of pathogens [6,7]. Fruit and vegetables with better sensory and nutritional attributes have a relevant economic value. Consequently, inadequate preservation practices, besides causing important losses in nutritional and quality characteristics, can have a detrimental economic impact along the entire supply chain, from growers to consumers. The Food and Agriculture Organization (FAO) estimated that 33% of the total food produced for human consumption is lost due to postharvest spoilage (Figure 1). Overall, 44% of losses occur in industrialized (developed) countries and 40% occur in developing countries [8].

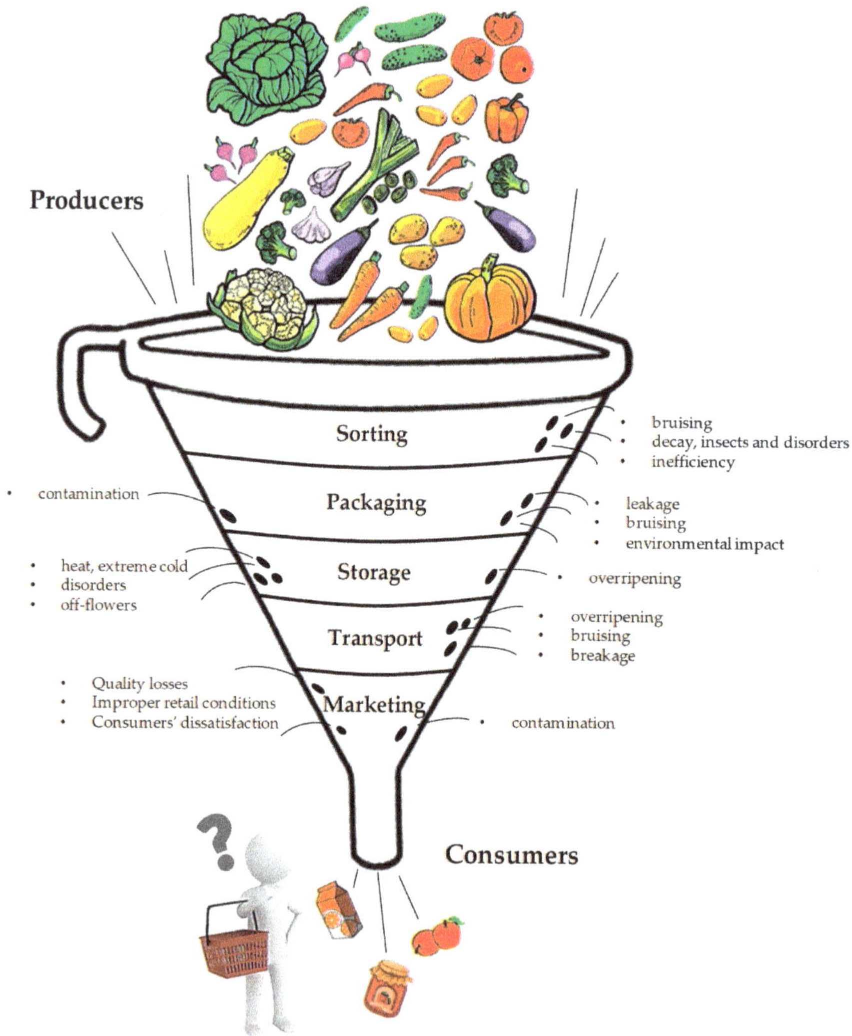

Figure 1. Causes of postharvest losses along the supply chain. (Adapted from [9]).

In a recent report released by the FAO [10], it was reported that fruit and vegetables are the food group with the second-highest value of losses and waste (about 22%), exceeded only by roots, tubers, and oil-bearing crops at all stages in the food supply chain (Figure 2), due to their highly perishable nature.

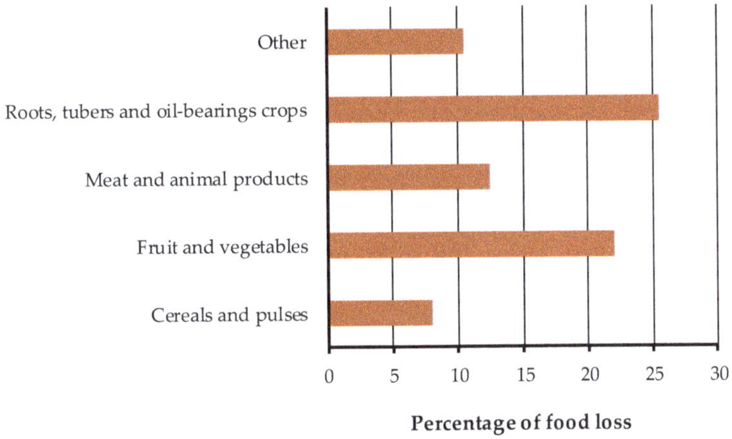

Figure 2. Food losses and waste along the supply chain (percentage for each food group).

The reduction of losses and wastage in fruit and vegetable production is important for improving security, promoting the sustainability of the environment, and reducing production costs.

Optimal postharvest handling, including storage time and temperature management, relative humidity, chemical and/or physical treatments, and packaging (i.e., a modified atmosphere) can slow down the biological processes caused by senescence and maturation, reduce or inhibit the development of physiological disorders, and minimize microbial growth and contamination [1,11,12]. For example, temperature is the most important and simplest postharvest factor that can delay the decay of the product [2,13]. Generally, there is a recommended storage temperature for each product, and optimum quality preservation can be achieved if the commodity is rapidly cooled to its best postharvest storage temperature as soon as possible after harvest [14].

Furthermore, the amounts of internationally traded agricultural products have generally increased with the progress of food globalization. Consequently, the distance and the duration of transport have been extended; therefore, prolonged exposure to non-optimal storage conditions can cause a rapid deterioration in the product quality, with the consequent losses and waste. For these reasons, increased product shelf-life and more advanced quality control technology over longer distances have become key issues [15].

Nowadays, there is much research about the effects of new postharvest physical and chemical treatments and about the biocontrol techniques used to preserve the quality, nutritional value, and safety of fresh produce, from harvest to consumer consumption. It has been widely demonstrated that these methods, whether alone or in combination with the appropriate management of storage temperature, could preserve the principal quality and nutritional traits of fruit and vegetables.

Conventional methods (sensory evaluations and analytical methodologies) used to evaluate fruit and vegetable quality are destructive, time-consuming, and labor- and cost-intensive. Moreover, these techniques are not suitable for in-line application in industrial or in market settings to give real-time information to the consumers on the quality of the product at hand. On the contrary, the emerging contactless and non-destructive technologies for the quality monitoring of fresh produce, including near-infrared spectroscopy, hyperspectral or multispectral imaging, image analysis, electronic noses, etc., present numerous advantages over conventional destructive methods [3]. These emerging techniques are normally used for external and internal quality evaluation and are principally based on the measurement of chemical or physical attributes that correlate with certain quality traits of agricultural products [16].

The aim of this review article is to report a detailed collection of the research works carried out in the last years about the principal application of new postharvest physical and chemical emerging technologies to preserve the quality and reduce postharvest losses and waste in fresh fruit and vegetables. Moreover, the effectiveness and advantages of quality evaluation via some contactless and non-destructive methodologies are described and compared to conventional approaches.

2. Postharvest Strategies to Extend the Shelf-Life of Fruit and Vegetables

2.1. Physical Treatments

Emerging non-thermal physical technologies have moved into the spotlight in the last few years, with the aim of replacing the traditional postharvest technologies based on thermal processing. Besides being highly water-consuming, the conventional methodologies can show deleterious effects on the fresh commodities' quality aspects. These novel technologies can reduce nutrient losses, increase consumer acceptability, promote food quality, and prolong shelf-life and freshness, guaranteeing the complete absence of chemical byproducts in the processed product, combined with a reduction in the environmental impact [17]. Among these emerging methods, microwave heating, high hydrostatic pressure, pulsed electric fields, high hydrostatic pressure, and cold plasma have been applied to reduce the microbial load, thus helping to preserve the freshness and quality characteristics of fruit and vegetables [18]. However, these techniques show numerous advantages and disadvantages that have begun to be investigated to achieve a suitable quality standard by adopting cost-effective methods [19].

This subsection aims to briefly collect the emerging physical technologies that have been applied to fresh and minimally processed or fresh horticultural products over the past few years, particularly focusing on the treatments' effectiveness in maintaining the quality and safety of fruit and vegetables.

2.1.1. Microwave

Generally, heating processes, such as hot water and hot air treatments, high-temperature/short-time treatments, and radio frequency can cause a reduction in the contents of essential nutrients and flavor-related compounds, due to heat application and its slow distribution in plant tissue by means of conduction or conduction processes. Microwaving was applied in this context as an alternative to conventional heating [20], with the aim of achieving a fast and effective increase in temperature without a temperature gradient. This technique briefly treats fruits and vegetables to manage microbial growth throughout the product's minimal processing, minimizing losses of quality and simultaneously guaranteeing the smallest effect on the environment and the absence of residues in the treated product [21].

However, little information is available in the literature related to the application of this innovative physical technique with the aim of minimizing quality loss during the postharvest phase (Table 1). Minimally processed carrots [22], apples, and bok choy (*Brassica campestris* L.) were subjected to high-power/short-time treatments, demonstrating promising results from a microbiological point of view [23,24]. An excessive duration or intensity of the microwave treatment, however, can induce excessive temperature increase, damaging the fresh tissue, due to non-uniform heating [25]. For this reason, there is a need to overcome several challenges to achieve a successful microwave process on an industrial scale for fresh and minimally processed products.

Table 1. The main applications of emerging physical postharvest treatments on fresh and minimally processed horticultural products, as reported in research published over the last five years.

Treatment	Experimental Conditions	Food Matrix	Effects	Reference
Microwave	454 W/5 s	Minimally processed bok choy	Microwaving decreases respiration rate, while retarding decay occurrence, and improves cell membrane integrity.	[23]
Microwave	300–100 W/35–10 s	Fresh cut apples	Application of the treatments at the highest intensity after minimal processing led to the best mesophiles and psychrophiles control.	[24]
Pulsed Electric Field	10,000 pulses at 0.5, 1.0 and 1.5 kV cm^{-1}	Fresh-cut lotus root	Sugar content was reduced after PEF application in fresh samples, thus lowering the browning index and reducing acrylamide.	[26]
Pulsed Electric Field	5 pulses of 350 kVm^{-1}	Carrot	Twelve hours after the treatment, high CO_2 and volatiles production was observed; after 24 h, the largest total phenolic increase occurred.	[27]
Pulsed Electric Field	0.8, 2 and 3.5 kV cm^{-1} and 5, 12 and 30 pulses	Carrot	Treatment did not affect color, while an increase in the phenolic and carotenoid content and softening was observed.	[28,29]
High hydrostatic pressure	Red cabbage: 150–200 MPa at 35–55 °C, 5–20 min/Radish: 100–200 MPa, at 20–40 °C, 5–10 min	Red cabbage leaves and radish tubers	Variations in the HHP treatments affected the integrity of the tissues and cell turgor.	[30]
High hydrostatic pressure	50–400 MPa for 3–60 min	Fresh-cut papaya fruit	After HHP treatment and storage, the enhancement of carotenoid precursors and carotene content was observed.	[31]
High hydrostatic pressure	100–600 MPa for 2 min	Fresh-cut pumpkin	Color parameters, firmness, electrical conductance, and pectin esterification were positively affected by HHP.	[32]
Mild high hydrostatic pressure	20–80 MPa for 10 min	Mango	HHP reduced the respiration rate, prevented tissue damage, and positively affected the content of bioactive compounds.	[33]
High pressure processing	100–300 MPa for 5–20 min	Minimally processed pineapple	HPP significantly affected firmness, flavonoids, polyphenols, vitamin C content, and colorimetric parameters.	[34]
High pressure treatments	200, 400, 600 MPa for 5 min	Pumpkin	After applying 400 MPa, the pectinmethylesterase enzyme was inactivated. Color parameters decreased and an increase in antioxidant activity was observed.	[35]
High hydrostatic pressures	400–600 MPa; 1–5 min	Blueberries	Treatment caused high tissue damage, thus resulting in the leakage of bioactive cellular components.	[36]
High-pressure treatments	400,600 MPa; 1, 5 min	Zucchini slices	The longest treatment led to more severe cell lysis, browning and dehydration occurrence.	[37]
Dielectric Barrier Discharge (DBD)	5 + 5 min on each side, 10 + 10, and 15 + 15; 30 and 60 min	Fresh cut apple	Plasma treatment caused less browning incidence and enhanced phenols and antioxidant activity after 10 min of treatment.	[38,39]
Cold plasma	40 kV/90 s	Fresh cut cantaloupe	Cold plasma treatment significantly reduced bacteria and mold development during storage. Final product showed higher quality, firmness, and sensory attributes.	[40]
Dielectric barrier discharge cold plasma	Plasma levels of 7% and 14% duty cycle for 5, 10, 20 min.	Strawberry	Plasma treatment of 20 minutes reduced the mesophilic bacteria and yeasts and molds, while not affecting texture and color.	[41]

Table 1. *Cont.*

Treatment	Experimental Conditions	Food Matrix	Effects	Reference
Atmospheric cold plasma (ACP)	ACP at 60 kV for 10, 15, 30 min	Strawberry	ACP treatment was able to prolong the product shelf-life in treated strawberries. A 15-minute treatment resulted in 2 log bacteria reduction, also enhancing the phenolic content and antioxidant activity. TSS, pH, and moisture were not affected.	[42]
Dielectric barrier discharge cold plasma	60 kV/5 min	Fresh cut pitaya	Total aerobic bacterial count was significantly reduced by treatment, while phenolic content and antioxidant activity increased.	[43]
Atmospheric double barrier discharge plasma	1 or 5 min at 45 and 65 kV	Fresh cut pears	Treatment effectively inhibited the growth of mesophiles and yeast and mold. The 65 kV/1 min treatment slowed respiration rate and maintained organoleptic properties and quality.	[44]
Cold atmospheric dielectric barrier discharge plasma	5, 10, 15 and 20 min	Ready-to-eat rocket leafy salad	A treatment length of 10 min was optimal for the adequate reduction of the microbial load while maintaining color and firmness.	[45]
Atmospheric cold plasma (ACP)	0, 5, 10, 15 and 20 min	Blueberries	ACP treatment inhibited microbial development and decay occurrence. Treatments of 5 and 10 min showed negligible effects on firmness, pH, ORP and anthocyanin concentration, but darkened the color.	[46]
Dielectric barrier discharge gas plasma and arc plasma-activated water (PAW)	20-min DBD treatment and 20 min PAW immersion	Shiitake mushroom	Treatment reduced the total bacterial load after 7 days of storage, also slowing down the overall color modification and positively affecting firmness.	[47]
Plasma-activated water (PAW)	7.0 kHz at 6 kV, 8 kV (PAW-8), 10 kV for 5 min	Fresh cut apple	PAW-8 treatment inhibited bacterial development and the reduced browning of the cut surface without affecting firmness, titratable acidity, radical scavenging activity, and antioxidant content.	[48]
Plasma activated water (PAW)	Activation times of 10, 20, 30, 45 and 60 min, 5 min washing	Fresh cut apple	PAW variably affected enzyme activities, while it did not show an effect on total phenolic content and antioxidant activity. Significant reductions in the aerobic bacteria and yeast and mold loads were observed.	[49]

2.1.2. Pulsed Electric Field

Recently, pulsed electric field (PEF) technology has become of the most interest because of its capability to obtain safe food with minimal heat production through the use of µs to ms-pulses of a high electric field of high intensity [50]. This technique has been widely used on liquid, semi-solid and solid foods, also including fresh fruit, vegetable smoothies, and juices. The PEF parameters to be optimized to obtain microbial and enzymatic inactivation in fresh products are represented by the strength of the electric fields, the treatment time, and the frequency, polarity, or shape of the pulses. Beneficial effects from PEF treatment were observed in the reduction of enzymatic activity with a consequent improvement in the quality parameters [27–29,51–53], as reported in Table 1. PEF-treated products better retained their fresh flavor, textural, and functional attributes, including a longer shelf-life and greater microbiological safety. However, PEF-induced metabolic stress could negatively affect the quality of the final product, thus limiting the application of PEF to fresh-cut products. Among the limited studies reporting the effects on the metabolism and characteristics of minimally processed produce, Li et al. [26] have reported that PEF treatment limited the browning index and acrylamide content in ready-to-eat lotus root during postharvest life.

2.1.3. High Hydrostatic Pressure

High hydrostatic pressure (HHP) technology is mostly used for microbial inactivation or reduction and for enzyme denaturation. However, high pressure, inducing injuries on microbial cellular structures, might show similar effects on the plant cells; thus, an in-depth study into treatment optimization in various fresh systems is required. Several results show that HHP significantly affects microbial load; however, it also influences the functionality of proteins, such as enzymes and tissue structure, specifically and differentially due to the wide variety of product types [30]. Moreover, the stimulation and accumulation of nutraceutical compounds were also observed. The effects of HHP application have been reported for different minimally processed horticultural commodities [31,32,34,37], whole produce [30,35,36,46], and juice [54], demonstrating great efficiency in improving food safety aspects and in maintaining quality.

2.1.4. Cold Plasma

The application of cold plasma is widely used in the whole and minimally processed fruit and vegetable industry as an innovative technology used to handle microbial development [55], with the aim of replacing conventional sanitation treatments, meanwhile preserving the nutritional and antioxidant aspects of food products. Several scientific studies reported the effectiveness of non-thermal plasma on different horticultural products (Table 1). Several fruit-based fresh-cut products have been subjected to plasma treatment with beneficial effects, in terms of the quality parameters and the inhibition of microbial growth. During the last few years, plasma-activated water (PAW) application has also been more and more widely studied. This technique allows the producer to avoid cell damage caused by direct exposure to cold plasma, representing a valuable alternative to the conventional solution of washing during fresh-cut processing for several products. To date, cold plasma and PAW washing applications have been described for strawberries [41,42,56–58], kumquat fruit [59,60] green leafy vegetables [45,57,61–64], blueberries [46,65–67], fresh-cut apples [38,39,48,49,68,69] pears [44,70], cantaloupe melons [40], mushrooms [47,71], tomatoes [72], kiwifruits [73], and red currants [74].

2.2. *Dipping and Vacuum Impregnation*

The quality of fresh-cut fruit and vegetables is widely affected by physiological changes, such as enzymatic browning caused by tissue damage and high respiration rates, and by physical factors, including mechanical injuries and the removal of outer protective coverings, which culminate in faster weight loss, shriveling, loss of color and appearance, and shorter shelf-life [75]. Hence, innovative food processing technologies, such as dipping and vacuum impregnation techniques, are being investigated and implemented to sanitize, reduce enzymatic browning, improve the texture, and use nutrients (vitamins, probiotics, minerals, organic acids, phenols, etc.) to fortify fresh-cut fruit and vegetables, in order to preserve and improve the quality and to extend the shelf-life of these products [76–78].

Dipping treatments consist of soaking the product, with or without mechanical agitation, followed by the removal of the excess solution. This method is commonly used on whole, peeled, shredded, and sliced commodities and on more perishable products, as it favors the dispersion of the solution, covering the maximum surface area of the product without any damage or stress [79]. One of the major advantages of these dipping treatments is the removal of the cellular exudates, which can have a detrimental effect on the postharvest quality of commodities. Depending on the food product, the variables of the dipping process to be optimized are the time of soaking, the frequency, the solute composition, the temperature, and the concentration of the solution. Several studies have tested dipping treatments with calcium (Ca) salts to extend the shelf-life of products. Ca enrichment, in fact, has several advantages, including the reduction of microbial growth due to a decrease in activity water, the improvement of texture, acceptability, and storability, and the prevention of browning due to the oxidation phenomena and the development of off-flavors in fresh-cut foods [80–83]. Giacalone and Chiabrando [84] compared the effect of dipping in Ca salts (chloride or propionate) and citric acid solutions on apple cubes, highlighting an improvement in firmness, color, sensory attributes, and browning inhibition on fruits treated with Ca chloride and citric acid. Albertini et al. [85] tested the immersion of papaya slices in a water solution containing cinnamaldehyde, Ca chloride, and their combinations, showing that a combination treatment increased the maintenance of firmness and had an effect on flavor and taste. Other studies have evaluated the dipping process in solutions of Ca chloride, combined with pectin methyl-esterase (PME), on firmness and some of the quality attributes of raspberry fruits during refrigerated storage, showing that the treatment improved the firmness and reduced the weight loss of the fruit [86]. The application of natural extracts for use as anti-browning agents (polyphenols, carotenoids, organic acids, and bioactive peptides) represents a recent approach that is used to improve the quality and extend the shelf-life of fresh-cut fruit and vegetables [87]. Many natural agents from tomato skin [88], pineapple juice [89], pomegranate peel [90], mango peel [91], aloe vera gel [92], and extracts from pumpkin, artichoke, grape, and broccoli, etc. [93] have been reported for browning control in fresh-cut fruits. Supapvanich et al. [94] tested the immersion of fresh-cut apple slices in coconut water and evaluated several quality attributes (visual quality, color, and enzymatic and antioxidant activity) during storage, reporting that coconut-liquid endosperms could inhibit the browning incidence of fresh-cut fruit for up to 9 days at 4 °C. Wessels et al. [93] evaluated the anti-browning effect of 36 plant extracts, applied as dipping solutions on the enzymatic activity and the color parameters of fresh-cut apple slices, suggesting that the inhibition of browning might be attributable to the antioxidant activity of secondary plant metabolites, especially the phenolic compounds.

Food vacuum impregnation (VI) (Figure 3) is a method that allows producers to directly introduce, dissolve, or suspend substances in the void fraction (i.e., the pores) of a food matrix in a controlled manner [95]. VI includes two main steps: (1) the pressure is reduced in the system (under vacuum), the native gases and liquids are removed, and the product pores are expanded under the action of pressure gradients until mechanical equilibrium is achieved; (2) the atmospheric pressure is restored (relaxation period) and, with the opposite pressure gradient, the external solution fills the pores while the tissues relax, until a new equilibrium has been reached. The hydrodynamic mechanism and deformation-relaxation phenomena take place during the vacuum impregnation process, leading to the flow of external solutions into the intracellular spaces of foods [96].

Before the application of VI treatment, it is necessary to consider the porosity, the tissue structure, the size and geometry of the food, the impregnation solution (concentration and type of solute), and the process parameters (vacuum pressure, exposure time, relaxation time at atmospheric pressure, temperature, product/solution relationship, and agitation). Fruit and vegetables have a substantial amount of intercellular space occupied by gas; therefore, the VI represents a suitable tool for incorporating compounds that allow producers to extend the shelf-life without modifying the cellular structure of food [97]. In the last few years, several authors have studied the application of VI to obtain foods enriched with antimicrobial, functional, and structural compounds. Kang and Kang [98] evaluated the effect of VI when applied to the washing process, with a malic acid solution used to remove microorganisms from the surface of broccoli. Yılmaz and Ersus Bilek [99] evaluated the simultaneous effect of an ultrasound-VI process, using natural phenolic compounds extracted from black carrot, on several quality attributes and on the inhibition of microorganism growth in fresh-cut apple discs. Santana Moreira et al. [100] enriched minimally processed fruit salad with β-carotene and lutein, which are natural pigments used to improve the appearance and color of food products, as well as to offer health benefits related to antioxidant and anticancer activity. Natural fruit and vegetable juices, rich in bioactive compounds, were used to impregnate solid fruits and generate products with high nutritional value [101,102]. Derossi et al. [103], using the VI technology, enriched fresh-cut apple slices to improve their antioxidant activity and healthy properties by filling their pores with an aloe vera gel extract; other authors used grape juice to increase the antioxidant activity of apple cubes [104]. A recent study highlighted the finding that the application of VI, using a solution of Ca chloride and pectin methyl-esterase (PME), had a positive effect on firmness, weight loss, soluble solids content, and vitamin C levels of jujube fruits [105], while other authors used VI with solutions of Ca lactate and PME, both alone and in combination, to improve the product's quality attributes, including texture profile and color, and to prolong the shelf-life of fresh-cut papaya cubes [106].

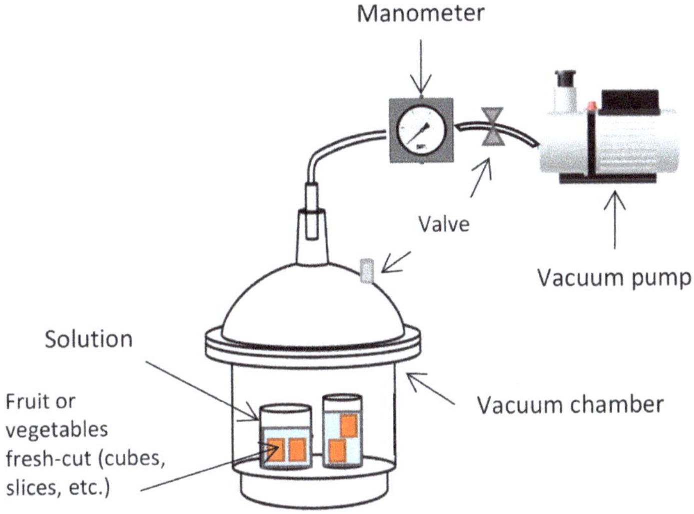

Figure 3. Schematic representation of the vacuum impregnation device, the arrows point at each system element.

2.3. Edible Active Packaging, Based on Natural Compounds

Edible active packaging consists of edible polymers with the incorporation of natural antioxidants. Edible coatings have proven to be an effective primary packaging material in delaying the ripening process, preserving the nutritional properties and preventing quality loss by decreasing several natural processes, including gaseous exchange and the respiration and transpiration rate [107]. Recently, it has been observed that the efficiency of edible coatings can be significantly improved by incorporating active natural components with antioxidant and/or antimicrobial properties. These packaging products are called "active" and are designed to interact with food by releasing components with biological properties. The integration of active compounds into biopolymer matrices enhances the oxidation stability of the food product and inhibits the growth of food-borne pathogens, providing additional safety features for the food products, even in the absence of cold storage [108].

An alginate and chitosan edible coating enriched with an extract from olive leaves has demonstrated the ability to increase the shelf-life of sweet cherries. Moreover, compared with the uncoated samples, a delay in the ripening process and the significant retention of phenolic and antioxidant components were also observed [109]. Robles-Sánchez et al. [110] demonstrated that an alginate coating, when incorporated with citric and acetic acid, allowed fresh-cut mangoes to retain several quality attributes, such as antioxidant activity, phenolic content, and color. However, the flavonoids and β-carotene could not be retained [110]. Moreover, an edible coating with 1% chitosan and 5% ascorbic acid inhibited browning, retained flesh firmness, delayed microbial growth, and maintained phenolic compounds throughout the storage period in fresh-cut apples [111]. Liu et al. [112] verified the effectiveness of gallic acid-grafted chitosan film as a novel active packaging material for the preservation of mushrooms (*Agaricus bisporus*). Compared to commercial polyethylene film, mushrooms packed with the active film showed significantly lower respiration rate, browning degree, malondialdehyde content, electrolyte leakage rate, superoxide anion production rate, and hydrogen peroxide content, with a potential increase in the antioxidant status of mushrooms, which, in turn, maintained the postharvest quality of the products. Carvalho et al. [113] showed that a 2% chitosan-based coating, enriched with 500 mg L^{-1} antimicrobial *trans*-cinnamaldehyde, preserved the quality of fresh-cut cantaloupe melons during their storage at 4 °C by preserving the content of total vitamin C and carotenoids, lightness, and firmness. Moreover, it was reported that carrots coated with turmeric and casein extended their shelf life by about 7 days, preserving the carotenoid content, the texture, and the antibacterial properties [114].

Imeneo et al. [115] demonstrated that a pectin-based coating, enriched with an extract of lemon by-product, ensured that fresh-cut carrots maintained stable structural integrity for 14 days of storage at 4 °C, due to a reduction in enzymatic bacterial activity. This kind of treatment also resulted in higher levels of carotenoids, phenolic compounds, and antioxidant activity.

Lemon essential oil, incorporated into a chitosan edible coating, has been reported to reduce the respiration rate of strawberries and improve the antifungal activity of chitosan against *Botrytis cinerea* [116]. Ghafoor et al. [117] investigated the effect of chitosan edible coatings with natural functional ingredients obtained from orange peel (OPE) and olive cake (OCE) on the quality attributes of fresh Barhi date fruit. When chitosan was mixed with OPE or OCE, a significant increase in phenolic content and in radical scavenging activity was observed with respect to the uncoated samples. Moreover, chitosan-coating significantly preserved the product's textural properties, particularly hardness, and inhibited mold growth, without any non-significant changes in the consumer acceptability of the fruit throughout the storage time.

Vieira et al. [118] placed active film pads made from chitosan, enriched by the ethanolic extracts of green tea and rosemary as natural antifungal agents, on the bottom of commercial trays for the preservation of raspberry fruits. This sustainable packaging successfully decreased raspberry fruit fungal incidence, preserved the overall quality during storage, and extended the shelf-life of fruits by up to 14 days.

2.4. Strategies of Biocontrol

The study of microbiota represents one of the hottest topics in agri-food research, and this trend is confirmed by the great interest in the microbiota associated with fruit and vegetables [119]. These complex microbial ecosystems encompass the diversity of naturally associated bacteria, yeasts, and filamentous fungi (molds) [120,121]. At harvest, this microbial diversity can play an extremely diversified role, modulating the quality and safety of the postharvest products [122,123]. This microbiota can be associated with undesired microorganism pathogens, the producers of toxins, bacteria-carrying antibiotic resistance genes, and/or with potential spoilage activity with respect to the food matrices of interest [124,125]. At the same time, these undesired microorganisms can arise from postharvest processes, operators, and environments [125]. These spoilage microbes can be controlled using different strategies that are often combined in the framework of hurdle technology applications [125]. Among the other approaches, biocontrol represents the key solution in the field of biological treatments [123,126]. Exploiting selected microbes as control agents, biocontrol is considered one of the more sustainable postharvest approaches to increasing the shelf-life of fruit and vegetables [126,127]. Bio-protection relies on the application of selected microbes that can limit the development of undesired microorganisms. The idea of having the microbiome of the fruit or vegetable as a target offers a privileged perspective in understanding the variables involved in biocontrol solutions, including the currently emerging interest in products that have been developed for preharvest applications but that also demonstrate interest in terms of postharvest biological control [123]. The presence of products on the market (Table 2) helps us to define the diversity of microorganisms that are actually valued in this sector.

Table 2. A non-exhaustive list of biocontrol-based products that are currently marketed for postharvest applications. Adapted with permission from Sellitto et al. [123]. Copyright 2021 MDPI.

Product	Active Ingredient	Country/Company	Fruit/Vegetable	Target
Bio-fungicides recommended for postharvest applications				
Biosave®	*Pseudomonas syringae*	Jet Harvest Solutions USA	Pome Fruit, Citrus, Strawberry, Cherry, Potato	*Penicillium*, *Botrytis*, *Mucor*
Nexy®	*Candida oleophila*	Lesaffre Belgium	Pome Fruit	*Botrytis*, *Penicillium*
BoniProtect® BlossomProtect® Botector®	*Aureobasisium pullulans* (2 strains)	Bio-ferm, Austria	Pome Fruit Grape	*Penicillium*, *Botrytis*, *Monilinia*
Noli	*Metschnikowia fructicola*	Koppert The Netherlands	Table Grape, Pome Fruit, Strawberry, Stone Fruit, Sweet Potato	*Botrytis*, *Penicillium*, *Rhizopus*, *Aspergillus*
Bio-fungicides developed for preharvest applications, also recommended for postharvest				
Serenade® Opti	*Bacillus subtilis*	Bayer	Grape, Berry Fruits, Potato	*Botrytis*, Silver scarf
Amylo-x®	*Bacillus amyloliquefaciens*	Biogard, Italy CBC-Europe, Germany	Grape, Apple, Pear, Kiwifruit	*Botrytis*, *Pseudomonas syringae*
Bio-protection agents developed for food processing, also recommended for postharvest				
Gaia™	*Metschnikowia fructicola*	IOC, France	Harvested Grape, Withering Grape, Grape Musts	*Botrytis*, non-*Saccharomyces* spoiling yeasts
Nymphea™	*Torulaspora delbrueckii*	ICV/ Lallemand, France	Harvested Grapes, Grape Musts	*Botrytis*, non-*Saccharomyces* spoiling yeasts

All the solutions currently on the market encompass the application of yeast and bacteria as control agents (Table 2). The successful application of yeast in postharvest biocontrol is related to their general predisposition toward physically colonizing the surfaces (also in response to the released exopolysaccharides), efficiently competing for the nutrients to overcome frequently used pesticides, release lytic enzymes, and induce the host resistance [128,129] (e.g., lytic enzymes, Figure 4).

Alongside a global aptitude for concretizing antagonism, there is a diversity of specific killer toxins that can be bio-produced by these eukaryotic microorganisms and that have specific cellular targets in the killer-sensitive microbial cell (e.g., membrane, cell wall, RNA, and replication) [128]. This dual aptitude for creating antagonism justifies the amount of interest in yeasts for postharvest control actions on a wide range of fresh fruit and vegetables (e.g., lemon, see Figure 5).

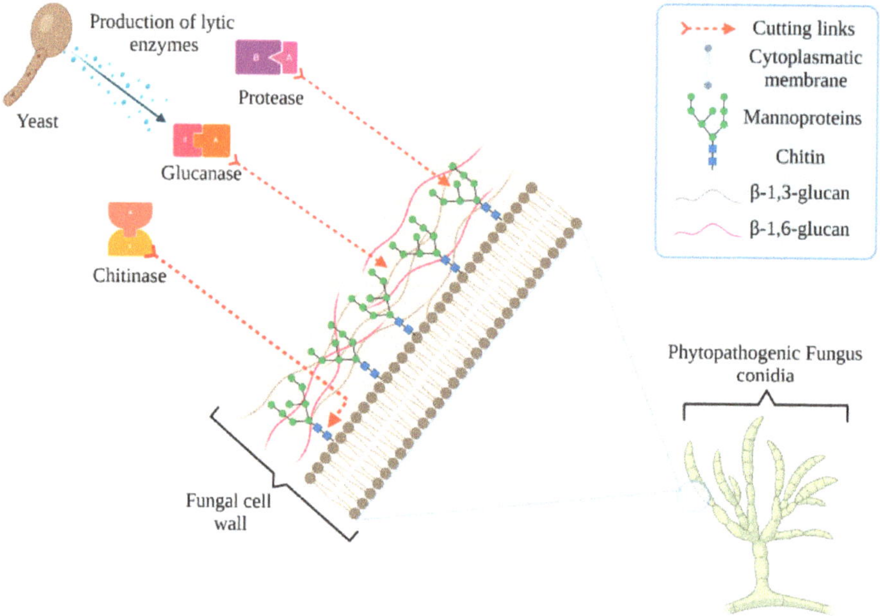

Figure 4. Enzyme production via selected yeasts and their lytic effect on the phytopathogenic fungus cell wall. Reprinted with permission from Hernandez-Montiel et al. [129]. Copyright 2021 MDPI.

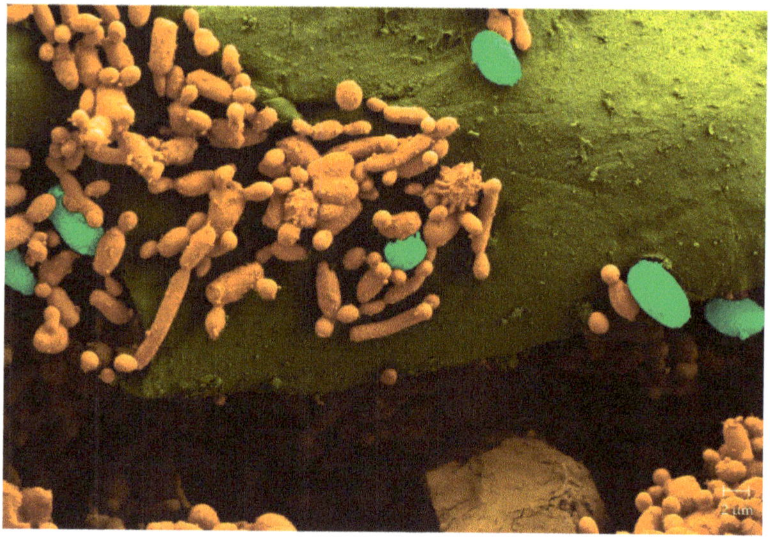

Figure 5. Colored scanning electron microscope (SEM) image of the surface of a lemon wound, inoculated with the killer yeast (orange) and ungerminated spores of citrus fungal postharvest phytopathogen (green). Reprinted with permission from Díaz et al. [128]. Copyright 2020 MDPI.

Even among the prokaryotic microorganisms, steric competition and the antagonism in terms of nutritional factors, as well as the release of lytic enzymes, biofilm formation, and resistance induction (in the host plant), represent the tools of the biological arsenal against postharvest pathogens [130]. To these aspects, we must add the producers of antifungal molecules (e.g., peptides and volatile organic compounds) and the siderophores that chelate iron [130]. The genera *Bacillus* and *Pseudomonas* are considered some of the most effective antagonists for postharvest control [131,132]. *Bacillus* is commonly recognized as a biological alternative to standard chemical fungicides/bactericides in agriculture [132]. *Pseudomonas* has a wide diffusion in the field, attesting to the resilience of this genus in the agricultural environment [131]. Among the different phenotypes of interest in the selected strains, the ability to survive harsh environmental conditions, fast growth (also with low nutrient availability), the capability to modulate the host response, and the production of lytic enzymes/antimicrobial compounds (including the endospore formation for *Bacillus*) have been widely explored [131–133].

In the field of prokaryotes, the heterogeneous group of lactic acid bacteria (LAB) represents a promising reservoir of potential bio-based solutions for sustainable agriculture, also in terms of postharvest applications [134,135]. LAB generally have a long history of safe use in food fermentations and, also due to this characteristic, several LAB species are recognized as having facilitated paths in defining their safe use in food, according to the different legislative environments [136,137]. The biological mechanisms exploited in the control of undesired microbes on fruits and vegetables after the harvest are multivariate and belong to classes that are widely studied in other food applications [135]. Among the others, this molecular arsenal includes organic acids, bacteriocins, and antimicrobial peptides [138].

All these biomolecules have been exploited to improve the shelf-life of fresh plant products during storage conditions [139–142], both through the inoculation of cells and/or of the cell-free supernatants [138,143]. There were several matrices, including leafy green, mixed salads, lettuce, potato, mushroom, tomato, melon, cabbage, apple, table grape, lotus root, litchi, strawberry, kiwifruit, and banana, on which the LAB have been successfully applied for this type of purpose [139,142]. The target microorganisms used in the studies are mainly pathogenic bacteria (e.g., *Listeria monocytogenes*, *Salmonella enterica*, and *Staphylococcus aureus*), but fungal spoilage bacteria (e.g., *Botrytis cinerea*, *Penicillium expansum*, and *Aspergillus flavus*) have also been employed [142–145]. It is possible to detect emerging trends regarding the treatment methods used by the selected LABs and/or with molecules having antimicrobial action. One process of particular interest is the use of fermented products, obtained by inoculating the strain of interest [141,145–147], or the use of edible coatings [139,148,149]. This is a rapidly evolving field in which there are numerous open problems: from the screening of new strains suitable for maximizing efficacy against a broad spectrum of undesirable microorganisms, to the strategy of applying bacteria to the product, without altering the product's sensory quality. In general, the application of supernatants and solutions based on the incorporation of the bacterium in a coating (in some cases, this is also of microbial origin) represent interesting solutions as they are able to improve the degree of standardization for these bio-based strategies [139,150,151]. Future perspectives in the field of biocontrol in the postharvest of fresh plant products show that it is crucial to highlight the need for a clear shared global regulatory environment (e.g., [138]) to speed up innovative actions and the assessment of biocontrol, in combination with other physical and chemical solutions (within the framework of hurdle technology) (e.g., [152]).

3. Innovative Non-Destructive Techniques for the Quality Monitoring of Fruit and Vegetables

3.1. Image Analysis through a Computer Vision System

Computer vision systems (CVSs) are part of an innovative, contactless, and non-destructive technology, based on traditional imaging in the visible range of the electromagnetic spectrum, which is widely used for the in-line grading of fruit and vegetables [153–159]. CVSs include novel technologies to automatically extract from an image the relevant visual information related to the visual quality of the product: they are used to classify and grade, to assess the quality, to detect defects, and to estimate the internal properties. Proper image analysis algorithms and regression or classification models can perform these tasks. The principal benefits of objective, consistent, and pervasive food control along the entire supply chain, from the producers to the final consumers, that are provided by this technology are a reduction in loss and waste, as well as increased consumer satisfaction.

A CVS typically acquires images using a setup composed of the combination of a digital camera, an illumination system, and a personal computer that extracts classification features and builds appropriate models, using statistical methods or machine learning approaches (Figure 6).

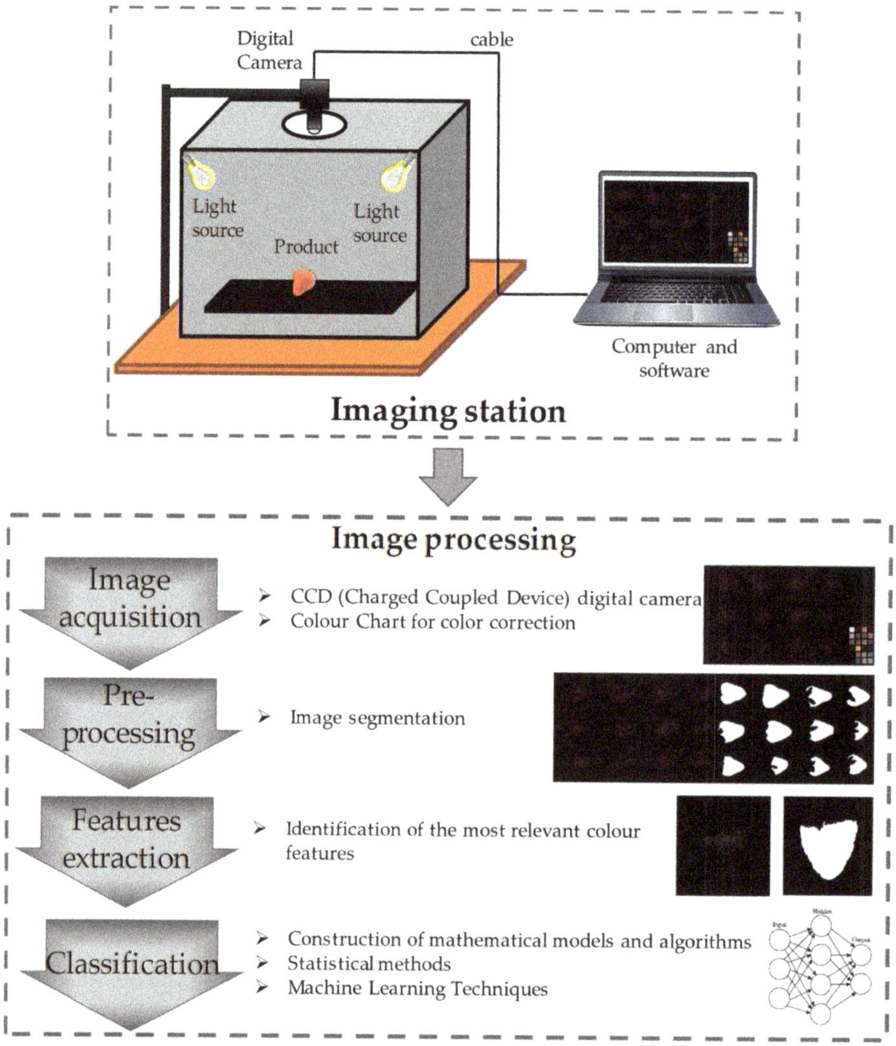

Figure 6. The imaging station and the general workflow of an image-processing system. Four main steps are foreseen in the image processing: Image acquisition, Pre-processing, Features extraction and Classification.

CVSs have been developed to evaluate the quality and marketability of several fresh commodities since color has proven to be crucial for market acceptance. CVS technology is able to estimate the color properties at the pixel level and to provide a more objective and consistent evaluation with respect to standard colorimeters, which have limited applicability at an industrial level and for in-line monitoring [160].

Recently, an innovative CVS has been developed to evaluate the quality level of rocket leaves and to non-destructively discriminate the cultivation approach using color information extracted from digital images. The proposed CVS permitted the authors to achieve an accuracy of about 95% in quality-level assessment and of about 65–70% in the discrimination of the cultivation approach by the use of a random forest model for the automatic selection of the relevant features for classification [161].

The use of CVS also represents an effective tool to monitor fresh-cut fruit quality along the entire supply chain, reducing food waste and ensuring freshness at the market level. In this context, several interesting applications have been reported in the literature on fresh-cut products: artichokes [162], nectarines [163], iceberg lettuce [153], radicchio [164], apples [165], and potatoes [166]. As regards packaged products, the identification of regions of images affected by the presence of shadows or highlights produced by the interaction of light with a plastic bag is a critical problem that needs to be solved for quality-level assessment to be achieved through the packaging material. A robust and powerful segmentation approach that selects the regions where colors can be measured properly is mandatory to achieve performances similar to the results obtained on unpackaged samples [167,168].

Besides their external appearance, CVS can be used for the evaluation of the inner quality of horticultural products. During postharvest storage, ripening or senescence processes can cause alterations in nutritional quality, leading to changes in the product's visual attributes, including color and/or texture [169]. In this regard, the yellowing of green leafy vegetables is strictly related to chlorophyll loss, while the browning processes that occur on the surfaces of fresh-cut products are caused by polyphenol oxidase and peroxidase activity on phenolic compounds. Moreover, the total soluble solids and pH values are statistically affected by the various levels of the ripening stages: when the particular fruit color is enhanced, the total soluble solids content increases, and the fruit acidity decreases.

Pace et al. [170], studying the relationships between antioxidant activity (AA) or total phenols (TP) and the color traits of a local landrace of yellow to purple pigmented carrots, developed CVS regression models that were able to estimate the AA and TP contents (R^2 = 0.97 and R^2 = 0.94, respectively).

Moreover, enforcement of the image-processing method, based on the JPEG images of plums at several maturity stages, highlighted the finding that the indices of an RGB color scale of sample images were correlated to the different chemical traits of plums. A robust correlation between fruit acidity (expressed as the total soluble solids content) and the mean intensity of the green color (R^2 = 0.99) and the R/G ratio (R^2 = 0.85) was obtained [155].

Recently, image analysis by CVS has been applied to predict the enzymatic activity of both polyphenol oxidase (PPO) and peroxidase (POD) on banana samples, evaluating the peel browning that occurred during 9 days of storage at 25 °C. The extraction of several color features (such as the average and variance of all RGB color parameters) from the images that were acquired by CVS allowed the obtaining of two equations using a genetic programming model that could predict PPO and POD activity during the browning process of banana peels. However, the correlation coefficients did not show significant differences between the predicted and measured values of the enzymatic activity of PPO and POD (R^2 = 0.98 and 0.97, respectively) [171].

Sabzi et al. [172] implemented an automatic CVS method able to predict with high accuracy (R^2 = 0.95) the pH value of oranges from images in the visible wavelengths of the external peel. The average time spent in estimating the pH values with the proposed non-destructive technique was 0.42 s, suitable for in-line industrial applications.

A CVS was recently adopted to discriminate between the ripening stages (half-red or red) of strawberries harvested at three different times. Among the chemical indicators of ripening, titratable acidity resulted in it being statistically correlated to the image data

(Pearson correlation coefficient = 1), providing a good indicator for the non-destructive evaluation of the ripening stage of strawberries [173].

Additionally, Palumbo et al. [168] proposed a combination of image analysis and the random forest model to forecast the contents of chlorophyll and ammonia, as objective indicators of senescence, for unpackaged and packaged rocket leaves. An irrelevant performance loss on packaged products (Pearson's linear correlation coefficient was 0.84 and 0.91 for chlorophyll and ammonia content, respectively) compared to unpackaged ones (0.86 for chlorophyll and 0.92 for ammonia) was reported. Moreover, the three partial least squares regression (PLSR) models, built to predict the visual quality level of fresh-cut rocket leaves, using as predictors the total chlorophyll and the ammonia contents obtained by destructive methods, were employed by CVS on packaged products and by CVS on unpackaged ones, highlighting the high performance in validation in terms of R^2 (0.70, 0.77 and 0.80 for destructive methods, CVS through packaging, and CVS without packaging, respectively).

The application of Image-processing techniques has been recently evaluated for the estimation of the total soluble solids and pH of strawberries. The RGB, HSV, and HSL color-space channels were used as input variables to develop multiple linear regression (MLR) and support vector machine regression (SVM-R) models. The results demonstrated that an SVM-R model working on the characteristics in the HSV color space performed better than an MLR model for total soluble solids and pH prediction (accuracy of 84.1% and 79.2% for total soluble solids and 78.8% and 72.6% for pH in the training and testing stages, respectively) [174].

Finally, one very interesting application of image analysis was reported by Li et al. [175], who provided innovative and smart technology to predict the shelf-life and the quality of kiwifruit during cold storage, via acquiring and calculating the RGB value extracted from photos taken by a smartphone camera. The results showed that the R to B ratio values (Central R/B) were negatively correlated with titratable acidity, vitamin C content, and firmness and were positively correlated with soluble solids content, total soluble sugars, and total plate counts. The smartphone has become an essential portable device nowadays; the methodology proposed by these authors, based on smartphone image analysis, is simpler and faster than the other reported prediction methods. The obtained rapid evaluation of the postharvest quality of kiwifruit may avoid losses and waste, especially for ordinary consumers or fruit retailers at the end of the supply chain.

3.2. E-nose

Odors and flavors, crucial sensory parameters for consumer acceptability, are strictly affected by organic volatile compounds (VOCs), which are the final products of the metabolism of plant-based food matrices [176]. The identification and semi-quantification of volatile metabolites via headspace solid-phase microextraction (HS-SPME) sampling, followed by gas chromatography–mass spectrometry (GC-MS), has become a well-established methodology for the assessment of the quality of fruit and vegetables on a molecular basis [177], but the concrete application of HS-SPME/GC-MS in the food industry is limited [178].

In the last few years, the electronic nose (E-nose) has become one of the most favorable sensing technologies as an alternative to conventional HS-SPME/GC-MS. From a practical point of view, it achieves the differentiation and classification of food matrices with different aroma signatures by evaluating the presence and the content of specific volatile metabolites in the headspace of the samples [177,179].

The E-nose is a sensing device supplied with a set of partial specific and broad-spectrum electronic chemical sensors that mimic human olfactory perception; it provides a digital VOC fingerprint that can be explored using suitable statistical tools. These electronic devices are generally composed of three parts: a sample-handling apparatus, a detector, and a system for data acquisition [48]. The use of the olfactometric methodology for the evaluation of the VOC profiles of horticultural food products offers several important advantages, including the low cost of analysis, ease of use, rapidity, non-destructiveness, no need for preliminary sample-preparation steps, environmentally friendly use, and automatic data handling [180,181]. Among the different sensors, the metal oxide semiconductor (MOS) sensors are most frequently used in E-nose applications, presenting the advantages of fast response, high sensitivity, and low cost [182].

Due to the complexity and size of the data matrices produced by the E-nose, it is necessary to treat the sensory responses with chemometric tools to extract information regarding food quality, safety, and fraud recognition and postharvest handling evaluations, as well as ripening assessments [173,183–186].

Numerous reports have been published on the application of E-noses in the analysis of different foods of plant origin [183]. In particular, according to some recent studies it can be indicated that an E-nose is a useful methodology for investigating the ripening stage of fruit [173,184,187].

In the study by Palumbo et al. [173], an E-nose, along with attenuated total reflection-Fourier transform infrared (ATR-FTIR) spectroscopy and image analysis (IA), were used as fast and non-destructive methodologies to discriminate between two ripening stages (half-red or red) of the strawberry variety "Sabrosa", harvested at three different times. In particular, the PCA carried out on the E-nose values indicated that there was a potential correlation among the E-nose signals and the fruit-ripening stage, as the E-nose device reacted sensitively and selectively to changes in the aroma profiles of strawberries at different maturity degrees. Moreover, the correlation analysis between E-nose data and the VOCs profiles, previously obtained by HS-SPME/GC-MS, demonstrated that E-nose responses were in line with HS-SPME/GC-MS analysis [186,188].

Aghilinategh et al. [184] explored the combination of the responses of a MOS gas sensor E-nose with suitable pattern recognition methods, including artificial neural networks (ANN), principal component analysis (PCA), and linear discriminant analysis (LDA) to determine the maturation stage of white berry and blackberry. Most consistent with the results, although all the statistical methods used were able to differentiate among the fruit, based on the ripeness grades, the ANN showed the best classification performance, with an accuracy of more than 88%.

Recently, Qiao et al. [189] investigated the fruit quality indexes, including titratable acidity, soluble sugar content, sugar–acid ratio, soluble solids amount, soluble protein level, and flavor profile via an E-nose among naturally and artificially ripe crab apples. The aroma patterns generated by 12 MOS sensors were treated by PCA, LDA, SVM, and random forest (RF) analyses. The data obtained indicated that the RF, with an average recognition accuracy of about 98%, can be considered the best algorithm for distinguishing between naturally or artificially ripe crab apples. On the other hand, the correlation analysis, performed by PLSR, among the E-nose values and the quality parameters allowed the authors to establish good predictive models with regression coefficients (R^2) that were higher than 0.91 [189].

There are several studies reporting the applicability of the E-nose, combined with chemometric strategies, in classifying fruit and vegetables in line with their geographical origin [183].

Li et al. [190] investigated the origin of 303 maca samples collected from more than 100 sites within the main growing area in China, using GC-MS and an MOS-based E-nose to detect the maca samples' volatile and odor fingerprints, respectively. Correlation and multi-regression analyses showed that all sensors had a significant association with specific maca volatiles. Moreover, using E-nose and a backpropagation (BP) neural network algorithm, a maca odor database was implemented that was able to trace the maca origin with predictability greater than 78%. This study suggested that the E-nose is a fast, reliable, and efficient method by which to predict the geographical provenance of Chinese maca [190].

In an interesting paper, data obtained via an MOS E-nose treated with statistical tools were used to discriminate table grapes, depending on the different agronomic practices (conventional vs. organic farming) and geographical origin [191]. PCA experiments performed on the E-nose responses showed poor clustering of the fruit samples according to the growing site or cultivation methods, while a supervised approach using LDA allowed promising prediction rates of 84% and 85% for the discrimination of the growing handling and geographic provenance, respectively [191].

In line with previous reports, the E-nose has been reported to provide effective and useful information for an overall evaluation of the freshness of fruit and vegetables [183,192].

Cozzolino et al. [186] explored the potentiality of the E-nose as a rapid technique in discriminating samples of the sweet cherry variety, "Ferrovia", packaged in a high-CO_2 (16% O_2 + 20% CO_2 + 64% N_2) or air (20% O_2 + 0.03% CO_2 + 80% N_2) environment, for up to 21 days. The projection to latent structures (PLS) methods that were applied to the E-nose data indicated that the fresh sample and the packaged or unpackaged fruit could be classified, based on both the storage conditions and the storage time. Furthermore, a correlation analysis among the E-nose responses and the overall VOCs, detected in a previous study via HS-SPME/GC-MS [187] on the same cherry samples, allowed the researchers to associate samples with specific flavor profiles, using one or more E-nose sensors. Specifically, the S10 sensor results were related to some VOCs that are thought to be possible markers of freshness and could, thus, be used for a rapid assessment of product quality. Finally, since the data of the rapid and the conventional approaches agreed, the E-nose is confirmed to be an appropriate method for the real-time physiological and quality evaluations of postharvest plant-based food.

Ghasemi-Varnamkhasti et al. [193] used an E-nose equipped with 8 MOS sensors to evaluate the freshness of strawberries stored in three different polymer packaging types, including polypropylene (PPP), ethylene vinyl alcohol (EVOH), and polyvinyl chloride (PVC). E-nose data, treated with pattern recognition approaches such as PCA, LDA, and SVM, allowed the researchers to properly classify the unpackaged and packaged samples and explore the effects of polymer packages on strawberry freshness. The response surface method (RSM) was considered for the choice of the optimized sensor array with regard to the contribution of each sensor in the sample classification. Sample headspace profiles were studied on days 1, 8, and 16. The results showed that the PCA explained 84% of the variance of the data, while the LDA categorized all the sensor responses with an accuracy of 86.4%. Moreover, the SVM method could accurately distinguish between the samples by 86.4% and by 50.6% in training and validation, respectively, by using a polynomial basis function (C-SVM) and could distinguish between them by 85.2% and 55.6% in training and validation, respectively, using a radial basis function (Nu-SVM). Finally, among the eight sensors used in the study, four of them were selected as the optimal sensors for adoption in an E-nose system.

Huang et al. [194] reported a rapid and non-destructive method to investigate the changes in freshness quality of postharvest spinach from 1 to 12 days of cold storage, using machine vision and an E-nose, combined with chemometrics. Ten trained panelists classified the spinach freshness during cold storage into four grades. K-nearest neighbors (KNN), SVM, and a backpropagation artificial neural network (BPNN) were used for the prediction of spinach freshness. The results obtained from applying the BPNN model related to machine vision showed the same output as the KNN approach, with a classification accuracy of 85.4% in the prediction of spinach freshness. On the other hand, the BPNN model, based on E-nose data, allowed obtaining a better result with respect to the SVM approach, with classification accuracies of 81.2% and 75.0%, respectively. Besides, the BPNN model, built on multisensory data fusion using machine vision and E-nose data, greatly improved the accuracy of the freshness evaluation of postharvest spinach by reaching a classification accuracy of 93.7%.

3.3. Near-Infrared Spectroscopy

Infrared (IR) spectroscopy is a very useful technique to recognize specific functional groups in a molecule (Table 3) and, thus, the chemical composition of a product by relating the vibrational properties of matter to certain internal features. Therefore, every sample has a specific IR spectrum; fruit and vegetable products with a similar spectrum also present similar bioactive compounds and nutritional value [195]. Among the IR methodologies, near-infrared (NIR) spectroscopy, which covers the magnetic spectrum range between 780 and 2500 nm, is a rapid, non-destructive, multi-analytical technique that is widely and effectively used in several sectors, such as the food industry, agriculture, chemicals, pharmaceuticals, textiles, polymers, cosmetics, and medical applications [196].

Table 3. Adsorption peaks of the functional groups. Adapted from [197].

		Adsorption Peaks	
	Functional Group	**Wavenumber (cm^{-1})**	**Wavelength (μm)**
O-H	aliphatic and aromatic	3600–3000	2.8–3.3
NH_2	amine	3600–3100	2.8–3.2
CH	aromatic	3150–3000	3.2–3.3
CH	aliphatic	3000–2850	3.3–3.5
C≡N	nitril	2400–2200	4.2–4.6
C≡C-	alkyne	2260–2100	4.4–4.8
COOR	ester	1750–1700	5.7–5.9
COOH	carboxylic acid	1740–1670	5.7–6.0
C=O	aldehydes and ketones	1740–1660	5.7–6.0
$CONH_2$	amide	1720–1640	5.8–6.1
C=C-	alkene	1670–1610	6.0–6.2
Ø-O-R	aromatic	1300–1180	7.7–8.5
R-O-R	aliphatic	1160–1060	8.6–9.4

According to the specific application, three setup modes are possible (Figure 7) to use. First, it is important to note that the penetration of NIR radiation into the tissues decreases with the depth, so it is necessary to select the measurement configuration. For example, thick-skinned fruit can compromise reflectance and interactance ability in detecting internal quality or injuries. On the other hand, transmittance can yield information about the skin and core of a fruit, but the process needs high light intensity that can damage the product (Figure 7).

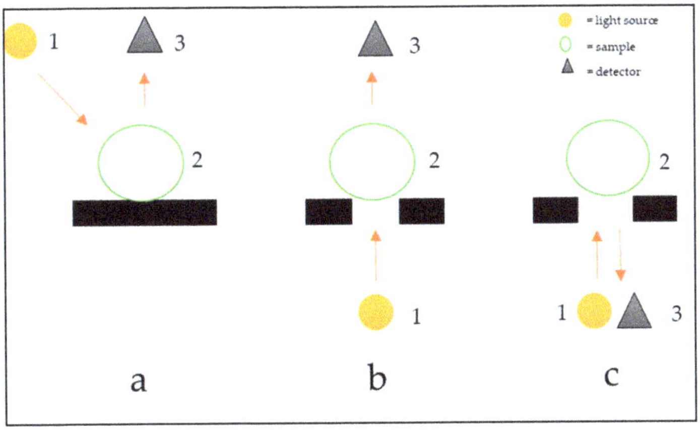

Figure 7. Setup for acquisition in reflectance (**a**), transmittance (**b**), and interactance (**c**). The arrows indicate the direction of light coming out of the lamp (1) hits the sample (2) and then reaches the detector (3).

The outcomes of NIR spectroscopy, combined with multivariate techniques and an image analyzer, can yield information about the solid soluble content (SSC), thc firmness, the dry matter (DM), the hardness, and, in some cases, the internal injuries of harvested produce.

Peirs et al. [198] examined the spectral variability of apples in reflectance mode and demonstrated that it depends mainly on the seasons and cultivars, and depends less on orchards. Besides, the data variability seems to be associated with the changes in the produce's texture properties and in its chemical composition. Actually, the sugar and acid content, as well as the cell size, the number of cells, and the number of intercellular spaces were affected by the different growing conditions. Conversely, Schaare and Fraser [199] compared the three NIR modes in estimating the SSC, density, and flesh color of kiwifruit, achieving a higher accuracy by using the interactance mode. The dry matter (DM), the SSC, and the flesh color of kiwifruit were evaluated by Clark et al. [200], applying an interactance modality in order to predict fruit-storage disorder. The model successfully classified the fruits according to their susceptibility to storage, based on the NIR profile. The incidence of postharvest storage disorders increased for fruit that was less mature, which contained less DM, lower SSC, and greener flesh color than their unaffected counterparts. McGlone and Kawano [201] also analyzed the kiwi, obtaining an excellent predictive model for SSC and DM, but not for firmness.

Several studies have focused their research on the internal quality of citrus, specifically. Liu et al. [202] demonstrated the NIR potential to measure the SSC of intact navel orange fruit. This suggests that it can be used as a valid tool for the determination of other internal quality indices in other thick-skinned fruits, such as tangerine and lemon. The results obtained by Gómez et al. [203] confirm NIR as a non-destructive technique for measuring mandarin quality profiling, especially in predicting the sugar content. Similar results were obtained by Lee et al. [204], who evaluated citrus using transmittance methods to yield information about the fruit flesh; in particular, a PLSR model was found to predict sugar content.

Among the NIR techniques, Fourier transform infrared spectroscopy (FTIR) is involved in several studies as it offers the advantages of high sensitivity, high resolution, and fast data acquisition speed. Conversely, FTIR is an expensive and complex tool. Schulz et al. [205] applied NIR-FT-Raman, ATR-IR, and NIR spectroscopy, in combination with chemometric algorithms for the rapid determination of pungency in black and white ground pepper and green whole pepper berries, demonstrating that vibrational spectroscopy methods can replace the conventional procedures for the effective quality control of peppercorns and pepper extracts and of pepper oil treated in the industry. Furthermore, the spectral data can be used to simultaneously classify the pepper oils according to their different monoterpene and sesquiterpene compositions. Amodio et al. [206] applied an FTIR spectrometer to discriminate among strawberries produced by three different fertility management systems. Data treated by chemometric tools allowed the sensor to obtain good classification models to successfully predict TSS, pH, and TA. FTIR, coupled with attenuated total reflectance (ATR-FTIR) techniques, has been demonstrated to present several advantages, including better signal-to-noise ratio, multiplexing, higher energy, and improved resolution [207]. In Palumbo et al. [173], the ATR-FTIR data have been significantly correlated to titratable acidity as a good indicator of the maturity stage of "Sabrosa" strawberries.

The NIR system is an effective non-invasive technique that can be used to analyze the chemical composition of fresh produce and quality changes during storage, promoting its application directly on the field and in the industrial line and as a valid approach for the traceability and authentication of agricultural produce.

4. Conclusions

This review paper provides an overview of the effects of advanced postharvest tools (active packaging, dipping, vacuum impregnation, pulsed electric field, high hydrostatic pressure, and cold plasma) and of biocontrol techniques to preserve the high nutritional value and safety of fresh produce after harvesting.

Physical treatments (such as microwaving, a pulsed electric field, high hydrostatic pressure, and cold plasma) and a biocontrol strategy proved to be useful for improving the safety of products and consequently extending their shelf-life. Technologies such as dipping, vacuum impregnation, and edible active packaging should be applied when the aim of producers is to preserve the quality and enhance the nutritional value of fresh fruits and vegetables. Moreover, the combination of more techniques, among those reported, might positively affect both the safety and the quality of the products. Therefore, in conclusion, the adoption of these technologies represents an innovation in the fruit and vegetable sector, in order to satisfy the consumer's needs. However, cost analysis is necessary to verify the real applicability.

Regarding the non-destructive techniques reported (image analysis, the E-nose, and near-infrared spectroscopy), the research in this field has made it possible to validate its effectiveness for the non-destructive evaluation of fresh and fresh-cut fruit and vegetables from "the field to the fork", with the aim of optimizing the process phases and limiting losses. Furthermore, the application of these techniques by the end user could increase the satisfaction of the players in the supply chain, improving the quality of the process.

To this end, portable tools to be applied on industrial lines or in the field are required. Further research is in progress, investigating new technologies to extend shelf life and enable the continuous monitoring of product quality at all stages of the supply chain.

Author Contributions: Conceptualization, M.P., R.C., B.P. and M.C.; literature search, M.P., V.C., R.C., A.C., M.L.V.d.C., I.R. and M.C.; investigation, M.P., R.C., B.P. and M.C.; writing—original draft preparation, M.P., V.C., R.C. and M.C.; writing—review and editing, M.P., G.A., V.C., R.C., A.C., M.L.V.d.C., B.P., S.P., I.R., R.R. and M.C.; supervision, M.P., G.A., V.C., R.C., M.L.V.d.C., B.P. and M.C.; project administration, B.P. and M.C.; funding acquisition, B.P. and M.C. All authors have read and agreed to the published version of the manuscript.

Funding: This research was funded by the project Prin 2017 "Multifunctional polymer composites based on grown materials (MIFLOWER)" from the Italian Ministry of Education University and Research Project (grant number: 2017B7MMJ5_001) and by Italian Ministry of University and Research (MUR), project "Conservabilità, qualità e sicurezza dei prodotti ortofrutticoli ad alto contenuto di servizio-ARS01_00640–POFACS", D.D. 1211/2020 and 1104/2021.

Data Availability Statement: Data is contained within the article.

Acknowledgments: The authors thank Massimo Franchi of CNR-ISPA for his technical support in the laboratory.

Conflicts of Interest: The authors declare no conflict of interest.

References

1. Mahajan, P.V.; Caleb, O.J.; Singh, Z.; Watkins, C.B.; Geyer, M. Postharvest Treatments of Fresh Produce. *Philos. Trans. R. Soc. Math. Phys. Eng. Sci.* **2014**, *372*, 20130309. [CrossRef] [PubMed]
2. Brasil, I.M.; Siddiqui, M.W. Postharvest Quality of Fruits and Vegetables: An Overview. In *Preharvest Modulation of Postharvest Fruit and Vegetable Quality*; Elsevier: Amsterdam, The Netherlands, 2018; pp. 1–40. ISBN 978-0-12-809807-3.
3. Chauhan, O.P.; Lakshmi, S.; Pandey, A.K.; Ravi, N.; Gopalan, N.; Sharma, R.K. Non-Destructive Quality Monitoring of Fresh Fruits and Vegetables. *Def. Life Sci. J.* **2017**, *2*, 103. [CrossRef]
4. Barth, M.; Hankinson, T.R.; Zhuang, H.; Breidt, F. Microbiological Spoilage of Fruits and Vegetables. In *Compendium of the Microbiological Spoilage of Foods and Beverages*; Sperber, W.H., Doyle, M.P., Eds.; Springer New York: New York, NY, USA, 2009; pp. 135–183. ISBN 978-1-4419-0825-4.
5. De Corato, U. Improving the Shelf-Life and Quality of Fresh and Minimally-Processed Fruits and Vegetables for a Modern Food Industry: A Comprehensive Critical Review from the Traditional Technologies into the Most Promising Advancements. *Crit. Rev. Food Sci. Nutr.* **2020**, *60*, 940–975. [CrossRef]
6. Berger, C.N.; Sodha, S.V.; Shaw, R.K.; Griffin, P.M.; Pink, D.; Hand, P.; Frankel, G. Fresh Fruit and Vegetables as Vehicles for the Transmission of Human Pathogens: Fresh Produce as Vehicles for Transmission of Human Pathogens. *Environ. Microbiol.* **2010**, *12*, 2385–2397. [CrossRef] [PubMed]
7. Srisamran, J.; Atwill, E.R.; Chuanchuen, R.; Jeamsripong, S. Detection and Analysis of Indicator and Pathogenic Bacteria in Conventional and Organic Fruits and Vegetables Sold in Retail Markets. *Food Qual. Saf.* **2022**, *6*, fyac013. [CrossRef]
8. Gustavsson, J.; Cederberg, C.; Sonesson, U. The Methodology of the FAO Study: "Global Food Losses and Food Waste—Extent, Causes and Prevention"—FAO. Available online: https://www.diva-portal.org/smash/get/diva2:944159/FULLTEXT01.pdf (accessed on 29 August 2022).

9. Bourne, M. Post Harvest Food Losses—The Neglected Dimension in Increasing the World Food Supply. In *Cornell International Agriculture Mimeograph 53*; Cornell University: Ithaca, NY, USA, 1977.
10. SOFA 2019—The State of Food and Agriculture in the World. Available online: https://www.fao.org/state-of-food-agriculture/2019/en/ (accessed on 1 October 2022).
11. Singh, V.; Hedayetullah, M.; Zaman, P.; Meher, J. Postharvest Technology of Fruits and Vegetables: An Overview, Journal of Post Harvest Technology. *J. Postharvest Technol.* **2014**, *2*, 124–135.
12. Ali, A.; Yeoh, W.K.; Forney, C.; Siddiqui, M.W. Advances in Postharvest Technologies to Extend the Storage Life of Minimally Processed Fruits and Vegetables. *Crit. Rev. Food Sci. Nutr.* **2018**, *58*, 2632–2649. [CrossRef]
13. Fatchurrahman, D.; Amodio, M.L.; Colelli, G. Quality of Goji Berry Fruit (*Lycium barbarum* L.) Stored at Different Temperatures. *Foods* **2022**, *11*, 3700. [CrossRef]
14. Kader, A.A. *Postharvest Technology of Horticultural Crops*; University of California Agriculture and Natural Resources: Oakland, CA, USA, 2002; ISBN 978-1-879906-51-8.
15. Fahmy, K.; Nakano, K. Effective Transport and Storage Condition for Preserving The Quality of 'Jiro' Persimmon in Export Market. *Agric. Agric. Sci. Procedia* **2016**, *9*, 279–290. [CrossRef]
16. Bhargava, A.; Bansal, A. Fruits and Vegetables Quality Evaluation Using Computer Vision: A Review. *J. King Saud Univ. - Comput. Inf. Sci.* **2021**, *33*, 243–257. [CrossRef]
17. García-Oliveira, P.; Fraga-Corral, M.; Pereira, A.G.; Prieto, M.A.; Simal-Gandara, J. Solutions for the Sustainability of the Food Production and Consumption System. *Crit. Rev. Food Sci. Nutr.* **2022**, *62*, 1765–1781. [CrossRef] [PubMed]
18. Fan, X.; Wang, W. Quality of Fresh and Fresh-Cut Produce Impacted by Nonthermal Physical Technologies Intended to Enhance Microbial Safety. *Crit. Rev. Food Sci. Nutr.* **2022**, *62*, 362–382. [CrossRef] [PubMed]
19. Ganesan, A.R.; Tiwari, U.; Ezhilarasi, P.N.; Rajauria, G. Application of Cold Plasma on Food Matrices: A Review on Current and Future Prospects. *J. Food Process. Preserv.* **2021**, *45*, e15070. [CrossRef]
20. Dar, A.H.; Shams, R.; ul Eain Hyder Rizvi, Q.; Majid, I. *Microwave and Ohmic Heating of Fresh Cut Fruits and Vegetable Products*; Elsevier Inc.: Amsterdam, The Netherlands, 2019; ISBN 9780128161845.
21. Usall, J.; Ippolito, A.; Sisquella, M.; Neri, F. Physical Treatments to Control Postharvest Diseases of Fresh Fruits and Vegetables. *Postharvest Biol. Technol.* **2016**, *122*, 30–40. [CrossRef]
22. Martínez-Hernández, G.B.; Amodio, M.L.; Colelli, G. Potential Use of Microwave Treatment on Fresh-Cut Carrots: Physical, Chemical and Microbiological Aspects. *J. Sci. Food Agric.* **2016**, *96*, 2063–2072. [CrossRef]
23. Song, L.; Luo, H.; Cheng, X.; Yan, F.; Yang, Z.; Yu, Z. Effects of Microwave Treatment on Physiology and Quality of Minimally Processed Bok Choy (*Brassica campestris* L.) during Storage at 5 °C. *J. Food Meas. Charact.* **2018**, *12*, 913–922. [CrossRef]
24. Colelli, G.; Amodio, M.L.; de Chiara, M.L.V. Operating Conditions for Microwave Application throughout Production Process to Reduce Microbial Load of Fresh-Cut Apples. *Acta Hortic.* **2021**, *1319*, 223–230. [CrossRef]
25. Vadivambal, R.; Jayas, D.S. Non-Uniform Temperature Distribution During Microwave Heating of Food Materials—A Review. *Food Bioprocess Technol.* **2010**, *3*, 161–171. [CrossRef]
26. Li, J.; Shi, J.; Huang, X.; Wang, T.; Zou, X.; Li, Z.; Zhang, D.; Zhang, W.; Xu, Y. Effects of Pulsed Electric Field Pretreatment on Frying Quality of Fresh-Cut Lotus Root Slices. *LWT* **2020**, *132*, 109873. [CrossRef]
27. López-Gámez, G.; Elez-Martínez, P.; Martín-Belloso, O.; Soliva-Fortuny, R. Pulsed Electric Fields Affect Endogenous Enzyme Activities, Respiration and Biosynthesis of Phenolic Compounds in Carrots. *Postharvest Biol. Technol.* **2020**, *168*, 111284. [CrossRef]
28. López-Gámez, G.; Elez-Martínez, P.; Martín-Belloso, O.; Soliva-Fortuny, R. Enhancing Phenolic Content in Carrots by Pulsed Electric Fields during Post-Treatment Time: Effects on Cell Viability and Quality Attributes. *Innov. Food Sci. Emerg. Technol.* **2020**, *59*, 102252. [CrossRef]
29. López-Gámez, G.; Elez-Martínez, P.; Martín-Belloso, O.; Soliva-Fortuny, R. Changes of Carotenoid Content in Carrots after Application of Pulsed Electric Field Treatments. *LWT* **2021**, *147*, 111408. [CrossRef]
30. Rux, G.; Gelewsky, R.; Schlüter, O.; Herppich, W.B. High Hydrostatic Pressure Treatment Effects on Selected Tissue Properties of Fresh Horticultural Products. *Innov. Food Sci. Emerg. Technol.* **2020**, *61*, 102326. [CrossRef]
31. Ramos-Parra, P.A.; García-Salinas, C.; Rodríguez-López, C.E.; García, N.; García-Rivas, G.; Hernández-Brenes, C.; Díaz de la Garza, R.I. High Hydrostatic Pressure Treatments Trigger de Novo Carotenoid Biosynthesis in Papaya Fruit (*Carica papaya* Cv. Maradol). *Food Chem.* **2019**, *277*, 362–372. [CrossRef]
32. Hu, X.; Ma, T.; Ao, L.; Kang, H.; Hu, X.; Song, Y.; Liao, X. Effect of High Hydrostatic Pressure Processing on Textural Properties and Microstructural Characterization of Fresh-cut Pumpkin (*Cucurbita pepo*). *J. Food Process Eng.* **2020**, *43*, e13379. [CrossRef]
33. Hu, K.; Peng, D.; Wang, L.; Liu, H.; Xie, B.; Sun, Z. Effect of Mild High Hydrostatic Pressure Treatments on Physiological and Physicochemical Characteristics and Carotenoid Biosynthesis in Postharvest Mango. *Postharvest Biol. Technol.* **2021**, *172*, 111381. [CrossRef]
34. Kundukulangara Pulissery, S.; Kallahalli Boregowda, S.; Suseela, S.; Jaganath, B. A Comparative Study on the Textural and Nutritional Profile of High Pressure and Minimally Processed Pineapple. *J. Food Sci. Technol.* **2021**, *58*, 3734–3742. [CrossRef]
35. Paciulli, M.; Rinaldi, M.; Rodolfi, M.; Ganino, T.; Morbarigazzi, M.; Chiavaro, E. Effects of High Hydrostatic Pressure on Physico-Chemical and Structural Properties of Two Pumpkin Species. *Food Chem.* **2019**, *274*, 281–290. [CrossRef]

36. Paciulli, M.; Medina Meza, I.G.; Rinaldi, M.; Ganino, T.; Pugliese, A.; Rodolfi, M.; Barbanti, D.; Morbarigazzi, M.; Chiavaro, E. Improved Physicochemical and Structural Properties of Blueberries by High Hydrostatic Pressure Processing. *Foods* **2019**, *8*, 272. [CrossRef]
37. Paciulli, M.; Ganino, T.; Meza, I.G.M.; Rinaldi, M.; Rodolfi, M.; Morbarigazzi, M.; Chiavaro, E. High Pressure and Thermal Processing on the Quality of Zucchini Slices. *Eur. Food Res. Technol.* **2021**, *247*, 475–484. [CrossRef]
38. Tappi, S.; Ramazzina, I.; Rizzi, F.; Sacchetti, G.; Ragni, L.; Rocculi, P. Effect of Plasma Exposure Time on the Polyphenolic Profile and Antioxidant Activity of Fresh-Cut Apples. *Appl. Sci.* **2018**, *8*, 1939. [CrossRef]
39. Tappi, S.; Ragni, L.; Tylewicz, U.; Romani, S.; Ramazzina, I.; Rocculi, P. Browning Response of Fresh-Cut Apples of Different Cultivars to Cold Gas Plasma Treatment. *Innov. Food Sci. Emerg. Technol.* **2019**, *53*, 56–62. [CrossRef]
40. Zhou, D.; Li, T.; Cong, K.; Suo, A.; Wu, C. Influence of Cold Plasma on Quality Attributes and Aroma Compounds in Fresh-Cut Cantaloupe during Low Temperature Storage. *LWT* **2022**, *154*, 112893. [CrossRef]
41. Ahmadnia, M.; Sadeghi, M.; Abbaszadeh, R.; Ghomi Marzdashti, H.R. Decontamination of Whole Strawberry via Dielectric Barrier Discharge Cold Plasma and Effects on Quality Attributes. *J. Food Process. Preserv.* **2021**, *45*, e15019. [CrossRef]
42. Rana, S.; Mehta, D.; Bansal, V.; Shivhare, U.S.; Yadav, S.K. Atmospheric Cold Plasma (ACP) Treatment Improved in-Package Shelf-Life of Strawberry Fruit. *J. Food Sci. Technol.* **2020**, *57*, 102–112. [CrossRef]
43. Li, X.; Li, M.; Ji, N.; Jin, P.; Zhang, J.; Zheng, Y.; Zhang, X.; Li, F. Cold Plasma Treatment Induces Phenolic Accumulation and Enhances Antioxidant Activity in Fresh-Cut Pitaya (Hylocereus Undatus) Fruit. *LWT* **2019**, *115*, 108447. [CrossRef]
44. Zhang, Y.; Zhang, J.; Zhang, Y.; Hu, H.; Luo, S.; Zhang, L.; Zhou, H.; Li, P. Effects of In-package Atmospheric Cold Plasma Treatment on the Qualitative, Metabolic and Microbial Stability of Fresh-cut Pears. *J. Sci. Food Agric.* **2021**, *101*, 4473–4480. [CrossRef]
45. Giannoglou, M.; Stergiou, P.; Dimitrakellis, P.; Gogolides, E.; Stoforos, N.G.; Katsaros, G. Effect of Cold Atmospheric Plasma Processing on Quality and Shelf-Life of Ready-to-Eat Rocket Leafy Salad. *Innov. Food Sci. Emerg. Technol.* **2020**, *66*, 102502. [CrossRef]
46. Hu, X.; Sun, H.; Yang, X.; Cui, D.; Wang, Y.; Zhuang, J.; Wang, X.; Ma, R.; Jiao, Z. Potential Use of Atmospheric Cold Plasma for Postharvest Preservation of Blueberries. *Postharvest Biol. Technol.* **2021**, *179*, 111564. [CrossRef]
47. Gavahian, M.; Sheu, F.; Tsai, M.; Chu, Y. The Effects of Dielectric Barrier Discharge Plasma Gas and Plasma-activated Water on Texture, Color, and Bacterial Characteristics of Shiitake Mushroom. *J. Food Process. Preserv.* **2020**, *44*, e14316. [CrossRef]
48. Liu, C.; Chen, C.; Jiang, A.; Sun, X.; Guan, Q.; Hu, W. Effects of Plasma-Activated Water on Microbial Growth and Storage Quality of Fresh-Cut Apple. *Innov. Food Sci. Emerg. Technol.* **2020**, *59*, 102256. [CrossRef]
49. Perinban, S.; Orsat, V.; Raghavan, V. Influence of Plasma Activated Water Treatment on Enzyme Activity and Quality of Fresh-Cut Apples. *Food Chem.* **2022**, *393*, 133421. [CrossRef] [PubMed]
50. Vanga, S.K.; Wang, J.; Jayaram, S.; Raghavan, V. Effects of Pulsed Electric Fields and Ultrasound Processing on Proteins and Enzymes: A Review. *Processes* **2021**, *9*, 722. [CrossRef]
51. González-Casado, S.; Martín-Belloso, O.; Elez-Martínez, P.; Soliva-Fortuny, R. Enhancing the Carotenoid Content of Tomato Fruit with Pulsed Electric Field Treatments: Effects on Respiratory Activity and Quality Attributes. *Postharvest Biol. Technol.* **2018**, *137*, 113–118. [CrossRef]
52. Sotelo, K.A.G.; Hamid, N.; Oey, I.; Pook, C.; Gutierrez-Maddox, N.; Ma, Q.; Ying Leong, S.; Lu, J. Red Cherries (*Prunus avium* Var. Stella) Processed by Pulsed Electric Field—Physical, Chemical and Microbiological Analyses. *Food Chem.* **2018**, *240*, 926–934. [CrossRef]
53. Ribas-Agustí, A.; Martín-Belloso, O.; Soliva-Fortuny, R.; Elez-Martínez, P. Enhancing Hydroxycinnamic Acids and Flavan-3-Ol Contents by Pulsed Electric Fields without Affecting Quality Attributes of Apple. *Food Res. Int.* **2019**, *121*, 433–440. [CrossRef]
54. Pokhrel, P.R.; Boulet, C.; Yildiz, S.; Sablani, S.; Tang, J.; Barbosa-Cánovas, G.V. Effect of High Hydrostatic Pressure on Microbial Inactivation and Quality Changes in Carrot-Orange Juice Blends at Varying PH. *LWT* **2022**, *159*, 113219. [CrossRef]
55. Bagheri, H.; Abbaszadeh, S. Effect of Cold Plasma on Quality Retention of Fresh-Cut Produce. *J. Food Qual.* **2020**, 8866369. [CrossRef]
56. Li, M.; Li, X.; Han, C.; Ji, N.; Jin, P.; Zheng, Y. Physiological and Metabolomic Analysis of Cold Plasma Treated Fresh-Cut Strawberries. *J. Agric. Food Chem.* **2019**, *67*, 4043–4053. [CrossRef]
57. Ziuzina, D.; Misra, N.N.; Han, L.; Cullen, P.J.; Moiseev, T.; Mosnier, J.P.; Keener, K.; Gaston, E.; Vilaró, I.; Bourke, P. Investigation of a Large Gap Cold Plasma Reactor for Continuous In-Package Decontamination of Fresh Strawberries and Spinach. *Innov. Food Sci. Emerg. Technol.* **2020**, *59*, 102229. [CrossRef]
58. Hozák, P.; Jirešová, J.; Khun, J.; Scholtz, V.; Julák, J. Shelf Life Prolongation of Fresh Strawberries by Nonthermal Plasma Treatment. *J. Food Process. Preserv.* **2022**, *46*, e16150. [CrossRef]
59. Puligundla, P.; Lee, T.; Mok, C. Effect of Intermittent Corona Discharge Plasma Treatment for Improving Microbial Quality and Shelf Life of Kumquat (Citrus Japonica) Fruits. *LWT* **2018**, *91*, 8–13. [CrossRef]
60. Guo, J.; Qin, D.; Li, W.; Wu, F.; Li, L.; Liu, X. Inactivation of Penicillium Italicum on Kumquat via Plasma-Activated Water and Its Effects on Quality Attributes. *Int. J. Food Microbiol.* **2021**, *343*, 109090. [CrossRef] [PubMed]
61. Tan, J.; Karwe, M.V. Inactivation of Enterobacter Aerogenes on the Surfaces of Fresh-Cut Purple Lettuce, Kale, and Baby Spinach Leaves Using Plasma Activated Mist (PAM). *Innov. Food Sci. Emerg. Technol.* **2021**, *74*, 102868. [CrossRef]

62. Laurita, R.; Gozzi, G.; Tappi, S.; Capelli, F.; Bisag, A.; Laghi, G.; Gherardi, M.; Cellini, B.; Abouelenein, D.; Vittori, S.; et al. Effect of Plasma Activated Water (PAW) on Rocket Leaves Decontamination and Nutritional Value. *Innov. Food Sci. Emerg. Technol.* **2021**, *73*, 102805. [CrossRef]
63. Silvetti, T.; Pedroni, M.; Brasca, M.; Vassallo, E.; Cocetta, G.; Ferrante, A.; De Noni, I.; Piazza, L.; Morandi, S. Assessment of Possible Application of an Atmospheric Pressure Plasma Jet for Shelf Life Extension of Fresh-Cut Salad. *Foods* **2021**, *10*, 513. [CrossRef]
64. Sudarsan, A.; Keener, K. Inactivation of Spoilage Organisms on Baby Spinach Leaves Using High Voltage Atmospheric Cold Plasma (HVACP) and Assessment of Quality. *Innov. Food Sci. Emerg. Technol.* **2022**, *79*, 103023. [CrossRef]
65. Dong, X.Y.; Yang, Y.L. A Novel Approach to Enhance Blueberry Quality During Storage Using Cold Plasma at Atmospheric Air Pressure. *Food Bioprocess Technol.* **2019**, *12*, 1409–1421. [CrossRef]
66. Ji, Y.; Hu, W.; Liao, J.; Jiang, A.; Xiu, Z.; Gaowa, S.; Guan, Y.; Yang, X.; Feng, K.; Liu, C. Effect of Atmospheric Cold Plasma Treatment on Antioxidant Activities and Reactive Oxygen Species Production in Postharvest Blueberries during Storage. *J. Sci. Food Agric.* **2020**, *100*, 5586–5595. [CrossRef]
67. Pathak, N.; Grossi Bovi, G.; Limnaios, A.; Fröhling, A.; Brincat, J.; Taoukis, P.; Valdramidis, V.P.; Schlüter, O. Impact of Cold Atmospheric Pressure Plasma Processing on Storage of Blueberries. *J. Food Process. Preserv.* **2020**, *44*, e14581. [CrossRef]
68. Segura-Ponce, L.A.; Reyes, J.E.; Troncoso-Contreras, G.; Valenzuela-Tapia, G. Effect of Low-Pressure Cold Plasma (LPCP) on the Wettability and the Inactivation of Escherichia Coli and Listeria Innocua on Fresh-Cut Apple (Granny Smith) Skin. *Food Bioprocess Technol.* **2018**, *11*, 1075–1086. [CrossRef]
69. Zhou, R.; Zhou, R.; Mai-Prochnow, A.; Zhang, X.; Xian, Y.; Cullen, P.J.; Ostrikov, K. (Ken) Surface Plasma Discharges for the Preservation of Fresh-Cut Apples: Microbial Inactivation and Quality Attributes. *J. Phys. Appl. Phys.* **2020**, *53*, 174003. [CrossRef]
70. Chen, C.; Liu, C.; Jiang, A.; Guan, Q.; Sun, X.; Liu, S.; Hao, K.; Hu, W. The Effects of Cold Plasma-Activated Water Treatment on the Microbial Growth and Antioxidant Properties of Fresh-Cut Pears. *Food Bioprocess Technol.* **2019**, *12*, 1842–1851. [CrossRef]
71. Zhao, Z.; Wang, X.; Ma, T. Properties of Plasma-Activated Water with Different Activation Time and Its Effects on the Quality of Button Mushrooms (Agaricus Bisporus). *LWT* **2021**, *147*, 111633. [CrossRef]
72. Lee, T.; Puligundla, P.; Mok, C. Intermittent Corona Discharge Plasma Jet for Improving Tomato Quality. *J. Food Eng.* **2018**, *223*, 168–174. [CrossRef]
73. Zhao, Y.; Chen, R.; Liu, D.; Wang, W.; Niu, J.; Xia, Y.; Qi, Z.; Zhao, Z.; Song, Y. Effect of Nonthermal Plasma-Activated Water on Quality and Antioxidant Activity of Fresh-Cut Kiwifruit. *IEEE Trans. Plasma Sci.* **2019**, *47*, 4811–4817. [CrossRef]
74. Limnaios, A.; Pathak, N.; Grossi, G.; Fr, A.; Valdramidis, P.; Taoukis, P.S.; Schlüter, O. Effect of Cold Atmospheric Pressure Plasma Processing on Quality and Shelf Life of Red Currants. *LWT* **2021**, *151*, 112213. [CrossRef]
75. Siddiqui, W.; Chakraborty, I.; Ayala-Zavala, J.F.; Dhua, R.S. Advances in Minimal Processing of Fruits and Vegetables: A Review. *J. Sci. Ind. Res.* **2011**, *70*, 12.
76. Escobedo-Avellaneda, Z.; García-García, R.; Valdez-Fragoso, A.; Mújica-Paz, H.; Welti-Chanes, J. Fruit Preservation and Design of Functional Fruit Products by Vacuum Impregnation. In *Fruit Preservation*; Rosenthal, A., Deliza, R., Welti-Chanes, J., Barbosa-Cánovas, G.V., Eds.; Food Engineering Series; Springer New York: New York, NY, USA, 2018; pp. 335–349. ISBN 978-1-4939-3309-9.
77. Chinnaswamy, S.; Rudra, S.G.; Sharma, R.R. Texturizers for Fresh-Cut Fruit and Vegetable Products. In *Fresh-Cut Fruits and Vegetables*; Elsevier: Amsterdam, The Netherlands, 2020; pp. 121–149. ISBN 978-0-12-816184-5.
78. Joshi, A.; Prajapati, U.; Sethi, S.; Arora, B.; Sharma, R.R. Fortification in Fresh and Fresh-Cut Horticultural Products. In *Fresh-Cut Fruits and Vegetables*; Elsevier: Amsterdam, The Netherlands, 2020; pp. 183–204. ISBN 978-0-12-816184-5.
79. Martín-Diana, A.B.; Rico, D.; Frías, J.M.; Barat, J.M.; Henehan, G.T.M.; Barry-Ryan, C. Calcium for Extending the Shelf Life of Fresh Whole and Minimally Processed Fruits and Vegetables: A Review. *Trends Food Sci. Technol.* **2007**, *18*, 210–218. [CrossRef]
80. Soliva-Fortuny, R.C.; Martín-Belloso, O. New Advances in Extending the Shelf-Life of Fresh-Cut Fruits: A Review. *Trends Food Sci. Technol.* **2003**, *14*, 341–353. [CrossRef]
81. Alzamora, S.M.; Salvatori, D.; Tapia, M.S.; López-Malo, A.; Welti-Chanes, J.; Fito, P. Novel Functional Foods from Vegetable Matrices Impregnated with Biologically Active Compounds. *J. Food Eng.* **2005**, *67*, 205–214. [CrossRef]
82. Mu, B.; Xue, J.; Zhang, S.; Li, Z. Effects of the Use of Different Temperature and Calcium Chloride Treatments during Storage on the Quality of Fresh-Cut "Xuebai" Cauliflowers. *Foods* **2022**, *11*, 442. [CrossRef] [PubMed]
83. Mola, S.; Uthairatanakij, A.; Srilaong, V.; Aiamla-or, S.; Jitareerat, P. Impacts of Sodium Chlorite Combined with Calcium Chloride, and Calcium Ascorbate on Microbial Population, Browning, and Quality of Fresh-Cut Rose Apple. *Agric. Nat. Resour.* **2016**, *50*, 331–337. [CrossRef]
84. Giacalone, G.; Chiabrando, V. Effect of Different Treatments with Calcium Salts on Sensory Quality of Fresh-Cut Apple. *J. Food Nutr. Res.* **2013**, *52*, 79–86.
85. Albertini, S.; Lai Reyes, A.E.; Trigo, J.M.; Sarriés, G.A.; Spoto, M.H.F. Effects of Chemical Treatments on Fresh-Cut Papaya. *Food Chem.* **2016**, *190*, 1182–1189. [CrossRef]
86. Yan, R.; Han, C.; Fu, M.; Jiao, W.; Wang, W. Inhibitory Effects of CaCl2 and Pectin Methylesterase on Fruit Softening of Raspberry during Cold Storage. *Horticulturae* **2021**, *8*, 1. [CrossRef]
87. Hamdan, N.; Lee, C.H.; Wong, S.L.; Fauzi, C.E.N.C.A.; Zamri, N.M.A.; Lee, T.H. Prevention of Enzymatic Browning by Natural Extracts and Genome-Editing: A Review on Recent Progress. *Molecules* **2022**, *27*, 1101. [CrossRef]

88. Martínez-Hernández, G.B.; Castillejo, N.; Artés-Hernández, F. Effect of Fresh–Cut Apples Fortification with Lycopene Microspheres, Revalorized from Tomato by-Products, during Shelf Life. *Postharvest Biol. Technol.* **2019**, *156*, 110925. [CrossRef]
89. Supapvanich, S.; Mitsang, P.; Srinorkham, P. Effects of 'Queen' and 'Smooth Cayenne' Pineapple Fruit Core Extracts on Browning Inhibition of Fresh-Cut Wax Apple Fruit during Storage. *Int. Food Res. J.* **2017**, *24*, 559–564.
90. Turrini, F.; Malaspina, P.; Giordani, P.; Catena, S.; Zunin, P.; Boggia, R. Traditional Decoction and PUAE Aqueous Extracts of Pomegranate Peels as Potential Low-Cost Anti-Tyrosinase Ingredients. *Appl. Sci.* **2020**, *10*, 2795. [CrossRef]
91. Jirasuteeruk, C.; Theerakulkait, C. Ultrasound-Assisted Extraction of Phenolic Compounds from Mango (Mangifera Indica Cv. Chok Anan) Peel and Its Inhibitory Effect on Enzymatic Browning of Potato Puree. *Food Technol. Biotechnol.* **2019**, *57*, 350–357. [CrossRef] [PubMed]
92. Supapvanich, S.; Mitrsang, P.; Srinorkham, P.; Boonyaritthongchai, P.; Wongs-Aree, C. Effects of Fresh Aloe Vera Gel Coating on Browning Alleviation of Fresh Cut Wax Apple (Syzygium Samarangenese) Fruit Cv. Taaptimjaan. *J. Food Sci. Technol.* **2016**, *53*, 2844–2850. [CrossRef] [PubMed]
93. Wessels, B.; Damm, S.; Kunz, B.; Schulze-Kaysers, N. Effect of Selected Plant Extracts on the Inhibition of Enzymatic Browning in Fresh-Cut Apple. *J. Appl. Bot. Food Qual.* **2014**, *87*, 16–23. [CrossRef]
94. Supapvanich, S.; Yimpong, A.; Srisuwanwichan, J. Browning Inhibition on Fresh-Cut Apple by the Immersion of Liquid Endosperm from Mature Coconuts. *J. Food Sci. Technol.* **2020**, *57*, 4424–4431. [CrossRef]
95. Gras, M.; Vidal-Brotóns, N.; Betoret, A.; Chiralt; Fito, P. The Response of Some Vegetables to Vacuum Impregnation. *Innov. Food Sci. Emerg. Technol.* **2002**, *3*, 263–269. [CrossRef]
96. Fito, P. Modelling of Vacuum Osmotic Dehydration of Food. *J. Food Eng.* **1994**, *22*, 313–328. [CrossRef]
97. Andrés, A.; Salvatori, D.; Albors, A.; Chiralt, A.; Fito, P. Vacuum Impregnation Viability of Some Fruits and Vegetables. In *Osmotic Dehydration & Vacuum Impregnation*; Fito, P., Chiralt, A., Barat, J.M., Spiess, W.E.L., Behsnilian, D., Eds.; CRC Press: Boca Raton, FL, USA, 2019; pp. 53–60. ISBN 978-0-429-13221-6.
98. Kang, J.-W.; Kang, D.-H. Enhanced Antimicrobial Effect of Organic Acid Washing against Foodborne Pathogens on Broccoli by Vacuum Impregnation. *Int. J. Food Microbiol.* **2016**, *217*, 85–93. [CrossRef]
99. Yılmaz, F.M.; Ersus Bilek, S. Ultrasound-Assisted Vacuum Impregnation on the Fortification of Fresh-Cut Apple with Calcium and Black Carrot Phenolics. *Ultrason. Sonochem.* **2018**, *48*, 509–516. [CrossRef]
100. Santana Moreira, M.; de Almeida Paula, D.; Maurício Furtado Martins, E.; Nascif Rufino Vieira, É.; Mota Ramos, A.; Stringheta, P.C. Vacuum Impregnation of β-Carotene and Lutein in Minimally Processed Fruit Salad. *J. Food Process. Preserv.* **2018**, *42*, e13545. [CrossRef]
101. Lech, K.; Michalska, A.; Wojdyło, A.; Nowicka, P.; Figiel, A. The Influence of Physical Properties of Selected Plant Materials on the Process of Osmotic Dehydration. *LWT* **2018**, *91*, 588–594. [CrossRef]
102. Peng, J.; Bi, J.; Yi, J.; Allaf, K.; Besombes, C.; Jin, X.; Wu, X.; Lyu, J.; Asghar Ali, M.N.H. Apple Juice Concentrate Impregnation Enhances Nutritional and Textural Attributes of the Instant Controlled Pressure Drop (DIC)-dried Carrot Chips. *J. Sci. Food Agric.* **2019**, *99*, 6248–6257. [CrossRef]
103. Derossi, A.; Ricci, I.; Fiore, A.G.; Severini, C. Apple Sclices Enriched With Aloe Vera By Vacuum Impregnation. *Ital. J. Food Sci.* **2017**, *30*, 256–257. [CrossRef]
104. González-Pérez, J.E.; Jiménez-González, O.; Ramírez-Corona, N.; Guerrero-Beltrán, J.A.; López-Malo, A. Vacuum Impregnation on Apples with Grape Juice Concentrate: Effects of Pressure, Processing Time, and Juice Concentration. *Innov. Food Sci. Emerg. Technol.* **2022**, *77*, 102981. [CrossRef]
105. Zhang, L.; Wang, P.; Chen, F.; Lai, S.; Yu, H.; Yang, H. Effects of Calcium and Pectin Methylesterase on Quality Attributes and Pectin Morphology of Jujube Fruit under Vacuum Impregnation during Storage. *Food Chem.* **2019**, *289*, 40–48. [CrossRef]
106. Yang, H.; Wu, Q.; Ng, L.Y.; Wang, S. Effects of Vacuum Impregnation with Calcium Lactate and Pectin Methylesterase on Quality Attributes and Chelate-Soluble Pectin Morphology of Fresh-Cut Papayas. *Food Bioprocess Technol.* **2017**, *10*, 901–913. [CrossRef]
107. Pandey, V.K.; Islam, R.U.; Shams, R.; Dar, A.H. A Comprehensive Review on the Application of Essential Oils as Bioactive Compounds in Nano-Emulsion Based Edible Coatings of Fruits and Vegetables. *Appl. Food Res.* **2022**, *2*, 100042. [CrossRef]
108. M. Rangaraj, V.; Rambabu, K.; Banat, F.; Mittal, V. Natural Antioxidants-Based Edible Active Food Packaging: An Overview of Current Advancements. *Food Biosci.* **2021**, *43*, 101251. [CrossRef]
109. Zam, W. Effect of Alginate and Chitosan Edible Coating Enriched with Olive Leaves Extract on the Shelf Life of Sweet Cherries (*Prunus avium* L.). *J. Food Qual.* **2019**, *2019*, 8192964. [CrossRef]
110. Robles-Sánchez, R.M.; Rojas-Graü, M.A.; Odriozola-Serrano, I.; González-Aguilar, G.; Martin-Belloso, O. Influence of Alginate-Based Edible Coating as Carrier of Antibrowning Agents on Bioactive Compounds and Antioxidant Activity in Fresh-Cut Kent Mangoes. *LWT - Food Sci. Technol.* **2013**, *50*, 240–246. [CrossRef]
111. Özdemir, K.S.; Gökmen, V. Effect of Chitosan-Ascorbic Acid Coatings on the Refrigerated Storage Stability of Fresh-Cut Apples. *Coatings* **2019**, *9*, 503. [CrossRef]
112. Liu, J.; Liu, S.; Zhang, X.; Kan, J.; Jin, C. Effect of Gallic Acid Grafted Chitosan Film Packaging on the Postharvest Quality of White Button Mushroom (Agaricus Bisporus). *Postharvest Biol. Technol.* **2019**, *147*, 39–47. [CrossRef]
113. Carvalho, R.L.; Cabral, M.F.; Germano, T.A.; de Carvalho, W.M.; Brasil, I.M.; Gallão, M.I.; Moura, C.F.H.; Lopes, M.M.A.; de Miranda, M.R.A. Chitosan Coating with Trans-Cinnamaldehyde Improves Structural Integrity and Antioxidant Metabolism of Fresh-Cut Melon. *Postharvest Biol. Technol.* **2016**, *113*, 29–39. [CrossRef]

114. Jagannath, J.H.; Nanjappa, C.; Gupta, D.D.; Bawa, A.S. Studies on the Stability of an Edible Film and Its Use for the Preservation of Carrot (Daucus Carota). *Int. J. Food Sci. Technol.* **2006**, *41*, 498–506. [CrossRef]
115. Imeneo, V.; Piscopo, A.; Martín-Belloso, O.; Soliva-Fortuny, R. Efficacy of Pectin-Based Coating Added with a Lemon Byproduct Extract on Quality Preservation of Fresh-Cut Carrots. *Foods* **2022**, *11*, 1314. [CrossRef] [PubMed]
116. Perdones, A.; Sánchez-González, L.; Chiralt, A.; Vargas, M. Effect of Chitosan–Lemon Essential Oil Coatings on Storage-Keeping Quality of Strawberry. *Postharvest Biol. Technol.* **2012**, *70*, 32–41. [CrossRef]
117. Ghafoor, K.; Al-Juhaimi, F.Y.; Babiker, E.E.; Mohamed Ahmed, I.A.; Shahzad, S.A.; Alsawmahi, O.N. Quality Attributes of Refrigerated Barhi Dates Coated with Edible Chitosan Containing Natural Functional Ingredients. *Foods* **2022**, *11*, 1584. [CrossRef]
118. Vieira, T.M.; Alves, V.D.; Moldão Martins, M. Application of an Eco-Friendly Antifungal Active Package to Extend the Shelf Life of Fresh Red Raspberry (*Rubus idaeus* L. Cv. 'Kweli'). *Foods* **2022**, *11*, 1805. [CrossRef]
119. Jurburg, S.D.; Eisenhauer, N.; Buscot, F.; Chatzinotas, A.; Chaudhari, N.M.; Heintz-Buschart, A.; Kallies, R.; Küsel, K.; Litchman, E.; Macdonald, C.A.; et al. Potential of Microbiome-Based Solutions for Agrifood Systems. *Nat. Food* **2022**, *3*, 557–560. [CrossRef]
120. Wassermann, B.; Müller, H.; Berg, G. An Apple a Day: Which Bacteria Do We Eat With Organic and Conventional Apples? *Front. Microbiol.* **2019**, *10*, 1629. [CrossRef]
121. Kłapeć, T.; Wójcik-Fatla, A.; Farian, E.; Kowalczyk, K.; Cholewa, G.; Cholewa, A.; Dutkiewicz, J. Mycobiota of Berry Fruits—Levels of Filamentous Fungi and Mycotoxins, Composition of Fungi, and Analysis of Potential Health Risk for Consumers. *Ann. Agric. Environ. Med.* **2022**, *29*, 28–37. [CrossRef]
122. Vermote, L.; Verce, M.; Mozzi, F.; De Vuyst, L.; Weckx, S. Microbiomes Associated With the Surfaces of Northern Argentinian Fruits Show a Wide Species Diversity. *Front. Microbiol.* **2022**, *13*, 872281. [CrossRef]
123. Sellitto, V.M.; Zara, S.; Fracchetti, F.; Capozzi, V.; Nardi, T. Microbial Biocontrol as an Alternative to Synthetic Fungicides: Boundaries between Pre- and Postharvest Applications on Vegetables and Fruits. *Fermentation* **2021**, *7*, 60. [CrossRef]
124. Zhang, H.; Zhang, Q.; Chen, S.; Zhang, Z.; Song, J.; Long, Z.; Yu, Y.; Fang, H. Enterobacteriaceae Predominate in the Endophytic Microbiome and Contribute to the Resistome of Strawberry. *Sci. Total Environ.* **2020**, *727*, 138708. [CrossRef] [PubMed]
125. De Simone, N.; Pace, B.; Grieco, F.; Chimienti, M.; Tyibilika, V.; Santoro, V.; Capozzi, V.; Colelli, G.; Spano, G.; Russo, P. Botrytis Cinerea and Table Grapes: A Review of the Main Physical, Chemical, and Bio-Based Control Treatments in Post-Harvest. *Foods* **2020**, *9*, 1138. [CrossRef] [PubMed]
126. De Simone, N.; Capozzi, V.; Amodio, M.L.; Colelli, G.; Spano, G.; Russo, P. Microbial-Based Biocontrol Solutions for Fruits and Vegetables: Recent Insight, Patents, and Innovative Trends. *Recent Pat. Food Nutr. Agric.* **2021**, *12*, 3–18. [CrossRef]
127. Capozzi, V.; Fragasso, M.; Bimbo, F. Microbial Resources, Fermentation and Reduction of Negative Externalities in Food Systems: Patterns toward Sustainability and Resilience. *Fermentation* **2021**, *7*, 54. [CrossRef]
128. Díaz, M.A.; Pereyra, M.M.; Picón-Montenegro, E.; Meinhardt, F.; Dib, J.R. Killer Yeasts for the Biological Control of Postharvest Fungal Crop Diseases. *Microorganisms* **2020**, *8*, 1680. [CrossRef]
129. Hernandez-Montiel, L.G.; Droby, S.; Preciado-Rangel, P.; Rivas-García, T.; González-Estrada, R.R.; Gutiérrez-Martínez, P.; Ávila-Quezada, G.D. A Sustainable Alternative for Postharvest Disease Management and Phytopathogens Biocontrol in Fruit: Antagonistic Yeasts. *Plants* **2021**, *10*, 2641. [CrossRef]
130. Carmona-Hernandez, S.; Reyes-Pérez, J.; Chiquito-Contreras, R.; Rincon-Enriquez, G.; Cerdan-Cabrera, C.; Hernandez-Montiel, L. Biocontrol of Postharvest Fruit Fungal Diseases by Bacterial Antagonists: A Review. *Agronomy* **2019**, *9*, 121. [CrossRef]
131. Aiello, D.; Restuccia, C.; Stefani, E.; Vitale, A.; Cirvilleri, G. Postharvest Biocontrol Ability of Pseudomonas Synxantha against Monilinia Fructicola and Monilinia Fructigena on Stone Fruit. *Postharvest Biol. Technol.* **2019**, *149*, 83–89. [CrossRef]
132. Bu, S.; Munir, S.; He, P.; Li, Y.; Wu, Y.; Li, X.; Kong, B.; He, P.; He, Y. Bacillus Subtilis L1-21 as a Biocontrol Agent for Postharvest Gray Mold of Tomato Caused by Botrytis Cinerea. *Biol. Control* **2021**, *157*, 104568. [CrossRef]
133. Wang, X.; Zhou, X.; Cai, Z.; Guo, L.; Chen, X.; Chen, X.; Liu, J.; Feng, M.; Qiu, Y.; Zhang, Y.; et al. A Biocontrol Strain of Pseudomonas Aeruginosa CQ-40 Promote Growth and Control Botrytis Cinerea in Tomato. *Pathogens* **2020**, *10*, 22. [CrossRef]
134. Raman, J.; Kim, J.-S.; Choi, K.R.; Eun, H.; Yang, D.; Ko, Y.-J.; Kim, S.-J. Application of Lactic Acid Bacteria (LAB) in Sustainable Agriculture: Advantages and Limitations. *Int. J. Mol. Sci.* **2022**, *23*, 7784. [CrossRef] [PubMed]
135. Linares-Morales, J.R.; Gutiérrez-Méndez, N.; Rivera-Chavira, B.E.; Pérez-Vega, S.B.; Nevárez-Moorillón, G.V. Biocontrol Processes in Fruits and Fresh Produce, the Use of Lactic Acid Bacteria as a Sustainable Option. *Front. Sustain. Food Syst.* **2018**, *2*, 50. [CrossRef]
136. EFSA Panel on Biological Hazards (BIOHAZ); Koutsoumanis, K.; Allende, A.; Alvarez-Ordóñez, A.; Bolton, D.; Bover-Cid, S.; Chemaly, M.; Davies, R.; De Cesare, A.; Hilbert, F.; et al. Update of the List of QPS-recommended Biological Agents Intentionally Added to Food or Feed as Notified to EFSA 13: Suitability of Taxonomic Units Notified to EFSA until September 2020. *EFSA J.* **2021**, *19*. [CrossRef]
137. Russo, P.; Spano, G.; Capozzi, V. Safety Evaluation of Starter Cultures. In *Starter Cultures in Food Production*; Speranza, B., Bevilacqua, A., Corbo, M.R., Sinigaglia, M., Eds.; John Wiley & Sons, Ltd: Chichester, UK, 2017; pp. 101–128. ISBN 978-1-118-93379-4.
138. Mani-López, E.; Arrioja-Bretón, D.; López-Malo, A. The Impacts of Antimicrobial and Antifungal Activity of Cell-free Supernatants from Lactic Acid Bacteria in Vitro and Foods. *Compr. Rev. Food Sci. Food Saf.* **2022**, *21*, 604–641. [CrossRef]

139. Agriopoulou, S.; Stamatelopoulou, E.; Sachadyn-Król, M.; Varzakas, T. Lactic Acid Bacteria as Antibacterial Agents to Extend the Shelf Life of Fresh and Minimally Processed Fruits and Vegetables: Quality and Safety Aspects. *Microorganisms* **2020**, *8*, 952. [CrossRef]
140. Tenea, G.N.; Olmedo, D.; Ortega, C. Peptide-Based Formulation from Lactic Acid Bacteria Impairs the Pathogen Growth in Ananas Comosus (Pineapple). *Coatings* **2020**, *10*, 457. [CrossRef]
141. Dong, A.; Malo, A.; Leong, M.; Ho, V.T.T.; Turner, M.S. Control of Listeria Monocytogenes on Ready-to-Eat Ham and Fresh Cut Iceberg Lettuce Using a Nisin Containing Lactococcus Lactis Fermentate. *Food Control* **2021**, *119*, 107420. [CrossRef]
142. De Simone, N.; Capozzi, V.; de Chiara, M.L.V.; Amodio, M.L.; Brahimi, S.; Colelli, G.; Drider, D.; Spano, G.; Russo, P. Screening of Lactic Acid Bacteria for the Bio-Control of Botrytis Cinerea and the Potential of Lactiplantibacillus Plantarum for Eco-Friendly Preservation of Fresh-Cut Kiwifruit. *Microorganisms* **2021**, *9*, 773. [CrossRef]
143. Yin, H.-B.; Chen, C.-H.; Colorado-Suarez, S.; Patel, J. Biocontrol of *Listeria Monocytogenes* and *Salmonella Enterica* on Fresh Strawberries with Lactic Acid Bacteria During Refrigerated Storage. *Foodborne Pathog. Dis.* **2022**, *19*, 324–331. [CrossRef]
144. Yap, P.-C.; MatRahim, N.-A.; AbuBakar, S.; Lee, H.Y. Antilisterial Potential of Lactic Acid Bacteria in Eliminating Listeria Monocytogenes in Host and Ready-to-Eat Food Application. *Microbiol. Res.* **2021**, *12*, 234–257. [CrossRef]
145. Luz, C.; D'Opazo, V.; Quiles, J.M.; Romano, R.; Mañes, J.; Meca, G. Biopreservation of Tomatoes Using Fermented Media by Lactic Acid Bacteria. *LWT* **2020**, *130*, 109618. [CrossRef]
146. Ranjith, F.H.; Muhialdin, B.J.; Yusof, N.L.; Mohammed, N.K.; Miskandar, M.H.; Hussin, A.S.M. Effects of Lacto-Fermented Agricultural By-Products as a Natural Disinfectant against Post-Harvest Diseases of Mango (*Mangifera indica* L.). *Plants* **2021**, *10*, 285. [CrossRef] [PubMed]
147. Ghosh, M.; Chouhan, D.; Kamra, A.; Sharma, V. Sustainable Utilization of Potato Industry Waste for Antifungal Biopolymer Production by Lactobacillus Helveticus and Its Application on Pomegranates (*Punica granatum* L.). *Circ. Econ. Sustain.* **2021**, *1*, 1297–1312. [CrossRef]
148. de Oliveira, K.Á.R.; Fernandes, K.F.D.; de Souza, E.L. Current Advances on the Development and Application of Probiotic-Loaded Edible Films and Coatings for the Bioprotection of Fresh and Minimally Processed Fruit and Vegetables. *Foods* **2021**, *10*, 2207. [CrossRef]
149. Fernandes, K.F.D.; de Oliveira, K.Á.R.; de Souza, E.L. Application of Potentially Probiotic Fruit-Derived Lactic Acid Bacteria Loaded into Sodium Alginate Coatings to Control Anthracnose Development in Guava and Mango During Storage. *Probiotics Antimicrob. Proteins* **2021**. [CrossRef]
150. Álvarez, A.; Manjarres, J.J.; Ramírez, C.; Bolívar, G. Use of an Exopolysaccharide-Based Edible Coating and Lactic Acid Bacteria with Antifungal Activity to Preserve the Postharvest Quality of Cherry Tomato. *LWT* **2021**, *151*, 112225. [CrossRef]
151. Rocchetti, M.T.; Russo, P.; Capozzi, V.; Drider, D.; Spano, G.; Fiocco, D. Bioprospecting Antimicrobials from Lactiplantibacillus Plantarum: Key Factors Underlying Its Probiotic Action. *Int. J. Mol. Sci.* **2021**, *22*, 12076. [CrossRef]
152. Khalil, O.A.A.; Mounir, A.M.; Hassanien, R.A. Effect of Gamma Irradiated Lactobacillus Bacteria as an Edible Coating on Enhancing the Storage of Tomato under Cold Storage Conditions. *J. Radiat. Res. Appl. Sci.* **2020**, *13*, 318–330. [CrossRef]
153. Pace, B.; Cefola, M.; Da Pelo, P.; Renna, F.; Attolico, G. Non-Destructive Evaluation of Quality and Ammonia Content in Whole and Fresh-Cut Lettuce by Computer Vision System. *Food Res. Int.* **2014**, *64*, 647–655. [CrossRef]
154. Mohammadi, V.; Kheiralipour, K.; Ghasemi-Varnamkhasti, M. Detecting Maturity of Persimmon Fruit Based on Image Processing Technique. *Sci. Hortic.* **2015**, *184*, 123–128. [CrossRef]
155. Kaur, H.; Sawhney, B.K.; Jawandha, S.K. Evaluation of Plum Fruit Maturity by Image Processing Techniques. *J. Food Sci. Technol.* **2018**, *55*, 3008–3015. [CrossRef]
156. Cavallo, D.P.; Cefola, M.; Pace, B.; Logrieco, A.F.; Attolico, G. Non-Destructive and Contactless Quality Evaluation of Table Grapes by a Computer Vision System. *Comput. Electron. Agric.* **2019**, *156*, 558–564. [CrossRef]
157. Fashi, M.; Naderloo, L.; Javadikia, H. The Relationship between the Appearance of Pomegranate Fruit and Color and Size of Arils Based on Image Processing. *Postharvest Biol. Technol.* **2019**, *154*, 52–57. [CrossRef]
158. Ireri, D.; Belal, E.; Okinda, C.; Makange, N.; Ji, C. A Computer Vision System for Defect Discrimination and Grading in Tomatoes Using Machine Learning and Image Processing. *Artif. Intell. Agric.* **2019**, *2*, 28–37. [CrossRef]
159. Fan, S.; Li, J.; Zhang, Y.; Tian, X.; Wang, Q.; He, X.; Zhang, C.; Huang, W. On Line Detection of Defective Apples Using Computer Vision System Combined with Deep Learning Methods. *J. Food Eng.* **2020**, *286*, 110102. [CrossRef]
160. Cavallo, D.P.; Cefola, M.; Pace, B.; Logrieco, A.F.; Attolico, G. Contactless and Non-Destructive Chlorophyll Content Prediction by Random Forest Regression: A Case Study on Fresh-Cut Rocket Leaves. *Comput. Electron. Agric.* **2017**, *140*, 303–310. [CrossRef]
161. Palumbo, M.; Pace, B.; Cefola, M.; Montesano, F.F.; Serio, F.; Colelli, G.; Attolico, G. Self-Configuring CVS to Discriminate Rocket Leaves According to Cultivation Practices and to Correctly Attribute Visual Quality Level. *Agronomy* **2021**, *11*, 1353. [CrossRef]
162. Amodio, M.L.; Cabezas-Serrano, A.B.; Peri, G.; Colelli, G. Post-Cutting Quality Changes of Fresh-Cut Artichokes Treated with Different Anti-Browning Agents as Evaluated by Image Analysis. *Postharvest Biol. Technol.* **2011**, *62*, 213–220. [CrossRef]
163. Pace, B.; Cefola, M.; Renna, F.; Attolico, G. Relationship between Visual Appearance and Browning as Evaluated by Image Analysis and Chemical Traits in Fresh-Cut Nectarines. *Postharvest Biol. Technol.* **2011**, *61*, 178–183. [CrossRef]
164. Pace, B.; Cavallo, D.P.; Cefola, M.; Colella, R.; Attolico, G. Adaptive Self-Configuring Computer Vision System for Quality Evaluation of Fresh-Cut Radicchio. *Innov. Food Sci. Emerg. Technol.* **2015**, *32*, 200–207. [CrossRef]

165. Subhashree, S.N.; Sunoj, S.; Xue, J.; Bora, G.C. Quantification of Browning in Apples Using Colour and Textural Features by Image Analysis. *Food Qual. Saf.* **2017**, *1*, 221–226. [CrossRef]
166. Hongyang, T.; Daming, H.; Xingyi, H.; Aheto, J.H.; Yi, R.; Yu, W.; Ji, L.; Shuai, N.; Mengqi, X. Detection of Browning of Fresh-cut Potato Chips Based on Machine Vision and Electronic Nose. *J. Food Process Eng.* **2021**, *44*. [CrossRef]
167. Cavallo, D.P.; Cefola, M.; Pace, B.; Logrieco, A.F.; Attolico, G. Non-Destructive Automatic Quality Evaluation of Fresh-Cut Iceberg Lettuce through Packaging Material. *J. Food Eng.* **2018**, *223*, 46–52. [CrossRef]
168. Palumbo, M.; Pace, B.; Cefola, M.; Montesano, F.F.; Colelli, G.; Attolico, G. Non-Destructive and Contactless Estimation of Chlorophyll and Ammonia Contents in Packaged Fresh-Cut Rocket Leaves by a Computer Vision System. *Postharvest Biol. Technol.* **2022**, *189*, 111910. [CrossRef]
169. Xia, Z.; Wu, D.; Nie, P.; He, Y. Non-Invasive Measurement of Soluble Solid Content and PH in Kyoho Grapes Using a Computer Vision Technique. *Anal. Methods* **2016**, *8*, 3242–3248. [CrossRef]
170. Pace, B.; Cefola, M.; Renna, F.; Renna, M.; Serio, F.; Attolico, G. Multiple Regression Models and Computer Vision Systems to Predict Antioxidant Activity and Total Phenols in Pigmented Carrots. *J. Food Eng.* **2013**, *117*, 74–81. [CrossRef]
171. Nadafzadeh, M.; Abdanan Mehdizadeh, S.; Soltanikazemi, M. Development of Computer Vision System to Predict Peroxidase and Polyphenol Oxidase Enzymes to Evaluate the Process of Banana Peel Browning Using Genetic Programming Modeling. *Sci. Hortic.* **2018**, *231*, 201–209. [CrossRef]
172. Sabzi, S.; Javadikia, H.; Arribas, J.I. A Three-Variety Automatic and Non-Intrusive Computer Vision System for the Estimation of Orange Fruit PH Value. *Measurement* **2020**, *152*, 107298. [CrossRef]
173. Palumbo, M.; Cozzolino, R.; Laurino, C.; Malorni, L.; Picariello, G.; Siano, F.; Stocchero, M.; Cefola, M.; Corvino, A.; Romaniello, R.; et al. Rapid and Non-Destructive Techniques for the Discrimination of Ripening Stages in Candonga Strawberries. *Foods* **2022**, *11*, 1534. [CrossRef]
174. Basak, J.K.; Madhavi, B.G.K.; Paudel, B.; Kim, N.E.; Kim, H.T. Prediction of Total Soluble Solids and PH of Strawberry Fruits Using RGB, HSV and HSL Colour Spaces and Machine Learning Models. *Foods* **2022**, *11*, 2086. [CrossRef] [PubMed]
175. Li, H.; Lv, S.; Feng, L.; Peng, P.; Hu, L.; Liu, Z.; Hati, S.; Bimal, C.; Mo, H. Smartphone-Based Image Analysis for Rapid Evaluation of Kiwifruit Quality during Cold Storage. *Foods* **2022**, *11*, 2113. [CrossRef] [PubMed]
176. Farneti, B.; Alarcón, A.A.; Papasotiriou, F.G.; Samudrala, D.; Cristescu, S.M.; Costa, G.; Harren, F.J.M.; Woltering, E.J. Chilling-Induced Changes in Aroma Volatile Profiles in Tomato. *Food Bioprocess Technol.* **2015**, *8*, 1442–1454. [CrossRef]
177. Zhu, D.; Ren, X.; Wei, L.; Cao, X.; Ge, Y.; Liu, H.; Li, J. Collaborative Analysis on Difference of Apple Fruits Flavour Using Electronic Nose and Electronic Tongue. *Sci. Hortic.* **2020**, *260*, 108879. [CrossRef]
178. Barbosa-Pereira, L.; Rojo-Poveda, O.; Ferrocino, I.; Giordano, M.; Zeppa, G. Assessment of Volatile Fingerprint by HS-SPME/GC-QMS and E-Nose for the Classification of Cocoa Bean Shells Using Chemometrics. *Food Res. Int.* **2019**, *123*, 684–696. [CrossRef]
179. Gaggiotti, S.; Mascini, M.; Pittia, P.; Della Pelle, F.; Compagnone, D. Headspace Volatile Evaluation of Carrot Samples—Comparison of GC/MS and AuNPs-HpDNA-Based E-Nose. *Foods* **2019**, *8*, 293. [CrossRef]
180. Applying Electronic Nose Based on Odour Classification and Identification Technology in Detecting the Shelf Life of Fresh Fruits. *Chem. Eng. Trans.* **2018**, *68*, 217–222. [CrossRef]
181. Bonah, E.; Huang, X.; Aheto, J.H.; Osae, R. Application of Electronic Nose as a Non-Invasive Technique for Odor Fingerprinting and Detection of Bacterial Foodborne Pathogens: A Review. *J. Food Sci. Technol.* **2020**, *57*, 1977–1990. [CrossRef]
182. Du, D.; Wang, J.; Wang, B.; Zhu, L.; Hong, X. Ripeness Prediction of Postharvest Kiwifruit Using a MOS E-Nose Combined with Chemometrics. *Sensors* **2019**, *19*, 419. [CrossRef]
183. Galvan, D.; Aquino, A.; Effting, L.; Mantovani, A.C.G.; Bona, E.; Conte-Junior, C.A. E-Sensing and Nanoscale-Sensing Devices Associated with Data Processing Algorithms Applied to Food Quality Control: A Systematic Review. *Crit. Rev. Food Sci. Nutr.* **2022**, *62*, 6605–6645. [CrossRef]
184. Aghilinategh, N.; Dalvand, M.J.; Anvar, A. Detection of Ripeness Grades of Berries Using an Electronic Nose. *Food Sci. Nutr.* **2020**, *8*, 4919–4928. [CrossRef] [PubMed]
185. Nategh, N.A.; Dalvand, M.J.; Anvar, A. Detection of Toxic and Non-Toxic Sweet Cherries at Different Degrees of Maturity Using an Electronic Nose. *J. Food Meas. Charact.* **2021**, *15*, 1213–1224. [CrossRef]
186. Cozzolino, R.; Cefola, M.; Laurino, C.; Pellicano, M.P.; Palumbo, M.; Stocchero, M.; Pace, B. Electronic-Nose as Non-Destructive Tool to Discriminate "Ferrovia" Sweet Cherries Cold Stored in Air or Packed in High CO2 Modified Atmospheres. *Front. Nutr.* **2021**, *8*, 720092. [CrossRef] [PubMed]
187. Cozzolino, R.; Martignetti, A.; Cefola, M.; Pace, B.; Capotorto, I.; De Giulio, B.; Montemurro, N.; Pellicano, M.P. Volatile Metabolites, Quality and Sensory Parameters of "Ferrovia" Sweet Cherry Cold Stored in Air or Packed in High CO2 Modified Atmospheres. *Food Chem.* **2019**, *286*, 659–668. [CrossRef]
188. Cozzolino, R.; Pace, B.; Palumbo, M.; Laurino, C.; Picariello, G.; Siano, F.; De Giulio, B.; Pelosi, S.; Cefola, M. Profiles of Volatile and Phenolic Compounds as Markers of Ripening Stage in Candonga Strawberries. *Foods* **2021**, *10*, 3102. [CrossRef]
189. Qiao, J.; Su, G.; Liu, C.; Zou, Y.; Chang, Z.; Yu, H.; Wang, L.; Guo, R. Study on the Application of Electronic Nose Technology in the Detection for the Artificial Ripening of Crab Apples. *Horticulturae* **2022**, *8*, 386. [CrossRef]
190. Li, A.; Duan, S.; Dang, Y.; Zhang, X.; Xia, K.; Liu, S.; Han, X.; Wen, J.; Li, Z.; Wang, X.; et al. Origin Identification of Chinese Maca Using Electronic Nose Coupled with GC-MS. *Sci. Rep.* **2019**, *9*, 12216. [CrossRef]

191. Longobardi, F.; Casiello, G.; Centonze, V.; Catucci, L.; Agostiano, A. Electronic Nose in Combination with Chemometrics for Characterization of Geographical Origin and Agronomic Practices of Table Grape. *Food Anal. Methods* **2019**, *12*, 1229–1237. [CrossRef]
192. Liu, K.; Zhang, C. Volatile Organic Compounds Gas Sensor Based on Quartz Crystal Microbalance for Fruit Freshness Detection: A Review. *Food Chem.* **2021**, *334*, 127615. [CrossRef]
193. Ghasemi-Varnamkhasti, M.; Mohammad-Razdari, A.; Yoosefian, S.H.; Izadi, Z.; Rabiei, G. Selection of an Optimized Metal Oxide Semiconductor Sensor (MOS) Array for Freshness Characterization of Strawberry in Polymer Packages Using Response Surface Method (RSM). *Postharvest Biol. Technol.* **2019**, *151*, 53–60. [CrossRef]
194. Huang, X.; Yu, S.; Xu, H.; Aheto, J.H.; Bonah, E.; Ma, M.; Wu, M.; Zhang, X. Rapid and Nondestructive Detection of Freshness Quality of Postharvest Spinaches Based on Machine Vision and Electronic Nose. *J. Food Saf.* **2019**, *39*. [CrossRef]
195. Nicolaï, B.M.; Beullens, K.; Bobelyn, E.; Peirs, A.; Saeys, W.; Theron, K.I.; Lammertyn, J. Nondestructive Measurement of Fruit and Vegetable Quality by Means of NIR Spectroscopy: A Review. *Postharvest Biol. Technol.* **2007**, *46*, 99–118. [CrossRef]
196. McClure, W.F. 204 Years of near Infrared Technology: 1800–2003. *J. Infrared Spectrosc.* **2003**, *11*, 487–518. [CrossRef]
197. Reusch, W. Table of Contents. Available online: https://www2.chemistry.msu.edu/faculty/reusch/VirtTxtJml/intro1.htm (accessed on 1 October 2022).
198. Peirs, A.; Tirry, J.; Verlinden, B.; Darius, P.; Nicolaï, B.M. Effect of Biological Variability on the Robustness of NIR Models for Soluble Solids Content of Apples. *Postharvest Biol. Technol.* **2003**, *28*, 269–280. [CrossRef]
199. Schaare, P.N.; Fraser, D.G. Comparison of Reflectance, Interactance and Transmission Modes of Visible-near Infrared Spectroscopy for Measuring Internal Properties of Kiwifruit (Actinidia Chinensis). *Postharvest Biol. Technol.* **2000**, *20*, 175–184. [CrossRef]
200. Clark, C.J.; McGlone, V.A.; De Silva, H.N.; Manning, M.A.; Burdon, J.; Mowat, A.D. Prediction of Storage Disorders of Kiwifruit (Actinidia Chinensis) Based on Visible-NIR Spectral Characteristics at Harvest. *Postharvest Biol. Technol.* **2004**, *32*, 147–158. [CrossRef]
201. McGlone, V.A.; Kawano, S. Firmness, Dry-Matter and Soluble-Solids Assessment of Postharvest Kiwifruit by NIR Spectroscopy. *Postharvest Biol. Technol.* **1998**, *13*, 131–141. [CrossRef]
202. Liu, Y.; Sun, X.; Ouyang, A. Nondestructive Measurement of Soluble Solid Content of Navel Orange Fruit by Visible–NIR Spectrometric Technique with PLSR and PCA-BPNN. *LWT - Food Sci. Technol.* **2010**, *43*, 602–607. [CrossRef]
203. Gómez, A.H.; He, Y.; Pereira, A.G. Non-Destructive Measurement of Acidity, Soluble Solids and Firmness of Satsuma Mandarin Using Vis/NIR-Spectroscopy Techniques. *J. Food Eng.* **2006**, *77*, 313–319. [CrossRef]
204. Lee, K.J.; Kim, G.Y.; Kang, S.W.; Son, J.R.; Choi, D.S.; Choi, K.H. Measurement of Sugar Contents in Citrus Using Near Infrared Transmittance. *Key Eng. Mater.* **2004**, *270–273*, 1014–1019.
205. Schulz, H.; Baranska, M.; Quilitzsch, R.; Schütze, W.; Lösing, G. Characterization of Peppercorn, Pepper Oil, and Pepper Oleoresin by Vibrational Spectroscopy Methods. *J. Agric. Food Chem.* **2005**, *53*, 3358–3363. [CrossRef] [PubMed]
206. Amodio, M.L.; Ceglie, F.; Chaudhry, M.M.A.; Piazzolla, F.; Colelli, G. Potential of NIR Spectroscopy for Predicting Internal Quality and Discriminating among Strawberry Fruits from Different Production Systems. *Postharvest Biol. Technol.* **2017**, *125*, 112–121. [CrossRef]
207. Braue, E.H.; Pannella, M.G. Consistency in Circle Cell FT-IR Analysis of Aqueous Solutions. *Appl. Spectrosc.* **1987**, *41*, 1057–1067. [CrossRef]

MDPI
St. Alban-Anlage 66
4052 Basel
Switzerland
Tel. +41 61 683 77 34
Fax +41 61 302 89 18
www.mdpi.com

Foods Editorial Office
E-mail: foods@mdpi.com
www.mdpi.com/journal/foods

www.ingramcontent.com/pod-product-compliance
Lightning Source LLC
Chambersburg PA
CBHW061814210926
51850CB00006B/702

9783036576114